POWER INTEGRATED CIRCUITS: Physics, Design, and Applications

POWER INTEGRATED CIRCUITS: Physics, Design, and Applications

Paolo Antognetti, *Editor*

Professor
Department of Electronics (DIBE)
University of Genoa
Genoa, Italy

McGraw-Hill Book Company

New York St. Louis San Francisco Auckland
Bogotá Hamburg Johannesburg London Madrid
Mexico Montreal New Delhi Panama Paris
São Paulo Singapore Sydney Tokyo Toronto

Library of Congress Cataloging-in-Publication Data
Main entry under title:

Power integrated circuits.

Includes index.
1. Power electronics. 2. Integrated circuits.
I. Antognetti, Paolo.
TK7881.15.P68 1986 621.381′73 85-23117
ISBN 0-07-002129-5

1234567890 DOCDOC 8932109876

ISBN 0-07-002129-5

The editors for this book were Richard Krajewski and James T. Halston, the designer was Mark E. Safran, and the production supervisor was Sara L. Fliess. It was set in Caledonia by University Graphics, Inc.

Printed and bound by R. R. Donnelley and Sons

Contents

List of Contributors *vii*
Foreword *ix*
Preface *xi*

CHAPTER 1
Power Integrated Circuits: An Introduction 1.1

CHAPTER 2
Bipolar Power Transistors and Their Characteristics 2.1

CHAPTER 3
Power MOS Devices in Discrete and Integrated Circuits 3.1

CHAPTER 4
Evolution of the Planar Process for PICs 4.1

CHAPTER 5
Improved Transistor Structures in Power Integrated Circuits 5.1

CHAPTER 6
Technological Limitations and Protection Techniques in Power ICs 6.1

CHAPTER 7
Packaging of PICs 7.1

CHAPTER 8
Power Integrated Circuit Reliability 8.1

CHAPTER 9
Audio Amplifiers 9.1

CHAPTER 10
Power Integrated Circuits for Industrial Applications 10.1

CHAPTER 11
Three-Terminal Voltage Regulators and Their Applications 11.1

CHAPTER 12
BiMOS Power Integrated Circuits 12.1

Index follows Chapter 12

List of Contributors

A. A. ADAMIAN, Fairchild Semiconductor, Mountain View, California
F. BERTOTTI, SGS Microelettronica, Milan, Italy
R. A. BLANCHARD, Siliconix, Inc., Santa Clara, California
F. R. BORRI, SGS Microelettronica, Milan, Italy
C. CINI, SGS Microelettronica, Milan, Italy
G. D'ESPINOSA, SGS Microelettronica, Milan, Italy
P. R. EMERALD, Sprague Electric, Worcester, Massachusetts
P. G. ERRATICO, SGS Microelettronica, Milan, Italy
M. FELICI, Thomson Semiconductors, Aix en Provence, France
S. GARUE, SGS Microelettronica, Milan, Italy
T. KIKKAWA, NEC Corporation, Kawasaki, Japan
B. MURARI, SGS Microelettronica, Milan, Italy
S. PALARA, SGS Microelettronica, Milan, Italy
J. D. PLUMMER, Stanford University, Stanford, California
C. ROVERE, SGS Microelettronica, Milan, Italy

Foreword

The economic importance of integrated circuits that perform functions involving large power on-chip power dissipation (power integrated circuit) has grown continually since the first such circuits were implemented in the late 1960s. The realization of low-level analog and digital processing circuitry on the same integrated circuit with power transistors handling large voltages and currents has resulted in the implementation of complete functions such as voltage regulators and audio amplifiers at lower cost than had been the case for discrete power transistor realizations. This growth will continue as improved technology and circuit design techniques allow higher levels of integration to be achieved in these circuits both in terms of their power handling capability and the complexity of the low-level signal-processing circuitry that can be included on the same chip.

The design of power integrated circuits has historically presented a special set of problems to the circuit designer. These problems involve technological, device, circuit, and packaging considerations resulting from the required high levels of voltage, current density, and power density. At the technology level, perhaps the most pervasive problem is the realization of power devices with sufficiently large current-handling capability and voltage capability in the same technology with standard integrated circuit components. A related important problem is the simultaneous implementation of digital circuitry of sufficient density and speed together with high voltage devices.

At the device level, the design of MOS and bipolar power devices that reliably achieve the necessary levels of power dissipation and safe area of operation while consuming the minimum possible silicon die area is a key problem. The problem of prediction a priori the detailed electrical behavior of a power transistor structure with complex electrical and thermal topology under transient conditions of high voltage, high current, and high level injection is still not fully solved.

At the circuit level, many unique problems present themselves. One is the problem of designing circuitry which protects the circuit from catastrophic failure under worstcase load-fault conditions. A second aspect that is unique to power ICs is the thermal interaction between the devices on the chip which dissipate large power and the low-level analog circuitry on the same chip. This problem can result if grossly degraded performance is not properly considered during the design and layout phase.

Finally, at the packaging level, the development of a physical structure that can be manufactured at low cost, which has a sufficient number of electrical connections, which is capable of conducting sufficient thermal energy from the die with a given maximum allowable temperature drop, and which is compatible with component insertion equipment for low cost of manufacture is a challenging task.

Because of these and other specialized problems, the design of power integrated circuits and devices has evolved into a separate specialty within the field of integrated circuit design. In comparison to the skills required to undertake the design of more conventional integrated circuits, the design of a power integrated circuit requires an understanding of a broader set of physical principles. These include thermodynamics, thermal properties of materials, a detailed understanding of the effects of temperature variations on devices, high-level injection effects in transistors, high-field mobility behavior in silicon, complex electro-thermal interaction phenomena, and so forth.

This book is the first attempt to compile a volume specifically focused on the problems of power integrated circuits. In it, contributions from leading experts in the field have been assembled. The result is a reference work that should be invaluable to those involved in the design of this highly challenging and interesting type of integrated circuit.

Paul R. Gray
University of California, Berkeley

Preface

As described by P. Gray in his Foreword, this book represents the first attempt to compile a volume specifically focused on power integrated circuits (PIC). After an introductory chapter, which gives an interesting historical perspective, technological problems related to the fabrication process are described, as well as the particular features of power devices. Both types of power devices are considered, bipolar and FET, and their treatment concerns theoretical and practical aspects. Subsequent chapters discuss other important topics, such as packaging, which represents a fundamental part of the design cycle of a PIC, and reliability, which enables the designer and the process engineer to improve the quality of their products. Finally the core of the book focuses on the design and applications of PIC's. General aspects of the design process for power circuits are described first, as opposed to the design of regular, low power IC's. In the second part of the book, five chapters are dedicated to the description of special designs for particular applications, such as audio amplifiers, PIC's for industrial applications, regulators, and interface circuits to the external world.

The book has been edited with contributions from leading experts in the field, from Europe, the United States, and Japan. However, the backbone of the book has been written by authors from the SGS company in Milan, Italy. SGS is generally considered a leader in the field of PIC's. The other authors represent the following companies: Fairchild, Sprague, Siliconix, and NEC. One author is affiliated with Stanford University.

The authors have assumed that the reader is familiar with the basic aspects of semiconductor devices and processing technology, as well as design criteria for low power IC's. The book is aimed at design engineers and process engineers in semiconductor houses, specialized in the fabrication of power circuits, and at application engineers in system houses, where power electronics is relevant in the design of an electronic system.

Each chapter contains many circuit schematics, die photos, plots and charts. This book will most certainly prove a valuable reference tool for designers in industrial research and application labs. The first part of the book covers in detail theoretical aspects of device physics, both bipolar and FET, and will probably appeal also to graduate students desiring a thorough treatment of power devices.

I would like to extend my gratitude to Dr. P. Pistorio, Chairman of SGS, for his initial and decisive encouragement and support; I would also like to thank Mr. B. Murari of SGS for his enthusiasm in supporting the initial idea of this book, and all the authors for the time spent working on their chapters, mostly taken from their leisure time; in particular I wish to thank two authors, Mr. T. Kikkawa of NEC and Dr. M. Felici, formerly of SGS, for their timeliness and precision.

P. ANTOGNETTI
UNIVERSITY OF GENOA

CHAPTER 1

Power Integrated Circuits: An Introduction

F. Bertotti
S. Garue
B. Murari
SGS Microelettronica
Milan, Italy

1.1 PRELIMINARIES 1.2
1.2 SEMICONDUCTOR TECHNOLOGIES 1.5
1.3 CIRCUITS 1.7
1.4 TRENDS 1.8

In the 1980s, integrated electronics is dominated by the acronym VLSI (*v*ery *l*arge *s*cale *i*ntegration). Without doubt this is a fascinating subject, and for some time, rapid and continuous development in the field has justified research aimed at identifying the physical limits of the process of integrating micrometric devices.

At the same time there has been an equally significant, though perhaps less visible, development in the field of *power integrated circuits* (PICs). This we can identify, for brevity and for an objective dualism, by the acronym VLAI (*v*ery *l*arge *a*rea *i*ntegration). The meaning that we intend to attribute to this acronym reflects the dimensions of the elementary device (a power transistor) and the dimensions of the resulting integrated circuit. With this the intention is to stress the increase in power density—in terms of both voltage and current—and in the total number of components included in a power integrated circuit.

Technological advances and circuit and systems development activities involved in the realization of a monolithic electronic power system have now reached a level of maturity that suggests the drafting of a volume that collects the contributions of numerous experts in the field. Such a volume should offer a reasonably complete review of past developments, of the state of the art, and of the emerging prospects in this sector of integrated electronics.

The guideline for development is obviously industrial needs, which can be expressed as reduction in costs, improved performance, and ever-improving reliability. The results achieved demonstrate the continuing interaction between circuits and technologies. This interaction spreads ever more frequently beyond the relatively restricted environment of the semiconductor manufacturer to create new developments in the architecture of electronic systems, which are essential for optimizing the industrial economy offered by the new designs. This confirms, among other things, the decidedly interdisciplinary character of integrated circuit engineering.

Moreover, technological developments in PICs often yield results which are of interest to the world of VLSI, and vice versa. It is the conviction of the authors that the electronic revolution fired by power integrated circuits will also bring results as far-ranging as those obtained so far in the undoubtedly fascinating field of submicrometric devices.

1.1 PRELIMINARIES

PIC engineering can be divided into three parts: semiconductor technology, circuits, and packages. In turn, each part presents a variety of problems which often impinge on other parts. In the semiconductor technology field, for example, the optimization of the most important elementary cell, the power transistor, entails modeling a circuit type such that maximum equalization of current distribution is possible when several elementary transistors are connected in parallel.

In the circuitry field, as we will see, a possible overcurrent protection network configuration exploits a minimum-resistance sensing element obtained by placing two Kelvin contacts on a suitable geometry. In the assembly field, certain operations (finishing of the back of the silicon wafer, for example, or the mechanical die-separation technique) determine the performance of the device and require an extensive knowledge of electrothermic problems.

Electrothermic effects in integrated circuits are known and are generally controllable with sufficient precision (thermal feedback in an operational amplifier is a typical example), but the fundamental problem in PICs becomes exquisitely thermodynamic, in three dimensions and in a medium anything but homogeneous. The determination of the temperature field in such a system constitutes a formidable problem both from the point of view of numerical analysis and from that of strict experiment.

A determining element in the economy of an integrated circuit is obviously the area of the silicon die. Current and voltage are not equal in their effect on the power density obtainable in an integrated circuit. We will briefly describe here the interrelationship between system application and silicon die area, using bipolar technologies. The current density is determined by the characteristics of the base-emitter junction, i.e., the doping profile in a relatively superficial region; however, the breakdown voltage required by the process is a determinant of the characteristics of the isolation junction and thus affects considerably the surface area of each isolated device. The power required from an integrated circuit often determines the individual voltage or current as a function of the characteristics of the user system so that the choice of technology for the semiconductor manufacturer is obligatory.

Bipolar devices, by definition, depend in their mode of operation on the transport, trapping, and/or recombination of the two types of carriers. Bipolar device operation takes place essentially in the bulk of the semiconductor. Indeed, the most famous member of this category, the bipolar transistor, was invented by Shockley and coworkers to avoid the problems connected with the surface. However, since no devices of infinite or semi-infinite extent exist, regions of transition from one conductivity type to another eventually also arrive at the surface. Thus, surface effects still play a role in the operation of such devices, often a crucial one in determining final IC performance. High-voltage bipolar transistors are, in fact, very complicated structures for both process and geometric design, and a deep knowledge of surface problems is required in order to avoid accidentally creating parasitic MOS devices.

Thus, to produce good PICs certain basic conditions have to be met: the availability of silicon technologies suitable for the integration of high-voltage and high-current devices, the availability of power packages capable of dissipating the heat generated, and the ability to design innovative circuits and systems. Mainly because of this last need, the design of PICs becomes every day less exclusively the domain of the semiconductor manufacturer and involves (already remarkably) the designer of electronic systems, which play a much more essential role than in the recent past.

Several times we have used the term *electronic system* and perhaps, at this point, it merits a few words of comment. Radios, telephones, television sets, automobiles, and typewriters, for example, are electronic systems in which the most natural division of functions can be illustrated: supply, transduction, processing, and actuation (see Fig. 1-1). In general, each function requires a dedicated technology. The natural application domain of power technologies comprises mainly the supply and the actuation functions. This does not exclude a more or less relevant part of the signal processing function from the power circuit. A recurring example is provided by parallel printer-driver circuits, where serial transmission of the data to be transferred permits a definite economy in the number of pins used but obviously requires a serial-parallel conversion inside the power circuit. Even this relatively simple problem requires new technological performance such as the coexistence of analog and digital functions and the minimization of the area of silicon necessary to implement digital functions at high voltages. A notable example in the bipolar field is provided by high-voltage technologies, which, with minimum process complications, allow the integration of I^2L structures.

Direct interfacing with microprocessors and peripherals and direct driving of actuation systems are a measure, as a function of the total cost, of the excellence of VLAI technology.

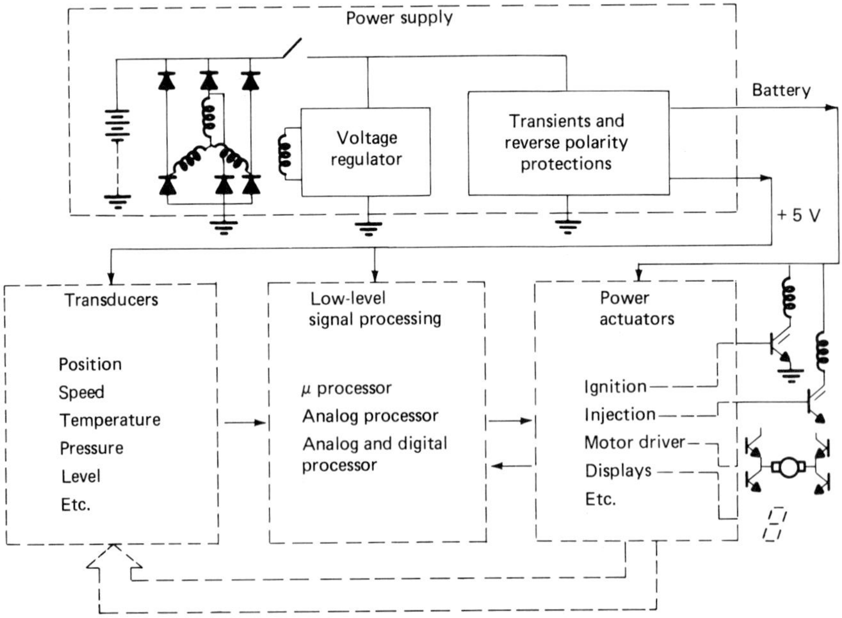

FIG. 1-1 A typical electronic system partition.

1.2 SEMICONDUCTOR TECHNOLOGIES

An important parameter in PIC process design is the maximum voltage that can be impressed on the collector with respect to the emitter (the worst-case situation is circuit operation with a high impedance at the base terminal). The resulting common-emitter breakdown voltage is the lowest expected breakdown configuration resulting from a current-gain multiplication of collector-base junction leakage. The collector avalanche voltage can be limited by several effects, including junction curvature at edges or, worse, at corners; one-dimensional breakdown calculations usually must be modified to take this into account.

The most significant limitation upon nominal *collector-base* junction breakdown voltage *is* one-dimensional in nature and so is directly related to vertical doping profile. If the collector depletion region reaches the n^+ buried layer before the collector-base junction breaks down as a result of another limitation (perhaps junction curvature), the electric field builds up at a higher rate because additional charge depletes at a slower rate. Within this so-called reach-through condition, the limitations arising from diffused-junction isolation or deep n^+ (to lower collector series resistance) can be realized by considering the amount of n^+ buried-layer outdiffusion caused by the large amount of post-epitaxy high-temperature processing. Variation in epitaxial thickness results in a direct variation in the post-epitaxy diffusion length required to reach the substrate and thus in a buried-layer outdiffusion that can consume almost one-half the width of the epitaxial layer.

Ion-implanted depositions allow the isolation and sinker regions to be produced from top and bottom and this results in a consistent reduction of post–epitaxial deposition high-temperature processing. This is one of the most attractive applications of ion implantation in high-voltage processes. It is not the only one. Vertical *pnp* transistors can also be obtained by using selectively placed dual buried layers with implanted boron, which will outdiffuse more rapidly than the antimony host layer. This technique is widely used also in VLSI technologies: I^2L upward operation and the VMOS active channel region give well-known examples.

Thus ion implantation (but also process and device computer-aided design, or CAD) reveals a strong influence in the evolution of PIC processes which is closely related to the corresponding evolution in MOS processes.

Polycrystalline silicon deposition is another technique widely used by bipolar process engineers: device isolation, field-plate surface protection, and deposited resistors are the most important applications in high-voltage technology. Furthermore, it should be remembered that, in order to obtain really effective surface protection, one has also to prevent thermal-aided gate movement due to oxide charge mobility. Problems of this kind are closely linked to overall IC reliability.

High-current devices for integrated circuits present quite the same design problems as in the case of discrete components; some further complications may arise because of the need to limit device size and to put the collector contact on the front of the chip. Nevertheless, a deep understanding of high-injection effects in bipolar transistor operation (e.g., base widening and current crowding) is required. Emitter efficiency and perimeter-to-area ratio are the most common technological parameters to be optimized, particularly when high-voltage processes are necessary and epitaxial-layer thickness and resistivity cannot be tailored for high-current operation. What is much more difficult to improve, in comparison with discrete devices, is transistor ruggedness against direct and reverse second breakdown failure. To improve direct second breakdown protection while maintaining acceptable device characteristics, a number of emitter "fingers" are used and provisions are made for so-called ballasting resistors in each finger in order to equalize currents and prevent hot spots from occurring. But second breakdown in power transistors will not cease being an actively discussed subject, and likewise in IC applications. What one has to explain, at least from a phenomenological point of view, is a failure that is preceded by a rapid drop in the voltage across the device and a rapid increase in the current.

There are many theories about second breakdown mechanisms and initiation processes, but we still have no unified global approach. Solutions can be applied locally upon further investigation after an a priori choice among contending theories.

Precision and matching of parameters are frequently essential in analog circuits. This is true, for example, of resistor values. Beyond the effects induced by the natural dispersion of process parameters, one of the determining mechanisms is found to be sensitivity to mechanical stresses. As is known, these modify the energy band structure of the semiconductor or modify to a lesser degree the mobility of carriers and their intrinsic concentration. Mechanical stresses arise during both the process phase and the assembly phase. In the first case, everything is traceable to the differing coefficients of thermal expansion of the materials used (silicon, silicon dioxide, silicon nitride, etc.). In the second case, chip soldering techniques and the mechanical deformation of the resin in the polymerization phase prove to be the most critical processes. As far as piezoresistive effects are concerned, in particular, the choice of ⟨100⟩ substrate allows an orientation of the critical layers with minimum sensitivity of stresses. In the first case again, copper, for example, has a thermal expansion coefficient significantly different from that of silicon. When thermal dissipation requirements compel the use of a copper frame, the chip soldering process requires the use of suitable materials capable of absorbing the stresses created at the copper-silicon interface to avoid, among other things, breakage of the chip.

1.3 CIRCUITS

PICs are mainly used in supply and actuation functions: the former, to convert energy from the primary source (battery, supply network, dynamo, alternator, solar cells, or whatever) into a supply suitable for the system; the latter, to drive motors, solenoids, loudspeakers, and other inductive loads after power amplication of low-level input signals.

The supply functions include rectifiers, voltage regulators and filters, dc-dc converters, inverters, and all circuits to suppress transients coming from the primary supply (the external supply), such as "crowbars," zener diodes, and clamp diodes. All these as a group must be able to handle high current, high voltage, or both together, and most of them can be produced using monolithic technology.

The first important limitation may arise from package thermal resistance (forced-air and liquid cooling are not taken into consideration, because of their high cost): switching regulation techniques, for example, are a circuit solution to the problem of producing stabilized supplies with high efficiency and low dissipated power.

On the other hand, a basic concept underlying actuation device design is that stabilized supplies must be avoided because regulation reduces the efficiency of the system; they must be designed, therefore, to operate in fairly wide voltage ranges.

Power packages have a relatively limited number of pins, and this poses another constraint, both for partitioning the system for sharing of functions between low-power and high-power technologies, on the one hand, and, once again, for the thermal resistance of the package.

PICs often have to drive inductive loads, e.g., loudspeakers or solenoid relays in general. The switching of high current in these loads gives rise to voltages which can exceed the supply and fall below the ground value.

The most simple solution for inductive load protection is to clamp the overvoltage, by means of clamp diodes, to one V_{BE} above positive supply voltage and one V_{BE} below ground value. Negative overvoltages are automatically clamped by the substrate-collector isolation junction of the lower power transistors (i.e., those with grounded emitter), which is always present and of sufficient dimensions to carry the peak current.

But there are many hidden details in this very simple explanation, such as parasitic effects due to unwanted lateral *npn*-transistor activation, latch-up due to firing of parasitic SCRs, and so on. The layout becomes very critical and the position of each element must be fixed with care.

Protection against positive overvoltage arising at the output can be provided by simply connecting a diode between the output pad and $+V_{CC}$ to withstand all supply voltages during normal operation and to clamp overvoltages to $V_{CC} + V_{BE}$. Unfortunately, the integrated diode for overvoltage limiting has a vertical *pnp* parasitic transistor intrinsic in its structure, which loses current downward to the substrate. Therefore new diode structures have been developed to circumvent this problem and to reduce the dissipated power too.

In the event of supply voltage reversal, the collector of the upper output transistor is forward-biased toward the substrate, and the bond wires for the V_{CC} and ground pins are forced to carry a high current which is limited only by the substrate resistance. The wires must therefore be capable of withstanding this current for a relatively long time, longer than, for example, the blowout time of a fast-blow 1-A fuse connected in series with the supply.

Furthermore, just as future digital VLSI circuits should contain the capability for analog functions on the same chip, analog-digital-compatible PICs every day are becoming more and more versatile in their applications. Entire-system integration often requires a single chip to receive weak signals and to deliver powerful currents or voltages. In regard to this requirement, MOS transistors are still noisy at low frequencies and can deliver only a small amount of current; thus, bipolar ICs until now have been the only candidate for providing sensitive amplifiers and high-power outputs in the same device.

It has been argued that I^2L offers the best opportunity for dense fast-digital-logic fabrication with compatible analog circuit devices. Standard I^2L, however, exhibits some limitations when pursuing both logic speed and analog-device breakdown voltage. Something like a figure of merit can be defined as a quantitative trade-off between the I^2L minimum delay time and *npn* power integrated circuit BV_{CBO}. A deep analysis shows that the product of I^2L maximum speed and PIC-device collector-base breakdown voltage cannot exceed $(1 \div \frac{1}{3})$ V/ns. Nevertheless, very low-power operation with high packing density has been demonstrated to be quite feasible with breakdwon voltages exceeding 60 V (BV_{CEO}).

1.4 TRENDS

We think that the usefulness of PIC theory is not restricted to engineers actively engaged in power systems design. In fact, it is our contention that almost every design engineer in the field of integrated electronics will have some opportunity to draw on this subject. PIC applications may be suggested whenever signal-processing results in communications, control, or instrumentation systems, for example, are handled in other than purely digital form. (Designing almost any system is not unlike jumping from boulder to boulder to cross a stream: each boulder, in this analogy, may represent a specific area for design innovation.)

If, as we believe, the past is a guide to the future, the major advances in power integrated circuits will be made by people who have a firm grasp of technologies, circuits, and packages. PIC engineers, like Caesar, have been exploring and expanding the frontiers of a vast region; the weapons of conquest, sometimes impressive and sometimes simple-minded, have been used to consolidate gains in a territory of great ferment and excitement. But describing the rapid advancement of PICs as an evolutionary process is overly simplistic. As well-established techniques are exploited, extended, and optimized, they also approach their basic limitations. These limitations, when coupled with the increased demands upon newer products, stimulate the development of better methods. Since old techniques are frequently used to advantage in the design of new solutions, the superposition of old upon new, an evolutionary-revolutionary trend, appears to be a better way to explain this accelerated growth.

Long-range predictions are tougher to make, but, looking ahead, it seems easy to project near-future trends. The key objectives, once again, will be to improve the performance, to lower the cost, to enhance the reliability, and to shorten the development and manufacturing cycle. This last (but by no means least) requirement arises because marketing considerations demand, more and more frequently, a minimal lapse of time between product definition and product availability.

We do not aim to become the very conscience of the community of integrated circuit engineers but, nevertheless, after giving a brief backward look at three great antecedents to VLAI, we can suppose what the next main step could be (see Fig. 1-2).

The early efforts toward PICs came to fruition around 1970 with the first commercially available 2-W audio amplifier. The second cornerstone, around 1975, was a 20-W amplifier for high-fidelity applications which resulted from process (high-voltage, low-noise), circuit (short-circuit protection), and package (copper frame) improvements. We should not pass over in silence the fact that, after this, the growing feeling among technicians was that the fertility of power increase was approaching exhaustion. But systems evolution (peripheral equipment, particularly) put new emphasis on higher power needs. All this resulted in a 200-W, switching-mode, dual full-bridge driver designed to accept

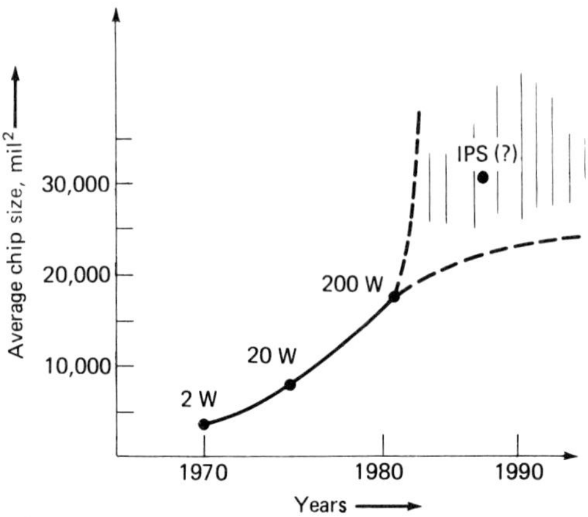

FIG. 1-2 The main steps toward VLAI.

standard logic levels and to drive inductive loads such as relays, solenoids, and dc and stepping motors.

Of course, new advances will keep coming in even as this book is being printed, and we cannot hope that this work will remain up to date for more than a few years. The circuit that is now given a starring role will quickly become a market standard.

We believe that it is easy to predict (and only to a certain extent, at that) only the general approach toward VLAI: the two disjoint domains of PIC application (supply and actuation) have to be unified in order to allow a high-efficiency direct connection to the external power supply. Such a monolithic interfaceless power system (IPS) will be undoubtedly the most significant departure from the present state of the art.

We would like at this point to give a second look at the evolution of PICs by assessing the wealth of their present *power*. We have already pointed out that the silicon area occupied by an integrated circuit is a growing function of the operating voltage, i.e., of the breakdown voltage to be guaranteed for each device. A reference silicon area may be expressed in terms of both the total number of components and the total power implemented inside the chip. One can compute a voltage-valued area for complexity, for current, or for both. This means, at first order, that a fixed number of pieces can be combined in order to obtain the required integrated circuit.

These considerations can be put into a pictorial form which closely resembles the so-called pseudosphere of analytical geometry (i.e., the volume enclosed by the surface of a rotating tractrix); see Fig. 1-3. This kind of representation may look artificial at first sight, but it is actually practical and even if it could not be supported by something like a *theorema egregium*, we do think this is the way PICs are evolving.

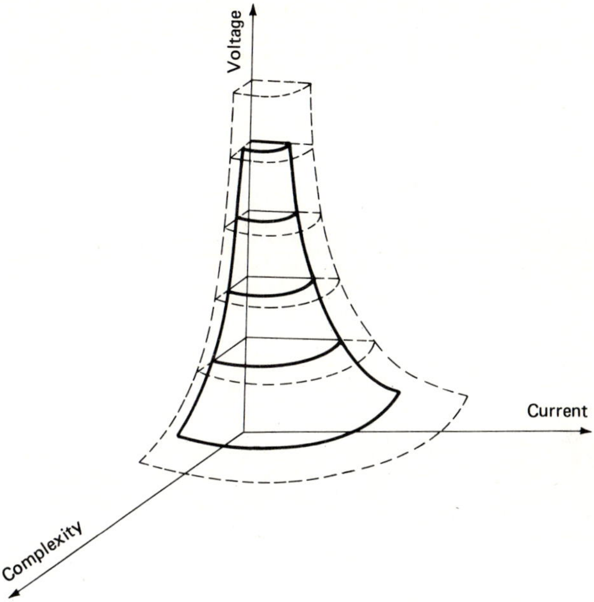

FIG. 1-3 PICs evolution represented by a pseudosphere.

In Fig. 1-3, different sections of the pseudosphere have been marked off to identify the discrete spectrum of voltage values corresponding to classes of industrially available processes. Thus, at a given voltage, it is possible to exploit a certain amount of silicon area to implement "current" or "complexity" or a suitable mixture of both. The "real estate" (throughout this book we will go over and over this point again) is voltage-limited.

The meaning of the evolution, expressed by the dashed pseudosphere, which describes the state of the art at a later time, is that, for a similar current-voltage-complexity relationship and the same silicon physical constraints, the usable real estate at a given voltage will be increased. All the efforts will thus fall within the ambit of what we have called very large area integration.

A less pictorial, and perhaps more vivid, description of the voltage-current-complexity relationship shown in Fig. 1-3 can be gained by comparing the amount of silicon area required for three basic functions that are often seen in circuit synthesis: an operational amplifier, an edge-triggered D flip-flop, and an *npn* power transistor.

We will consider two remarkable processes identified by an *npn* BV_{CEO} of 10 V and 60 V, respectively. The 10-V process operates in an analog-digital LSI environment, while the 60-V process operates in an analog-digital power environment. The comparison will leave out of

TABLE 1-1 PIC Chip Area for Various Circuit Functions, mil^2

Process	Function: Operational amplifier	Edge-triggered D flip-flop	1-A power *npn* transistor
$BV_{CEO} > 10$ V	400	100	100
$BV_{CEO} > 60$ V	1000	400	300

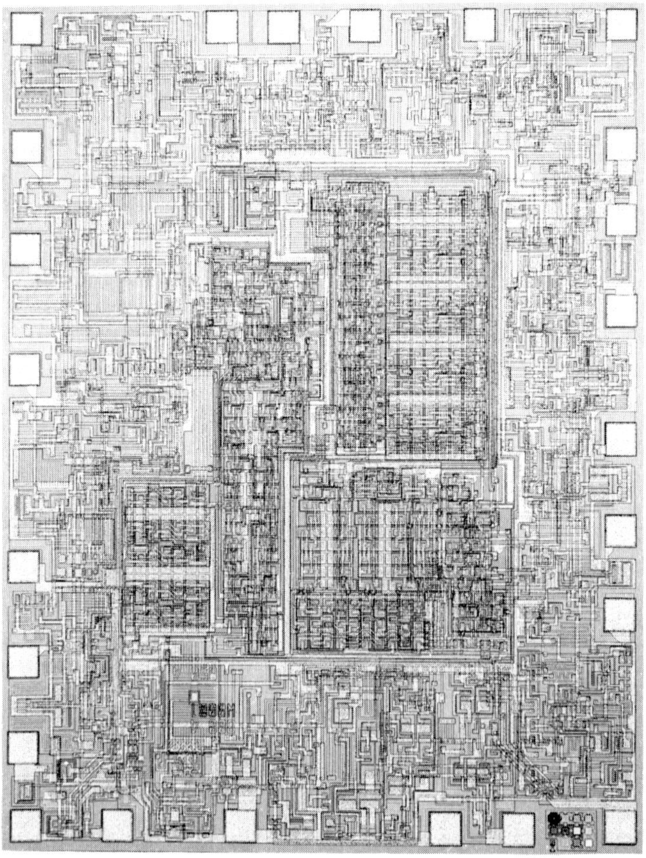

FIG. 1-4 Signal-processing bipolar PIC, realized with the 10-V process.

consideration the performance modification which arises when one integrates the same circuit configuration using two different processes (e.g., the op-amp gain-bandwith product, the maximum toggle frequency of the D flip-flop, and the saturation voltage of a power transistor at a given current). Table 1-1 illustrates the real estate of the selected functions in the two processes mentioned above; in both cases the digital functions exploit I^2L compactness.

Let us assume that the maximum chip size is almost the same in the two processes, i.e., that the process yields are practically at the same level. This rough assumption allows us to argue that a scaling factor exists for current and complexity capability, at a given chip size, in response to a process BV_{CEO} modification. This is fully equivalent to

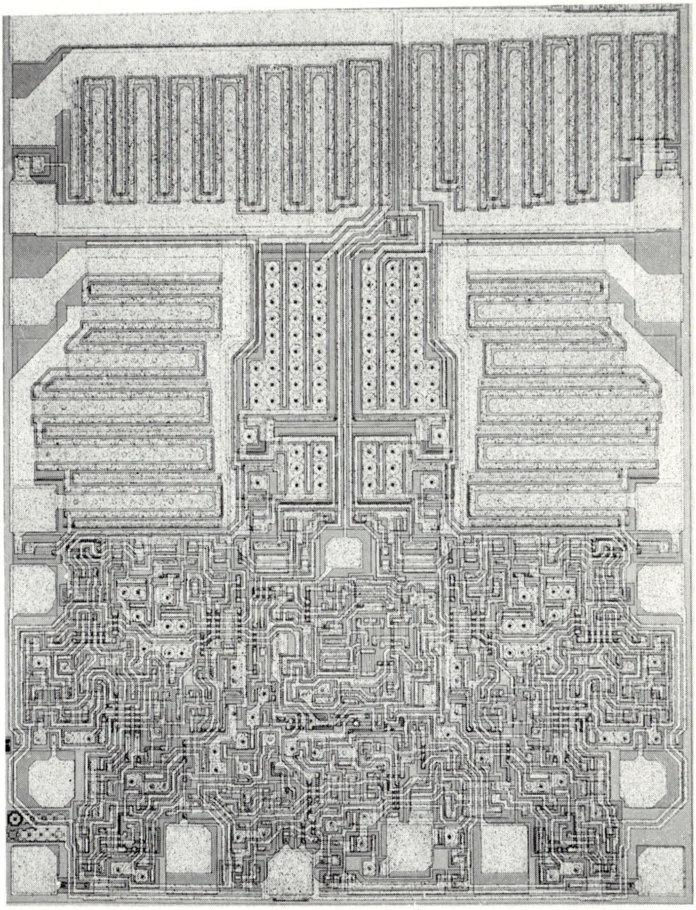

FIG. 1-5 Switch-mode solenoid driver, realized with the 60-V process.

stating that a given mix of current and complexity capability can be preserved at higher voltage by increasing the silicon area and that, beyond the maximum chip size which can be reliably handled, an increased BV_{CEO} forces a scaling-down of the current-complexity mix.

Besides these arguments, one can have a look at the integrated circuits shown in Figs. 1-4 and 1-5, which have exactly the same area (18,500 mil^2) and have been realized with the 10-V (Fig. 1-4) and the 60-V (Fig. 1-5) processes, respectively.

The first circuit provides signal processing functions for horizontal and vertical deflection in television sets and monitors; it works with a standard 5-V supply voltage and has a complexity expressed by 310 triple-output I^2L gates, 265 bipolar transistors, 10 capacitors of 20 pF each on the average, and 220 base resistors for a total value of 1.3 MΩ.

The second circuit is a dual–switch mode solenoid driver able to deliver 2.5 A per channel with a supply voltage of 50 V; it contains 164 bipolar transistors, 72 base resistors, and 4 Darlington output stages.

But the world of PICs is in ferment, and more, much more than this, could probably be shown in a book about this subject written in the next decade.

CHAPTER 2

Bipolar Power Transistors and Their Characteristics

M. Felici
Thomson Semiconductors
Aix en Provence, France

2.1 VOLTAGE RATINGS 2.2
2.1.1 Collector-to-Base Avalanche Breakdown (Plane Unlimited Junction) 2.3
2.1.2 Collector-to-Base Avalanche Breakdown (Plane Limited Junction) 2.7
2.1.3 Collector-to-Base Avalanche Breakdown (Planar Junction) 2.10
2.1.4 Leakage Current 2.11
2.1.5 Collector-to-Emitter Sustaining Voltage 2.12
2.2 CURRENT RATINGS 2.17
2.2.1 Low Current Densities (Low Injection) 2.18
2.2.2 High Injection in the Base 2.22
2.2.3 Early Effect 2.23
2.2.4 High Injection in the Collector (Saturation and Quasi Saturation) 2.24
2.2.5 Extensions of the Model 2.28
2.2.6 Lifetime 2.33
2.2.7 Emitter Gummel Number 2.36
2.3 THERMAL RATINGS 2.38
2.3.1 Maximum Temperature 2.38
2.3.2 Thermal Resistance 2.38
2.3.3 Maximum Power 2.40
2.3.4 Thermal Resistance Measurement 2.42

2.4 SECOND BREAKDOWN 2.43
2.4.1 Forward Second Breakdown 2.43
2.4.2 Reverse Second Breakdown 2.45
2.5 SWITCHING BEHAVIOR 2.47
2.5.1 Electrical Switching Times 2.48
2.5.2 Physical Switching Times 2.49
2.5.3 Base and Collector Transit Times 2.51
2.5.4 One-Dimensional Theory 2.52
2.5.5 Two-Dimensional Effects 2.55

Power integrated transistors must be designed to withstand high voltage, current, and power. Wide operation areas in both the forward and the reverse base-driving condition and good switching speed are also often required.

In this chapter the physical phenomena that determine power integrated transistor performance are discussed. The aim is not to develop the *perfect* theory. An approximate and simple picture will be given, sufficiently complete and accurate to provide the right algebra to the designer and to describe device behavior to both the designer and the user.

Basic semiconductor and PN junction theory is assumed as a background [1–6]. NPN bipolar transistors are the ones considered in the text, unless otherwise indicated. The subject matter developed is applicable to integrated as well as discrete bipolar power transistors.

The physical constants used in the text and the most important properties of silicon are summarized in Tables 2-1 and 2-2.

2.1 VOLTAGE RATINGS

Bipolar transistor voltage ratings can be explained by starting from two main physical phenomena: *avalanche breakdown* and *sustaining.* Lim-

TABLE 2-1 Important Physical Constants

Constant	Symbol	Magnitude
Boltzmann's constant	k	1.38×10^{-23} J/K
Electron charge	q	1.6×10^{-19} C
Electronvolt	eV	1.6×10^{-19} J
Electron rest mass	m_0	9.11×10^{-28} g
Permittivity of free space	ϵ_0	8.86×10^{-14} F/cm
Thermal voltage at 295 K	V_T	25.4 mV

TABLE 2-2 Properties of Silicon

Property	Magnitude
Atomic number	14
Atomic weight	28.06
Crystal structure	diamond
Lattice constant	5.431 Å
Density	2.33 g/cm^3
Dielectric constant K	11.7
Energy gap* E_g	1.126 eV
Energy gap at 0 K, E_{g0}	1.206 eV
Intrinsic carrier concentration* n_i	9.82×10^9 cm^{-3}
Intrinsic mobility*	
Electron	1350 cm^2/(V·s)
Hole	480 cm^2/(V·s)
Limiting velocity in bulk Si for electron and hole v_l	1×10^7 cm/s
Diffusivity for minority carriers:*†	
Electron D_n	29 cm^2/s
Hole D_p	11 cm^2/s
Thermal conductivity* σ_{th}	1.5 W/(K·cm)
Coefficient of thermal expansion*	2.33×10^{-6} K^{-1}

*At T = 295 K.
†For background dopant concentration = 1×10^{16} cm^{-3}.

iting factors are *punch-through, reach-through,* and *junction curvature.* Emitter-base breakdown voltage [7], which is seldom relevant, is not discussed.

2.1.1 Collector-to-Base Avalanche Breakdown (Plane Unlimited Junction)

An *abrupt junction* is considered, with constant acceptor dopant concentration N_A on the P side and constant donor dopant concentration N_D on N side. In this section the P and N regions are assumed to be *unlimited* and their metallurgical boundary to be *plane.*

When a reverse voltage V_0 is applied, an electric field E is generated, which sweeps out free carriers near the junction, inducing a *space charge region,* or *depletion layer,* where acceptors and donors are ionized.

For a plane junction, the one-dimensional *Poisson equation* applies:

$$\frac{d^2V}{dx^2} = -\frac{\rho(x)}{\epsilon} \tag{2-1}$$

where $\rho(x)$ = charge density
$\epsilon = K\epsilon_0$
K = dielectric constant (11.7 for silicon)
ϵ_0 = permittivity of free space (8.86×10^{-14} F/cm)

The two following assumptions are made (Fig. 2-1*a*): (1) all dopant atoms in the depletion layer are *ionized* and (2) the *neutrality condition,* given by

$$N_A x_{dP} = N_D x_{dN} \tag{2-2}$$

applies, where x_{dP}, x_{dN} are the depletion widths in the P and N regions. Poisson's equation is thus reduced to

$$\frac{dE}{dx} = -\frac{d^2V}{dx^2} = -\frac{qN_A}{\epsilon} \quad \text{(P region, } -x_{dP} \le x \le 0) \tag{2-3a}$$

$$\frac{dE}{dx} = -\frac{d^2V}{dx^2} = \frac{qN_D}{\epsilon} \quad \text{(N region, } 0 \le x \le x_{dN}) \tag{2-3b}$$

where q = the electron charge (1.6×10^{-19} C). The boundary conditions are

$$E(x_{dN}) = E(-x_{dP}) = 0 \tag{2-4a}$$

$$V(x_{dN}) = V_0 \tag{2-4b}$$

$$V(-x_{dP}) = 0 \tag{2-4c}$$

By simple integration of Eqs. 2-3*a* and 2-3*b*,

$$E_0 \equiv |E(0)| = \frac{qN_A x_{dP}}{\epsilon} = \frac{qN_D x_{dN}}{\epsilon} \tag{2-5}$$

$$V_P = \frac{qN_A x_{dP}^2}{2\epsilon} \tag{2-6}$$

$$V_N = \frac{qN_D x_{dN}^2}{2\epsilon} \tag{2-7}$$

$$V_0 = V_P + V_N \tag{2-8}$$

$$x_{dP} = \left(\frac{2\epsilon V_P}{qN_A}\right)^{1/2} \tag{2-9}$$

$$x_{dN} = \left(\frac{2\epsilon V_N}{qN_D}\right)^{1/2} \tag{2-10}$$

The situation is shown in Fig. 2-1*b*.

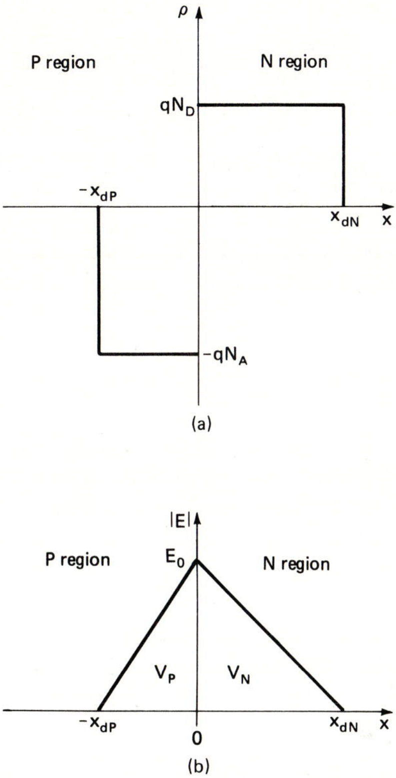

FIG. 2-1 Two-sided PN junction: (*a*) charge density and (*b*) electric field.

A *transition capacitance* is associated with a reverse-biased PN junction:

$$C = \frac{\epsilon A_j}{x_{dP} + x_{dN}} \tag{2-11}$$

where A_j = junction area.

The junction is called *one-sided* when dopant concentrations are strongly different in magnitude ($N_A \gg N_D$ or $N_D \gg N_A$). Depletion-layer width is very small in the heavily doped side, which gives a negligible contribution to total voltage drop across the junction.

The reverse breakdown of a PN junction may be ascribed either to the *zener* or to the *avalanche* effect.

Zener Effect [8]. At very high electric fields ($E > 10^6$ V/cm) electrons can be excited from the valence to the conduction band through the depletion layer, by the quantum-mechanical tunneling effect. Zener breakdown voltages are below about 5 V and only concern junctions heavily doped on both sides.

Avalanche Effect. When a junction is reverse-biased, a normally small leakage current is found to flow (see Sec. 2.1.4) as electrons and holes cross the depletion layer. These free carriers are accelerated from the field, between each two consecutive collisions with silicon lattice atoms. There is a finite probability for such carriers to acquire an energy equal to or greater than the *ionization energy* of silicon, creating an electron-hole pair: a silicon atom is ionized, sending an electron to the conduction band, and is neutralized by taking an electron from the valence band, where a new hole is created. The electron *multiplication coefficient* M_n is defined as the ratio between electron currents after and before crossing the depletion layer. The hole multiplication coefficient M_p is defined as the symmetric opposite of M_n. When the junction is reverse-biased, both M_n and M_p are greater than unity.

At some value of junction field $E_0 = E_{cr}$, impact ionization strongly increases and reverse junction current ceases being limited. For $E_0 = E_{cr}$, M_n (or M_p) is found to be infinitely large, and breakdown is caused by high-energy electrons (or holes). This situation, which appears as an abrupt increase of reverse current, is known as *avalanche breakdown*. *Breakdown voltage* V_{br} is expressed by

$$V_{br} = \frac{\epsilon E_{cr}^2}{2q}\left(\frac{1}{N_D} + \frac{1}{N_A}\right) \tag{2-12}$$

According to Eqs. 2-3*a* and 2-3*b*, the electric field E decreases, from the maximum value E_0 at the junction, with a slope proportional to the dopant concentration. For a fixed E_0, a lower slope (indicating lower N_D or N_A) determines a wider high-field region, where more ionizations of lattice atoms occur; higher M_n and M_p are then found. As a consequence, a lower E_{cr} is associated with lower dopant concentrations. It must be pointed out, however, that only the region close to the junction gives a major contribution to M_n and M_p, the probability for ionization being a very strong function of the electric field. Thus, the variation of E_{cr} with dopant concentrations, though not negligible, is not very pronounced.

For dopant concentrations in the range 10^{13} to 10^{16} cm^{-3}, the following approximation allows a direct calculation of E_{cr} with an error less than 6 percent [9]:

$$E_{cr} = \left(\frac{K_D}{N_D} + \frac{K_A}{N_A}\right)^{-1/8} \tag{2-13}$$

where $K_D = 2.405 \times 10^{-29}\ \text{V}^{-8}\cdot\text{cm}^5$ and $K_A = 1.966 \times 10^{-29}\ \text{V}^{-8}\cdot\text{cm}^5$. (Equation 2-13 is also applicable to one-sided P^+N or N^+P junctions, by letting N_A or $N_D \to \infty$. For PICs a P^+N junction normally has to be considered.) In the range $N_D = 10^{15}$ to $10^{16}\ \text{cm}^{-3}$, a better fit allows the determination of E_{cr} within 4 percent [9]:

$$E_{cr} = \left[\frac{N_D\ (\text{cm}^{-3})}{2 \times 10^{-29}}\right]^{1/8} \quad \text{V/cm} \tag{2-14}$$

$$V_{br} = \left[\frac{1.77 \times 10^{18}}{N_D\ (\text{cm}^{-3})}\right]^{3/4} \quad \text{V} \tag{2-15}$$

2.1.2 Collector-to-Base Avalanche Breakdown (Plane Limited Junction)

In actual cases the dimensions of the P- and N-type regions of a junction are obviously limited, and two phenomena may occur: *reach-through* and *punch-through.*

Reach-Through. For vertical NPN transistors, the collector-base junction can be considered one-sided (P^+N), base doping being much greater than collector doping. *Collector width* W_C is the vertical distance between the junction and the N^+ buried layer (BL). The overall structure is well-modeled as a P^+NN^+ abrupt diode (Figs. 2-2 and 2-3) with $W_C < x_{dN}$. The electric field at breakdown is shown in Fig. 2-4.

When $W_C > 0.2x_{dN}$, a condition always verified in practical cases, E_{cr}

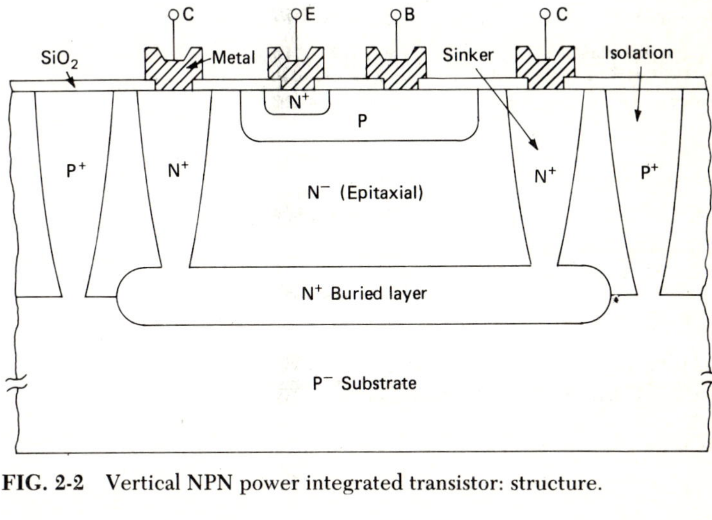

FIG. 2-2 Vertical NPN power integrated transistor: structure.

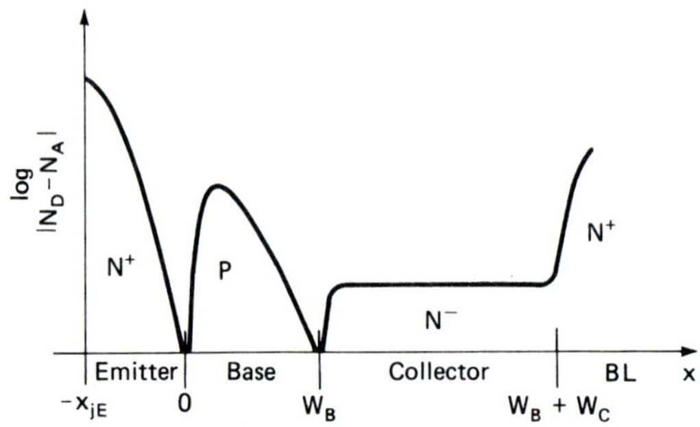

FIG. 2-3 Vertical NPN power integrated transistor: dopant concentration profile.

is found not to vary appreciably with respect to the unlimited junction value, given by Eqs. 2-13 and 2-14 [9]. The rapid decrease of E in the N^+ buried layer, however, determines lower values for V_{br}. Assuming (1) an abrupt transition between the collector and the buried layer and (2) a much higher dopant concentration in the buried layer than in the collector, the breakdown voltage is reduced to V'_{br}:

$$V'_{br} = V_{br}\left[1 - \left(1 - \frac{W_C}{x_{dN}}\right)^2\right] \tag{2-16}$$

V_{br} can be derived by Eq. 2-15, and x_{dN} by Eq. 2-10 with $V_N = V_{br}$. The condition $W_C < x_{dN}$, normally known as reach-through, often occurs in transistor collector-base junctions as well as in diodes.

Punch-Through. This is a condition possible only for transistor structures, which can be found especially in lateral PNP integrated transistors (Figs. 2-5 and 2-6). The depletion region extends mainly into the

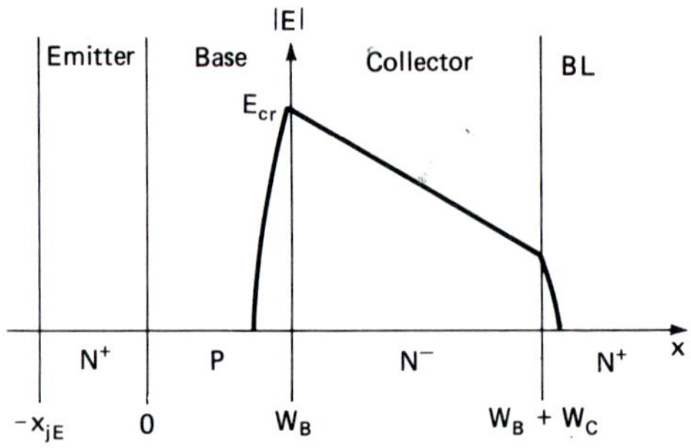

FIG. 2-4 Electric field at breakdown for a vertical NPN transistor.

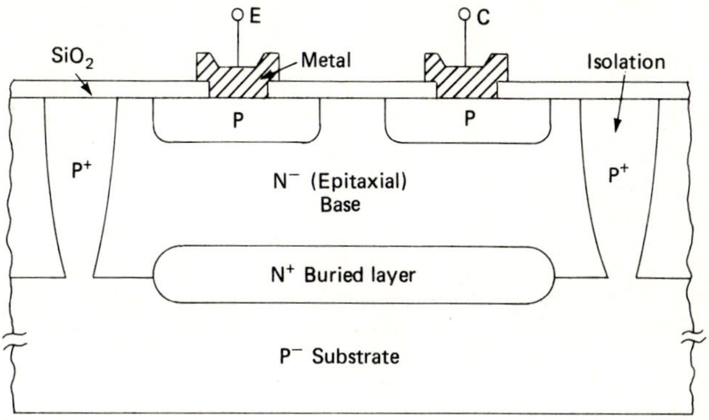

FIG. 2-5 Lateral PNP integrated transistor: structure (base contact not shown).

base of the lateral PNP. For low values of emitter-collector spacing W_B and low base dopant concentrations N_D, the depletion layer may *touch* the emitter at some value of junction field $E_{pt} < E_{cr}$ (Fig. 2-7). An emitter-collector short circuit via the depleted base occurs, allowing an unlimited current flow between collector and emitter. The electric field cannot be raised over E_{pt}, and a *pseudo breakdown* is verified at a voltage V_{pt}, which can be considerably lower than V_{br}. In this case the collector-base junction breaks down via the emitter-base junction. V_{pt} is given by

$$V_{pt} = \frac{\epsilon E_{pt}^2}{2qN_D} \tag{2-17}$$

and

$$BV_{CBO} = V_{pt} + BV_{EBO} \tag{2-18}$$

Punch-through may appear also in vertical NPN transistors, when base doping under the emitter is set at very low values (normally with

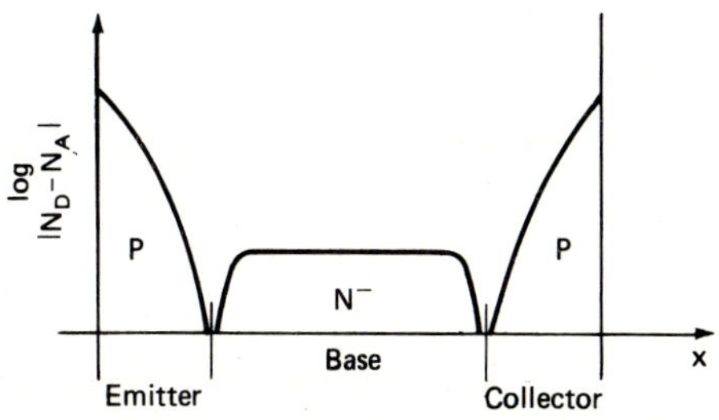

FIG. 2-6 Lateral PNP integrated transistor: dopant concentration profile.

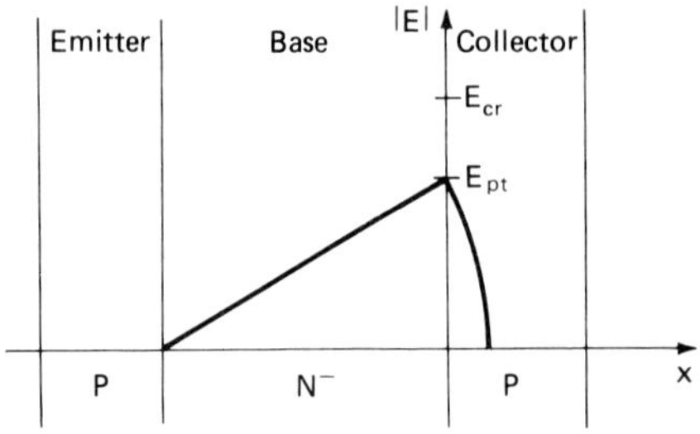

FIG. 2-7 Electric field at punch-through for a lateral PNP transistor.

the purpose of increasing device current gain). The charge neutrality principle, which in the case of a two-sided step junction is expressed by Eq. 2-2, states a minimum value for Q_B (*total base doping* under the emitter, per unit surface), in order to prevent punch-through:

$$Q_B \geq \left(\frac{2\epsilon V_{br} N_D}{q}\right)^{1/2} = \frac{\epsilon E_{cr}}{q} \tag{2-19}$$

For a 100-V power integrated circuit, E_{cr} is about 32 V/μm, and from Eq. 2-19, $Q_B \geq 2 \times 10^{12}$ cm^{-2} is calculated. Equation 2-19 states the minimum base charge density for NPN vertical transistors, or the minimum $W_B N_D$ product for PNP lateral transistors. For PNP transistors, with $N_D \simeq 2.5 \times 10^{15}$ cm^{-3}, punch-through is avoided when $W_B \geq 8.5$ μm. Larger values of W_B are required for higher-voltage PICs, according to the more general relation, obtained by combining Eqs. 2-14 and 2-19,

$$W_B \geq \left[\frac{2.86 \times 10^{16}}{N_D\ (\text{cm}^{-3})}\right]^{7/8} \quad \mu\text{m} \tag{2-20}$$

2.1.3 Collector-to-Base Avalanche Breakdown (Planar Junction)

Integrated transistor collector-base junctions are normally *planar*. In the curve region more space is available in the external N layer than in the internal P layer. In order to have an overall charge of zero, the depletion-layer extension on the N side must be lower in the curve region than in the plane portion [10]. Equipotential lines are then more closely spaced, the electric field is higher, and E_{cr} is reached at a voltage V_{br}^P lower than V_{br}.

In order to derive an exact solution, Poisson's equation must be solved in at least two dimensions. This implies the use of numerical techniques, such as *finite difference* [11–14] or *finite element* [15] methods.

Temple and Adler [16] derived a general empirical expression for the ratio r_E between electric field at breakdown in the plane portion and E_{cr}:

$$r_E = 0.871 + 0.125 \ln \frac{x_{jB}}{x_{dN}} \tag{2-21}$$

where x_{dN} is the depletion width at $V_N = V_{br}$, as given in Eq. 2-10. V_{br}^P is found by integrating Poisson's equation in the plane portion of the junction, where it is reduced to the one-dimensional form (Eq. 2-1):

$$V_{br}^P = \frac{\epsilon (r_E E_{cr})^2}{2qN_D} \tag{2-22}$$

E_{cr} can be derived from Eq. 2-14.

Reach-through occurs when

$$W_C < x_{dN}^P = \left(\frac{2\epsilon V_{br}^P}{qN_D} \right)^{1/2} \tag{2-23}$$

and can be calculated by inserting V_{br}^P and x_{dN}^P into Eq. 2-16.

Several techniques have been adopted which reduce the planar junction effect and allow r_E to approach unity. The principal ones are *floating P^+ limiting rings,* contouring the transistor base region [17, 18], and *field plates,* which are extensions of the base electrode beyond the junction, over the N region [19].

2.1.4 Leakage Current [20–22]

The reverse *junction current I_{CBO}* (emitter open) is the sum of *surface, generation,* and *diffusion* currents.

Surface Current I_S. This current is due to the formation of *depleted* or *inverted* regions at the surface, as a consequence of *ionic charges* which may be present at the surface or outside the semiconductor. I_S is rather insensitive to temperature variations and can be limited to very low values (nanoamps) by correct processing.

Generation Current I_G. Electron-hole pairs generated in the collector-base depletion layer cross the junction when they are accelerated by the electric field:

$$I_G = \frac{qn_i x_{dN} A_j}{2\tau_{sc}} \tag{2-24}$$

where τ_{sc} = space charge generation lifetime and n_i = free carrier concentration in intrinsic silicon [23, 24]:

$$n_i = aT^{3/2}e^{-T_0/T} \tag{2-25}$$

where $a = 3.80 \times 10^{16}\ \text{cm}^{-3}\cdot\text{K}^{-3/2}$
$T_0 = E_{g0}/2k$ (=6991 K for silicon)
E_{g0} = energy gap at 0 K (=1.206 eV for silicon)

Diffusion Current I_D. Electron-hole pairs are also generated outside the collector-base depletion layer. Some of these carriers diffuse to the depletion layer boundaries and then cross the junction, accelerated by the electric field. For an NPN transistor with a P^+N collector-base junction, I_D is essentially given by

$$I_D = \frac{qD_{pC}n_i^2A_j}{N_DW_C} \tag{2-26}$$

where D_{pC} = hole diffusivity in the collector. The contribution of I_D to I_{CBO} is relevant only for high-voltage discrete transistors ($BV_{CBO} > 500$ V) at high temperatures (over 100°C).

The variation of I_{CBO} versus T in integrated transistors is mainly determined by the variation of I_G versus T, when correct processing assures low I_S. From Eqs. 2-24 and 2-25,

$$\frac{d(\ln I_G)}{dT} = \frac{3}{2T} + \frac{T_0}{T^2} \tag{2-27}$$

Linearizing Eq. 2-27 around an average operation temperature, for instance, 330 K, somewhere between room temperature and the maximum junction temperature generally rated, results in a doubling of I_G (I_{CBO}) for each 10-K temperature increment. This is a useful rule of thumb to determine the variation of I_{CBO} with temperature.

2.1.5 Collector-to-Emitter Sustaining Voltage

Current distribution in a transistor structure is summarized in Fig. 2-8, where solid lines represent electron currents and dashed lines hole currents. $\alpha_F = I_C/I_E$ (<1) is the *common-base current gain.*

The electron current I_{no} injected at the emitter contact is reduced to $\alpha_F I_{no}$ at the depletion layer boundary and a hole base current $(1 - \alpha_F)I_{no}$ is injected from the base contact. Electrons crossing the depletion region create electron-hole pairs by impact ionization, which determines an electron current $M_n\alpha_F I_{no}$ at the lower end of the depletion layer and a hole current $\alpha_F(M_n - 1)I_{no}$ at the upper end. In V_{CEO} polar-

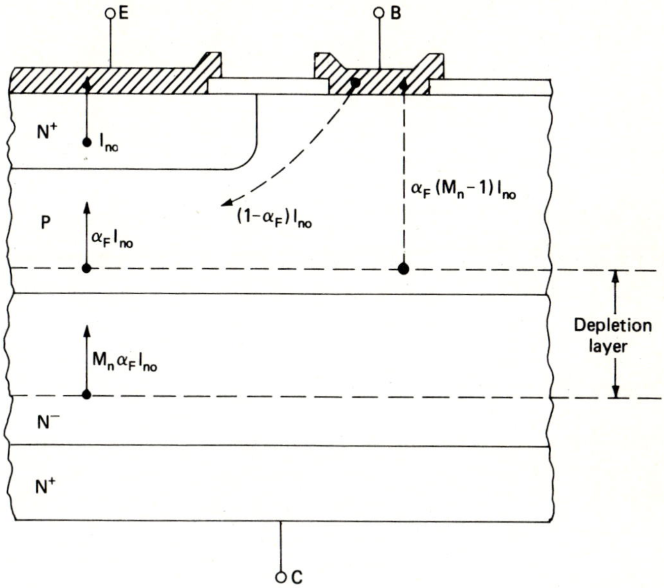

FIG. 2-8. Electron (solid lines) and hole (dashed lines) currents for an NPN transistor, including multiplication effect.

ization the base terminal is disconnected and then $I_B = 0$. The transistor can then sustain a current between collector and emitter only if

$$\alpha_F(M_n - 1)I_{no} = (1 - \alpha_F)I_{no} \tag{2-28}$$

or

$$M_n = \frac{1}{\alpha_F} \tag{2-29}$$

The *sustaining field* is defined as the junction field for which Eq. 2-29 holds. The *sustaining voltage* is

$$V_S = \frac{\epsilon E_S^2}{2q}\left(\frac{1}{N_D} + \frac{1}{N_A}\right) \tag{2-30}$$

E_S is lower than E_{cr}, and hence V_S is considerably lower than V_{br}; V_S depends on doping levels as well as on transistor current gain. Equation 2-29 clearly states that higher current gains imply lower M_n for sustaining, and therefore lower E_S and V_S. The variation of transistor current gain with I_C is reflected in sustaining characteristics, as shown in Fig. 2-9.

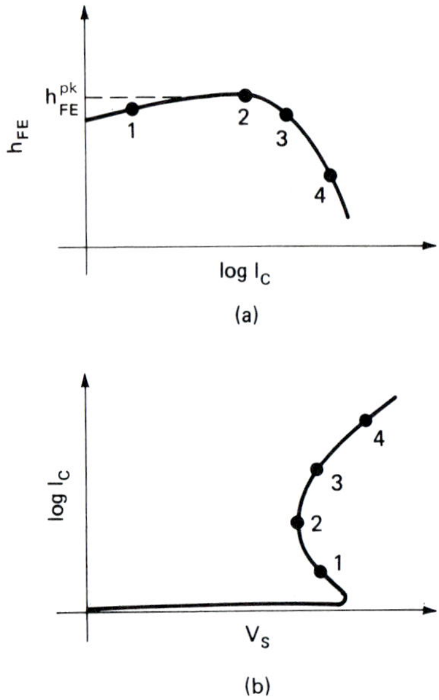

FIG. 2-9 Characteristics of (*a*) common-emitter current gain and (*b*) sustaining voltage. Corresponding points are marked by the same number.

Calculated V_S values show a good agreement with a modified *Miller-type* relation [25, 26]:

$$V_S = f\,V_{br}(1 + h_{FE})^{-1/n} \tag{2-31}$$

where f and n are empirical parameters fitted to exactly calculated values and h_{FE} is the *common-emitter current gain,* given by

$$h_{FE} = \frac{I_C}{I_B} = \frac{\alpha_F}{1 - \alpha_F} \tag{2-32}$$

In Figs. 2-10 and 2-11, f and n are plotted for vertical NPN and lateral PNP transistors, as a function of donor concentration N_D in the epitaxial layer.

To obtain actual V_{CEO} in vertical NPN transistors, it may be necessary to correct for reach-through:

$$V_{CEO} = V_S\left[1 - \left(1 - \frac{W_C}{x_{dN}^S}\right)^2\right] \tag{2-33}$$

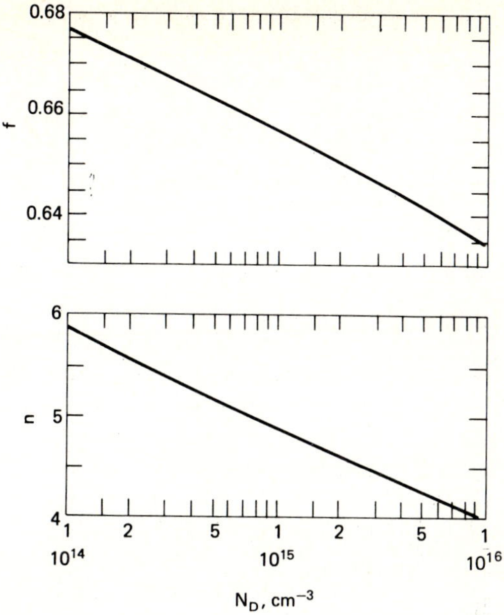

FIG. 2-10 Miller parameters f and n for vertical NPN transistors vs. N^- collector donor concentration N_D.

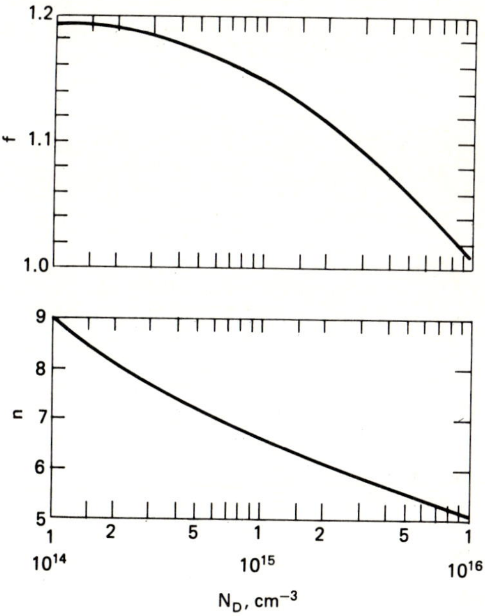

FIG. 2-11 Miller parameters f and n for lateral PNP transistors vs. N^- base donor concentration N_D.

where
$$x^S_{dN} = \left(\frac{2\epsilon V_S}{qN_D}\right)^{1/2} \tag{2-34}$$

If $V_S < V_{pt}$, punch-through occurs at lower field strengths than sustaining, and $V_{CEO} = V_{pt}$.

Curvature effects do not in theory influence V_S, because the active part of the transistor, where currents flow, is well within the plane portion of the junction, and Eq. 2-31 is always valid. It would be wrong to use BV_{CBO} or V^P_{br} instead of V_{br} in Eq. 2-31. However, when $r_E E_{cr} < E_S$, BV_{CBO} would be lower than V_{CEO}; in this case V_{CEO} turns out to be equal to the sum of BV_{CBO} and forward V_{BE}, that is, almost equal to BV_{CBO}.

Other interesting transistor configurations are V_{CER}, V_{CEV}, and V_{CES}, which will be discussed briefly.

In the V_{CER} condition (Fig. 2-12), a resistor R connects base and emitter. An approximate picture can be obtained defining a forward $V_{BE,0}$, or *thresholdlike* voltage. Below $V_{BE,0}$ ($\sim$0.4 V), current flow across the emitter-base junction is neglected. Two situations are possible:

1. $I_{CE} \leq V_{BE,0}/R$; no current flows across the emitter-base junction and

$$V_{CER} = BV_{CBO} + I_{CE}R \simeq BV_{CBO} \tag{2-35}$$

2. $I_{CE} > V_{BE,0}/R$; the V_{CER} sustaining condition is approximately

$$M_n = \frac{1/\alpha_F}{1 - V_{BE}(I_{CE})/I_{CE}R} \tag{2-36}$$

In the V_{CEV} condition a constant-voltage generator—V_A is added in series with R between the emitter and the base (Fig. 2-13). We have again two possible situations:

1. $I_{CE} \leq (V_{BE,0} + V_A)/R$; no current flows across the emitter-base junction and

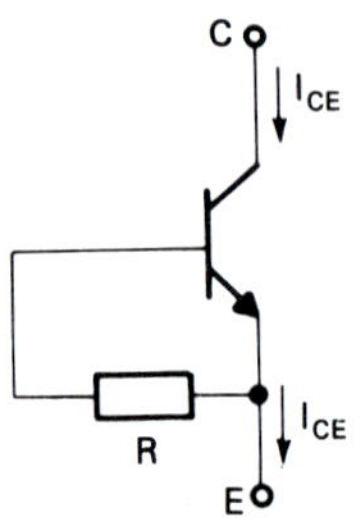

FIG. 2-12 V_{CER} transistor connection.

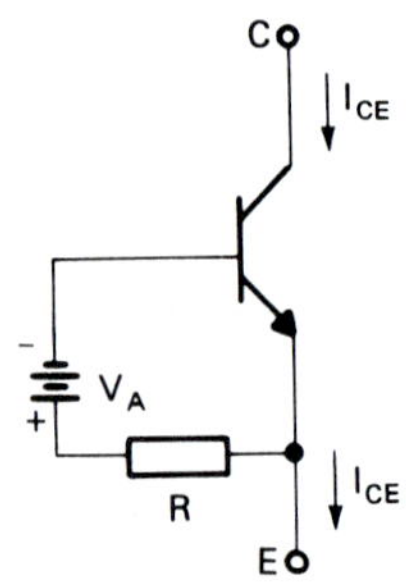

FIG. 2-13 V_{CEV} transistor connection.

$$V_{CEV} = BV_{CBO} + I_{CE}R - V_A \tag{2-37}$$

2. $I_{CE} > (V_{BE,0} + V_A)/R$; the sustaining condition is approximately

$$M_n = \frac{1/\alpha_F}{1 - [V_{BE}(I_{CE}) + V_A]/I_{CE}R} \tag{2-38}$$

V_{CER} and V_{CEV} can be calculated from Eq. 2-31, Fig. 2-10, and Fig. 2-11, by simple substitution of $1/M_n$, as given by Eq. 2-36 or 2-38, in place of α_F in Eq. 2-32.

In the V_{CES} condition (base shorted to emitter), sustaining normally does not occur, because the emitter-base junction cannot be forward-biased; thus, $V_{CES} = BV_{CBO}$. At rather high breakdown currents, internal device resistances can induce lateral voltage drops and some part of the junction could happen to be forward-biased, giving rise to a sustaining condition, much like the V_{CER} condition. If the device is punch-through-limited, $V_{CES} = V_{pt}$.

In Fig. 2-14 the voltage characteristics of an NPN transistor not punch-through-limited are qualitatively drawn.

2.2 CURRENT RATINGS

A typical transistor h_{FE} versus I_C curve is plotted in Fig. 2-9. In a wide region h_{FE} is almost constant and equal to its peak value h_{FE}^{pk}. The

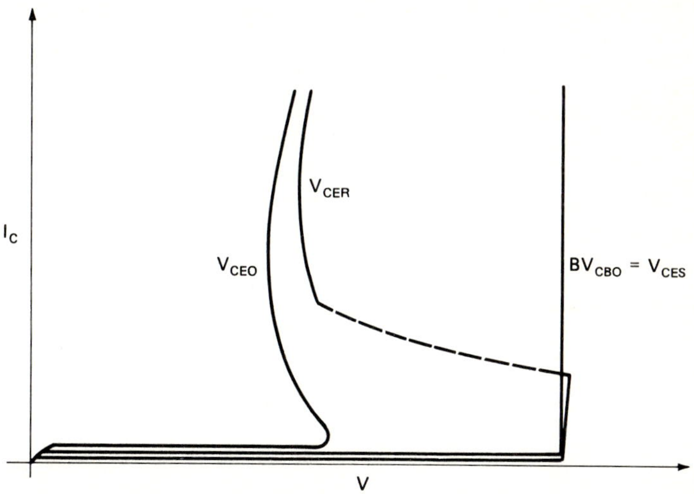

FIG. 2-14 Transistor limiting voltages BV_{CBO}, V_{CES}, V_{CEO}, and V_{CER} for a device not punch-through-limited.

decrease at low current levels, usually slight for correct wafer processing, does not substantially affect power transistor performance and will not be analyzed. An approximate theory, which proved to fit experimental data well, is proposed. Limitations and possible extensions of the model are discussed in Sec. 2.2.5. The last two subsections of Sec. 2.2 concern two parameters of primary importance for power transistors: *lifetime* and *emitter Gummel number.*

2.2.1 Low Current Densities (Low Injection)

An NPN vertical transistor is considered (Fig. 2-2), with a forward voltage V_{BE} and a reverse voltage V_{CB} applied to the junctions. Both a constant current density throughout the emitter and a collector current perpendicular to the device surface will be assumed; the two assumptions permit a *one-dimensional approximation.* V_{CB} is considered much lower than BV_{CBO}; multiplication effects are then negligible and $M_n \simeq M_p \simeq 1$.

The main currents involved in transistor action are described as follows:

1. Electron current I_e and hole current $I_{h\gamma}$, which cross the forward-biased emitter-base junction, are the primary currents.
2. While crossing the base region, electrons are recombined with holes injected at the base contact: current I_e is thus reduced to I'_e and a hole current $I_{h\tau}$ flows. Transport factor α_T is defined as

$$\alpha_T = \frac{I'_e}{I_e} = \frac{I_e - I_{h\tau}}{I_e} \tag{2-39}$$

The terminal currents are

$$I_E = I_e + I_{h\gamma} \tag{2-40}$$

$$I_B = I_{h\gamma} + I_{h\tau} \tag{2-41}$$

$$I_C = I_e - I_{h\tau} \tag{2-42}$$

The following condition usually holds for integrated as well as for discrete transistors:

$$W_B \ll L_{nB} \equiv (D_{nB}\tau_{nB})^{1/2} \tag{2-43}$$

where L_{nB}, τ_{nB} = electron diffusion length and lifetime in the base (Sec. 2.2.6). Under this condition, $I_{h\tau}$ turns out to be negligible with respect to I_e and the following approximation results:

$$I_C = I_e - I_{h\tau} \simeq I_e \tag{2-44}$$

The electron current I_e in the base region is the sum of a *drift term*, proportional to the drift field in the base E, and of a *diffusion term*, proportional to the gradient of the electron charge:

$$I_e = -qA_E\left(\mu_{nB}nE + D_{nB}\frac{dn}{dx}\right) \tag{2-45}$$

where A_E = emitter area
n = electron concentration
μ_{nB} = electron mobility in the base
D_{nB} = electron diffusivity in the base

In a similar way, a hole current I_h, always zero, can be defined. I_h must not be confused with $I_{h\gamma}$, determined by holes injected from the base contact into the emitter.

$$I_h = qA_E\left(\mu_{pB}pE - D_{pB}\frac{dp}{dx}\right) = 0 \tag{2-46}$$

where p = hole concentration
μ_{pB} = hole mobility in the base
D_{pB} = hole diffusivity in the base

When currents flow, the semiconductor is in nonequilibrium and the condition $pn = n_i^2$ is no longer valid. The *quasi-neutrality condition* must be applied:

$$p(x) = n(x) + N_A(x) \tag{2-47}$$

where $N_A(x)$ = acceptor dopant concentration in the base. At low current densities, the *low-injection approximation* can be used:

$$p(x) \simeq N_A(x) \tag{2-48}$$

In the case of a constant acceptor concentration (homogeneous base), from Eq. 2-48, $dp/dx = 0$ and thus, from Eq. 2-46, $E = 0$. Collector current in the base turns out to be only diffusion current. From Eqs. 2-44 and 2-45,

$$I_C = -qA_ED_{nB}\frac{dn}{dx} \tag{2-49}$$

A constant dn/dx and a triangular $n(x)$ distribution in the base (Fig. 2-15) are found and Eq. 2-49 becomes

$$I_C = \frac{qA_ED_{nB}n(0)}{W_B} \tag{2-50}$$

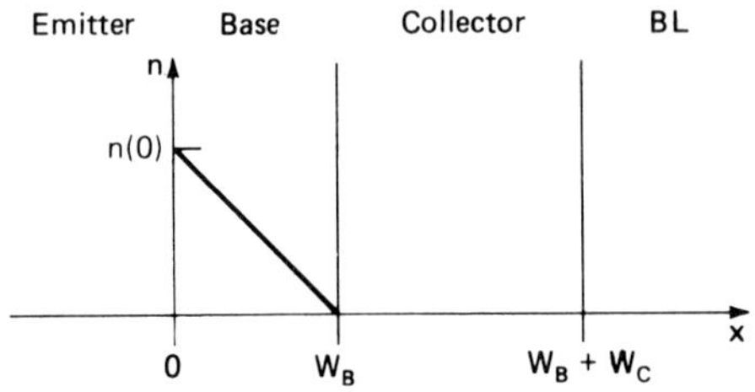

FIG. 2-15 Electron charge stored in low injection for a transistor with constant acceptor dopant concentration N_A in the base.

The semiconductor theory for nonequilibrium conditions gives

$$n(0) = \frac{n_i^2}{N_A} e^{V_{BE}/V_T} \tag{2-51}$$

where $V_T \equiv kT/q$ = thermal voltage (25.4 mV for T = 295 K) and k = Boltzmann's constant (1.38×10^{-23} J/K). Finally

$$I_C = \frac{qA_E D_{nB} n_i^2}{N_A W_B} e^{V_{BE}/V_T} \tag{2-52}$$

In the case of a *graded base,* the drift field is no longer negligible. Under low-injection conditions (Eq. 2-48), making use of Eq. 2-46,

$$E = \frac{D_{pB}}{\mu_{pB}} \frac{1}{p} \frac{dp}{dx} \simeq \frac{D_{pB}}{\mu_{pB}} \frac{1}{N_A(x)} \frac{dN_A(x)}{dx} \tag{2-53}$$

Taking into account *Einstein's relationship*

$$\frac{D}{\mu} = V_T \tag{2-54}$$

from Eqs. 2-44, 2-45, and 2-53,

$$I_C = -qA_E D_{nB} \left[\frac{n(x)}{N_A(x)} \frac{dN_A(x)}{dx} + \frac{dn(x)}{dx} \right] \tag{2-55}$$

$$\frac{d[n(x)N_A(x)]}{dx} = -\frac{I_C}{qA_E D_{nB}} N_A(x) \tag{2-56}$$

$$n(0)N_A(0) = \frac{I_C}{qA_E} \int_0^{W_B} \frac{N_A(x)}{D_{nB}[N_A(x)]} dx \tag{2-57}$$

The *base Gummel number* G_B [27] is defined as

$$G_B = \int_0^{W_B} \frac{N_A(x)}{D_{nB}[N_A(x)]}\,dx = \frac{1}{D_B}\int_0^{W_B} N_A(x)\,dx = \frac{Q_B}{D_B} \tag{2-58}$$

D_{nB} is a weak function of N_A, and D_B is some average value in the base. $n(0)$, similarly to Eq. 2-51, is given by

$$n(0) = \frac{n_i^2}{N_A(0)}\,e^{V_{BE}/V_T} \tag{2-59}$$

and finally

$$I_C = \frac{qA_E n_i^2}{G_B}\,e^{V_{BE}/V_T} \tag{2-60}$$

The hole current crossing the emitter-base junction can be derived in a similar way, by analyzing the hole current in the emitter side:

$$I_{h\gamma} = \frac{qA_E n_i^2}{G_E}\,e^{V_{BE}/V_T} \tag{2-61}$$

where G_E = *emitter Gummel number* (it depends on emitter structure and is discussed in Sec. 2.2.7).

The transport factor is given by

$$\alpha_T = 1 - \frac{W_B^2}{2\eta L_{nB}^2} \tag{2-62}$$

The η parameter [28] is equal to 1 for constant base doping; higher values are found for graded bases, as a consequence of the electric field associated with the dopant concentration gradient.

Substituting Eq. 2-62 into Eq. 2-39 gives

$$I_{h\tau} = I_e(1 - \alpha_T) \simeq \frac{I_C W_B^2}{2\eta L_{nB}^2} \tag{2-63}$$

The expressions for current gain can now be derived:

$$h_{FE0}^{-1} = \frac{I_B}{I_C} = \frac{I_{h\gamma} + I_{h\tau}}{I_C} = h_{FE0\gamma}^{-1} + h_{FE0\tau}^{-1} \tag{2-64}$$

$$h_{FE0\gamma} = \frac{I_C}{I_{h\gamma}} = \frac{G_E}{G_B} \tag{2-65}$$

$$h_{FE0\tau} = \frac{I_C}{I_{h\tau}} = \frac{2\eta L_{nB}^2}{W_B^2} = \frac{2\eta D_B \tau_{nB}}{W_B^2} \tag{2-66}$$

Equations 2-64 to 2-66 are valid in the region where h_{FE} is approximately constant with I_C. For integrated NPN vertical transistors, W_B is normally small and $h_{FE}^{pk} \simeq h_{FE0\gamma}$. The electron charge stored in the base is

$$Q_n = \int_0^{W_B} n(x)\, dx \tag{2-67}$$

In low injection, Q_n is given by

$$Q_n = \frac{I_C W_B^2}{2\eta q A_E D_B} \tag{2-68}$$

2.2.2 High Injection in the Base

When I_C is increasing, the electron charge stored in the base also increases, until Eq. 2-48 is no longer applicable and the more general Eq. 2-47 must be considered. Making no assumption regarding $N_A(x)$, Eq. 2-53 becomes

$$E(x) = \frac{D_{pB}}{\mu_{pB}} \frac{1}{N_A(x) + n(x)} \frac{d[N_A(x) + n(x)]}{dx} \tag{2-69}$$

By the same procedure followed in Sec. 2.2.1 we find

$$n(0)[n(0) + N_A(0)] = \frac{I_C}{qA_E D_B}(Q_B + Q_n) \tag{2-70}$$

In high injection,

$$N_A(0) \ll n(0) = n_i e^{V_{BE}/2V_T} \tag{2-71}$$

and so

$$I_C = \frac{qA_E D_B n_i^2}{Q_B + Q_n} e^{V_{BE}/V_T} \tag{2-72}$$

In order to evaluate Q_n, E is assumed constant and equal to $E(0)$. Thus, Eq. 2-71 holds and

$$E \simeq \frac{D_{pB}}{\mu_{pB}} \frac{1}{n} \frac{dn}{dx} \tag{2-73}$$

The collector current and the stored base charge turn out to be

$$I_C \simeq -2qA_ED_B\frac{dn}{dx} = \frac{2qA_ED_Bn(0)}{W_B} \qquad (2\text{-}74)$$

$$Q_n \simeq \frac{n(0)W_B}{2} \simeq \frac{I_CW_B^2}{4qA_ED_B} \qquad (2\text{-}75)$$

A comparison with Eq. 2-68 shows that the low-injection expression is still valid, provided η is set equal to 2, regardless of the dopant distribution in the base.

It is now possible to derive the expressions for current gain under high-injection conditions:

$$h_{FE\gamma}^{-1} = \frac{I_{h\gamma}}{I_C} = \frac{Q_B + Q_n}{D_BG_E} = h_{FE0\gamma}^{-1} + \frac{W_B^2I_C}{4qA_ED_B^2G_E} \qquad (2\text{-}76)$$

$$h_{FE\tau}^{-1} = \frac{I_{h\tau}}{I_C} = \frac{W_B^2}{4D_B\tau_{nB}} \qquad (2\text{-}77)$$

$$h_{FE}^{-1} = h_{FE\gamma}^{-1} + h_{FE\tau}^{-1} \qquad (2\text{-}78)$$

It can be noted that Q_n adds to Q_B in Eq. 2-76. This means that h_{FE} is inversely proportional to the total number of holes (per surface unit) in the base region.

2.2.3 Early Effect [29]

At high V_{CB} ($\simeq V_{CE}$) values, the extension of the depletion region into the base must be considered. Depleted-base charge and thickness must be subtracted from the low-voltage values in order to obtain the effective base charge Q_B and base width W_B, both functions of V_{CE}:

$$Q_B'(V_{CE}) = Q_B - \frac{\epsilon E_0}{q} = Q_B - x_{dN}N_D \simeq Q_B - \left(\frac{2\epsilon N_D V_{CE}}{q}\right)^{1/2} \qquad (2\text{-}79)$$

$$W_B'(V_{CE}) = W_B - x_{dP} \qquad (2\text{-}80)$$

All the expressions for current gain derived in the previous sections remain valid, provided the substitutions given in Eqs. 2-79 and 2-80 are made. The Early effect is observed as an increase of h_{FE} with voltage, which is more consistent when the total base charge Q_B is small. It is interesting to observe that when punch-through occurs (Eq. 2-19), Q_B' becomes zero, current gain becomes infinite, and there is an unlimited current flow.

2.2.4 High Injection in the Collector (Saturation and Quasi Saturation)

An approximate but effective model will be developed to describe *saturation* and *quasi saturation.* Transistor I_C-V_{CE} characteristics are divided into three regions, as shown in Fig. 2-16.

1. The *nonsaturation region,* at the right of the line marked $R_P + R_{SC}$. R_{SC} is the *collector series resistance.*

$$R_{SC} = \frac{\rho_C W_C}{A_E} \tag{2-81}$$

where ρ_C = collector resistivity. R_P is the parasitic series resistance, which comprises in this approximate theory the internal device resistances and the lead and connection resistances, as well as the voltage difference between V_{BE} and V_{BC}, both positive, since the emitter-base and collector-base junctions are both forward-biased in saturation.

2. The *quasi-saturation region,* between R_P and $(R_P + R_{SC})$.
3. The *saturation region,* formed by the points lying on R_P.

Transistor behavior in the three regions will be discussed for constant collector current I_C and decreasing V_{CE}.

Nonsaturation Region. At point A (Fig. 2-16), V_{CE} is the sum of the voltage drop in the depletion region and the ohmic drops in the undepleted collector and across R_P:

$$V_{CE} = \frac{\epsilon(E_0 - E_C)^2}{2qN_D} + E_C W_C + I_C R_P \tag{2-82}$$

where

$$E_C = \frac{\rho_C I_C}{A_E} \tag{2-83}$$

Figure 2-17 is a representation of electric field and electron charge distribution. V_{CE} is given by the dashed area (which includes the $I_C R_P$ drop). In nonsaturation the results obtained in Secs. 2.2.1 and 2.2.2 are valid.

Quasi-Saturation Region. At point B in Fig. 2-16,

$$V_{CE} = I_C(R_P + R_{SC}) \tag{2-84}$$

The depletion layer disappears and V_{CE} is determined only by the ohmic drop on R_{SC}. The electron distribution is the same as in the nonsaturation region, while the electric field varies according to Fig. 2-18.

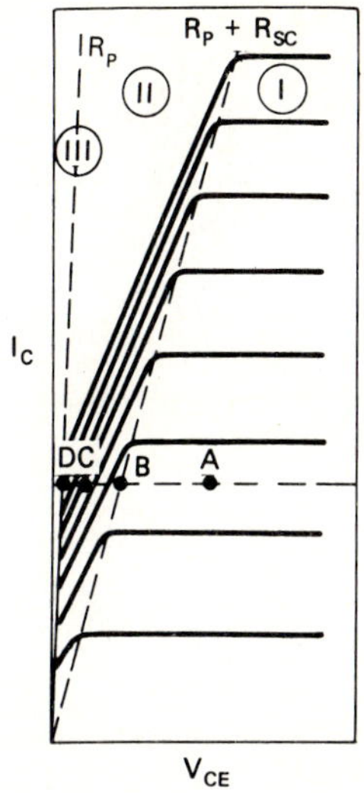

FIG. 2-16 Transistor I_C-V_{CE} characteristics, showing (I) nonsaturation, (II) quasi-saturation, and (III) saturation regions.

When V_{CE} is further decreased (point C in Fig. 2-16), the following phenomena occur [30–34]:

1. The collector-base junction becomes forward-biased.
2. Holes are injected from the base into some part of the collector region (from $x = W_B$ to $x = W_{CIB}$), which behaves as an extended base (Fig. 2-19). The sum of this extended base and of the metallurgical base is usually called the *c*urrent *i*nduced *b*ase.
3. In the extended base the quasi-neutrality condition must be applied and Eq. 2-47 is valid by simply substituting $-N_D$ for $N_A(x)$.
4. The normally small N_D value is negligible with respect to $n(x)$ and the extended base can be considered a real base region with $N_A = 0$.

For these reasons, the theory developed for high injection in the base (Sec. 2.2.2) is readily applicable to this *base-extension* or *base-widening* case. Q_n, h_{FE_γ}, and h_{FE_T} are still given by Eqs. 2-76 to 2-78, simply by substituting W_{CIB} for W_B. Total h_{FE} decreases with respect to the nonsaturation region, and a higher I_B is found (Fig. 2-16). At fixed I_C and I_B, W_{CIB} can be calculated by using h_{FE} expressions. V_{CE} is the sum of ohmic drops on the residual collector and on R_P:

$$V_{CE} = I_C R_P + \frac{\rho_C(W_B + W_C - W_{CIB})}{A_E} I_C \qquad (2\text{-}85)$$

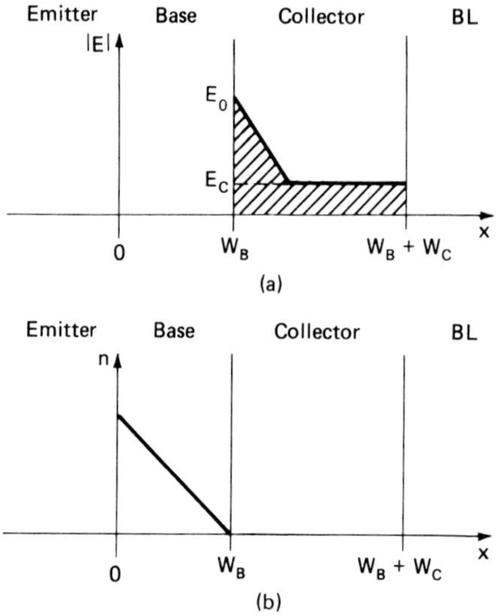

FIG. 2-17 Nonsaturation region: *(a)* electric field and *(b)* electron stored charge.

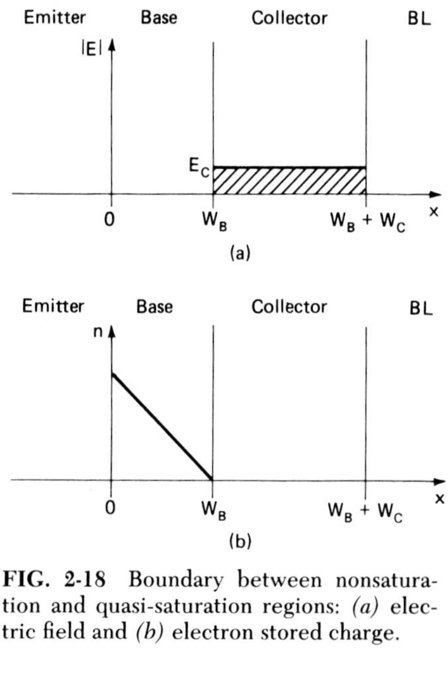

FIG. 2-18 Boundary between nonsaturation and quasi-saturation regions: *(a)* electric field and *(b)* electron stored charge.

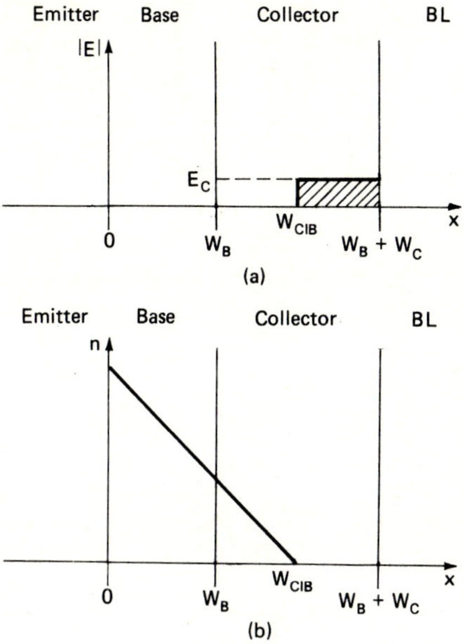

FIG. 2-19 Quasi-saturation region: *(a)* electric field and *(b)* electron stored charge.

Saturation Region. When the condition $V_{CE} = I_C R_P$ holds, the collector voltage drop is zero and $W_{CIB} = W_B + W_C$. The electron distribution, corresponding to point *D* in Fig. 2-16, is given by the solid line in Fig. 2-20 for $I_B = I_B^{sat}$, which is the minimum I_B required for the transistor to be in saturation. *Saturated gain* is defined as the ratio I_C/I_B^{sat}.

The expressions derived in Sec. 2.2.2 still hold, if $W_B + W_C$ is substituted for W_B:

$$Q_n^{sat} = \frac{I_C(W_B + W_C)^2}{4qA_E D_B} \tag{2-86}$$

$$h_{FE\gamma}^{sat\text{-}1} = \frac{Q_B + Q_n^{sat}}{D_B G_E} \tag{2-87}$$

$$h_{FE\tau}^{sat\text{-}1} = \frac{(W_B + W_C)^2}{4D_B \tau_{nB}} \tag{2-88}$$

$$h_{FE}^{sat\text{-}1} = \frac{I_B^{sat}}{I_C} = h_{FE\gamma}^{sat\text{-}1} + h_{FE\tau}^{sat\text{-}1} \tag{2-89}$$

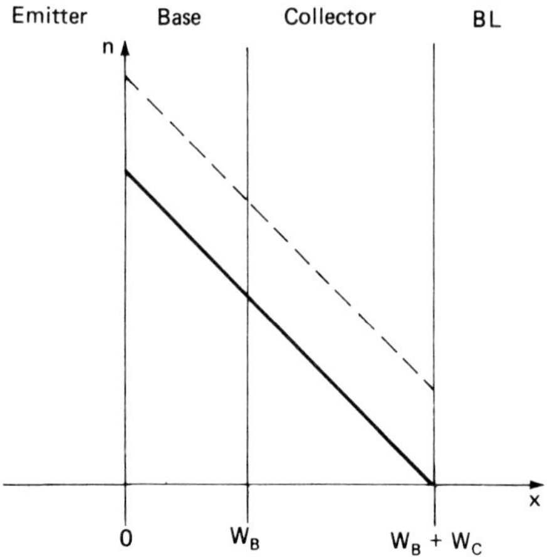

FIG. 2-20 Electron stored charge in saturation (solid line, $I_B = I_B^{sat}$) and in oversaturation (dashed line, $I_B > I_B^{sat}$).

When the two following conditions are met

$$h_{FE}^{sat} \ll h_{FE_T}^{sat} \tag{2-90}$$

and

$$h_{FE}^{sat} \ll \frac{G_E}{G_B} \tag{2-91}$$

a simple expression relating I_C and I_B^{sat} is derived:

$$\frac{I_C^2}{I_B^{sat}} \simeq \frac{I_C D_B G_E}{Q_n^{sat}} = \frac{4qA_E D_B^2 G_E}{(W_B + W_C)^2} \tag{2-92}$$

which clearly states the influence of physical parameters on the saturation characteristics. Integrated power transistors usually verify the conditions established by Eqs. 2-90 and 2-91, and then Eq. 2-92 is applicable.

When the base current exceeds I_B^{sat}, *oversaturation* occurs (dashed line in Fig. 2-20). The stored charge in oversaturation is obtained from Eq. 2-92:

$$Q_n \simeq Q_n^{sat} \frac{I_B}{I_B^{sat}} \qquad I_B > I_B^{sat} \tag{2-93}$$

2.2.5 Extensions of the Model

Some assumptions previously taken are discussed and indications are given to improve the current gain model developed.

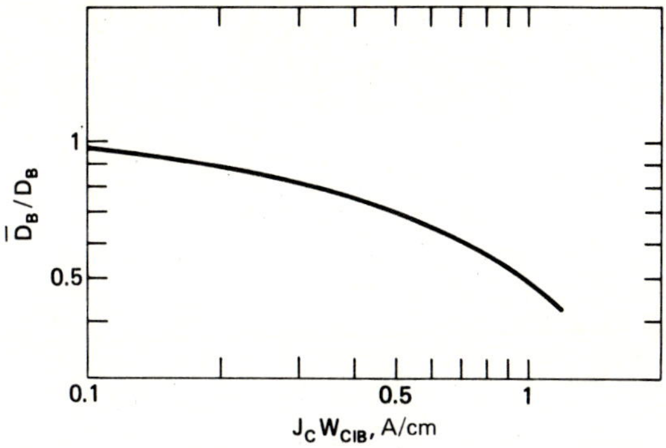

FIG. 2-21 Normalized diffusivity in high injection. (From Conti and Corda, Alta Frequenza, *46*, 122–62E, 1977.)

Base Diffusivity D_B. Diffusivity is known to decrease at high injection, because of *hole-electron scattering*. An effective diffusivity $\overline{D}_B$ can be defined and used instead of the low-injection value D_B. $\overline{D}_B$ is a function of the average carrier concentrations $\overline{p} \simeq \overline{n} = Q_n/W_{CIB} \propto J_C W_{CIB}/\overline{D}_B$. $\overline{D}_B$ is plotted versus the product $J_C W_{CIB}$ in Fig. 2-21 [35].

Collector Mobility. Electron mobility in the collector decreases at high electric fields E_C, due to *electron-electron scattering*. Mobility variations determine resistivity variations, according to

$$\rho = \frac{1}{q\mu N_D} \tag{2-94}$$

modifying voltage drops in the collector and quasi-saturation V_{CE}. Collector current and electric field are related by

$$E_C = \rho J_C = \rho \frac{I_C}{A_E} \tag{2-95}$$

Ryder [36] derived a fair approximation of the dependence of velocity v on E, shown in Fig. 2-22. The Ryder equations in the three regions are

$$\mu = \mu_0 \qquad E \leq E_1, \text{ region } A \tag{2-96a}$$

$$\mu = \mu_0 \left(\frac{E_1}{E}\right)^{1/2} \qquad E_1 \leq E \leq E_2, \text{ region } B \tag{2-96b}$$

$$v = v_l \qquad E \geq E_2, \text{ region } C \tag{2-96c}$$

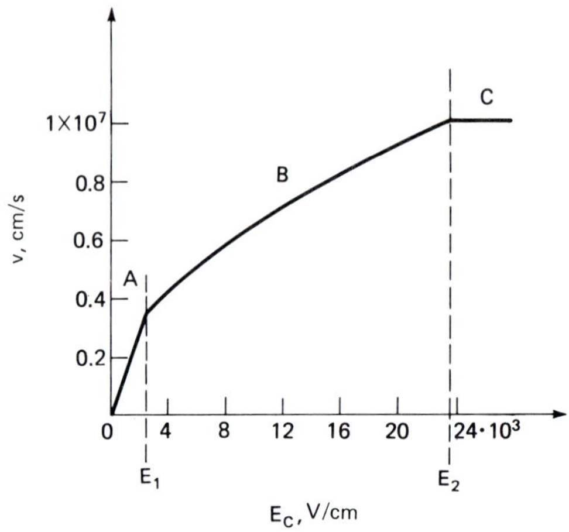

FIG. 2-22 Electron velocity vs. electric field, according to Ryder approximation [36].

where $E_1 = 2.5 \times 10^3$ V/cm for electrons
$E_1 = 5 \times 10^3$ V/cm for holes
$v_l = 1 \times 10^7$ cm/s for both electrons and holes [37, 38]
$E_2 = (v_l/\mu_0)^2/E_1$

The three regions must be considered equally for transistor I_C-V_{CE} characteristics [39].

Constant mobility (region A). The condition $E_C \leq E_1$ is equivalent to

$$I_C \leq \frac{E_1 A_E}{\rho_C} = q\mu_0 N_D E_1 A_E \tag{2-97}$$

In this case $\mu = \mu_0$ and $\rho = \rho_C$ (low field values). The results obtained in Sec. 2.2.4 apply, with $D_B = \mu_0 V_T$ (Eq. 2-54).

Intermediate (region B). The condition $E_1 \leq E_C \leq E_2$ is equivalent to

$$\frac{E_1 A_E}{\rho_C} \leq I_C \leq qA_E N_D v_l \tag{2-98}$$

Inserting Eqs. 2-94 and 2-96*b* into Eq. 2-95 gives

$$E_C = \frac{\rho_C^2 I_C^2}{E_1 A_E^2} \tag{2-99}$$

Quasi-saturation V_{CE} is modified to

$$V_{CE} = I_C R_P + \frac{\rho_C^2 I_C^2}{E_1 A_E^2} (W_B + W_C - W_{CIB}) \qquad (2\text{-}100)$$

Constant velocity (region C). The condition $E_C \geq E_2$ corresponds to compensation for fixed donors in the collector by traveling charges, namely, electrons for NPN transistors:

$$n = \frac{I_C}{q A_E v_l} \geq N_D \qquad (2\text{-}101)$$

An approximate model is obtained by fixing $E = E_2$ at $x = W_{CIB}$ and solving Poisson's equation in the residual collector region. Quasi-saturation V_{CE} turns out to be

$$V_{CE} = I_C R_P + \frac{q}{\epsilon} \left(\frac{I_C}{q A_E v_l} - N_D \right) \frac{(W_B + W_C - W_{CIB})^2}{2} + \frac{v_l^2}{\mu_0^2 E_1} (W_B + W_C - W_{CIB}) \qquad (2\text{-}102)$$

W_{CIB} in Eqs. 2-100 and 2-102 can be derived by quasi-saturation current gain (Sec. 2.2.4) when I_C and I_B are fixed. The I_C-V_{CE} characteris-

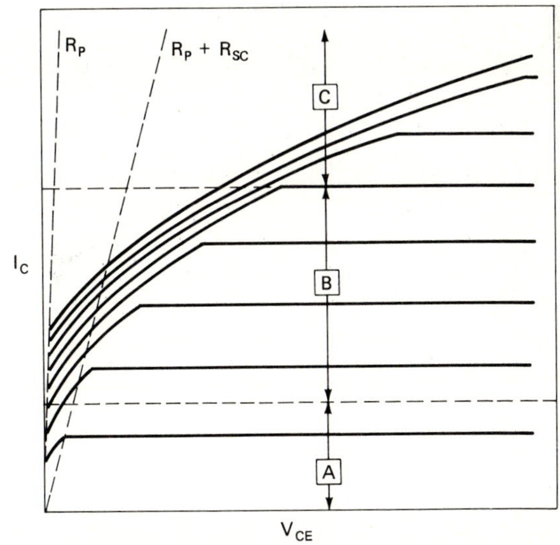

FIG. 2-23 Transistor I_C-V_{CE} characteristics, taking into account the reduction of electron mobility in the collector at high electric fields.

tics, as obtained by the use of nonconstant collector mobility, are shown in Fig. 2-23. The effect of mobility reduction is quite evident at large collector current densities: the quasi-saturation region spreads out well beyond the line marked $R_P + R_{SC}$.

Two-Dimensional Effects. The main two-dimensional effects in power transistor static operation are as follows:

1. *Enlargement of the cross-sectional area* of current flow passing from emitter to buried layer. A higher value of A_E with respect to *geometrical* emitter area had to be considered in current-gain theory. This effect is not very important in integrated power transistors, where $W_B + W_C$ values are small if compared with horizontal geometry dimensions. The situation is different for high-voltage discrete transistors, where this effect is found to be important.
2. *Current crowding (pinch-off)* [40], due to the *transverse base resistance* $r_{bb'}$, between the active base and the base contact (Fig. 2-24). As a result of the voltage drop caused by base current through $r_{bb'}$,

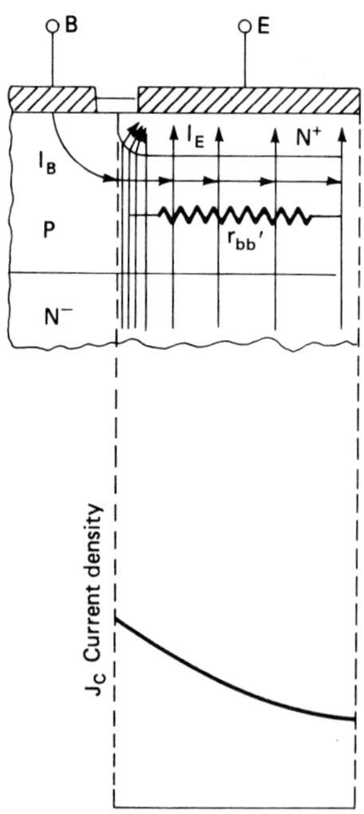

FIG. 2-24 Transistor pinch-off: voltage drops of base current on r_{bb}' determine higher current densities at emitter edge than at emitter center.

the forward V_{BE} at the emitter center is found to be lower than at the emitter edge. The current density increases exponentially with V_{BE}, according to Eq. 2-60, and consequently can vary considerably across the emitter. Interdigitated or cellular transistor geometries reduce this effect, which is less important in saturation and quasi saturation because, in these high-injection conditions, the base conductivity is enhanced by stored charges and Q_n adds to Q_B.

3. *Thermal crowding* may occur when the transistor has to withstand high power levels, determined by relatively low I_C and high V_{CE} [41]. In this case the emitter edge, surrounded by the cold base region, is cooled down better than the emitter center (where the temperature attains a maximum) due to lateral thermal conduction. As a consequence, higher current densities at the center are observed.

2.2.6 Lifetime

Electron lifetime τ_{nB}, assumed constant throughout this chapter, actually varies with the injection level. In this section the process of recombination will be examined and a technique to measure high-injection lifetime will be described.

The recombination of excess electrons in the base (metallurgical or current-induced) is described by the following equation.

$$-\frac{dn}{dt} = a_1 n + a_2 n^2 + a_3 n^3 \qquad \text{(2-103)}$$

The first term in Eq. 2-103 refers to *indirect recombination*, the second to *direct recombination*, and the third to *Auger recombination*.

Direct Recombination. There is a finite probability that an excess electron will make a *direct transition* from the conduction to the valence band. But silicon, like germanium and unlike GaAs, is an *indirect-gap* semiconductor, which means that the maximum energy level of the valence band and the minimum level of the conduction band do not occur for the same value of wave vector **k**. A momentum change, in addition to an energy change, is thus required. The result is that the probability of direct transition is very low and the term $a_2 n^2$ in Eq. 2-103 can be neglected.

Indirect Recombination. Heavy metals (such as copper, iron, or gold) or defects in the silicon lattice determine the presence of *deep energy levels*, situated approximately halfway between the valence and the conduction band. The termination of the lattice at the surface also determines interband levels. Indirect recombination corresponds to a *two-step transition* of an electron from conduction to valence band, via an intermediate energy level. The probability for such a process is much higher than for direct recombination; the term $a_I n$ is thus the most important in Eq. 2-103. The electron lifetime due to indirect recombination $\tau_{I,n} \equiv 1/a_1$ is found to be [42]

$$\tau_{I,n} = \tau_{n0} \text{ (low-injection)} \tag{2-104}$$

and

$$\tau_{I,n} = \tau_{n0} + \tau_{p0} \text{ (high-injection)} \tag{2-105}$$

where τ_{n0} = lifetime for electrons in P^+ regions and τ_{p0} = lifetime for holes in N^+ regions. It is interesting to observe that lifetime increases from low to high injection and that electron and hole lifetimes are equal in high-injection conditions.

Auger Recombination. In very-high-injection conditions the number of electrons and holes is very large and a *direct recombination*, with energy transfer to a *third carrier*, may occur. This is *Auger recombination*, which corresponds to the term $a_3 n^3$ in Eq. 2-103. The electron lifetime due to Auger recombination is

$$\tau_{A,n} = \frac{1}{a_3 n^2} \tag{2-106}$$

where $a_3 \simeq 2 \times 10^{-31}$ cm^6/s [24].

The transistor lifetime τ_{nB} is almost coincident with $\tau_{I,n}$ in low- and high-injection conditions, where Eqs. 2-104 and 2-105, respectively, apply. In very-high-injection conditions, Auger recombination cannot be neglected and

$$\frac{1}{\tau_{nB}} = \frac{1}{\tau_{I,n}} + \frac{1}{\tau_{A,n}} \tag{2-107}$$

where $\tau_{I,n}$ and $\tau_{A,n}$ are respectively given by Eqs. 2-105 and 2-106. By inserting the average electron concentration $\overline{n}$ into Eq. 2-106 and taking into account the variation of diffusivity versus the product $J_C W_{CIB}$ (Fig. 2-21), τ_{nB} is found to decrease appreciably for Auger recombination when $J_C W_{CIB} > 1$ A/cm, which corresponds to very high current densities, rarely used in the operation of integrated transistors.

Measurement of High-Injection Lifetime. To measure high-injection electron and hole lifetime τ, a simple method may be used, based on the natural voltage decay of the forward-biased collector-base diode. A procedure similar to the one developed in Sec. 2.2.2 leads to the following expressions, relating forward current I_F, forward voltage V_F, and stored charge Q:

$$Q = \frac{I_F W_C^2}{4qA_j D_{pC}} \tag{2-108}$$

$$I_F = \frac{2qA_j D_{pC} n_i}{W_C} e^{V_F/2V_T} = I_0 e^{V_F/2V_T} \tag{2-109}$$

$$V_F = 2V_T \ln \frac{I_F}{I_0} \tag{2-110}$$

A current I_F is applied to the collector-base diode. By opening the circuit, the natural decay of I_F and V_F can be observed (Fig. 2-25):

$$\frac{dQ}{dt} = -\frac{Q}{\tau} \tag{2-111}$$

$$Q(t) = Q(0)e^{-t/\tau} \tag{2-112}$$

From Eq. 2-108,

$$I_F(t) = I_F(0)e^{-t/\tau} \tag{2-113}$$

and, by inserting into Eq. 2-110,

$$\Delta V_F = V_F(t_2) - V_F(t_1) = -2V_T \frac{t_2 - t_1}{\tau} \tag{2-114}$$

The decay of V_F has the form shown in Fig. 2-26. An initial transient (region I) is followed by high-injection decay (region II) and by low-injection decay (region III). τ can be measured as the time required for V_F to decrease by an amount $2V_T$ in region II. It must, however, be

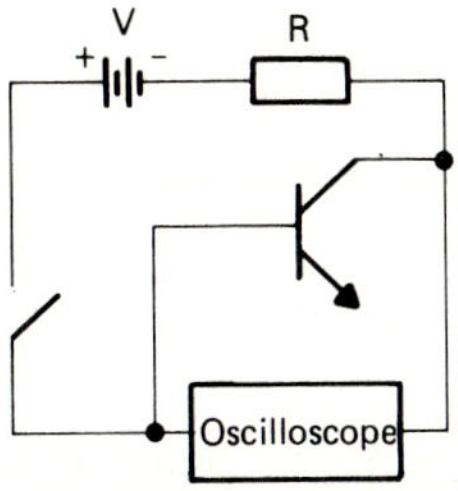

FIG. 2-25 Circuit for the measurement of high-injection lifetime.

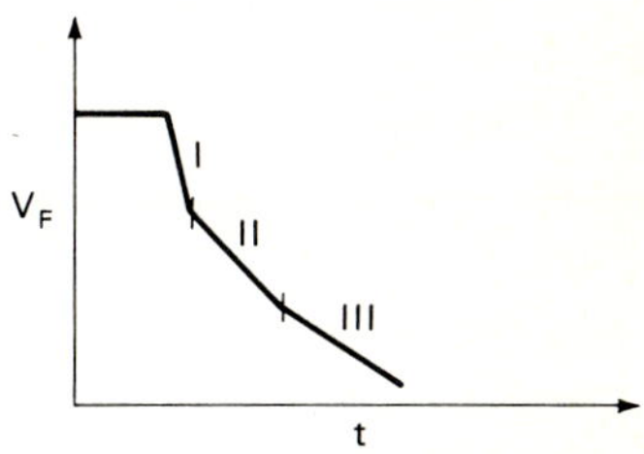

FIG. 2-26 Idealized V_F decay with time, resulting from the circuit shown in FIG. 2-25.

noted that, for vertical NPN integrated transistors, the condition $h_{FE} \simeq h_{FE\gamma}$ is normally verified, disregarding the degree of saturation, for lifetime values not less than about 1 μs, easily obtained in normal device processing.

2.2.7 Emitter Gummel Number

Emitter Gummel number G_E is a very important parameter for power transistors, because its magnitude strongly affects device current ratings. An analysis of hole current $I_{h\gamma}$ in the emitter region, parallel to the analysis of electron current I_e in the base region, developed in Sec. 2.2.1, would give the following results:

$$I_{h\gamma} = \frac{qA_E n_i^2}{G_E} e^{V_{BE}/V_T} \tag{2-61}$$

and

$$G_E = \int_{-x_{jE}}^{0} \frac{N_{DE}(x)}{D_{pE}[N_{DE}(x)]} dx \tag{2-115}$$

where D_{pE} = hole diffusivity and N_{DE} = donor concentration in the emitter region. Equation 2-115, derived symmetrically to Eq. 2-58, is *not valid* however, for the following two reasons:

1. For heavy emitter doping, a *degeneracy* of those donor levels lying in the band gap near the conduction band edge occurs. The effect is a *lowering* of the conduction band edge by some amount ΔE_g, known as *band-gap* narrowing [43–45]. ΔE_g is a function of N_{DE} and affects the value of the *pn* product, which in equilibrium conditions is no longer equal to n_i^2, but to n_{iE}^2, defined as

$$n_{iE}^2 = n_i^2 e^{\Delta E_g/kT} \tag{2-116}$$

2. There is no certainty about the values of hole diffusion length L_{pE} in the emitter, but it seems that for normal emitters (some micrometers deep), unlike for the base, the following condition can be stated [44]:

$$L_{pE} \equiv (D_{pE}\tau_{pE})^{1/2} < x_{jE} \tag{2-117}$$

where τ_{pE} = lifetime for holes in the emitter and x_{jE} = emitter junction depth.

In order to keep Eq. 2-61 valid, Eq. 2-115 must be modified into

$$G_E \simeq \int_{-L_{pE}}^{0} \frac{N_{DE}(x)}{D_{pE}[N_{DE}(x)]} \frac{n_i^2}{n_{iE}^2[N_{DE}(x)]} dx \tag{2-118}$$

An increase in emitter doping determines an increase in n_{iE}^2, and may then result in a lower G_E. Equation 2-118 may be simplified as

$$G_E \simeq \frac{Q_{E,\text{eff}}}{D_E} e^{-\Delta E_{g,\text{eff}}/kT} \tag{2-119}$$

where $Q_{E,\text{eff}}$ is the *effective emitter donor dopant charge* per surface unit, which takes into account the variation of the lower integration limit from Eq. 2-115 to Eq. 2-119; and $\Delta E_{g,\text{eff}}$ is the *effective band-gap narrowing,* the overall result of the normally nonconstant donor dopant concentration in the emitter.

For diffused-base transistors, the *temperature dependence* of nonsaturation, low-injection current gain can be derived from Eq. 2-119 as

$$h_{FE}^{pk} \simeq h_{FE0\gamma} = \frac{G_E}{G_B} \propto e^{-\Delta E_{g,\text{eff}}/kT} \tag{2-120}$$

Measurements of h_{FE}^{pk} at different temperatures give a good indication of the effective band-gap narrowing, normally in the range of a few tens of meV.

G_E measurements are easily accomplished as follows:

1. At rather low current densities, from Eq. 2-60,

$$G_B = \frac{qA_E n_i^2}{I_C} e^{V_{BE}/V_T} \tag{2-121}$$

It is necessary only to ensure that the I_C value chosen for V_{BE} measurement is in a range where Eq. 2-121 applies: a tenfold increase in I_C must correspond to a 60-mV increment of V_{BE} at room temperature.

2. Assuming $h_{FE}^{pk} \simeq h_{FE0\gamma}$, G_E is given simply by

$$G_E = G_B h_{FE}^{pk} \tag{2-122}$$

Care must be taken to select devices with rather flat h_{FE}-I_C characteristics, to measure V_{BE} and h_{FE}^{pk} at low V_{CE} to minimize the Early effect, and to monitor the temperature when measuring V_{BE}. n_i^2 is, in fact, a strong function of temperature, as is evident from linearizing Eq. 2-25 around room temperature:

$$\frac{\Delta G_E}{G_E} \simeq \frac{\Delta n_i^2}{n_i^2} \simeq 0.16\Delta T, \text{ in degrees Kelvin} \tag{2-123}$$

G_E values usually range between 1×10^{13} and 1×10^{14} cm$^{-4}\cdot$s.

2.3 THERMAL RATINGS

The main parameters related to the thermal performance of a power component are maximum junction temperature, junction-to-case thermal resistance, and maximum power. These parameters are in principle referred to the total power integrated circuit, but may be referred to the power transistor alone, which is the hottest region of the integrated circuit, where the major part of the total power is applied.

2.3.1 Maximum Temperature

Maximum storage temperature $T_{s,\max}$ is defined as the temperature which cannot be exceeded when the device is not operating. It depends mainly on the materials used for the die-to-package soldering. $T_{j,\max}$ refers to integrated circuit operating conditions and may be lower than $T_{s,\max}$, because electrical circuit performance may be correct only in a narrower range with respect to material limitations.

2.3.2 Thermal Resistance

Assuming one-dimensional heat flow, Fourier's heat equation has the form

$$P = \sigma_{\text{th}} A \frac{dT}{dx} \tag{2-124}$$

where P = power applied to the device
σ_{th} = thermal conductivity
A = section of the device normal to heat flow

By neglecting the slight variation of σ_{th} with T, Eq. 2-124 can be easily integrated. The result is

$$PR_{\text{th}} = \Delta T \tag{2-125}$$

where R_{th} is the *thermal resistance,* defined as

$$R_{\text{th}} = \frac{x}{\sigma_{\text{th}} A} \tag{2-126}$$

Equations 2-125 and 2-126 are formally identical to Ohm's laws. In addition, it is useful to recall the definition of *thermal capacitance* C_{th}:

$$C_{\text{th}} = \frac{dq}{dT} \tag{2-127}$$

with
$$q = \int_0^t P\,dt \tag{2-128}$$

From Eqs. 2-124 to 2-128 a correspondence can be established between thermal and electrical parameters (see Table 2-3), which allows the algebra used for electric circuits to become readily applicable to heat flow problems.

The heat conduction between the silicon die or the *junction* and the external package of the device, or the *case*, can be represented by an *equivalent thermal circuit*, given in Fig. 2-27.

According to Eq. (2-125), the *junction-to-case thermal resistance* $R_{th,jc}$ is defined as the ratio between a continuously applied power P and the resulting temperature difference between junction and case. From the equivalent thermal circuit (Fig. 2-27),

$$R_{th,jc} = R_{th,1} + R_{th,2} + R_{th,3} \tag{2-129}$$

where the suffixes, 1, 2, and 3, respectively, refer to silicon die, solder material, and package thermal mass, which is normally copper for power packages. $R_{th,1}$ is determined by silicon die thickness and area; $R_{th,2}$ by solder material, solder uniformity, and interface quality; and $R_{th,3}$ by thermal mass material, shape, and thickness.

One-dimensional models are unsuitable for $R_{th,jc}$ calculations: highly conductive materials have a consistent lateral conduction, and heat spreads over a larger section, thereby reducing total $R_{th,jc}$. Correct calculations must be at least two-dimensional. As for the solution of Poisson's equation (see Sec. 2.1.3), finite element or finite difference methods are suitable.

In normal operating conditions, the device is fixed to a mounting base, or *heat sink*. Case temperature is not fixed, and ambient temperature must be considered as a reference. *Junction-to-ambient thermal*

TABLE 2-3 Correspondence between Thermal and Electrical Quantities

Thermal			Electrical		
Symbol	Quantity	Unit	Symbol	Quantity	Unit
P	power	W	I	current	A
T	temperature	K	V	voltage	V
R_{th}	thermal resistance	K/W	R	resistance	Ω
σ_{th}	thermal conductivity	W/(K·cm)	σ	conductivity	$\Omega^{-1}\cdot cm^{-1}$
C_{th}	thermal capacitance	J/K	C	capacitance	F
q	heat	J	$\hat{Q}$	charge	C

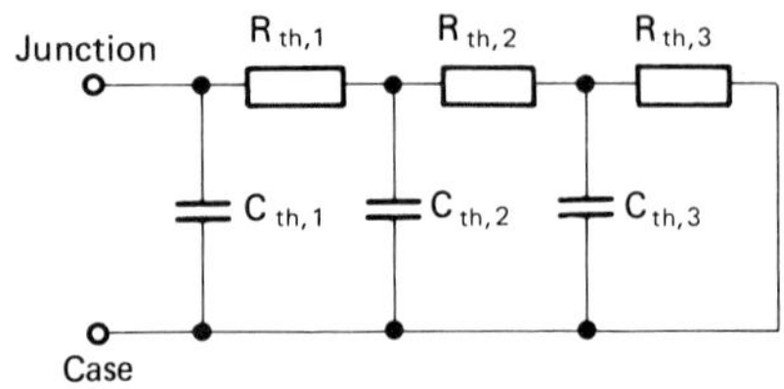

FIG. 2-27 Equivalent thermal circuit for a semiconductor device.

resistance $R_{th,ja}$ is the sum of junction-to-case ($R_{th,jc}$) and case-to-ambient ($R_{th,ca}$) contributions:

$$R_{th,ja} = R_{th,jc} + R_{th,ca} \tag{2-130}$$

$R_{th,ca}$ is determined by heat sink characteristics and by the quality of thermal contact between device and mounting base. Maximum power is found to be

$$P_{max} = \frac{T_{j,max} - T_a}{R_{th,ca}} \tag{2-131}$$

Equation 2-131 is still valid for power pulses, provided *thermal impedance* Z_{th}, derived from the equivalent thermal circuit, is substituted for $R_{th,ja}$.

2.3.3 Maximum Power

The power applied to a transistor is essentially given by

$$P = V_{CB} I_C \tag{2-132}$$

which must not exceed P_{max}, derived in the previous section. This simple picture would be complete if the collector current density were constant across the device's active (emitter) surface and equal to its average value I_C/A_E.

To go into greater detail, the influence of T on V_{BE} and J_C must be analyzed. For a constant J_C, from Eqs. 2-25 and 2-60,

$$\frac{dV_{BE}}{dT} = \frac{V_{BE}}{T} - \frac{k}{q}\left(3 + \frac{2T_0}{T}\right) \tag{2-133}$$

V_{BE} can be easily measured at the ambient temperature T_a. $V_{BE}(T_a) = V_{BE,a}$ is assumed as boundary condition and Eq. 2-133 can be integrated by separating variables. The solution is

$$V_{BE} = \frac{E_{g0}}{q} + \frac{T}{T_a}\left(V_{BE,a} - \frac{E_{g0}}{q}\right) - \frac{3kT}{q}\ln\frac{T}{T_a} \tag{2-134}$$

Equation 2-134 can be derived with respect to T, yielding

$$\phi \equiv -\frac{dV_{BE}}{dT} = \frac{E_{g0}/q - V_{BE,a}}{T_a} + \frac{3k}{q}\left(1 + \ln\frac{T}{T_a}\right) \tag{2-135}$$

For $V_{BE,a} = 0.6$ V a fairly constant value for $\phi \equiv -dV_{BE}/dT$ is found, ranging from 2.3 to 2.4 mV/K between 25 and 200°C. At constant J_C, V_{BE} decreases almost linearly with rising temperature. This implies that, for fixed V_{BE}, J_C increases exponentially with temperature:

$$J_C = J_0 \exp\frac{qV_{BE} + \phi(T - T_a)}{kT_a} \tag{2-136}$$

where
$$J_0 = \frac{qn_i^2(T_a)}{G_B} \tag{2-137}$$

The total power applied to the device is the integral over the active surface of power density Π dissipated by single elemental areas:

$$\Pi = \frac{dP}{dA} = V_{CB}J_C \tag{2-138}$$

J_C in real situations is not constant over the active surface, for the following reasons:

1. Nonuniformity of V_{BE} at different points of the emitter-base junction due to voltage drops across base and emitter metallizations, which act as low, but not negligible, ohmic resistances. It must be recalled that a few millivolts of variation in V_{BE} induces a strong variation in J_C.
2. Nonuniformity of V_{BE} due to voltage drops caused by base current flow across internal resistance $r_{bb'}$ (Sec. 2.2.5).
3. Nonuniformity of T_j due to microscopic variations in $R_{th,jc}$ (for instance, solder voids). Thermal crowding may also induce high temperatures in the central portions of the emitter pattern (Sec. 2.2.5).
4. Nonuniformity of J_0 due to crystal defects, which may alter doping profiles, and thus local G_B.

All these effects induce a nonconstant power density across the device surface, which results in a nonconstant T_j distribution. Accurate device design and processing allow relative equalization of temperature distribution across the silicon die. However, in order to allow safe operation, $T_{j,\max}$ in Eq. 2-131 must be regarded as the *peak temperature* in the silicon die.

2.3.4 Thermal Resistance Measurement [46]

Temperature distribution across the device when it is submitted to a power pulse may be observed through an infrared microscope or by coating the silicon die with a thermographic phosphor [47]. These operations are neither simple nor quick. They can be considered valid as a powerful tool for analysis, but too complicated for routine measurements. For this purpose, a temperature-sensitive transistor parameter is required: V_{BE} is the best choice for its linear variation with T over a wide temperature range (Sec. 2.3.3). V_{CB} could equally well be used, but V_{BE} is preferred, because it indicates a value nearer peak temperature than does V_{CB}.

The recommended method [48, 49] is now described.

1. The device is mounted on a performing heat sink, so that $R_{\text{th},ca} \ll R_{\text{th},jc}$.
2. The device is permanently biased at the fixed V_{CE}.
3. A collector current pulse I_M is applied and $V_{BE,1}$ is measured.
4. A power pulse $P = I_C V_{CE}$ is applied, for a duration long enough to establish thermal equilibrium.
5. A new current pulse I_M is applied quickly and $V_{BE,2}$ is measured.

The temperature increase, as a consequence of power application, is

$$\Delta T = \frac{V_{BE,1} - V_{BE,2}}{\phi} \tag{2-139}$$

and the thermal resistance is found to be

$$R_{\text{th},jc} = \frac{\Delta T}{P} \tag{2-140}$$

Care must be taken in choosing I_M. A good rule is to fix I_M corresponding to $V_{BE} \simeq 0.6$ V, which is neither too low a V_{BE} value, for which the validity of Eq. 2-60 would be doubtful, nor too high a V_{BE} (and I_M) value, for which voltage drops on emitter and base series resistances would no longer be negligible. A correct ϕ parameter would not be definable in these cases and Eq. 2-139 would no longer be applicable.

The method described is also applicable to short power pulses, to measure thermal impedance.

2.4 SECOND BREAKDOWN

Maximum voltage, current, and power do not completely define the *s*afe *o*perating *a*rea (SOA) of bipolar transistors. Second *b*reakdown (SB) is known to occur well within absolute maximum ratings, and the SOA must be reduced accordingly. The term *second breakdown* is not associated with a single mechanism. A clear distinction must be made between two different conditions:

1. *Forward SB,* occurring in forward base-driving conditions ($I_B > 0$), i.e., normal transistor operation. The *maximum collector current* is limited to I_{SB}, a function of voltage V_{CE} and of power pulse duration.
2. *Reverse SB,* occurring in reverse base-driving conditions ($I_B < 0$), normally found when switching off with inductive or resistive loads. A *maximum energy* E_{SB} is sometimes rated, as a function of the external circuit.

2.4.1 Forward Second Breakdown

The bipolar transistor's forward SOA defines the region of allowed operation on the graph of log I_C versus log V_{CE} (Fig. 2-28). The boundaries of the SOA are determined by the following parameters:

1. Maximum I_C, limited by wire bonding or metal interconnections (horizontal lines in Fig. 2-28). For short power pulse durations, a higher I_C is allowed than in dc operation.
2. Maximum power (45° contours in Fig. 2-28), which depends on thermal impedance and is higher for short pulses.
3. Forward second breakdown (the oblique contours).
4. Specified V_{CEO} (the vertical line).

A qualitative but correct description of forward second breakdown is obtained by deriving thermal stability conditions under constant V_{BE} and V_{CE} [50]. When a power pulse $P \simeq V_{CE}I_C$ is applied to the transistor, two opposite effects occur:

1. The junction temperature rises. This determines an increment in I_C, and thus in power:

$$dP_1 = V_{CE}\, dI_C \tag{2-141}$$

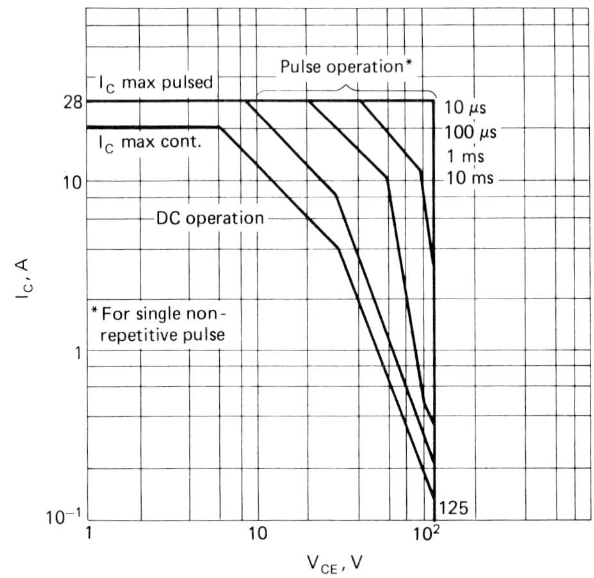

FIG. 2-28 Forward safe operating area (SOA) for a discrete NPN power transistor (BUX40).

2. Heat is dissipated, because of thermal exchange between device and ambient. The power dissipated is

$$dP_2 = -\frac{dT}{Z_{\text{th}}} \tag{2-142}$$

When $dP_1 \leq -dP_2$, thermal equilibrium is established. Otherwise, the device temperature tends to increase without limitation *(thermal runaway)*, with consequent transistor destruction. A *thermal stability factor* S may be defined as

$$S = \left|\frac{dP_1}{dP_2}\right| = Z_{\text{th}} V_{CE} \frac{dI_C}{dT} \tag{2-143}$$

For $S \leq 1$ the device is thermally stable. To better analyze the influence of device parameters on Eq. 2-143, the expression dI_C/dT must be derived. Adding emitter R_E and base R_B series resistances in Eq. 2-60 and taking into account the h_{FE} versus T dependence from Eq. 2-120,

$$\frac{dI_C}{dT} = \frac{I_C}{T} \frac{3 + \dfrac{2T_o}{T} + \ln \dfrac{qA_E n_i^2}{I_C G_B} + \dfrac{\Delta E_{g,\text{eff}}}{qV_T} \dfrac{I_C R_B/h_{FE}}{V_T}}{1 + I_C\,(R_E + R_B/h_{FE})/V_T} \tag{2-144}$$

The great effectiveness of *ballasting* [51], i.e., intergrating series emitter and base resistances, in reducing dI_C/dT is evident. It must, however, be noted that, while R_E is always effective, R_B appears divided by h_{FE} and is also present at the numerator of Eq. 2-144. Base ballasting is effective for small h_{FE} (for low-gain transistors or high currents) and small band-gap narrowing $\Delta E_{g,\text{eff}}$ (lightly doped emitters). Voltage drops on ballast resistors and band-gap narrowing must be compared to thermal voltage V_T to establish their weight in Eq. 2-144.

The simple physical picture presented allows a quantitative calculation of forward SOA, fitting experimental data fairly well. For a more accurate picture, current density nonhomogeneities must be considered. Current crowding, described in Sec. 2.2.5, tends to decrease the I_{SB} rating, the device being actually in less favorable conditions with respect to the ideal constant J_C of the simple one-dimensional model presented.

Solder voids between silicon die and thermal mass can induce high current densities as a consequence of local overheating, and thus can lead to poor forward SOA. Nonhomogeneous solder quality across the die back surface may be modeled as a parallel of many elemental transistors, perfectly identical in all the characteristics but differing as regards thermal resistance. Current is focalized in the element where R_{th} is maximum, which in turn sets a limit on the performance of the overall structure.

2.4.2 Reverse Second Breakdown [52–55]

Reverse second breakdown can occur during the turn-off of switching transistors, when a reverse base current is applied: high I_C and V_{CE} values may appear simultaneously, particularly in the case of an inductive load. At rather high values of V_{CE}, the electric field in the collector E_C is sufficiently high to determine carrier velocity to be almost equal to v_l (Sec. 2.2.5). The traveling electron concentration is then

$$n = \frac{J_C}{qv_l} \tag{2-145}$$

and Poisson's equation in the collector is modified from Eq. 2-3*b* to

$$\frac{dE_C}{dx} = \frac{q(N_D - n)}{\epsilon} \tag{2-146}$$

For low current densities $n \ll N_D$, and Eq. 2-146 is almost coincident with Eq. 2-3*b* (J_1 in Fig. 2-29). When, setting V_{CE} constant, J_C increases, n becomes comparable to N_D and the slope of E_C versus x varies (J_2).

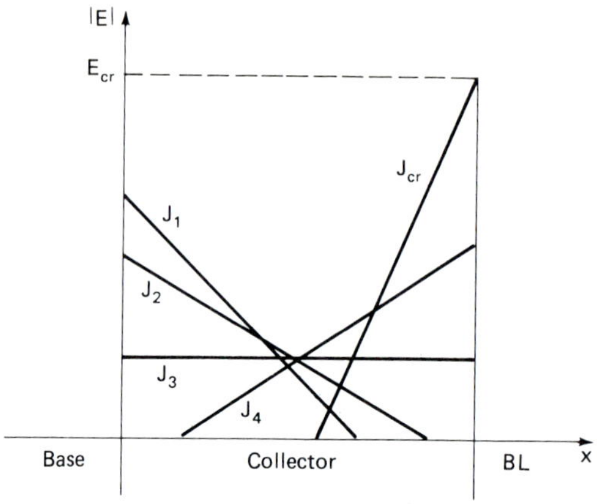

FIG. 2-29 Electric field in the collector at constant V_{CE} as a function of x and collector current density ($J_1 < J_2 < J_3 < J_4$).

At J_3, $n = N_D$ and E_C becomes constant. At higher current densities (J_4), the E versus x slope changes sign. If J_C is further increased to J_{cr}, E_{cr} is reached at the collector–buried-layer interface and ordinary avalanche breakdown occurs. It must be noted that this current-induced avalanche breakdown can occur at *any* V_{CE} voltage, provided J_C is sufficiently high. J_{cr} may be derived by combining Eqs. 2-145, 2-146, 2-12, and 2-13:

$$J_{cr} = qv_l N_D + \left(\frac{\epsilon}{2V_{CE}}\right)^{4/3} \frac{v_l}{(qK_A)^{1/3}} \tag{2-147}$$

or

$$J_{cr}\,(\mathrm{A/cm^2}) = (1.6 \times 10^{-12}) N_D\,(\mathrm{cm^{-3}}) + \frac{2.84 \times 10^6}{[V_{CE}(V)]^{4/3}} \tag{2-148}$$

The depletion region extension, starting from the collector–buried-layer interface, is given by

$$x_d = \left(\frac{2\epsilon V_{CE}}{J_{cr}/v_l - qN_D}\right)^{1/2} \tag{2-149}$$

J is calculated by Eq. 2-148. This value is inserted into Eq. 2-149 and x_d is calculated. If $x_d > W_C$ results, the value obtained for J_{cr} is higher than the actual value, because in this case reach-through occurs, and Eqs. 2-147 and 2-148 are no longer valid.

For an integrated transistor having $BV_{CBO} = 100$ V, a donor concentration $N_D \simeq 2 \times 10^{15}$ cm^{-3} is required. At $V_{CE} = 50$ V, $J_{cr} = 1.9 \times$

10^4 A/cm^2 is calculated and the condition $x_d < W_C$ is normally satisfied, because x_d turns out to be only 2.6 μm. J_{cr} in this case is well beyond the average current densities of operation, as given by the ratio I_C/A_E. However, in the case of reverse base driving, current nonhomogeneities are much more relevant than in the forward base current condition (Sec. 2.2.5). Ohmic drops on internal resistance $r_{bb'}$ change sign and the current tends to crowd in the emitter center *(pinch-in):* active area dramatically decreases and peak current density greatly exceeds the I_C/A_E ratio. It is very difficult to guess the effective emitter area, which also depends on reverse driving conditions, such as inductance magnitude, external base series resistance value, and reverse voltage applied to the emitter-base junction to turn off the device. Correlation between calculated J_{cr} and actual I_C-V_{CE} capability is thus very difficult, but J_{cr} maintains at least the meaning of a figure of merit for device reverse ruggedness.

An internal transistor parameter having some influence on reverse second breakdown performance is the transverse base resistance $r_{bb'}$, which must be as low as possible to maximize effective emitter area. High Q_B values (low peak h_{FE}) and narrow emitter patterns are required.

To keep J_{cr} values high, fairly high collector thicknesses must be used, particularly in high-voltage devices. This solution, however, adversely affects high-current performance (Sec. 2.2).

In discrete transistors a powerful tool to increase J_{cr}, and hence to improve reverse SOA, is the introduction of one or more buffer collector layers, between the N^- collector and the N^+ substrate, with intermediate doping levels [55]. This solution is unfortunately impossible for power integrated transistors, because a strong reduction of N^- collector–P^+ isolation breakdown voltage would occur. Improvements could be achieved by a graded buried-layer concentration profile, which, however, would reduce the N^- collector thickness and the device's voltage performance.

2.5 SWITCHING BEHAVIOR

Bipolar power transistors, discrete as well as integrated, are often used as switching devices.

An ideal switch has two states of operation: the *On state,* corresponding to large current flow with small voltage drop, and the *Off state,* with negligible current flow and high voltage applied. A rather obvious correspondence can be established between the bipolar transistor and the ideal switch: full saturation (forward V_{BE}) for the On state, and cutoff region (reverse V_{BE}) for the Off state. This describes *saturated-mode* bipolar transistor switching, which will be discussed in this section; other modes are possible (e.g., current and avalanche modes) but are rarely used. Another feature of an ideal switch is instantaneous transition between the two states; therefore, transistor switching times must be as short as possible.

The switched power is delivered to a load, which is normally resistive or inductive in its essence, rarely capacitive. *Resistive* loads will be considered in this analysis.

Voltage ratings (applicable to the Off state) and high current saturation performance (relevant to the On state) have been discussed in Secs. 2.1 and 2.2, respectively. Now, switching times will be examined, starting with a definition of *electrical* and *physical* parameters. A one-dimensional theory, including a theory of the quasi-saturation region (a must for power transistors), will be developed and a qualitative description of two-dimensional effects will be given.

2.5.1 Electrical Switching Times

A basic transistor switching circuit is shown in Fig. 2-30. A collector voltage V_{CC} is applied permanently. In the On state the base is driven by a constant-current generator I_{B1} and collector current is almost equal to $I_0 = V_{CC}/R_L$, since transistor V_{CE} in saturation is normally negligible with respect to V_{CC}. In the Off state a reverse $V_{BE(\mathrm{Off})}$ is applied, collector current is reduced to the very small leakage value, and transistor V_{CE} is almost equal to V_{CC}. The input base current and output collector current are shown in Fig. 2-31.

Electrical switching times are normally defined as follows:

1. t_d' (delay time): from the application of forward I_{B1} to starting current conduction ($I_C = 0.1I_0$)
2. t_r' (rise time): from starting ($I_C = 0.1I_0$) to almost full ($I_C = 0.9I_0$) current conduction

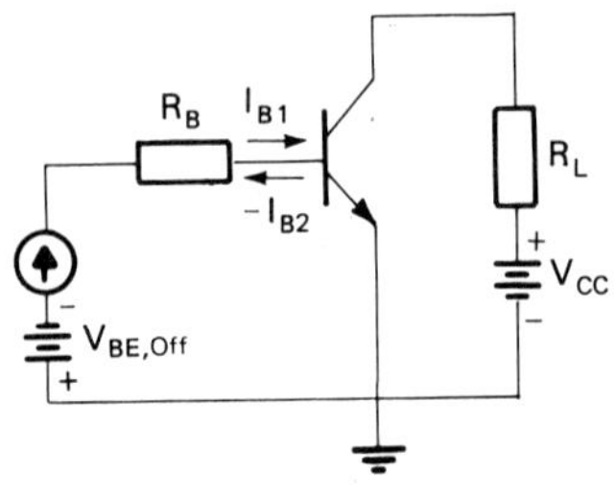

FIG. 2-30 Basic circuit for transistor switching with resistive load.

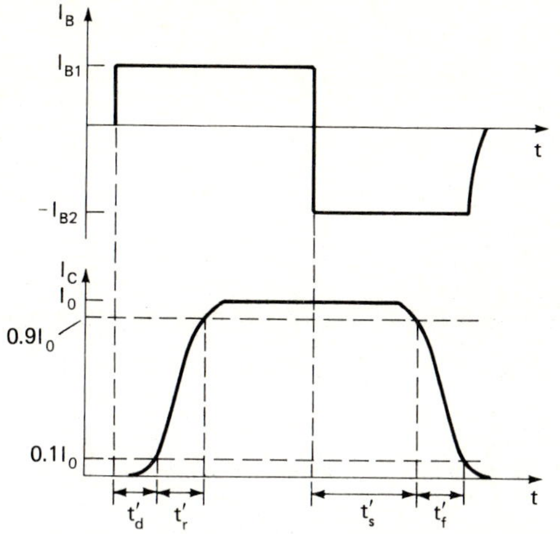

FIG. 2-31 Idealized I_B and I_C switching waveforms and electrical switching times.

3. t_s' (storage time): from the application of reverse $-I_{B2} = -V_{BE(\text{Off})}/R_B$ to the start of current decrease ($I_C = 0.9I_0$)
4. t_f' (fall time): from the start of current decrease ($I_C = 0.9I_0$) to negligible collector current ($I_C = 0.1I_0$)

On and Off times are defined as

$$t'_{\text{On}} = t'_d + t'_r \tag{2-150}$$

$$t'_{\text{Off}} = t'_s + t'_f \tag{2-151}$$

2.5.2 Physical Switching Times

Transistor characteristics, according to the model developed in Sec. 2.2, and switching load line are shown in Fig. 2-32. For simplicity R_P is assumed to be 0; this assumption does not affect the derivation of correct expressions for switching times.

Physical switching times, corresponding to transitions among different *physical states* of transistor operation, are defined as follows:

1. t_d *(delay time):* the charging time of the collector-base and emitter-base capacitances at $I_C = 0$. Leakage current is assumed negligible.
2. t_r *(rise time):* from $I_C = 0$ to the boundary between the nonsaturation and the quasi-saturation region ($I_C = I'$).
3. t_{qr} *(quasi rise time):* from $I_C = I'$ into full saturation ($I_C = I_0$).

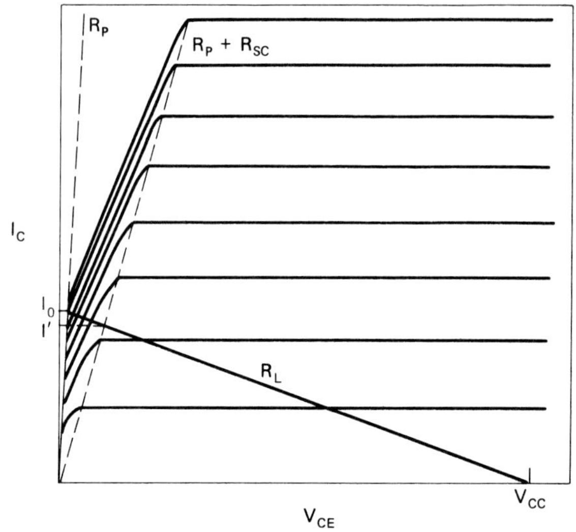

FIG. 2-32 Transistor I_C-V_{CE} characteristics and resistive load line.

4. t_s *(storage time)*: from oversaturation to full saturation. Collector current keeps constant and equal to I_0.
5. t_{qs} *(quasi storage time)*: from $I_C = I_0$ to $I_C = I'$.
6. t_f *(fall time)*: from $I_C = I'$ to $I_C = 0$.

Furthermore,

$$t_{On} = t_d + t_r + t_{qr} \tag{2-152}$$

$$t_{Off} = t_s + t_{qs} + t_f \tag{2-153}$$

t_{On} is longer than t'_{On}, which extends up to $I_C = 0.9I_0$. t_{Off} is longer than t'_{Off}, which extends down to $I_C = 0.1I_0$. Moreover, it can be noted that t'_d is the sum of t_d plus some contribution of t_r, t'_r is the residual major part of t_r plus some part of t_{qr}, t'_s is the sum of t_s plus some part of t_{qs}, and t'_f is residual t_{qs} plus the major part of t_f.

Electrical switching times could be derived from a physical analysis, but we believe this beyond our purpose, because physical switching times are sufficient to give an excellent picture of device switching performance. Current I' is the cross point between load line and R_{SC} (which is given by Eq. 2-81):

$$I' = \frac{V_{CC} - V_{CE}(I')}{R_L}$$
$$V_{CE}(I') = R_{SC}I' \tag{2-154}$$

Then
$$I' = \frac{I_0}{1 + \psi} \tag{2-155}$$

where
$$\psi = \frac{I_0 R_{SC}}{V_{CC}} \tag{2-156}$$

2.5.3 Base and Collector Transit Times [56]

Electron *transit time* t_{tr} from emitter to collector is the sum of several contributions:

1. *Emitter-base and collector-base capacitance charging,* through emitter and collector series resistances. This contribution is usually negligible.
2. *Collector depletion-layer transit time.* This contribution is also negligible, because free carriers are accelerated from the electric field to very high velocities, not far from v_l.
3. *Time required to cross the base region* (metallurgical and current-induced), where electron storage occurs. This is by far the most important term, which will be assumed coincident with t_{tr}.

The following equation holds

$$I_C = A_E\, qv(x)n(x) \tag{2-157}$$

and so
$$t_{tr} = \int_0^{W_{CIB}} \frac{dx}{v(x)} = \int_0^{W_{CIB}} \frac{qA_E}{I_C} n(x)\, dx = \frac{qA_E}{I_C} Q_n \tag{2-158}$$

In the nonsaturation region $W_{CIB} = W_B$, and t_{tr} is given by

$$t_{tro} = \frac{W_B^2}{2\eta D_B} \tag{2-159}$$

In quasi saturation Eq. 2-159 is still valid, simply by substituting W_{CIB} for W_B and setting $\eta = 2$.

In saturation,

$$t_{tr}^{sat} = \frac{(W_B + W_C)^2}{4D_B} \tag{2-160}$$

Current gain h_{FE} is known to decrease with frequency. *Cutoff frequency* f_T, also called the *current-bandwidth product*, is defined as the frequency at which h_{FE} becomes unity. f_T is simply related to transit time by

$$2\pi f_{T0} = \frac{1}{t_{tro}} = \frac{2\eta D_B}{W_B^2} \tag{2-161}$$

and

$$2\pi f_T^{sat} = \frac{1}{t_{tr}^{sat}} = \frac{4D_B}{(W_B + W_C)^2} \tag{2-162}$$

These parameters appear in the expressions for switching times shown in the next section.

2.5.4 One-Dimensional Theory [57]

Base current balance will be used to derive switching times: external base current must be equal to the sum of the static I_B (according to the device's instantaneous working point) and the time derivative of total electric charge Q_n stored in the base. This approach is equivalent to *charge control* models [58], but seems more directly related to device physics. $h_{FE} = h_{FE}\gamma$ is assumed throughout, an excellent approximation for integrated NPN power transistors.

Delay Time. t_d is mainly due to the charging time of the emitter-base junction capacitance from $V_{BE} = V_{BE(Off)}$ to $V_{BE} = 0$. It is normally small and negligible in switching time calculations.

Rise Time. During the rise time the transistor is in nonsaturation and collector current rises from 0 to I'. The base driving current is constant and equal to I_{B1}. The electric charge stored in the base is easily derived from Eq. 2-68:

$$\hat{Q}_n = qA_E Q_n = \frac{I_C W_B^2}{2\eta D_B} \tag{2-163}$$

Base current balance gives

$$I_{B1} = \frac{I_C}{h_{FE}} + \frac{d\hat{Q}_n}{dt} \tag{2-164}$$

From Eqs. 2-161 and 2-163

$$\frac{d\hat{Q}_n}{dt} = \frac{1}{2\pi f_{T0}} \frac{dI_C}{dt} \tag{2-165}$$

Inserting Eq. 2-165 into Eq. 2-164, separating variables, and integrating gives

$$t_r = \frac{1}{2\pi f_{T0}} \int_0^{I'} \frac{dI_C}{I_{B1} - I_C/h_{FE0}} = \frac{h_{FE0}}{2\pi f_{T0}} \ln \frac{I_{B1}}{I_{B1} - I'/h_{FE0}} \tag{2-166}$$

$h_{FE} \simeq h_{FE0}$ has been assumed, which is a very good approximation for the nonsaturation region. Defining $h_{FEF} = I_0/I_{B1}$ as the *externally forced current gain,*

$$t_r = \frac{h_{FE0}}{2\pi f_{T0}} \ln \frac{1}{1 - h_{FEF}/[h_{FE0}(1 + \psi)]} \tag{2-167}$$

The derivative of t_r with respect to h_{FE0} is always negative. A high h_{FE0} therefore improves t_r, until $h_{FE0} \gg h_{FEF}$ (see Fig. 2-33). In this case, often verified, Eq. 2-167 reduces to:

$$t_r \simeq \frac{h_{FEF}}{2\pi f_{T0}(1 + \psi)} \qquad (h_{FE0} \gg h_{FEF}) \tag{2-168}$$

Low h_{FEF} (high I_{B1}) and high f_{T0} (low W_B) are the most important conditions for a reduced t_r.

Quasi Rise Time. In the quasi rise time the transistor is in quasi saturation and the collector current rises from I' to I_0 (full saturation). The electrical charge stored in the base is

$$\hat{Q}_n = \frac{I_C W_{CIB}^2}{4D_B} \tag{2-169}$$

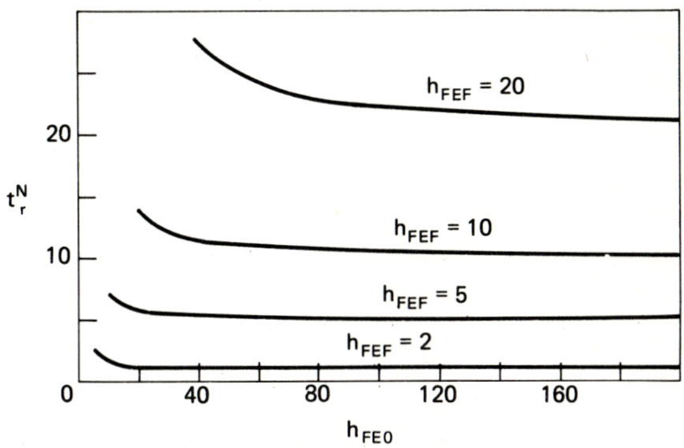

FIG. 2-33 Normalized rise time t_r^N as a function of common-emitter peak current gain h_{FE0} and forced gain h_{FEF}.

The base current balance equation can be integrated, but a rather complicated expression for t_{qr} results. If W_C is assumed to be greater than W_B, t_{qr} turns out to be only slightly sensitive to variations in ψ, because charge stored in saturation is much larger than charge stored in nonsaturation and is independent of the value of I'. A good approximation for t_{qr} can be obtained:

$$t_{qr} = \frac{h_{FE}^{sat}}{2\pi f_T^{sat}} \left(1 - \frac{h_{FE}^{sat}}{h_{FE0}} \right)^{-1} \ln \frac{1 - h_{FEF}/h_{FE0}}{1 - h_{FEF}/h_{FE}^{sat}} \tag{2-170}$$

When $h_{FE0} \gg h_{FE}^{sat} \gg h_{FEF}$ (i.e., under conditions of hard base driving), t_{qr} is well approximated by

$$t_{qr} \simeq \frac{h_{FEF}}{2\pi f_T^{sat}} \qquad (h_{FE}^{sat} \gg h_{FEF}) \tag{2-171}$$

Low h_{FEF} (high I_{B1}) and high f_T^{sat} (low $W_B + W_C$) are required to reduce t_{qr}.

Storage Time. When in the On condition, the transistor is oversaturated and the base current is $I_{B1} > I_B^{sat}$. t_s is the time required to bring the transistor from oversaturation to saturation. The collector current keeps constant and equal to I_0. The base driving is constant and equal to $-I_{B2}$. From Eq. 2-93, oversaturation stored charge is

$$\hat{Q}_n = \hat{Q}_n^{sat} \frac{I_B}{I_B^{sat}} \tag{2-172}$$

The base current balance equation is

$$-I_{B2} = I_B + \frac{d\hat{Q}_n}{dt} \tag{2-173}$$

Making use of Eq. 2-172,

$$\frac{d\hat{Q}_n}{dt} = \frac{d\hat{Q}_n}{dI_B}\frac{dI_B}{dt} = \frac{h_{FE}^{sat}}{2\pi f_T^{sat}}\frac{dI_B}{dt} \tag{2-174}$$

Separating variables and integrating Eq. 2-173 gives

$$t_s = \frac{h_{FE}^{sat}}{2\pi f_T^{sat}} \int_{I_B^{sat}}^{I_{B1}} \frac{dI_B}{I_{B2} + I_B} \tag{2-175}$$

Finally

$$t_s = \frac{h_{FE}^{sat}}{2\pi f_T^{sat}} \ln \frac{1 + (I_0/I_{B2})/h_{FEF}}{1 + (I_0/I_{B2})/h_{FE}^{sat}} \tag{2-176}$$

The derivative of t_s with respect to the ratio h_{FEF}/h_{FE}^{sat} turns out to be negative: as expected, t_s increases when the device is deeply oversaturated. High I_{B2}/I_{B1} ratios, high f_T^{sat} (low $W_B + W_C$), and low h_{FE}^{sat} are required to reduce t_s.

Quasi Storage Time. During the quasi storage time the transistor is in quasi saturation and collector current decreases from I_0 to I'. Base driving is constant and equal to $-I_{B2}$. t_{qs} is derived exactly as t_{qr} was, simply by substituting $-I_{B2}$ for I_{B1}:

$$t_{qs} = \frac{h_{FE}^{sat}}{2\pi f_T^{sat}}\left(1 - \frac{h_{FE}^{sat}}{h_{FE0}}\right)^{-1} \ln \frac{1 + (I_0/I_{B2})/h_{FE}^{sat}}{1 + (I_0/I_{B2})/h_{FE0}} \tag{2-177}$$

When $h_{FE0} \gg h_{FE}^{sat} \gg I_0/I_{B2}$ (under hard base driving conditions), t_{qs} is well approximated by

$$t_{qs} \simeq \frac{I_0/I_{B2}}{2\pi f_T^{sat}} \qquad \left(h_{FE}^{sat} \gg \frac{I_0}{I_{B2}}\right) \tag{2-178}$$

Low I_0/I_{B2} ratios and high f_T^{sat} are required to reduce t_{qs}.

Fall Time. During fall time the transistor is in nonsaturation and collector current decreases from I_0 to I'. Base driving is constant and equal to $-I_{B2}$. t_f is derived exactly as t_r is, simply by substituting $-I_{B2}$ for I_{B1}:

$$t_f = \frac{h_{FE0}}{2\pi f_{T0}} \ln\left[1 + \frac{I_0/I_{B2}}{h_{FE0}(1 + \psi)}\right] \tag{2-179}$$

Unlike that of t_r, the derivative of t_f with respect to h_{FE0} is always positive, and thus a low h_{FE0} improves t_f (see Fig. 2-34). When $h_{FE0} \gg I_0/I_{B2}$, t_f is well approximated by

$$t_f \simeq \frac{I_0/I_{B2}}{2\pi f_{T0}\,(1 + \psi)} \qquad \left(h_{FE0} \gg \frac{I_0}{I_{B2}}\right) \tag{2-180}$$

Low I_0/I_{B2} ratios reduce t_f and so would a high f_{T0} (low W_B), which, however, is often associated with a high h_{FE0}.

2.5.5 Two-Dimensional Effects

Two-dimensional effects cannot always be neglected when modeling transistor turn-off. Charges stored in the base and the collector are more rapidly removed from the region adjacent to the base contact than under the center of the emitter pattern [59, 60]. An additional time, not found in one-dimensional analysis, can be required to decrease collector current, and it appears as an electrical tail following

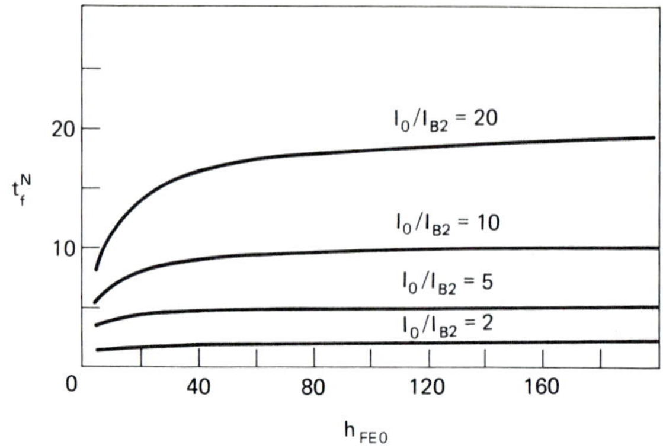

FIG. 2-34 Normalized fall time t_f^N as a function of common-emitter peak current gain h_{FE0} and the ratio I_0/I_{B2}.

fall time. To reduce this effect, the distance from each point of the emitter pattern to the base contact must be kept as small as possible. This can be achieved in two ways: reducing emitter finger width or eliminating emitter diffusion under the emitter bonding area. In the latter case an insulator, such as silicon dioxide, is required between metal and silicon, to avoid emitter-base shorts.

REFERENCES

1. A. J. Dekker, *Solid State Physics*, Prentice-Hall, Englewood Cliffs, N.J., 1957.
2. J. L. Moll, *Physics of Semiconductors*, McGraw-Hill, New York, 1964.
3. S. M. Sze, *Physics of Semiconductor Devices*, Wiley, New York, 1969.
4. A. B. Phillips, *Transistor Engineering*, McGraw-Hill, New York, 1962.
5. A. S. Grove, *Physics and Technology of Semiconductor Devices*, Wiley, New York, 1967.
6. E. S. Yang, *Fundamentals of Semiconductor Devices*, McGraw-Hill, New York, 1978.
7. P. R. Wilson, The Emitter-Base Breakdown Voltage of Planar Transistors, *Solid-State Electronics*, **17**:465 (1974).
8. S. M. Sze, *Physics of Semiconductor Devices*, Wiley, New York, 1969, pp. 110–111.
9. M. Felici, *SGS-ATES Internal Tech. Rep. SCD/59* (1976), unpublished.
10. O. Leistiko and A. S. Grove, Breakdown Voltage of Planar Silicon Junctions, *Solid-State Electronics*, **9**:847 (1966).

11. R. L. Davies and F. E. Gentry, Control of Electric Field at the Surface of PN Junctions, *IEEE Trans. Electron Devices,* **ED-11:**313 (1964).

12. J. Cornu, Field Distribution Near the Surface of Beveled PN Junctions in High-Voltage Devices, *IEEE Trans. Electron Devices,* **ED-20:**347 (1973).

13. M. Bakowski and K. I. Lundstrom, Depletion Layer Characteristics at the Surface of Beveled High-Voltage PN Junctions, *IEEE Trans. Electron Devices,* **ED-20:**550 (1973).

14. S. Yasuda and M. Kurata, Two-Dimensional Field Distribution Analysis of Reverse Biased PN Junction Devices, *Solid-State Electronics,* **23:**1077 (1980).

15. A. Savini, E. Dallago, M. Felici, M. Panigada, and F. Versiglia, Edge Effects in Glass-Passivated Collector-Base Junctions of High-Voltage Transistors, *IEE Proc., Pt. I,* **129:**116 (1982).

16. V. A. K. Temple and M. S. Adler, Calculation of the Diffusion Curvature Related Avalanche Breakdown in High-Voltage Planar PN Junctions, *IEEE Trans. Electron Devices,* **ED-22:**910 (1975).

17. Y. C. Kao and E. D. Wolley, High-Voltage Planar PN Junctions, *Proc. IEEE,* **55:**1409 (1967).

18. M. S. Adler, V. A. K. Temple, A. P. Ferro, and R. C. Rustay, Theory and Breakdown Voltage for Planar Devices with a Single Field Limiting Ring, *IEEE Trans. Electron Devices,* **ED-24:**107 (1977).

19. F. Conti and M. Conti, Surface Breakdown in Silicon Planar Diodes Equipped with Field Plate, *Solid-State Electronics,* **15:**93 (1972).

20. W. Shockley, The Theory of PN Junctions in Semiconductors and PN Junction Transistors, *Bell Syst. Tech. J.,* **28:**435 (1949).

21. C. T. Sah, R. N. Noyce, and W. Shockley, Carrier Generation and Recombination in PN Junctions and PN Junction Characteristics, *Proc. IRE,* **45:**1228 (1957).

22. J. L. Moll, The Evolution of the Theory for the Current-Voltage Characteristics of PN Junctions, *Proc. IRE,* **46:**1076 (1958).

23. J. W. Slotboom and H. C. de Graaff, Measurements of Bandgap Narrowing in Si Bipolar Transistors, *Solid-State Electronics,* **19:**857 (1976).

24. M. S. Adler and G. E. Possin, Achieving Accuracy in Transistor and Thyristor Modeling, *IEEE Trans. Electron Devices,* **ED-28:**1053 (1981).

25. S. L. Miller, Avalanche Breakdown in Germanium, *Phys. Rev.,* **99:**1234 (1955).

26. J. L. Moll, J. L. Su, and A. C. M. Wang, Multiplication in Collector Junctions of Silicon NPN and PNP Transistors, *IEEE Trans. Electron Devices,* **ED-17:**420 (1970).

27. H. K. Gummel and H. C. Poon, An Integral Charge Control Model of Bipolar Transistors, *Bell Syst. Tech. J.,* **49:**827 (1970).

28. J. L. Moll and I. M. Ross, The Dependence of Transistor Parameters on the Distribution of Base Layer Resistivity, *Proc. IRE,* **44:**72 (1956).

29. J. M. Early, Effects of Space Charge Layer Widening in Junction Transistors, *Proc. IRE,* **40:**1401 (1952).

30. C. T. Kirk, A Theory of Transistor Cutoff Frequency Falloff at High Current Densities, *IRE Trans. Electron Devices,* **ED-9:**164 (1962).

31. R. J. Whittier and D. A. Tremere, Current Gain and Cutoff Frequency Falloff at High Currents, *IEEE Trans. Electron Devices,* **ED-16:**39 (1969).

32. J. A. Pals and H. C. de Graaff, On the Behaviour of the Base-Collector Junction of a Transistor at High Collector Current Densities, *Philips Res. Rep.,* **24:**53 (1969).

33. H. C. de Graaff, Collector Models for Bipolar Transistors, *Solid-State Electronics,* **16:**587 (1973).

34. D. L. Bowler and F. A. Lindholm, High Current Regimes in Transistor Collector Regions, *IEEE Trans. Electron Devices,* **ED-20:**257 (1973).

35. M. Conti and G. Corda, One-Dimensional Models and Collector Characteristics of High Voltage Bipolar Transistor, *Alta Frequenza,* **46:**117–57 E (1977).

36. E. J. Ryder, Mobility of Holes and Electrons in High Electric Fields, *Phys. Rev.,* **90:**766 (1953).

37. J. F. Gibbons, Carrier Drift Velocities in Silicon at High Electric Field Strengths, *IEEE Trans. Electron Devices,* **ED-14:**37 (1967).

38. P. M. Smith, J. Frey, and P. Chatterjee, High-Field Transport of Holes in Silicon, *Appl. Phys. Lett.,* **39:**332 (1981).

39. G. Rey, F. Dupuy, and J. P. Bailbe, A Unified Approach to the Base Widening Mechanisms in Bipolar Transistors, *Solid-State Electronics,* **18:**863 (1975).

40. J. R. Hauser, The Effects of Distributed Base Potential on Emitter-Current Injection Density and Effective Base Resistance for Stripe Transistor Geometries, *IEEE Trans. Electron Devices,* **ED-11:**238 (1964).

41. D. Navon and R. E. Lee, Effect of Non-Uniform Emitter Current Distribution on Power Transistor Stability, *Solid-State Electronics,* **13:**981 (1970).

42. J. Cornu, R. Sittig, and W. Zimmermann, Analysis and Measurement of Carrier Lifetimes in the Various Operating Modes of Power Devices, *Solid-State Electronics,* **17:**1099 (1974).

43. H. J. J. De Man, The Influence of Heavy Doping on the Emitter Efficiency of a Bipolar Transistor, *IEEE Trans. Electron Devices,* **ED-18:**833 (1971).

44. P. J. Kannam, Effect of Emitter Doping on Device Characteristics, *IEEE Trans. Electron Devices,* **ED-20:**845 (1973).

45. M. S. Adler, B. A. Beatty, S. Krishna, V. A. K. Temple, and M. L. Torreno,

Limitations on Injection Efficiency in Power Devices, *IEEE Trans. Electron Devices,* **ED-23:**858 (1976).

46. F. K. Oettinger, D. L. Blackburn, and S. Rubin, Thermal Characterization of Power Transistors, *IEEE Trans. Electron Devices,* **ED-23:**831 (1976).
47. D. J. Brenner, A Technique for Measuring the Surface Temperature of Transistors by Means of Fluorescent Phosphors, *NBS Tech. Note 591,* July 1971.
48. W. M. Bullis, Methods of Measurement for Semiconductor Materials, Process Control, and Devices (Quarterly Report for Period Apr. 1 to June 30, 1972), *NBS Tech. Note 743,* pp. 34–35.
49. Thermal Resistance Measurements of Conduction Cooled Power Transistors, *EIA Recommended Std. RS-313-B* (revision of *RS-313-A*), October 1975.
50. P. L. Hower and P. K. Govil, Comparison of One- and Two-Dimensional Models of Transistor Thermal Instability, *IEEE Trans. Electron Devices,* **ED-21:**617 (1974).
51. R. P. Arnold and D. S. Zoroglu, A Quantitative Study of Emitter Balasting, *IEEE Trans. Electron Devices,* **ED-21:**385 (1974).
52. S. Krishna and P. L. Hower, Second Breakdown of Transistors during Inductive Turn-Off, *Proc. IEEE,* **61:**393 (1973).
53. B. A. Beatty, S. Krishna, and M. S. Adler, Second Breakdown in Power Transistors Due to Avalanche Injection, *IEEE Trans. Electron Devices,* **ED-23:**851 (1976).
54. D. L. Blackburn and D. W. Berning, An Experimental Study of Reverse-Bias Second Breakdown, *Tech. Dig. 1980 IEEE IEDM,* p. 297.
55. *SGS-ATES Design Note DN329,* August 1979.
56. R. Kumar and L. P. Hunter, Prediction of f_T and h_{fe} at High Collector Currents, *IEEE Trans. Electron Devices, ED-22:1031* (1975).
57. M. Felici, Thomson Internal Tech. Rep. TPU/1 (1985), unpublished.
58. R. Beaufoy and J. J. Sparkes, The Junction Transistor as a Charge-Controlled Device, *ATE J.,* **13:**310 (1957).
59. G. P. Chiavarotti, M. Conti, and S. Luciani, The Turn Off Time of High Voltage Expitaxial Transistors, *Rev. de Physique Appliquée,* **13:**737 (1978).
60. P. L. Hower, A Model for Turn-Off in Bipolar Transistors, *Tech. Dig. 1980 IEEE IEDM,* p. 289.

CHAPTER 3

Power MOS Devices in Discrete and Integrated Circuits

J. D. Plummer
Integrated Circuits Laboratory
Stanford University
Stanford, California

R. A. Blanchard
Siliconix, Inc.
Santa Clara, California

3.1 HISTORICAL DEVELOPMENT OF POWER MOSFETs 3.2
3.2 DOUBLE-DIFFUSED MOS OR DMOS FIELD-EFFECT TRANSISTORS 3.8
3.3 MAXIMIZING THE BREAKDOWN VOLTAGE OF POWER MOSFET STRUCTURES 3.14
3.4 LATCHBACK IN POWER FET STRUCTURES 3.27
3.5 MODELING AND MINIMIZATION OF ON RESISTANCE IN POWER MOSFETs 3.34
3.6 COMPOSITE MOS-BIPOLAR POWER DEVICE STRUCTURES 3.42
3.7 HIGH-VOLTAGE AND POWER MOS DEVICES IN INTEGRATED CIRCUITS 3.48
3.8 CONCLUSIONS AND FUTURE PROJECTIONS 3.54

Until quite recently, the term *power device* in the semiconductor industry was synonymous with a bipolar device. Both three-layer (NPN or PNP) bipolar transistors and four-layer regenerative structures (SCRs, triacs, etc.) have been available for many years with quite impressive power ratings. MOS technology has been largely reserved for low-voltage, low-current structures and has to some extent become synonymous with complex very large scale integrated circuits (VLSI).

Within the past five years, however, a dramatic change has taken place in the power semiconductor industry. Through the use of new or improved device structures and technology, the MOS transistor has been reshaped to make it competitive at high currents and at high voltages with the more traditional power devices. This has in many ways revolutionized traditional applications of power semiconductors and has made possible new applications as well. A further recent development which may also have great impact in the future is the incorporation of high-voltage or high-current MOS devices in integrated circuits which also contain MOS logic or control functions. This has made possible the realization of complex systems on a silicon chip along with output devices capable of controlling significant electrical loads.

This chapter begins by tracing the evolution from low-current, low-voltage MOS transistors to the high-power device structures which have become available today. The physics and technology of these new device structures are first discussed in discrete form, where we consider the optimization of the devices for high voltage and high current (low On resistance). We then consider discrete power structures which are really hybrids of traditional bipolar devices and the new MOS structures; these hybrids allow one to make use of the desirable features of both types of devices. Finally, we consider the extension of these ideas to integrated circuits containing power MOS devices and discuss some of the problems of integrating both types of structures on the same monolithic substrate. The chapter concludes with projections of both fundamental and practical limits to further progress in this field.

3.1 HISTORICAL DEVELOPMENT OF POWER MOSFETs

The device evolution that led to the eventual development of contemporary power MOSFETs started from the conventional low-voltage, low-current logic structure shown in Fig. 3-1. The physics of operation and the electrical characteristics of this device are adequately covered in numerous textbooks on this topic [1, 2, for example]; we present here

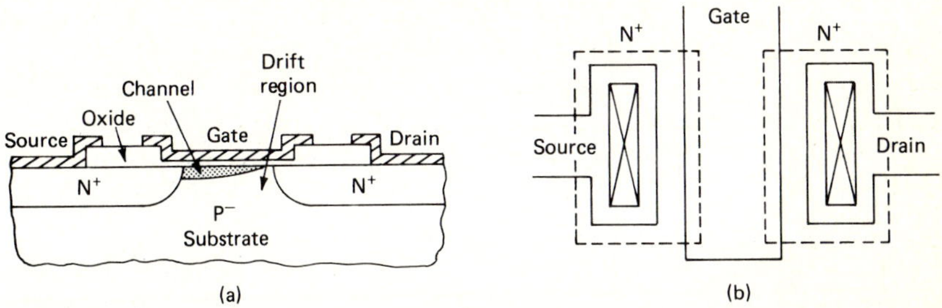

FIG. 3-1 Conventional low-voltage, low-power, metal-gate NMOS transistor: (*a*) cross section; (*b*) top view. The channel region near the drain is shown in part *a* in the pinched-off condition.

only a summary of the essential features which will be useful later in the chapter.

The device fundamentally operates through control of the surface conductivity under the gate. Negative voltage on the gate produces surface accumulation; i.e., the surface is made more P-type than the bulk. Under these conditions, essentially no current flows in the structure since all junctions are normally reverse-biased.

When the gate potential is made positive with respect to the source, the surface potential in the region under the gate rises, causing first depletion of mobile carriers in the surface region and, finally, inversion of the surface to N-type. Figure 3-2 illustrates this process with the well-known energy band diagram describing the MOS system. When the gate potential is sufficiently positive to cause surface inversion, a conducting channel of electrons exists between source and drain and the device is considered to be turned on. The surface conditions of accumulation, depletion, and inversion are represented in Fig. 3-2 by the position of the Fermi level E_f at the Si/SiO_2 interface. With E_f close to the semiconductor valence band, the surface is more P-type than the bulk and is therefore accumulated. As E_f moves up in energy, farther from the valence band edge, the hole concentration at the surface drops, producing depletion. As E_f crosses the middle of the band gap (E_i), the surface becomes N-type, producing a surface inversion layer.

It is worth pointing out that as soon as E_f crosses the midband position (E_i), the surface is N-type, and an inversion layer exists. Until E_f is relatively close to the conduction band, however, the mobile electron concentration at the surface is small and hence the device currents are small. By convention, when $\phi_s = 2\phi_f$, that is, when E_f is as far above E_i at the surface as it is below E_i in the bulk, the structure is considered to be strongly inverted. This is the classical definition of the threshold

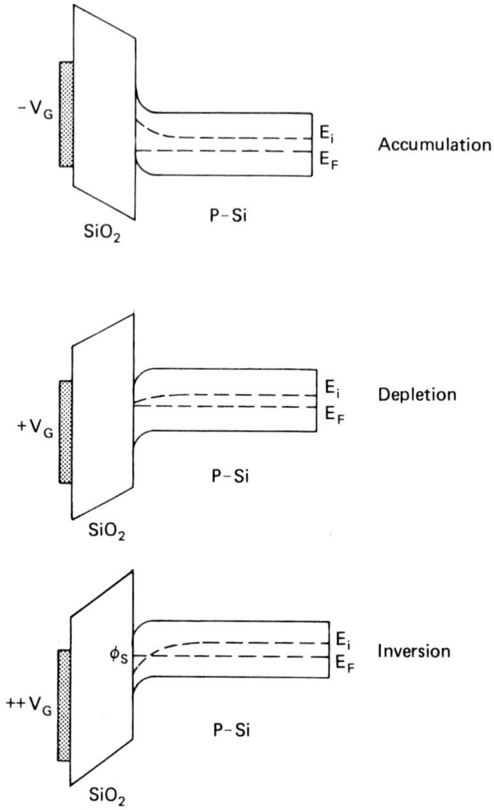

FIG. 3-2 Energy band diagram of the MOS structure under (*a*) accumulation, (*b*) depletion, and (*c*) inversion conditions.

voltage V_{TH}, or turn-on voltage. This definition of threshold voltage leads to the following equation:

$$V_{TH} = \phi_{MS} - \frac{Q_f}{C_o} + 2\phi_f + \frac{\sqrt{2q\epsilon_{Si}\epsilon_0 N_A(2\phi_f + V_{SSub})}}{C_o} \tag{3-1}$$

where ϕ_{MS} = metal-semiconductor work function difference
Q_f = fixed oxide charge at the Si/SiO_2 interface
C_o = gate oxide capacitance
ϵ_{Si} = dielectric constant of silicon (11.7)
ϵ_0 = permittivity of free space (8.86×10^{-14} F/cm)
N_A = substrate doping concentration
V_{SSub} = source-to-substrate voltage

When the gate voltage V_{GS} is less than V_{TH}, but positive enough that ϕ_s is above the midband position, the electron concentration at the surface varies exponentially with ϕ_s, and hence with the gate voltage V_g. This low-current or subthreshold region of device operation thus results in an exponential dependence of I_{DS} on V_g as follows [3, 4]:

$$I_{DS} = Ke^{qV_{GS}/nkT}(1 - e^{-qV_{DS}/kT}) \tag{3-2}$$

where V_{GS} = gate-to-source voltage $\quad V_{DS}$ = drain-to-source voltage

$$n \simeq 1 + \frac{C_D}{C_o}$$

and K depends on device geometry and material parameters. In the definition of n, C_D and C_o are the depletion region and gate oxide capacitances, respectively, and are given by

$$C_D \simeq \sqrt{\frac{q\epsilon_{\text{Si}}\epsilon_0 N_A}{2(\phi_s + V_{\text{SSub}})}} \tag{3-3}$$

$$C_o \simeq \frac{\epsilon_{0x}\epsilon_0}{t_{0x}} \tag{3-4}$$

where ϵ_{0x} = dielectric constant of SiO_2 (3.9)
t_{0x} = gate oxide thickness

This I-V relationship is similar to that in a bipolar transistor except that the slope is less steep in an MOS device ($n > 1$) because of the loss of some of the applied gate voltage across the gate oxide. That is, an MOS device is fundamentally less efficient than a bipolar transistor.

Once the strong inversion condition is reached, the surface potential is essentially fixed at $2\phi_f$ and a different set of I-V relationships holds. In power devices, these are generally of more interest because we are concerned with high-current applications. Figure 3-3 and the following equations summarize the well-known characteristics of the MOS device under strong inversion conditions. These equations should be regarded as first-order only; they will have to be modified later to more accurately represent power MOS devices.

Linear region:

$$I_{DS} \simeq \frac{W}{L}\mu C_o\left(V_{GS} - V_{\text{TH}} - \frac{V_{DS}}{2}\right)V_{DS} \tag{3-5}$$

$$R_{\text{On}} \triangleq \frac{\partial V_{DS}}{\partial I_{DS}} \simeq \frac{1}{(W/L)\mu C_o(V_{GS} - V_{\text{TH}})} \tag{3-6}$$

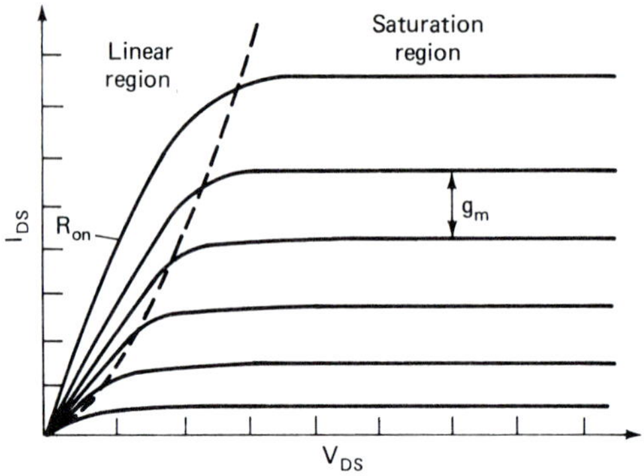

FIG. 3-3 Typical *I-V* characteristics of an MOS transistor operating above threshold. The different curves are for increasing gate voltages $V_{GS} > V_{TH}$.

where W = device channel width
L = device channel length
μ = electron mobility in the inversion layer

Saturation region:

$$I_{DS} \simeq \frac{\mu C_o W}{2L}(V_{GS} - V_{TH})^2 \tag{3-7}$$

$$g_m \triangleq \frac{\partial I_{DS}}{\partial V_{GS}} = \frac{1}{R_{On}} \simeq \frac{W}{L}\mu C_o(V_{GS} - V_{TH}) \tag{3-8}$$

If one examines the above equations and the conventional MOS structure in Fig. 3-1, several limitations of this technology become apparent:

1. The channel length L is determined by photolithographic tolerances, varying from 1 to 2 μm for extremely short-channel, low-voltage devices to 15 to 25 μm for higher-voltage, long-channel devices. The tolerance in the value L is set by the ability to accurately reproduce the drain-to-source line width.
2. The reverse-biased body-to-drain junction depletes further into the body region than the drain region because the drain region is more heavily doped than the body. The depletion of this junction back into the body as a result of large source-to-drain voltages necessitates large values of L if high voltages are to be attained.

Beginning about 15 years ago, modifications to the basic structure in Fig. 3-1 have been proposed, which circumvent some of these limitations. The earliest attempts centered around the introduction of a lightly doped drain region as illustrated schematically in Fig. 3-4. Numerous techniques for achieving such a structure were proposed between 1965 and 1975 [5–8]. These structures allow some of the drain depletion region to spread into the N^- drain region and therefore increase the device voltage capability, but they do not achieve the short effective channel lengths which are necessary to significantly improve both current and voltage capability. A structure which does accomplish simultaneously high-voltage and short-channel operation began to appear in the literature in the early 1970s, and it was this structure—the double-diffused or DMOS device—which made possible the revolution in discrete power MOS devices.

Before turning to the topic of DMOS devices, it is worthwhile to also point out an additional technology change which also occurred in the early 1970s. This was the replacement of metal-gate MOS technology by silicon gate structures. This new configuration is illustrated in Fig. 3-5. The change occurred for three main reasons:

1. The layer of polycrystalline silicon used as the gate also provides another conductive layer that can be used for interconnection.
2. The "self-aligned" feature of silicon-gate transistors allows for both smaller devices and correspondingly smaller capacitances, which result in higher-frequency transistors.
3. Silicon-gate technology eliminates aluminum metallization above the gate region. The inadvertent deposition of mobile ions, primarily sodium, during metal deposition can result in unstable device

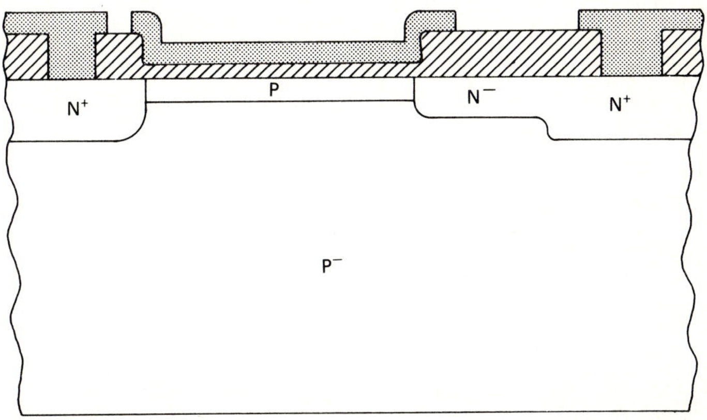

FIG. 3-4 Schematic representation of the addition of a lightly doped (usually ion-implanted) N^- drain region which partically supports the applied drain potential and increases overall device voltage capability.

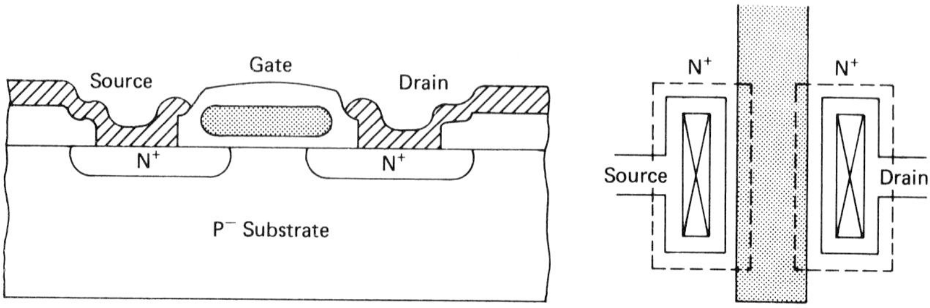

FIG. 3-5 Cross section and topological layout of basic NMOS silicon gate transistor. In this structure the polysilicon is deposited before the N^+ diffusions, resulting in a self-aligned structure since the edges of the N^+ regions are defined by the polysilicon gate.

performance, particularly as the electric field strength in the structure increases. Polysilicon deposition is normally accomplished with chemical vapor deposition, which is inherently a cleaner process.

3.2 DOUBLE-DIFFUSED MOS OR DMOS FIELD-EFFECT TRANSISTORS

An MOS structure that eliminates many of the voltage and On-resistance limitations of conventional MOS transistors is the double-diffused or DMOS technology [9]. Double-diffused MOS structures utilize the difference in the diffusion of sequentially introduced impurities from a common edge or boundary to define the channel length. The channel length is analogous to the base width W_B of a conventional double-diffused bipolar transistor. The control of the channel length and the peak dopant concentration in the channel or body are obtained, as in double-diffused bipolar technology, by proper control of the amount of dopant introduced at each step, and by the subsequent diffusion cycle. The resulting channel length and dopant profile for a lateral DMOS, or LDMOS, transistor are shown in Fig. 3-6. A DMOS transistor differs from a conventional MOS transistor in two significant ways:

1. Since the channel length L is determined by the difference between two sequential diffusions moving in the same direction from a common point of origin, L can be reproducibly controlled to values in the 1- to 2-μm range.
2. The body region is more heavily doped than the N^- drain region, resulting in a junction that depletes further into the drain region

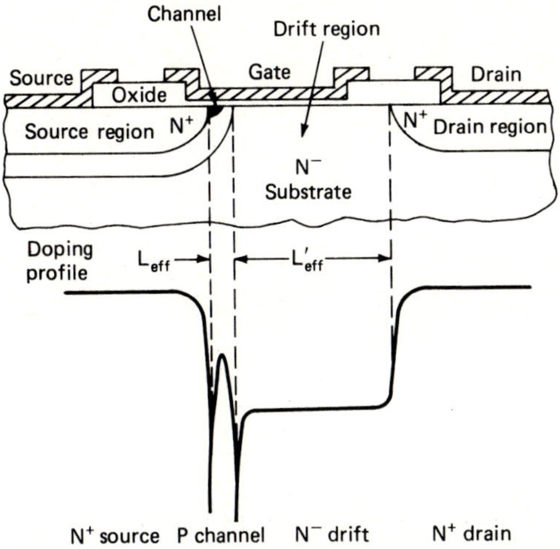

FIG. 3-6 Cross section and doping profile along the surface for an LDMOS transistor. The channel length L is determined by the difference in lateral diffusion of the P and N^+ regions.

than into the body region when a reverse bias is placed across the drain-to-body junction. This difference allows significantly higher voltages to be placed across the body-to-drain junction without markedly affecting the electrical channel length of the transistor.

These two differences result in MOS transistors with both a short channel length and the ability to withstand high drain-to-source voltages because of the separation of the active or channel region of the device from the region of the device that sustains the drain-to-source voltage. This structure is exactly analogous to that used in double-diffused bipolar transistors for many years. In these devices, a narrow, moderately doped base region controls device electrical characteristics. A lightly doped N^- collector region is used to support applied potentials.

Three variations of the basic DMOS structure have been used in the fabrication of high-current, high-voltage MOS transistors. These three variations include the LDMOS structure used to introduce the DMOS concept, the V-groove DMOS or VMOS structure [10], and the vertical DMOS or VDMOS structure [11]. The structures of the VMOS and VDMOS transistors are shown in Figs. 3-7 and 3-8. These devices are conceptually similar to the LDMOS transistor except that the drain N^+ contact is on the back side of the wafer and current flow in them is largely vertical rather than horizontal.

LDMOS transistors have several disadvantages in discrete device

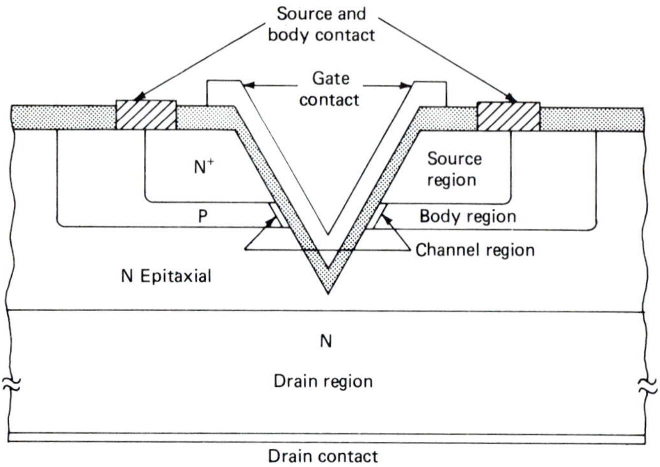

FIG. 3-7 Cross section of a VMOS transistor. Current flows from the N^+ source, along the side of the V groove in an inversion layer and finally in the N^-/N^+ drain region.

applications which have led to the dominance of the vertical structures in these situations. These disadvantages include the following:

1. The drain contact is made on the top surface, reducing the layout efficiency. A bottom drain contact as found in discrete bipolar tran-

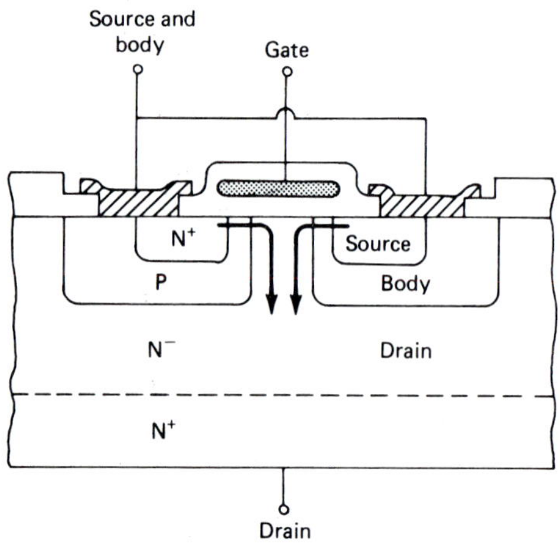

FIG. 3-8 Cross section of the VDMOS transistor. Current flow is initially lateral as in the LDMOS device, through a surface inversion layer, but then turns downward in the N^-/N^+ drain region.

sistors should provide a more efficient structure. (The LDMOS top-side drain contact may provide a desirable feature in integrated circuits using high-voltage DMOS devices, however.)

2. The depletion region of the body-to-drain diode spreads laterally, consuming significant surface area. A depletion region spreading vertically into the wafer would utilize considerably less surface area.
3. The geometrical constraints necessitated by having the source, gate, and drain regions on the surface preclude the use of many of the field shaping techniques that allow extremely high voltages to be reached.

These three disadvantages strongly suggest the investigation of DMOS transistor structures that utilize the vertical aspect of the substrate.

The structure shown in Fig. 3-7 utilizes the anisotropic etching characteristics of ⟨100⟩ silicon to achieve vertical current flow [12]. The V-shaped groove results from the preferential etching of silicon along the crystal plane when a chemical such as potassium hydroxide or hydrazine is used as a silicon etchant. The etch rate reduces to essentially zero when the V-shaped groove is formed, resulting in a groove whose depth is determined by the width of the opening at the surface. The sides of the V groove form a 54.7° angle with the surface and are ⟨111⟩ planes of the silicon crystal.

The channel region of the VMOS transistor is first formed by the sequential diffusion of the body and source regions from the surface of the wafer. The channel region is next exposed with a preferential etch sufficiently deep to reach through the body region to the drain. A gate oxidation cycle followed by metal or polysilicon electrode deposition results in the VMOS structure of Fig. 3-7.

Carriers moving from the source to the drain flow through a channel at the surface of the V groove induced by a voltage on the gate. This carrier flow differs from that of an LDMOS transistor in several ways:

1. For the same diffusion schedule, the VMOS channel length is larger by the geometrical factor associated with the 54.7° angle of the V groove. The difference is approximately 1.5 times the channel length of an LDMOS transistor.
2. The mobility and the scattering-limited velocity of electrons in the inversion layer of a ⟨111⟩ surface are less than those of electrons along a ⟨100⟩ surface [13].
3. The fixed oxide charge or Q_f is approximately 3 times higher on a ⟨111⟩ surface than on a ⟨100⟩ surface [14]. This difference in Q_f results in a higher N-channel threshold voltage for the same body dopant profile.

The carrier flow in a V-groove structure results in additional On resistance in the device due to the constriction in the carrier flow at the "notch" of the groove. The origin of this additional resistance is shown in Fig. 3-9, and is described in more detail in Refs. 15 and 16.

The use of a truncated V groove or U groove as shown in Fig. 3-10 helps to avoid the carrier-crowding phenomenon at the bottom of the groove [15]. This structure, however, requires careful control of the V-groove etching process because the etch must be stopped prior to its natural termination at the bottom of the V groove where the ⟨111⟩ planes intersect. VMOS power transistors, like LDMOS structures, may be fabricated using either metal-gate or silicon-gate technology.

The LDMOS structure may be modified in another fashion (Fig. 3-8) to obtain vertical carrier flow without introducing nonplanar processing requirements. Horizontal injection of the carriers into the drain region is retained while vertical carrier flow is added in the VDMOS transistor structure. The carriers flow from the source through the channel at the surface of the body region to the lightly doped drain region under the influence of voltages on the gate and drain. However, when the electrons reach the lightly doped drain region below the gate, they flow away from the surface of the silicon toward the more heavily doped drain region. In addition to retaining the advantages of a planar LDMOS structure, a vertical structure also displays the advantages discussed above. As with the LDMOS and the VMOS transistor, both metal-gate and silicon-gate VDMOS transistors may be fabricated.

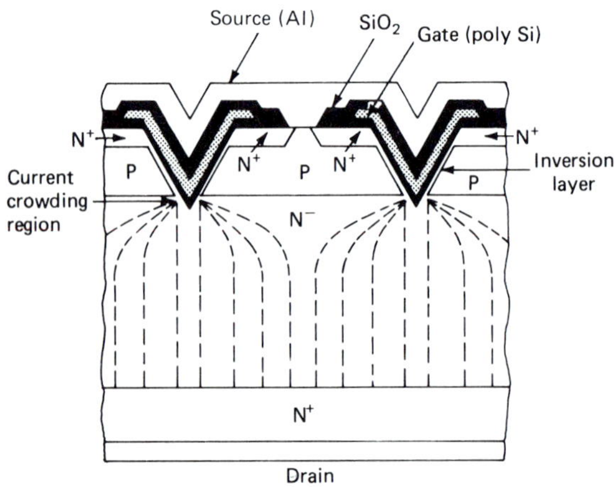

FIG. 3-9 Current flow in the VMOS structure may be constricted at the apex of the V grooves, resulting in additional resistance in the device.

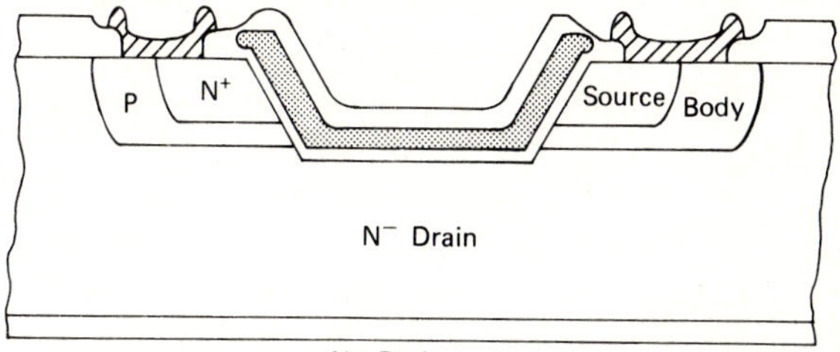

FIG. 3-10 Cross section of the UMOS or truncated V-groove device to minimize the current crowding illustrated in Fig. 3-9.

The two types of MOS power transistors which have found widespread commercial acceptance in discrete form are the LDMOS and VDMOS devices. Of these two the VDMOS is by far the more common. However, because the LDMOS device does have some advantages in specific applications (primarily lower-voltage situations), we shall discuss both structures in more detail in subsequent sections.

DMOS transistors exhibit some differences in the relationship between drain current I_{DS} and gate-to-source voltage V_{GS}. The simplest equation describing conventional MOS devices was previously given in Eq. 3-7:

$$I_{DS} \simeq \frac{\mu C_o W}{2L} (V_{GS} - V_{\mathrm{TH}})^2 \tag{3-7}$$

This equation indicates a square-law relationship between I_{DS} and V_{GS}. Such a dependence is actually observed in long-channel-length MOS transistors, under moderate gate voltages.

However, this relationship assumes that μ is a constant, independent of $V_{GS} - V_{\mathrm{TH}}$. It is well known [17] that large vertical electrical fields across the gate of the device decrease μ because of increased carrier scattering at the Si/SiO_2 interface. This can be modeled in the above equation by a mobility μ which is a function of $V_{GS} - V_{\mathrm{TH}}$.

The lateral electric field accelerating the carriers along the channel is given by

$$E = \frac{1}{L} (V_{GS} - V_{\mathrm{TH}}) \tag{3-9}$$

In devices with short channel lengths and at higher gate voltages, E fields can easily reach values greater than 10^4 V/cm or 1 V/μm. As the accelerating E fields increase, carrier velocity increases, since

$$v = \mu E \tag{3-10}$$

At sufficiently high electric fields ($\simeq 10^4$ V/cm), carrier velocity saturates at a value which is characteristic of the material and carrier type [13]. When this happens, Eq. 3-7 is no longer valid, because μ is effectively no longer a constant. A simplified expression for drain current under these conditions has been derived [9], and is given by

$$I_{DS} = C_o W v_{\text{sat}}(V_{GS} - V_{\text{TH}}) \tag{3-11}$$

where $v_{\text{sat}} \simeq 8 \times 10^6$ cm/s and is the limiting velocity for electrons in silicon [13].

Note that Eq. 3-11 predicts a linear relationship between I_{DS} and V_{GS} rather than the square-law relationship of Eq. 3-7. In addition, the device transconductance is now given by

$$g_m \triangleq \frac{\partial I_{DS}}{\partial V_{GS}} = C_o W v_{\text{sat}} \tag{3-12}$$

which is a constant, independent of both V_{GS} and channel length L. This is to be compared with Eq. 3-8, which holds for long-channel-length devices.

3.3 MAXIMIZING THE BREAKDOWN VOLTAGE OF POWER MOSFET STRUCTURES

The economics of silicon IC manufacturing demand that a given function, whether a VLSI chip or a discrete power transistor, be realized in as small a physical area as possible. This is because the functional yield of good devices on a wafer decreases exponentially with increased area, and hence minimum cost generally implies minimum area. In discrete power transistors, it is generally of interest to maximize the breakdown voltage and to minimize the On resistance. This latter criterion is equivalent to maximizing the current-carrying capability. We will deal with these topics in sequence and consider first the question of achieving the highest possible breakdown voltage in a particular power MOS device structure. The objective generally is to do this with a minimal increase in overall chip area.

The types of breakdown that occur in power MOSFETs are fundamentally the same as those found in other semiconductor devices. However, to obtain the lowest On resistance and the greatest current per unit area in a power MOSFET structure, the breakdown mechanisms and their causes must be understood. In the device structures of interest, the breakdown mechanisms that can occur are divided into the following four categories:

1. *Avalanche breakdown:* The electric field in the semiconductor accelerates carriers into atoms having bound electrons with sufficient energy to generate electron-hole pairs. These carriers are in turn accelerated by the field, resulting in further electron-hold pair production and essentially unlimited current.
2. *Zener breakdown:* The electric field is sufficiently high (on the order of 10^6 V/cm) that carriers gain enough energy to tunnel through the junction potential barrier. These field strengths are only reached in heavily doped N^+-P^+ junctions, and the resulting breakdown voltages are ≤ 6 V [18].
3. *Punch-through breakdown:* The depletion region of a reverse-biased junction spreads to another junction in the vicinity. When the depletion region reaches this junction, the two junctions are connected by a continuous region, and current flows from one junction to the other.
4. *Dielectric breakdown:* The electric field in a dielectric layer exceeds its dielectric strength, resulting in a large current flow and permanent damage to the dielectric.

A properly designed power MOS transistor avoids breakdown mechanisms 2, 3, and 4. Zener breakdown is avoided by keeping heavily doped N-type and P-type regions sufficiently far apart when an appreciable voltage is placed across the resulting junction. Dielectric-layer thicknesses are chosen to keep the electric field well below the maximum allowable field strength. The punch-through condition is avoided by controlling the doping concentration in the body region of the power MOSFET. The amount of immobile ionized charge in the body region is sufficiently large that the body-to-drain voltage does not deplete through the body to the source with the voltage applied. The remaining breakdown mechanism, avalanche breakdown, is the focus of this section, although we will discuss some aspects of punch-through breakdown as well. Avalanche breakdown is the mechanism that determines the drain-to-source breakdown or BV_{DSS} of a well-designed MOS power transistor.

The avalanche breakdown voltage of a plane junction has been studied in detail by many authors [19–22]. The general results of their investigations are:

1. The avalanche breakdown voltage of a plane junction is determined, to first order, by the doping concentration on the lightly

doped side. The carrier mobility, and hence the carrier velocity in an applied field, is higher on the side of the junction with the lower doping concentration. The carriers accelerated to the highest velocity are the ones capable of producing additional hole-electron pairs.

2. The breakdown voltage depends on the dopant profile near the junction. Two limiting approximations for junctions are the one-sided step junction and the linearly graded junction. These two approximations will be discussed in more detail to provide additional background material.

In the one-sided step junction, the doping concentration on one side of the junction is much greater than the doping concentration on the other side. The depletion region under applied bias spreads primarily into the lightly doped side and is given by

$$X_d = \sqrt{\frac{2\epsilon_{\mathrm{Si}}\epsilon_0(\phi_{bi} + V_R)}{qN_A}} \tag{3-13}$$

where X_d = depletion region width
ϕ_{bi} = junction built-in voltage
V_R = applied potential across the junction

The electric field peaks at the metallurgical junction, and its maximum value is given by

$$E_{\mathrm{max}} = \frac{qN_A X_d}{\epsilon_{\mathrm{Si}}\epsilon_0} \tag{3-14}$$

The breakdown voltage of the junction is reached when E_{max} is equal to the critical field E_{cr} for avalanche breakdown in silicon ($\simeq 10^5$ V/cm). Thus we have

$$\mathrm{BV} \simeq \frac{\epsilon_{\mathrm{Si}}\epsilon_0 E_{\mathrm{cr}}^2}{2qN_A} \tag{3-15}$$

This equation illustrates that the breakdown voltage BV is inversely proportional to the doping on the lightly doped side of the junction. This is illustrated in Fig. 3-11, which also shows the behavior of the critical field E_{cr} as a function of doping in silicon [23].

The linearly graded junction corresponds to a gradual transition from P to N doping and is characterized by the impurity concentration gradient a at the junction:

$$a = \left.\frac{dc}{dx}\right|_{x=x_j} \quad \mathrm{cm}^{-4} \tag{3-16}$$

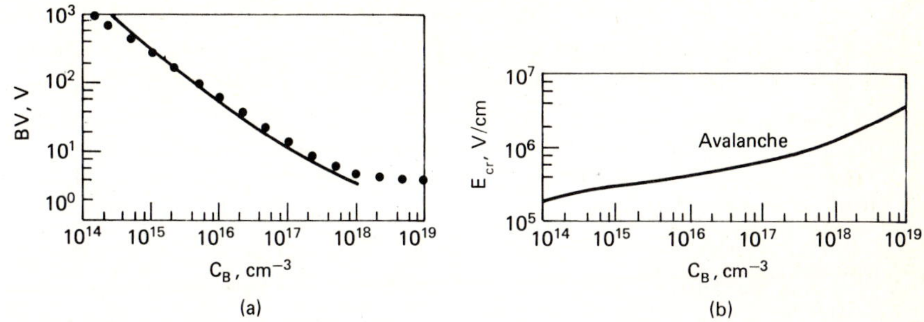

FIG. 3-11 (*a*) Breakdown voltage of a plane junction vs. doping on the lightly doped side. (*b*) Critical field in silicon for avalanche breakdown vs. doping [2].

In this case, the breakdown voltage of the junction is well known to be inversely related to the square root of a and is given by [18]

$$\mathrm{BV} = \sqrt{\frac{32\epsilon_{\mathrm{Si}}\epsilon_0 E_{\mathrm{cr}}^3}{9qa}} \tag{3-17}$$

In the fabrication of high-voltage power MOSFETs, the process should be tailored for the highest breakdown voltage for a given background concentration. The breakdown voltages of diffused plane PN junctions have been calculated on the basis of ionization rates, as is shown in Fig. 3-12; as a function of N_A, the background impurity concentration; and as a function of a, the impurity concentration gradient at the junction [20]. Figure 3-12 shows that as a decreases, the break-

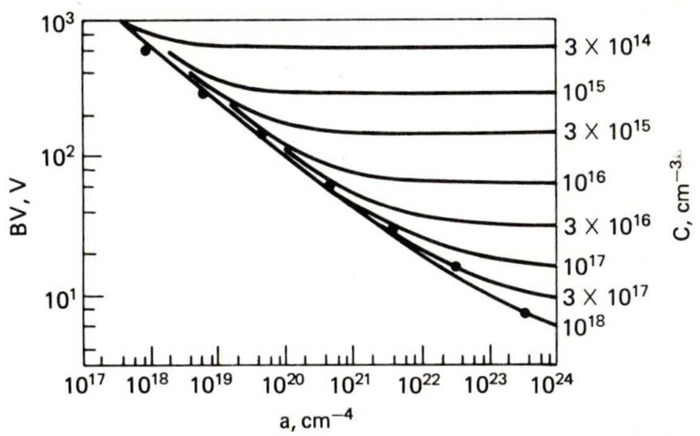

FIG. 3-12 Breakdown voltage of diffused plane junctions in silicon as a function of doping gradient a at the junction and the background doping on the lightly doped side of the junction [20].

down voltage increases from the minimum value determined by the background doping concentration to a higher value. The high values of a on the right-hand side of the figure correspond to the step-junction approximation ($a \simeq \infty$).

In normal silicon device processing, junctions are diffused from the surface of the wafer, resulting in planar, rather than plane, junctions. The curvature of the junction at the corners reduces the breakdown voltage below that of a corresponding plane junction. This reduction in BV is illustrated in Fig. 3-13 [23]. It is clear from this figure that very deep junctions are required to achieve high breakdown voltages if a simple planar junction is used. In fact, such deep junctions are not compatible with minimizing overall chip size and hence an alternative approach must be found to reduce the effects of junction curvature on breakdown voltage. We shall return to this topic shortly.

Junctions are often fabricated in a lightly doped epitaxial layer that has been deposited on a more heavily doped substrate. This structure minimizes the series resistance between the active region on the front of the wafer and the contact on the back of the wafer. If the lightly doped layer is thinner than the depletion width would be at the breakdown voltage, the depletion region will "reach through" to the heavily doped substrate under high reverse bias conditions. The reach-through phenomenon is illustrated in Fig. 3-14. The relationship between the epitaxial layer thickness, the depletion width, and the junction depth at reach-through is, simply, $x_j + W = W_{epi}$.

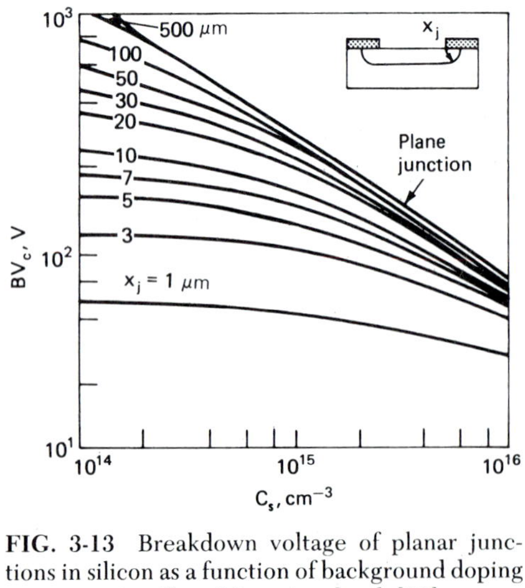

FIG. 3-13 Breakdown voltage of planar junctions in silicon as a function of background doping concentration C_S and junction depth [23].

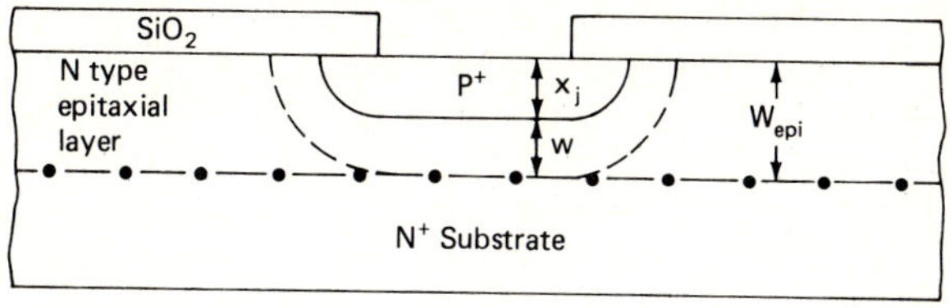

FIG. 3-14 Illustration of the "reach-through" phenomenon. Sufficient voltage is applied across the PN junction that the N^- region is completely depleted.

Once reach-through has occurred, the depletion region essentially stops spreading into the substrate because of the high concentration of ionized donor atoms which terminate the electric field. The breakdown voltage is therefore less than that of a junction in just a lightly doped substrate. This behavior is illustrated in Fig. 3-15 [2].

The highest breakdown voltage for a planar junction in epitaxial material or in the vicinity of heavily doped regions is obtained by keeping the spacing between the high-voltage junction and the heavily doped region at least as large as the depletion region width at planar breakdown. This condition again, however, may not be compatible with other device design objectives such as minimizing On resistance.

The goal of obtaining high breakdown voltage in a planar junction may be approached using a variety of techniques. However, all techniques have a common requirement—distributing the voltage in such a way that the electric field is as uniform as possible, and well below

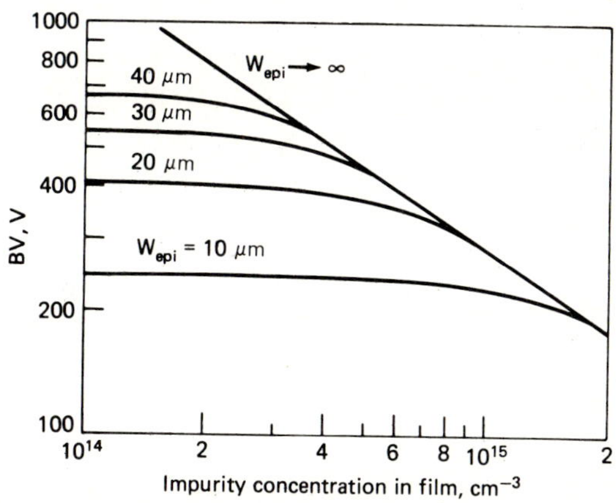

FIG. 3-15 Breakdown voltage of reach-through-limited $P^+N^-N^+$ structures as a function of N^- thickness and doping level [2].

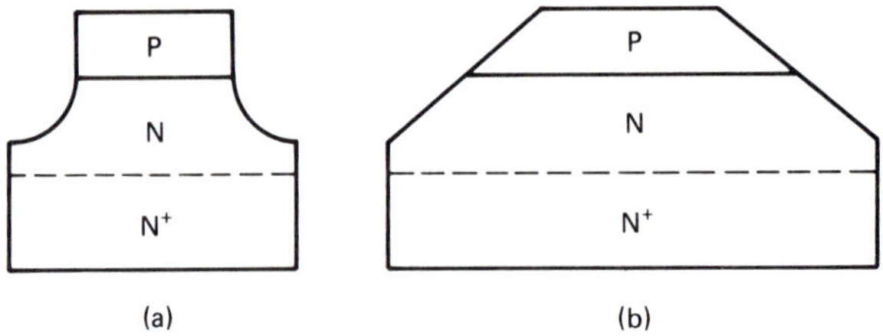

FIG. 3-16 Nonplanar techniques which minimize edge effects in planar junctions: (*a*) moat-etched structure; (*b*) beveled structure.

the field strength needed to cause current to flow in either the silicon or the associated dielectric layer. Two broad classifications of techniques that are used to obtain high voltages are planar and nonplanar, as described below.

Nonplanar technology utilizes etching or grinding techniques to shape the surface of the device at the perimeter of the high-voltage junction. Two of these techniques are shown in Fig. 3-16. When the sidewall of the resulting surface is essentially perpendicular to the original wafer surface, the term used is *moat* etch technology [24, 25]. In this instance, the plane breakdown characteristics of the junction are obtained by etching away the planar portion of the junction.

A junction not perpendicular to the high-voltage junction is a *beveled* junction [24]. A beveled junction is capable of withstanding higher voltages than a moat-etched or mesa junction because the nonperpendicular angle allows the depletion region to spread along the surface in a more controlled fashion.

Other etch contours have also been developed for obtaining high breakdown voltages. It is possible to take advantage of the anisotropic etch characteristics of ⟨100⟩ silicon to obtain a junction with a controlled etch angle at its perimeter [26]. While each of these technologies can be used to obtain breakdown voltages approaching those of plane junctions, and all have been used in high-voltage triacs and SCRs for many years, the nonplanar aspect of these technologies increases fabrication complexity, produces junctions without the low-leakage characteristics of SiO_2 passivated planar junctions, and results in structures that are difficult or impossible to integrate. Accordingly, only planar approaches to high-voltage MOSFETs are discussed in detail in this chapter.

The desire to make a planar junction appear to be a plane junction at its perimeter suggests several approaches for obtaining high breakdown voltages:

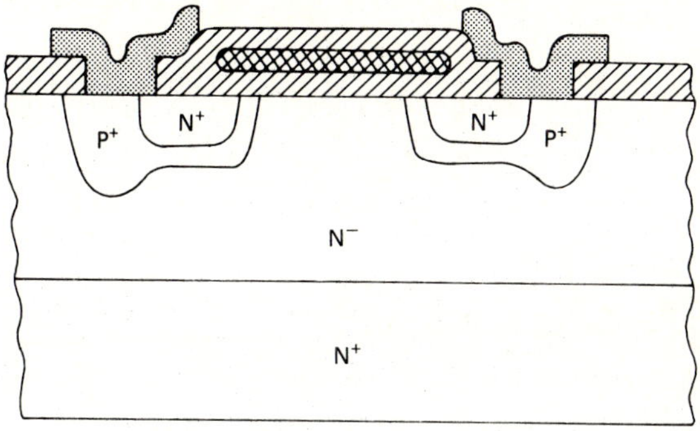

FIG. 3-17 Improved breakdown voltage structure which makes use of a deep perimeter diffusion to increase the radius of curvature of the PN diode at the device edge. The principle is illustrated for the VDMOS device.

1. The diffusion of the high-voltage junction to a larger depth to obtain larger radius of curvature, and, hence, a higher breakdown voltage, as shown in Fig. 3-17 [23]. This does not in general result in an optimum trade-off between On resistance and breakdown voltage, because the N^- epitaxial-layer thickness must be large enough to prevent reach-through breakdown under the deep P^+ regions. This adds series resistance in the central portion of the device where the current actually flows when the device is turned on. We shall not consider this approach further in this section.
2. The use of a field plate at the surface of the high-voltage junction to distribute the voltage along the device perimeter, as shown in Fig. 3-18 [27]. The field plate increases the effective radius of curvature of the depletion region, increasing the breakdown voltage. A variation on the use of a field plate is the utilization of a slightly

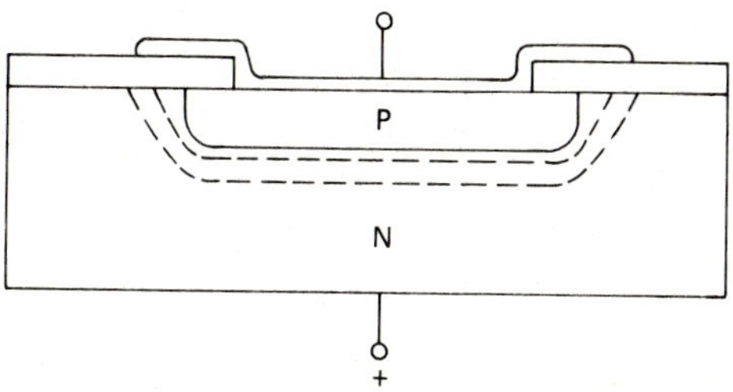

FIG. 3-18 Improved breakdown voltage structure which makes use of a field plate to widen the depletion region at the surface.

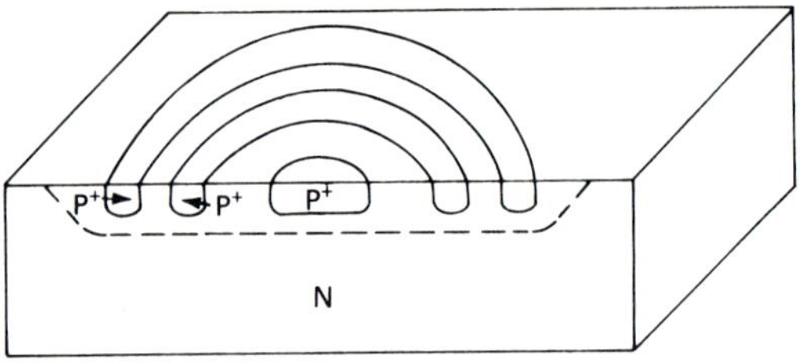

FIG. 3-19 Improved breakdown voltage structure which makes use of floating guard rings or "field limiting rings," at the junction perimeter.

conductive film such as semi-insulating polycrystalline silicon (SIPOS) [28, 29]. The use of such films has two effects on junction breakdown voltage:

a. The highly resistive film produces a continuous voltage drop between the regions at different potentials, eliminating any inhomogeneities in the surface electric field.

b. The film also serves as a barrier preventing either moisture or sodium ions from reaching the silicon surface.

3. The use of "field limiting rings" at the perimeter of the high-voltage junction to set the potential at the edge of the planar junction, as shown in Fig. 3-19 [30]. These rings also act to spread out the depletion region at the junction edges, effectively increasing the radius of curvature of the depletion region.

The field-plate technique has been investigated by a number of workers, particularly Grove et al. [31]. They show that for the gate-

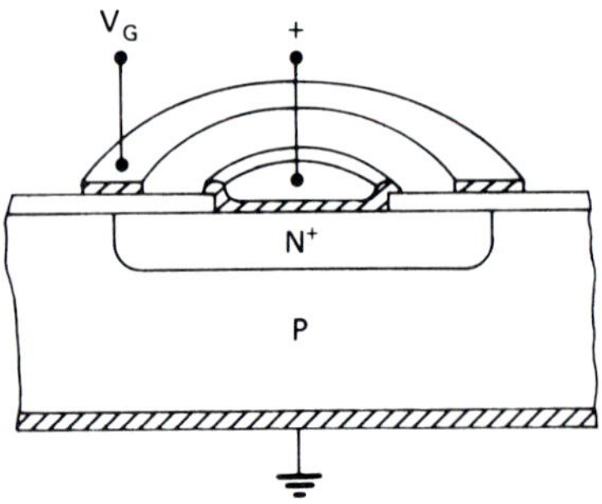

FIG. 3-20 Gate-controlled diode structure used to investigate the effect of a field plate on planar junction breakdown.

controlled planar diode of Fig. 3-20 over a considerable voltage range, the breakdown voltage is given by

$$BV = mV_G + \text{constant} \tag{3-18}$$

where m is a constant.

This behavior and the behavior of the breakdown voltage for large positive and large negative voltages on the gate are explained using the diagrams of Fig. 3-21. Of particular interest in the present context are the diagrams in Fig. 3-21*b* and *c*. From these diagrams, one sees that as the bias on the N^+ side of the diode goes positive, the depletion

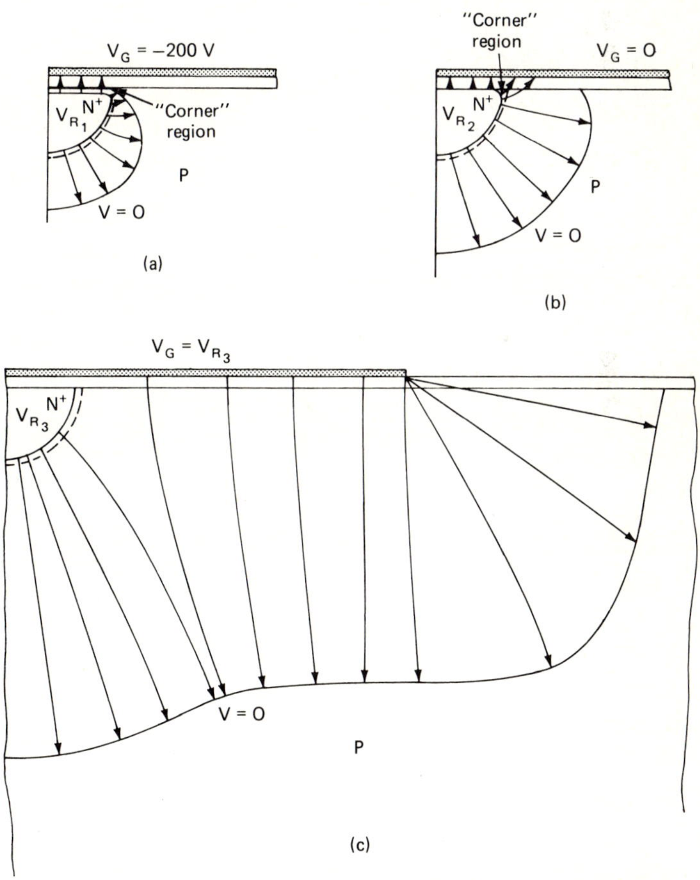

FIG. 3-21 Behavior of the surface depletion region at the edge of the gate-controlled diode sturcture of Fig. 3.20. (*a*) Large negative gate voltage. (*b*) Grounded gate. (*c*) Large positive voltage on the gate ($V_G = V_{N+}$).

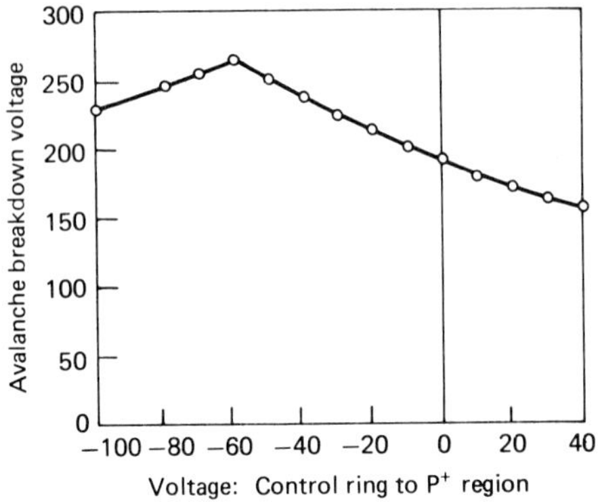

FIG. 3-22 Avalanche breakdown voltage of the structure in Fig. 3.20 as a function of the gate potential [32].

region curvature increases, goes through a maximum, then decreases with an even larger bias. The maximum breakdown voltage of the structure will occur when the radius of curvature of the depletion region induced by the field plate matches the lateral depletion width of the junction. This behavior is consistent with earlier observations that the avalanche breakdown voltage as a function of gate voltage increases, reaches a maximum, and then decreases, as shown in Fig. 3-22 [32].

First-order theory of the effect of field limiting rings (FLRs) explains the increase in junction breakdown voltage in terms of the electric field at the surface of the device and the width of the depletion region of the diffused junction. As shown in Fig. 3-23, a field limiting ring is spaced with respect to an inner junction so that the depletion region "reaches through" to the ring at a voltage lower than the inner junction breakdown voltage. For the first ring, this lateral punch-through voltage is $V_{pt,R1}$. As the punch-through to the first FLR occurs, the depletion region width greatly increases to encompass the diffused FLR, increasing the radius of curvature of the depletion region and the correspond-

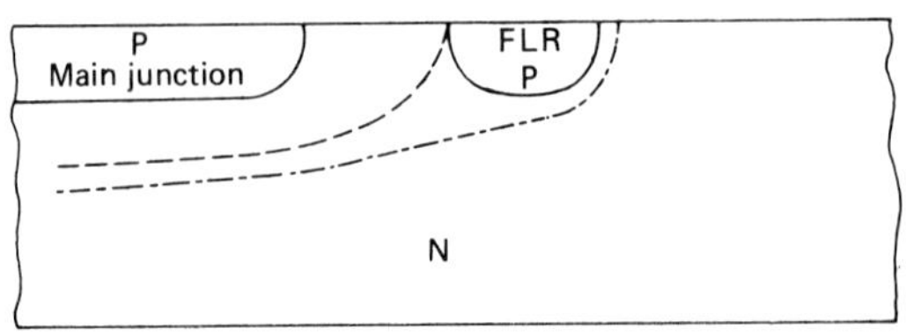

FIG. 3-23 Effect of a floating guard ring or field-limiting ring on the shape of the depletion region at the edge of a PN juction.

ing breakdown voltage. As additional voltage is applied to the main junction, the FLR becomes a voltage divider, with additional voltage beyond $V_{pt,R1}$ producing a depletion region spreading away from the first FLR, as shown in Fig. 3-23.

The theoretical behavior of planar diodes with a single field limiting ring has been investigated using computer techniques [33], resulting in modification to this one-dimensional model of the effect of FLRs. The two-dimensional nature of the structure results in an increased voltage between the main junction and the first ring.

The rate of voltage increase is approximately 0.35 for $V_R > V_{pt,R1}$, so under these conditions $V_{R1} = V_{pt,R1} + 0.35(V_R - V_{pt,R1})$ for $V_R > V_{pt,R1}$, where V_{R1} is the voltage on the first field limiting ring.

It follows that if the spacing of the first ring from the main junction is such that the depletion region reaches it just prior to breakdown, any increases in voltage will cause this breakdown to occur. As the number of FLRs increases, and as the voltage across the device increases, the distance between the device junction and the first FLR must decrease. This behavior is in marked contrast with standard depletion theory, where we expect the depletion region to spread as V to the ½ to ⅓ power corresponding to step or linear graded junctions.

Comparative studies of the field limiting ring and field plate techniques have been made. One such study is summarized in Table 3-1 [16]. The table shows that the FLR technique should be used to obtain maximum breakdown voltage for doping concentrations less than 5×10^{14} cm^{-3}, while field plates are more effective for higher doping concentrations. The crossover point in breakdown voltage occurs at approximately 250 V. This table also shows the amount of chip perimeter that is used in each of the termination techniques. (A 20-μm spac-

TABLE 3-1 Summary of Edge Termination Techniques versus Breakdown Voltage

	Edge termination technique		
Voltage	Field plate	Field limit rings	Edge width consumed by field plate or rings (μm)
100	X		34
150	X		38
200	X		44
250	X		50
300		X	100
400		X	120
500		X	140

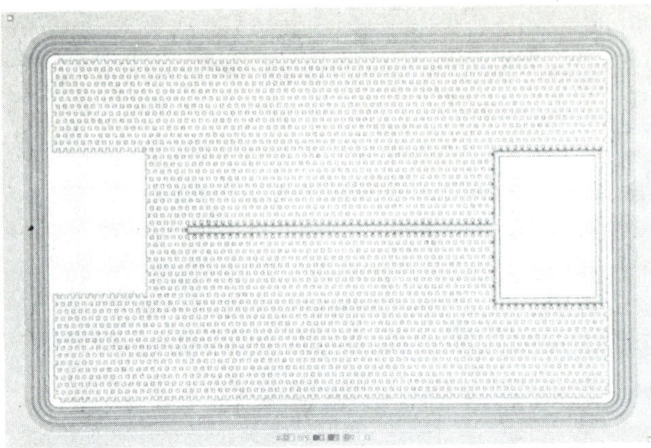

FIG. 3-24 Photomicrograph of a VDMOS power transistor which uses FLRs around its perimeter to maximize breakdown voltage. This discrete device is capable of supporting 450 V and carrying a current of 5 A. *(Photo courtesy of Siliconix, Inc.)*

ing from the field plate or the FLR to the edge of the die is assumed.) In terms of area consumed, a field plate is much more efficient than field limiting rings, but cannot be used for voltages greater than approximately 250 V, because of the much higher-resistivity substrate that is required. The amount of area needed for the high-voltage perimeter varies with the breakdown voltage. The active chip area is reduced by both the scribe line area and the area required for field shaping to obtain the required breakdown voltage. In actual practice, most commercially available power MOS devices today that have voltage ratings greater than 200 V use the FLR technique to control the fields at the

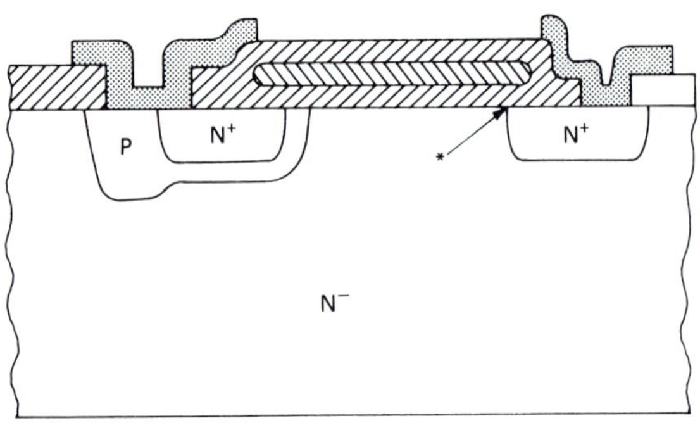

FIG. 3-25 Avalanche breakdown mechanism in the LDMOS device when the gate is allowed to overlap the N^+ drain region.

device perimeter. Figure 3-24 shows an example of a discrete power MOS device with such rings around its perimeter.

One final type of avalanche breakdown can occur in the LDMOS device when the gate electrode is allowed to overlap or come close to the N^+ drain region. The situation is illustrated in Fig. 3-25. In the region of the drain under the gate (the asterisk in Fig. 3-25), very high electric fields can exist which may cause the critical E field to be exceeded and hence induce avalanche breakdown [34]. This mechanism can be avoided by offsetting the gate electrode from the drain by a few micrometers (μm).

3.4 LATCHBACK IN POWER FET STRUCTURES

The last section was devoted to maximizing the breakdown voltage of the DMOS structure using planar techniques. Of equal importance for a power MOSFET is its ability to withstand overvoltage conditions for a period of time. In all but the most conservatively designed circuits, the voltage across a circuit component may exceed the maximum specified operating voltage for a variety of reasons. In particular, in switching circuits containing any inductance, the voltage may greatly exceed the supply voltage when switching occurs due to the collapsing of the flux in the inductor. Under these overvoltage conditions, an ideal power FET will not sustain any damage if its power dissipation limit is not exceeded. There are three methods of achieving this degree of overvoltage protection in a circuit using power MOSFETs:

1. Use FETs with a breakdown voltage so high that an overvoltage is never reached even under transient conditions.
2. Add components such as clamping diodes to steer potentially damaging currents through less fragile devices.
3. Use power FETs designed to withstand the overvoltage conditions even if the breakdown voltage is exceeded. This solution may be realized by integrating protective devices within the power MOSFET itself.

The first solution proposed above is not cost-effective, because the size of a device with a specified On resistance increases as the breakdown voltage to the 2.6 power [35]. In certain circumstances this alternative may provide a fast engineering solution, with the added cost being only the added price of a higher-voltage device. The second solution also provides a rapid solution to the problem, but with added circuit cost and complexity. In the majority of cases the third solution, the use of power MOSFETs designed to withstand overvoltage conditions, is preferable. This section investigates the behavior of power FET structures under overvoltage conditions. It will be shown that latchback and its prevention are the key phenomena to consider in designing rugged power FET devices. Latchback is destructive because a device suddenly switches from one blocking voltage to another, lower blocking voltage. The current that flows through the device, and hence the power being dissipated in it, may increase substantially when this transition occurs.

In the LDMOS and VDMOS structures, the body is contacted some distance from the channel region of the FET. The body contact also contacts the source region of the FET. This metallization shorts the emitter and base of the parasitic NPN transistor formed by the source, body, and drain regions of the FET. The diffused source and body regions are actually a distributed series of NPN bipolar transistors with their bases separated by the incremental resistance of the pinched body region beneath the source, as shown in Fig. 3-26 for the LDMOS structure. A similar equivalent circuit could be drawn for the VDMOS device. In breakdown, the FET can be modeled to first order by considering only the two NPN transistors at the extremes of the body region, the body resistor between these two transistors, and the diode formed by the body-drain junction in addition to the normal FET and the series resistance of the body region contact [16]. The equivalent circuit thus becomes that shown in Fig. 3-27. The circuit elements shown in this figure are:

R_b the body resistance between the contact at the surface and the body region beneath the source

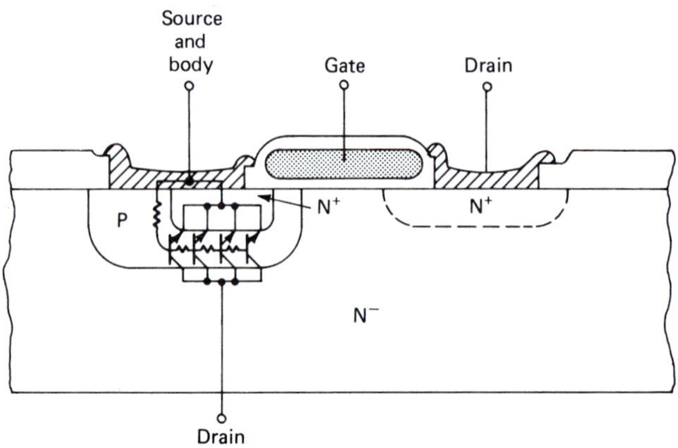

FIG. 3-26 Physical origin of the parasitic bipolar transistors that are intrinsic to DMOS devices, illustrated here for the LMDOS structure.

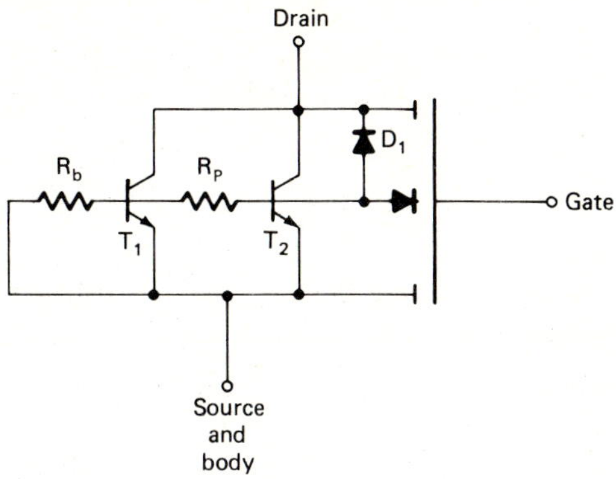

FIG. 3-27 Simplified equivalent circuit of the LDMOS device plus parasitic bipolar devices illustrated in Fig. 3-26.

D_1 the region of the body-to-drain diode where avalanche breakdown first occurs

T_1 the parasitic NPN transistor physically closest to the body contact

T_2 the parasitic NPN transistor adjacent to the surface near the gate

R_P the resistor formed by the pinched body region beneath the source

Modeling of the FET with the two bipolar transistors at the ends of the source-body junction provides circuit simplification while retaining the devices that largely determine the FET's behavior in breakdown.

With the gate of the FET biased to turn the MOS transistor on, current flows in the channel at the surface of the body region. No lateral current flows through the body resistor R_b. With the gate biased so no channel is present, only leakage currents flow through the device. For a typical junction in a large power device, these leakages are in the microampere range, the exact value depending upon the physical size of the junction. In the extreme case, this leakage current can be thought of as occurring at a single diode D_1. By replacing D_1 with a current source as shown in Fig. 3-28, the behavior of a power FET under reverse bias conditions can be explained.

The current supplied by the current source flows to the negative supply through the emitter-to-base diodes of T_1 and T_2, and through the resistors R_b and R_P. The relative impedance of these three paths changes with the voltage from the base to the emitter of transistor T_2. The impedances from the base of T_2 to ground seen by the currents I_1, I_2, and I_3 are

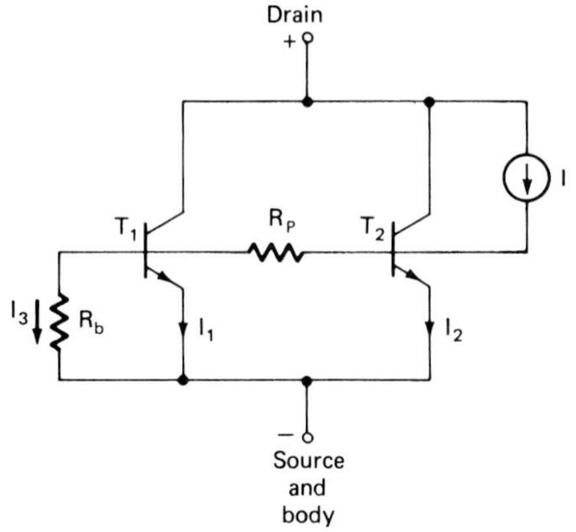

FIG. 3-28 Simplified equivalent circuit for the LDMOS in breakdown. *I* represents the current flow due to the avalanche process.

$$Z_1 = R_P + \frac{V_{BE1}}{I_{S1}} (e^{qV_{BE1}/kT} - 1)^{-1} \tag{3-19}$$

$$Z_2 = \frac{V_{BE2}}{I_{S2}} (e^{qV_{BE2}/kT} - 1)^{-1} \tag{3-20}$$

$$Z_3 = R_P + R_b \tag{3-21}$$

The relative impedances Z_2 and Z_3 are shown in Fig. 3-29 for a diode fabricated in silicon. The rapid change in impedance as a function of voltage is caused by the exponential nature of diode current with volt-

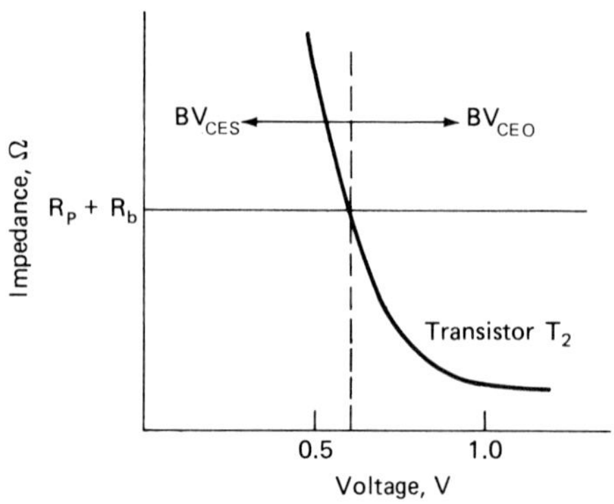

FIG. 3-29 Relative values of impedances $R_P + R_b$ and Z_2 (Eq. 3-21). As T_2 turns on it effectively carries all of the current.

age. For small values of current I, the voltage across the emitter-base diode of T_2 is small, and the majority of the current flows to the negative lead through the two resistors. The voltage from base to emitter in T_2 is small, so it is reasonable to assume that the base is shorted to the emitter and that the transistor is operating in the BV_{CES} mode. As the current I increases, the voltage across the resistors R_P and R_b increases, and the impedance of T_2 decreases. A larger fraction of the total current flows through the base-emitter diode of T_2. The relative impedance of T_2 becomes small compared to $R_P + R_b$, and the transistor is now operating in the BV_{CEO} mode. From the device model perspective, the transition from transistor BV_{CES} to BV_{CEO} occurs when the relative impedances of the two paths to ground are about equal. This point is the intersection of the two lines in Fig. 3-29. Using this information, the I-V curve of a power FET is shown in Fig. 3-30. The currents and voltages of importance are:

- BV_{CES} the collector-to-emitter (drain-to-source) breakdown voltage of the parasitic NPN transistor with the base shorted to the emitter
- BV_{CEO} the collector-to-emitter breakdown voltage of the parasitic NPN transistor with the base open
- I_T the current that triggers the transition from the BV_{CES} to the BV_{CEO} mode of operation

These three parameters are both layout- and process-dependent. Looking at these three parameters in detail:

- BV_{CES} This value is determined by the background doping concentration and the radius of curvature of the junction at its periphery, as has been discussed earlier, if the transistor is

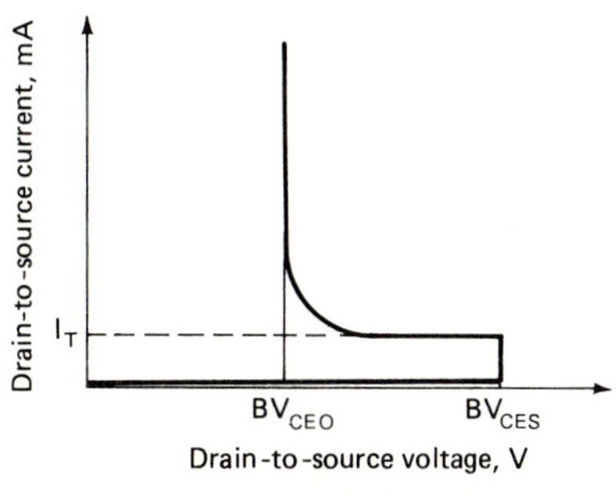

FIG. 3-30 Latchback phenomenon observed in power FET structures when transistor T_2 turns on.

not punch-through-limited. Punch-through is not present if $BV_{CES} = BV_{CBO}$ in the transistor [2].

BV_{CEO} In a bipolar transistor that is not punch-through-limited, the BV_{CEO} and the BV_{CES} are related by the equation [2]

$$\frac{BV_{CEO}}{BV_{CES}} = \frac{1}{\sqrt[n]{h_{fe}}} \tag{3-22}$$

where h_{fe} = the small-signal common-emitter current gain.
n = an experimentally determined value typically between 3 and 6 for a double-diffused NPN transistor [36]. It depends on the specifics of the substrate and the device processing.

I_T From the investigation of the impedance of the base-emitter junction shown in Fig. 3-29, the transition from the BV_{CES} mode of operation to the BV_{CEO} mode occurs at $V_{BE2} \simeq 0.6$ V. This voltage is reached when

$$(R_P + R_b)I_T \simeq 0.6 \text{ V} \tag{3-23}$$

or

$$I_T \simeq \frac{0.6 \text{ V}}{(R_P + R_b)} \tag{3-24}$$

When the avalanche current reaches this value, the transition from BV_{CES} to BV_{CEO} is triggered.

The goal of understanding the latchback characteristics in power FET structures is to develop techniques for preventing the unwanted latchback. Based on the model of Fig. 3-27, there are at least two methods of achieving this result:

1. Direct the current flow in the breakdown mode so that it bypasses the pinched resistor R_P, thus never turning on T_2.
2. Reduce the value $R_P + R_b$ so that any lateral current flow does not result in the 0.6-V drop necessary for the majority of this current to flow through the emitter-base diode.

We will consider each of these alternatives separately.

An example of the first approach is an integrated body-resistor-bypass diode which can be implemented by introducing a deeper, more heavily doped P^+ diffusion at the region where the contact to the body is made. A cross section of this structure and the equivalent circuit is shown in Fig. 3-31, for the VDMOS structure. In this structure, the

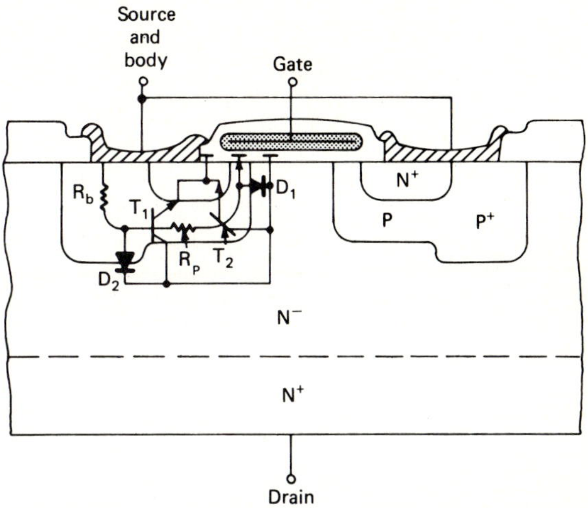

FIG. 3-31 Prevention of latchback in the VDMOS device through the use of a more heavily doped P^+ region at the device perimeter to control the breakdown location.

deeper, heavily doped P^+ diffusion (D_2) goes into avalanche breakdown before the lightly doped body region (D_1), diverting the avalanche current around the pinched resistor R_P. The higher doping concentration of D_2 somewhat reduces the breakdown voltage of the structure, but prevents the latchback phenomenon that may occur in its absence.

The second general method of reducing the latchback problem in power MOS structures is the reduction of $R_P + R_b$. This can be accomplished in either of two ways:

1. Minimize the distance from the source and body contact to the body region adjacent to the channel. This approach can reduce the resistance by a modest amount, but device geometries are such that a factor of 2 reduction in distance (and hence, in value of R_P) is not possible.
2. Include P^+ shorting bars from the source-body region to the body region adjacent to the channel. These shorting bars can be regions of the body that do not have a source diffusion. This technique reduces the amount of injecting source perimeter, and only increases the current level at which latchback occurs. It does not completely eliminate the latchback problem.

An example of this latter approach is shown in Fig. 3-32 for the LDMOS device. A P^+ region around the perimeter of the active device

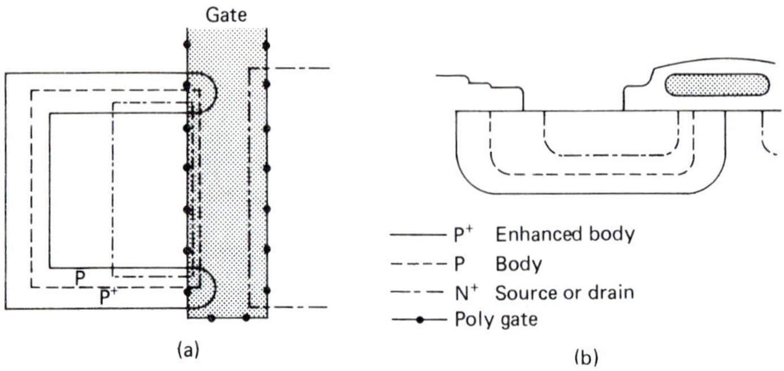

FIG. 3-32 Reduction of latchback in the LDMOS device through the use of P^+ shorting bars to bypass R_P. A variety of geometries can be used to accomplish this objective. (*a*) Top view; (*b*) side view.

provides a low-resistance path from the drain region to the source-body contact. This effectively bypasses the R_P in the equivalent circuit of Fig. 3-27 and hence markedly increases the triggering current necessary to initiate latchback [16].

In summary, the latchback phenomenon can be minimized in power MOSFET structures by proper device design. While some early commercial designs suffered from reliability problems due to this mechanism, more recent designs have effectively eliminated the problem by using the techniques described above, and power MOSFETs today have excellent reliability even in transient overvoltage situations. It should also be pointed out, however, that the parasitic bipolar device which is responsible for the latchback phenomenon is inherent in power MOSFET devices. In fact we will consider some modified structures later in the chapter which actually make use of this parasitic device.

3.5 MODELING AND MINIMIZATION OF ON RESISTANCE IN POWER MOSFETs

In most applications of discrete power MOSFETs, the device is used as a switch. Under these conditions, a crucial device parameter is the resistance in the On state or R_{On}. This parameter was defined for a simple MOS transistor in Eq. 3-6 and in Fig. 3-3. This simple definition, however, models only the resistance of the channel inversion layer. Particularly in high-voltage structures, this may be only a small component of the total device resistance. We shall consider in this section, then, more accurate models for device On resistance, specifically for the LDMOS and VDMOS devices.

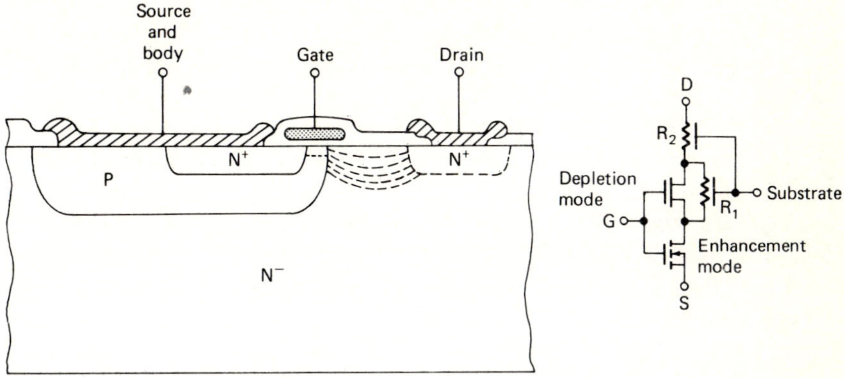

FIG. 3-33 Cross section and approximate equivalent circuit for the LDMOS device to calculate R_{On} [15].

Figures 3-33 and 3-34 illustrate approximate equivalent circuits for the LDMOS and VDMOS devices, respectively [15]. Each of the structures consists basically of an enhancement-mode transistor in series with a parasitic bulk resistance. A more careful look at the equivalent circuits shown in the figures will reveal some basic differences, however.

The LDMOS device consists of an enhancement-mode transistor (active channel) in series with a depletion-mode transistor (surface accumulation region), itself in parallel with a bulk resistance R_1, in series with a second bulk resistor R_2 [37]. The middle two components

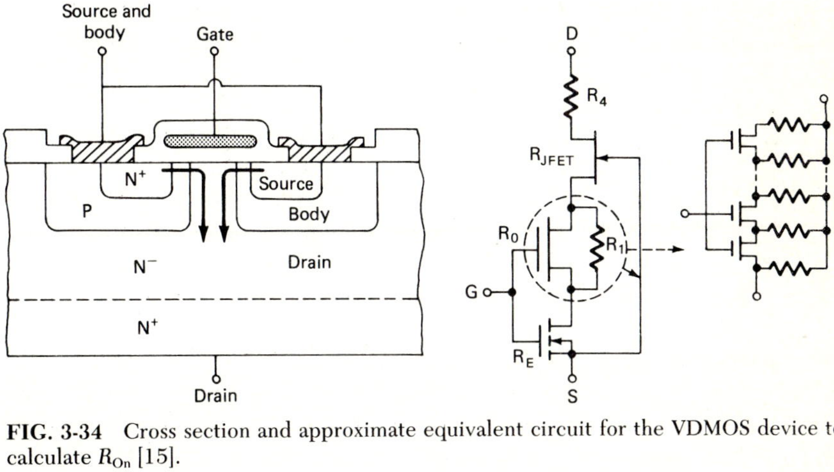

FIG. 3-34 Cross section and approximate equivalent circuit for the VDMOS device to calculate R_{On} [15].

model the gate-controlled accumulation layer in parallel with a bulk resistor; the current through the device will divide between these two in a manner determined by the gate voltage (i.e., the conductance of the accumulation layer). The third series component,* R_2, models the region of the device between the gate electrode and the drain contact; such a region would be missing in a low-voltage structure in which the gate conductor could extend to the N^+ region [34]. In a junction-isolated device, the substrate PN^- junction modulates the conductivity of R_1 and R_2 as indicated in the figure.

The VDMOS device model (Fig. 3-34) is distinguished from the LDMOS model principally in two ways. First, the depletion-mode transistor and its parallel bulk resistor must be regarded as distributed devices because the current flow changes in direction from horizontal to vertical along the length of the accumulated surface between the source regions. The distributed model will be simplified here to a single depletion-mode transistor. A correction factor will be introduced into the modeling later to account for the two-dimensional nature of the current flow. The second major factor distinguishing the VDMOS from the LDMOS is the presence of the series JFET in the equivalent circuit. This arises physically because of the pinching of the current between the adjacent P bulk diffusions. The spacing between the P diffusions (channel width of the JFET) can be of great importance in determining the fraction of the device's total On resistance contributed by the series JFET.

Analyses of the total R_{On} of these devices based upon these equivalent circuits have been performed [15, 16]. Clearly, approximations are involved in proceeding in this fashion because two-dimensional structures are being simplified to lumped equivalent circuits. The motivation for doing this is simply that clearer physical insight results than would be the case if more accurate computer techniques were used. The results described below largely follow the analysis given by Sun and Plummer [15].

The LDMOS equivalent circuit shown in Fig. 3-33 contains a total of four components. However, in high-voltage structures the lateral extent of the thin gate oxide over the N^- drift region and the lateral extent of the metal gate after it steps up over the thick oxide over the N^- drift region (see Fig. 3-33) are both generally small compared to the total extent of the N^- region [38]. This configuration is necessary in order to maximize the device voltage capability since the gate metal acts as a field plate [34, 38]. Because of this, the contributions of the depletion-mode device and the bulk resistor R_1 to the total device On resistance are generally small. We can, therefore, neglect the contribution of the depletion-mode device and include R_1 in the calculation of R_2. In lower-voltage devices in which this is not a valid approximation, the formulation given later for the VDMOS device can be directly extended to the LDMOS structure. With this simplifying assumption, the equivalent circuit becomes simply an enhancement-mode transistor in series with a bulk resistor.

The enhancement-mode channel resistance is given by the equation for R_{On} presented as Eq. 3-6,

$$R_E = \frac{1}{(W/L)\mu_E C_o(V_{GS} - V_{\mathrm{TH}_E})} \tag{3-25}$$

but here the subscript E has been added to create μ_E and V_{TH_E} to indicate the enhancement device in the equivalent circuit. In actually substituting values into this equation, the fact that μ_E varies both with doping concentration in the channel and with the vertical gate electric field must be taken into account [15].

To model the bulk resistor R_2 in the LDMOS equivalent circuit, the current is considered to flow from a line source of radius r_1, at the end of the channel, to a line sink with radius r_2 at the N^+ contact. This model is analogous to the electrostatic problem of an image charge in a conducting plane [39] and has been treated in detail by Pocha [40]. From Ref. 40, the resulting resistance of the bulk epitaxial region is

$$R_2 = \frac{\rho}{W\pi}\left(\ln\frac{L' - r_1}{r_1} + \ln\frac{L' - r_2}{r_2}\right) \tag{3-26}$$

where L' is the effective length of the N^- epitaxial resistor, r_1 is the effective radius of the current source at the end of the channel, and r_2 is the effective radius of the current sink at the N^+ contact.

L' is found directly from the mask dimensions of the device (the spacing between the P channel region and the N^+ drain contact minus lateral diffusion). r_1 and r_2 are likely dependent on the specific impurity profiles and doping levels used, but are typically 0.5 to 2.0 μm [15].

Calculation of On resistance for the VDMOS device is considered in Fig. 3-35, in which several dimensions to be used below are defined. In this device, the thin gate oxide normally extends all the way between adjacent source and channel diffusions. As a result, the surface accumulation layer (region 2 in the figure) plays an important role in total device On resistance and cannot be neglected. Physically, this region collects the channel current and distributes it over the region between adjacent P diffusions. It therefore reduces current crowding at the end of the channel and helps to reduce total device On resistance. This effect would be expected to be most efficient at high gate voltages, which maximize the conductivity of the accumulation layer.

In addition to this factor, a junction field-effect transistor exists between the adjacent P diffusions, as shown in the equivalent circuit in Fig. 3-34. Depending on the ratio of channel width (*a* in the figure) to the P diffusion depth in this JFET, it can play an important role in overall device On resistance.

Neglecting contact resistances and the resistance of the bulk N^+ substrate, the VDMOS On resistance is thus given by

$$R_{\text{OnVDMOS}} = R_E + R_D + R_{\text{JFET}} + R_4 \tag{3-27}$$

where R_E = enhancement-mode device On resistance
R_D = depletion-mode device On resistance (surface accumulation layer)
R_{JFET} = JFET On resistance
R_4 = bulk epitaxial-layer resistance

R_E is given by the same expression (see Eq. 3-25) as was used in the modeling of the LDMOS device. The depletion-mode device On resistance is given by an expression analogous to Eq. 3-25:

$$R_D = \frac{1}{12} \frac{1}{(W/L')\mu_D C_o (V_{GS} - V_{\text{TH}_D})} \tag{3-28}$$

where μ_D is the electron (majority-carrier) accumulation-layer mobility and L' is the effective depletion-mode device channel length.

The factor of 1/12 in the expression for R_D appears because of the two-dimensional nature of the current flow from the accumulation layer into the bulk N^- region [41, 42]. In essence, it models the conversion of current flow from lateral to vertical. More accurately, as shown in Fig. 3-34, the depletion-mode device should be modeled as a distributed structure; the factor of 1/12 has, however, provided reasonable agreement with experiment [15].

The JFET component of the VDMOS equivalent circuit in Fig. 3-34 is modeled with the geometry shown in Fig. 3-35*a*. The PN^- junction is modeled as a portion of a circle with the origin offset from the original mask edge by 0.15 X_j to account for the difference in lateral and vertical diffusion. The current is assumed to originate uniformly from the surface accumulation layer. Under these conditions, the resistance of the JFET region has been shown to be [15]

$$R_{\text{JFET}} = \frac{2\rho}{W}\left[\frac{1}{\sqrt{1-(2X_j/L)^2}}\left(\tan^{-1} 0.414 \sqrt{\frac{L+2X_j}{L-2X_j}}\right) - \frac{\pi}{8}\right] \tag{3-29}$$

Note that this formulation does not take into account any depletion regions which may exist in the N^- region. As expressed in Eq. 3-29, then, it applies to the case of low V_{DS}.

The JFET region of the VDMOS model is considered to end at the point where $\theta = 45°$. Below this boundary in the N^- material, a bulk resistor bounded by the trapezoidal geometry in Fig. 3-35*a* represents the series resistance of the epitaxial layer. The resistance of this trapezoidal volume is found to be [15]

$$R_4 = \frac{\rho}{W} \frac{1}{\tan \alpha} \ln \left(1 + 2 \frac{h}{a} \tan \alpha \right) \tag{3-30}$$

a and h are easily found from device geometry. The spreading angle α is analogous to the thermal-impedance spreading angle that has been

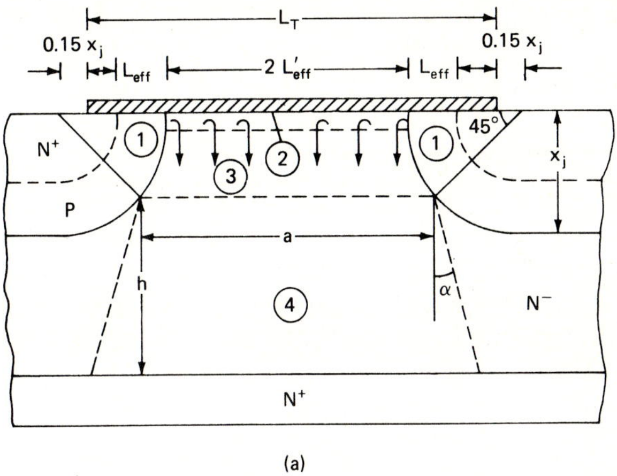

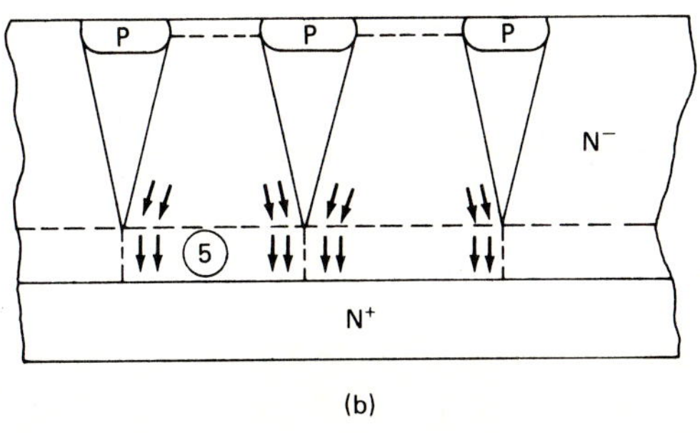

FIG. 3-35 Model used to calculate R_{On} for the VDMOS device [15]. (*a*) Top; (*b*) bottom.

calculated by David [43]. His exact calculations can be approximated within ±5 percent by the simple expressions shown below [15]:

$$\alpha = \begin{cases} 28° - \dfrac{h}{a} & \text{if } h \geq a \\ 28° - \dfrac{a}{h} & \text{if } h < a \end{cases} \tag{3-31}$$

If the epitaxial layer is very thick or if the surface packing density is very high, adjacent trapezoids may overlap, as shown in Fig. 3-35*b*. In this case, the resistance of region 5 in the figure is easily found by assuming uniform current flow in the merged region.

The formulas given above for R_{On} for the LDMOS and VDMOS devices provide reasonable agreement with experiment [15]. They also allow one to estimate the relative contributions of the various resistor components to the total R_{On}. An example of this is shown in Fig. 3-36, taken from Ref. 15, for VDMOS devices. The horizontal axis is the epitaxial-layer resistivity.

The calculations produce not unexpected results. For low-voltage devices ($\rho < 1\ \Omega\cdot$cm) channel resistance dominates, while in high-voltage devices ($\rho > 10\ \Omega\cdot$cm) the bulk contributions of the JFET and the epitaxial layer dominate. In fact, for epitaxial-layer resistivities greater than about 8 to 10 Ω·cm, the total device On resistance can be reasonably calculated from just the bulk terms $R_{JFET} + R_4$. The contribution of the depletion device R_D is not plotted explicitly in Fig. 3-36; it is simply the difference between 100 percent and the sum of the other two terms.

In the actual design of any discrete power MOSFET device, the

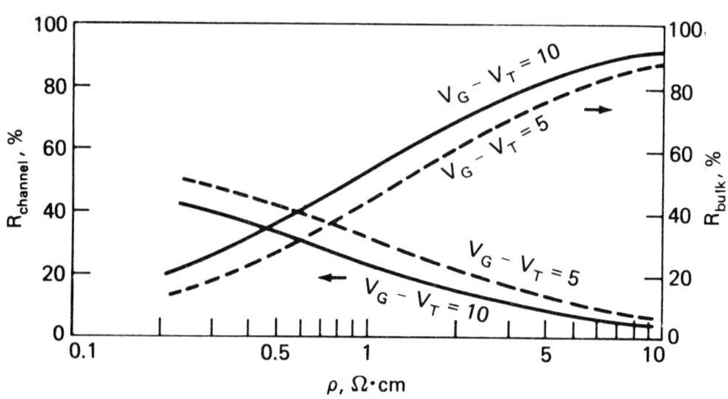

FIG. 3-36 Relative contributions of the channel resistance R_E and the bulk resistance $R_{JFET} + R_4$ for the VDMOS device as a function of the N^- region resistivity [15].

actual parameter of interest is the device R_{On} for a given chip area, not just the R_{On} of a unit cell in the structure as described above. Unfortunately, calculating R_{On} for a given chip area requires knowledge of specific layout and design rules, which vary from one manufacturer to another. It also demands knowledge of the geometric arrangement of the source, gate, and drain electrodes on the chip surface.

It has generally been found that the LDMOS device is area-efficient, in terms of R_{On} per unit area, only in low-voltage applications ($BV_{DSS} \leq$ 40 V). At these low voltages, the gate electrode can extend all the way to the drain N^+ region over the thin gate oxide, and the drift-region resistance is markedly reduced by surface accumulation. At higher voltages, this is not possible, and the large surface areas which must be allowed for laterally spreading depletion regions, coupled with the layout difficulties of having all three electrodes on the top surface, generally mean that the vertical VDMOS is preferred.

The VDMOS device allows considerable flexibility in surface layout since only the gate and source electrodes are on this surface. Most modern devices employ enclosed source geometries (circles, squares, hexagons, etc.) and two layers of interconnect. The lower layer is polysilicon and forms the gate electrode; the upper layer, which usually covers the whole surface, is aluminum and contacts the source regions individually. There is no universal agreement on the optimum source geometry. Figure 3-37 illustrates some of the commonly used arrangements.

	Square on square grid	Circle on square grid	Hexagon on square grid	Square on hexagonal grid	Circle on hexagonal grid	Hexagon on hexagonal grid
Source geometry and grid	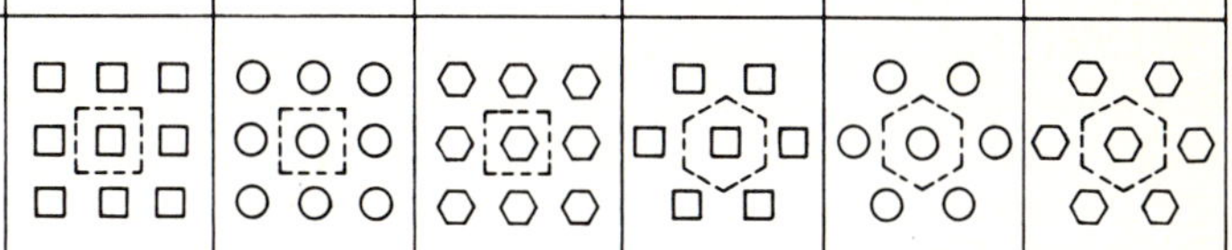					
Unit cell	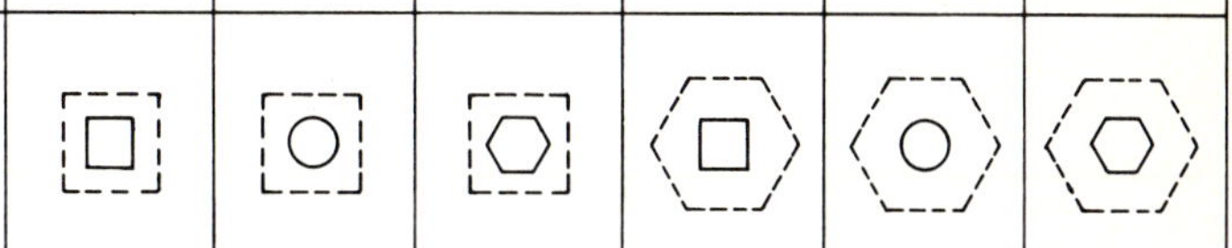					
Coefficient of cellular geometries	1.0	0.8862	0.9306	1.0746	0.9523	1.0

FIG. 3-37 Some possible surface geometries for VDMOS devices along with their respective relative efficiencies in terms of active source perimeter per unit surface area [44].

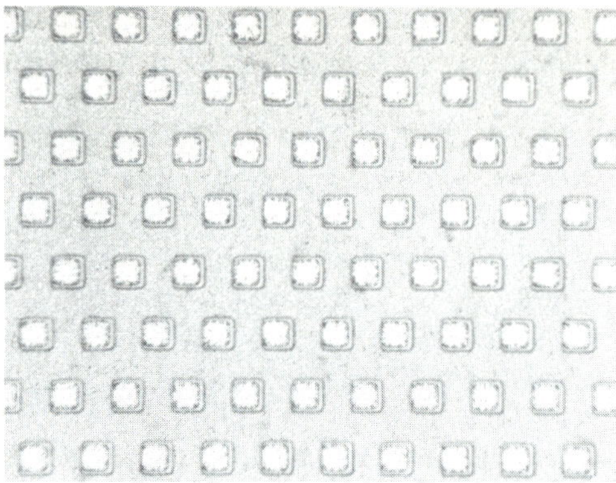

FIG. 3-38 Photomicrograph of a portion of the device illustrated in Fig. 3-24. The square source cells on a hexagonal grid are apparent in this photo. *(Photo courtesy of Siliconix, Inc.)*

The relative efficiencies of each in terms of source perimeter per unit surface area are compared in the figure, taken from Ref. 44. It is clear that all of the geometries are within ± 10 percent of each other. These differences are easily overshadowed by On resistance variations resulting from other process variables, and help to explain why there is no clear agreement on the optimum geometry to use. Figure 3-38 is a higher-magnification view of the device in Fig. 3-25, showing the particular source cell geometry used in this device.

3.6 COMPOSITE MOS-BIPOLAR POWER DEVICE STRUCTURES

Other chapters in this book deal with discrete power bipolar transistors. The advantages of such devices in comparison with power MOS devices are well known [45]. The principal differences in the two devices have to do with the influence of minority carriers in bipolar structures.

The On resistance of a bipolar transistor can be significantly less than that of an MOS device with a similar structure. This is primarily because the collector (or drain) resistance, R_4 in Eq. 3-27, is reduced as a result of conductivity modulation in a bipolar transistor. In power bipolar devices, if the collector-base junction is allowed to forward-bias so that the device enters the saturation region, holes are injected from the base (the P region in Fig. 3-35) into the N^- collector region. A corresponding increase in the electron concentration must occur to pre-

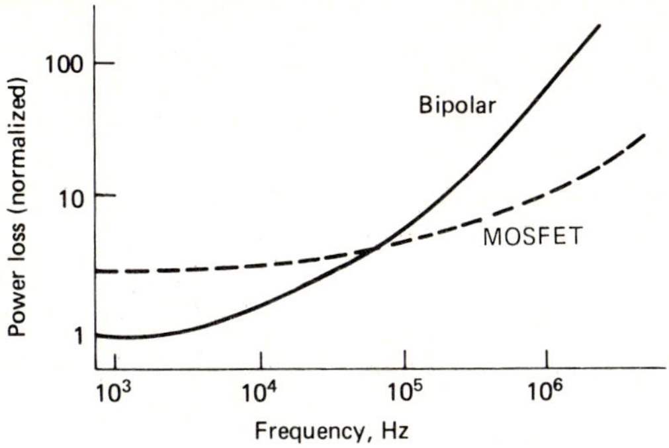

FIG. 3-39 Relative switching efficiencies of bipolar and MOS power transistors vs. switching frequency [45].

serve charge neutrality. This increases the conductivity of the N^- region (R_4) and therefore gives the bipolar transistor lower R_{On}.

This benefit is of course not obtained without a price. The same minority carriers which reduce R_{On} must be removed when the bipolar device is switched off. This requires time (the so-called storage time) and generally implies slower switching speeds for bipolar devices. This general conclusion is borne out by Fig. 3-39, which compares power bipolar and MOS devices [45]. At low frequencies the bipolar device produces smaller switching losses on account of its lower R_{On}. At higher frequencies, the MOS device generally wins because of the stored charge and therefore higher switching losses in the bipolar device.

A number of composite structures have been proposed recently which attempt to take advantage of the best features of both device types. None of these has as yet found widespread acceptance, but they are briefly discussed here because it is believed that ideas along these lines may be important in future device evolution.

The Darlington configuration with one bipolar transistor driving another can be modified to take advantage of the characteristics of an MOS transistor. Substitution of an MOS device for the input transistor as shown in Fig. 3-40 produces a composite device with some of the advantages of both transistor types [46]. The high input impedance of the MOS transistor reduces the circuit drive requirements. In addition, all of the current through both the MOS transistor and the bipolar transistor flows through the load. This configuration retains the current-handling capability and the low voltage drop of the bipolar transistor, but without the slow switching times. The MOS transistor across the base-collector junction of the bipolar transistor keeps the bipolar out of full saturation. The majority-carrier characteristics of the MOSFET

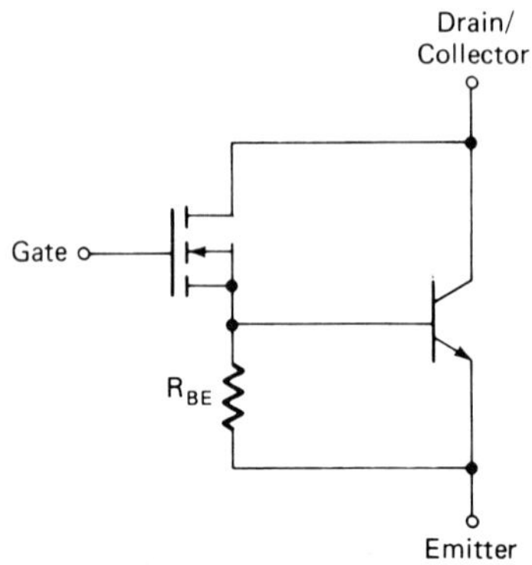

FIG. 3-40 Darlington configuration using an input MOS device and an output bipolar power transistor.

allow the bipolar transistor to switch rapidly in this configuration. Trade-offs in the base-emitter resistor value and the On resistance of the MOS transistor can be used to optimize the composite device for a particular application [46]. The ruggedness of this configuration is determined by the characteristics of the bipolar transistor. This configuration may be implemented by the circuit designer using discrete

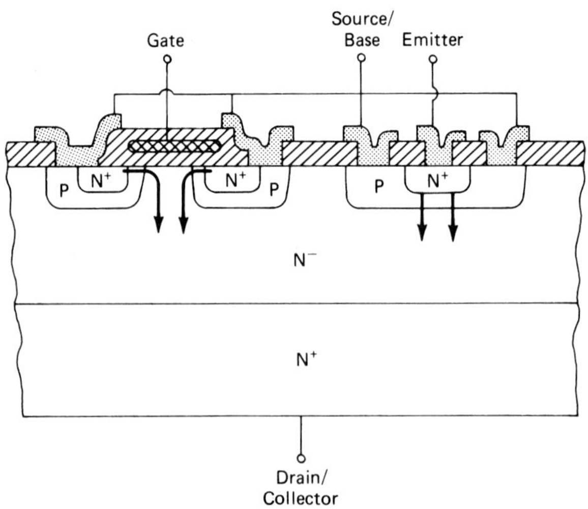

FIG. 3-41 Cross section of integrated version of the composite structure of Fig. 3-40. The resistor R_{BE} is not shown but may be integrated using the polysilicon layer or added externally.

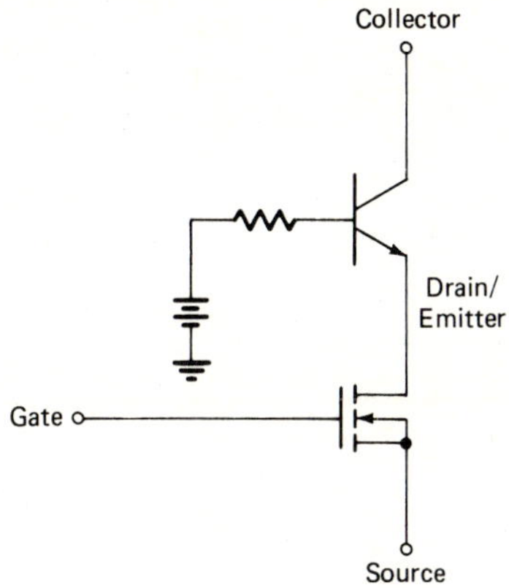

FIG. 3-42 Cascode composite device using power bipolar and MOS devices.

devices. It is also possible to fully integrate the Darlington configuration as shown in Fig. 3-41.

The MOS bipolar cascode configuration of Fig. 3-42 produces a combined device with the collector-base breakdown voltage of the bipolar transistor and the fast switching characteristics of the MOS transistor. Since the MOS transistor only sees the base-emitter breakdown voltage, a high-current, low–On resistance device may be used. The voltage drop is determined by the saturation voltage of the bipolar transistor and the On resistance of the MOSFET. This configuration also has the safe operating area of the common-base bipolar transistor. The cascode configuration may be realized using discrete components. It is not easily integrated.

The use of an MOS transistor in parallel with a bipolar transistor as shown in Fig. 3-43 has advantages that are not readily apparent. By properly sequencing the drive signal to the gate and the base, the ruggedness and fast switching speed of the MOS transistor may be combined with the low voltage drop of the bipolar transistor. The MOSFET is first turned on, allowing the load current to flow while absorbing any turn-on transient. The bipolar transistor is then turned on and the voltage across the parallel structure drops to its $V_{CE(sat)}$. To turn the parallel combination off, the bipolar transistor base drive is first removed and the load current flows through the MOSFET. After the current through the bipolar transistor reaches zero, the MOS transistor is turned off, absorbing the associated transient. The parallel combination of a bipolar and a MOS transistor may be optimized by using discrete devices, but, as with the Darlington configuration, it is possible to integrate this

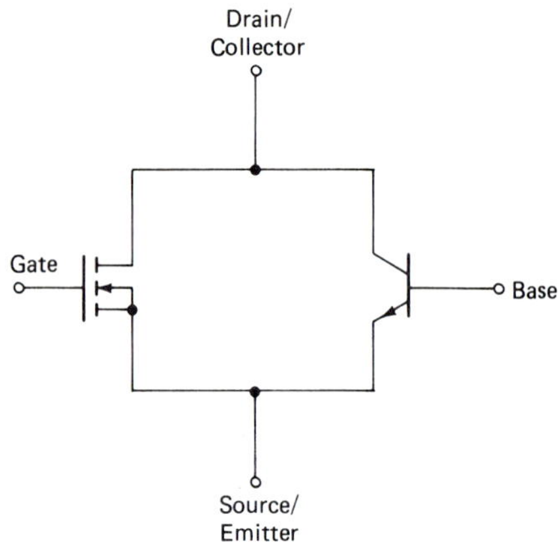

FIG. 3-43 Composite power device using a parallel arrangement of power bipolar and MOS devices.

structure. A cross section showing one monolithic implementation is shown in Fig. 3-44 [47].

One final interesting merged MOS-bipolar structure has recently appeared in the literature [48, 49]. This structure, shown in Fig. 3-45, essentially merges two LDMOS devices in a common N-drain region.

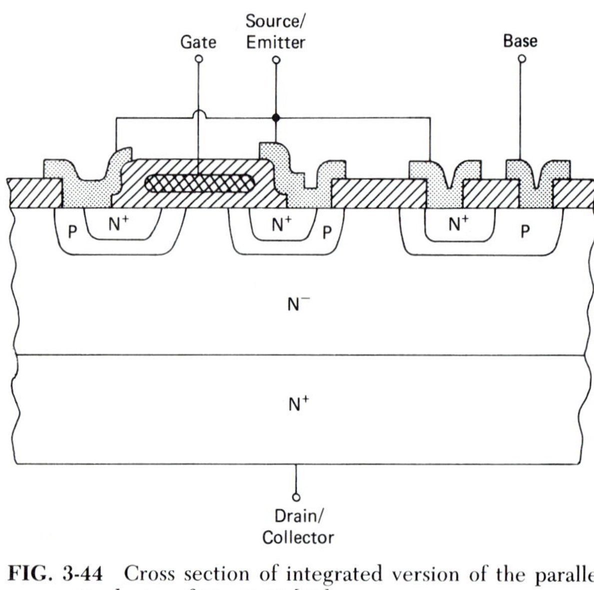

FIG. 3-44 Cross section of integrated version of the parallel composite device of Fig. 3-43 [47].

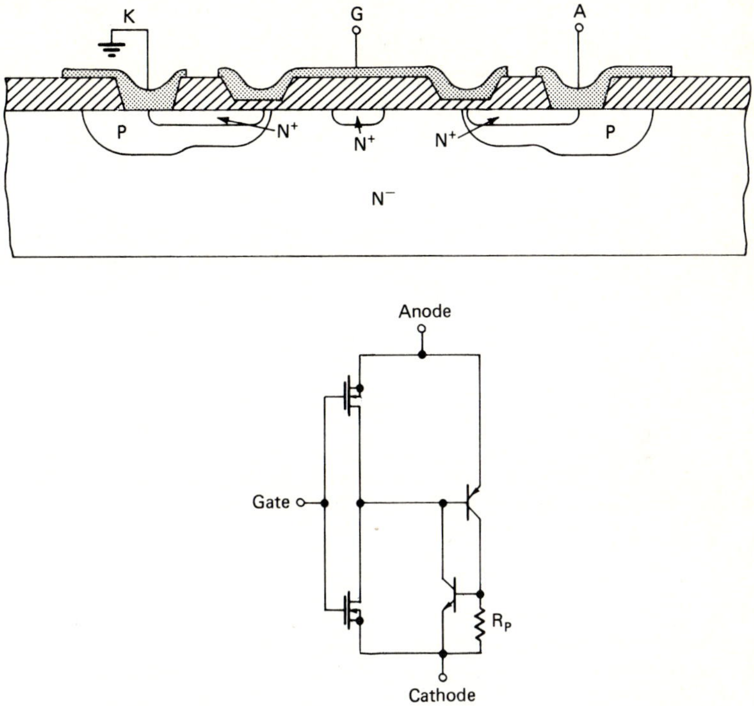

FIG. 3-45 Cross section and equivalent circuit of a merged bipolar/DMOS structure which performs the function of an insulated-gate SCR or triac [48, 49].

This makes possible regenerative operation as in a triac, but in this case, the device is switched on by the MOS input rather than by forward-biasing a PN junction as is the case in a conventional SCR or triac.

The equivalent circuit in Fig. 3-45 can be used to understand the device operation. The overall structure contains the two stacked LDMOS devices in series and merged PNP and NPN devices in parallel with the DMOS devices. The bipolar transistors are interconnected in the classic SCR configuration. For applied drain voltages less than $\simeq 1$ V, the bipolar devices are essentially not on and the total circuit current flows through the LDMOS devices under control of the gate potential V_g. The overall device acts just like two LDMOS devices in series.

For increasing drain potentials ($V_D \gtrsim 1.5$ V), sufficient voltage exists to forward-bias the PN junctions in the bipolar devices, and some of the current begins to flow through the PNP-NPN pair. As this current increases, the loop gain of the bipolar pair soon exceeds unity and the regenerative action seen in ordinary PNPN structures occurs. The PNP drives the NPN and vice versa, so that both devices become saturated, clamping the total structure voltage to $\simeq$1.5 V (1 diode drop + $V_{CE(sat)}$) and allowing very large currents to flow. The composite device is now latched into its On state, which can be changed only by interrupting the main current in the structure, as in a conventional SCR or triac.

The overall structure thus behaves as an MOS-controlled triac or SCR and allows a high-impedance input gate to control large amounts of dc or ac power. Other variations on this basic idea have been proposed [50], although at the present time none have reached commercial products. By way of overall summary of this section, it does seem clear that merging bipolar and MOS structures offers interesting possibilities for the long term and it is likely that some applications of this concept will prove to be commercially viable.

3.7 HIGH-VOLTAGE AND POWER MOS DEVICES IN INTEGRATED CIRCUITS

There are an increasing number of system applications which can benefit from the functional integration of low-voltage logic devices with high-voltage or high-power devices, on the same integrated circuit chip. These applications range from cross-point switches in telephone, communications, and testing systems, to display driver chips and to special-purpose chips for more limited markets, like medical electronics. In this section we shall consider some of the approaches to this problem which have recently appeared and also discuss some of the inherent difficulties that arise in such integration. We shall restrict ourselves to MOS logic and MOS-type power devices, since bipolar circuits and structures are covered elsewhere in this book.

Two general classes of high-voltage MOS integrated circuits exist [51]. The first is structured around a conventional NMOS or PMOS technology and by minor modifications to these processes produces output devices with voltage capabilities of 50 to 300+ V. These output devices use ion-implanted extended-drain regions, offset or stacked gates, or lightly doped drain regions to achieve high-voltage operation. An example of such a structure is shown schematically in Fig. 3-46, based, in this case, on a conventional NMOS silicon-gate technology.

In this example, a lightly doped ion-implanted N^- drift region is placed between the active device channel and the N^+ drain contact. The gate electrode extends over this N^- region on top of thick oxide, helping to shape the surface fields in the structure, thus acting as a field plate. Such devices depend for their voltage capability on the N^- region's supporting most of the applied voltage. A number of examples

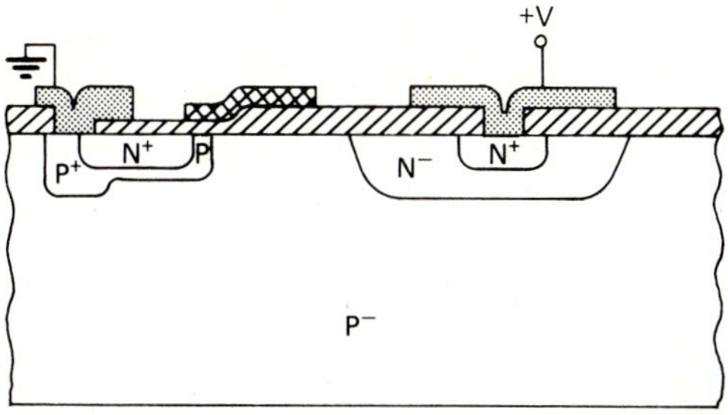

FIG. 3-46 Cross section of a high-voltage output MOS transistor compatible with low-voltage NMOS or DMOS circuits.

of this approach have appeared in the literature [52, 53] based upon either NMOS or DMOS devices as the active switching element. An example is shown in Fig. 3-47. Conceptually similar approaches, based upon thin N^- epitaxial layers, have also recently been proposed [54, 55]. These latter structures have been named *resurf* devices, after the

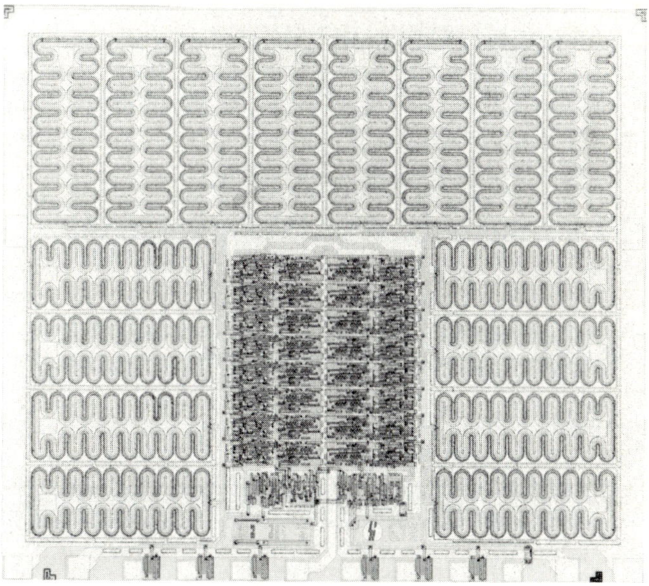

FIG. 3-47 Photomicrograph of an integrated circuit containing 16 output devices of the type shown in Fig. 3-46. The output devices are the large interdigitated devices around the three sides of the chip and can each support 250 V. The logic in the center in this case is made with DMOS devices. *(Photo courtesy of Supertex, Inc.)*

manner in which electric fields are controlled; they operate, however, in a similar fashion to the device illustrated in Fig. 3-46. Operation of all of these devices at the highest voltage levels implies that the N^- region under the extended gate electrode must be completely depleted in the Off state. This shifts the region of highest electric field away from the surface and into the bulk. Several empirical and computer-simulated studies of the optimum N^- implant dose in these devices [52, 54] have independently arrived at doses of 0.5 to 2×10^{12} cm^{-2}, reflecting the need for complete N^- depletion if the highest possible voltages are to be achieved.

Such devices generally operate in a grounded-source configuration because they are inherently asymmetric in their voltage capability. They also generally utilize an enclosed geometry with the drain bonding pad in the center in order to avoid parasitic field inversion problems. (Such enclosed bonding pads can be seen in the chip photomicrograph in Fig. 3-47.) As a consequence of the lightly doped N^- region, these devices are generally fairly high–On resistance structures. This type of technology has been used primarily for display-driving applications in which the grounded source and relatively low current capability of the output devices are acceptable. The basic NMOS or DMOS process used to build the devices makes it straightforward to implement complex logic functions on the same chip with the high-voltage output devices. Alternative approaches to high-voltage output devices have utilized offset [5, 6] or stacked gate electrodes with generally the same limitations on grounded-source, low-current applications.

A second general type of high-voltage MOS integrated circuit is based upon DMOS technology [38]. Such devices are generally fabricated in thick N^- epitaxial layers on P^- substrates and use junction isolation. Lateral DMOS devices are employed because they provide access to all three device terminals on the top wafer surface for easy interconnection to other devices. The combination of a short active channel and a lightly doped drift region provides high voltage, high speed, and low On resistance in the DMOS device. Such devices are directly compatible with low-voltage PMOS devices to implement logic functions, as illustrated schematically in Fig. 3-48.

An example of an integrated circuit fabricated with the technology of Fig. 3-48 is shown in Fig. 3-49. The large-area devices are high-voltage (125-V), high-current (0.3-A), low–On resistance ($<50\ \Omega$) analog switches, implemented with DMOS devices. The central portion of the chip contains PMOS logic for selecting 1 out of 8 of the analog switches and level shifting circuits to control the On or Off state of the high-voltage channels. Each of the analog channels can be selected in less than 3 μs and can switch analog or digital signals up to 10 MHz and 125 V. The overall chip size is 4 mm $\times$ 4 mm [51].

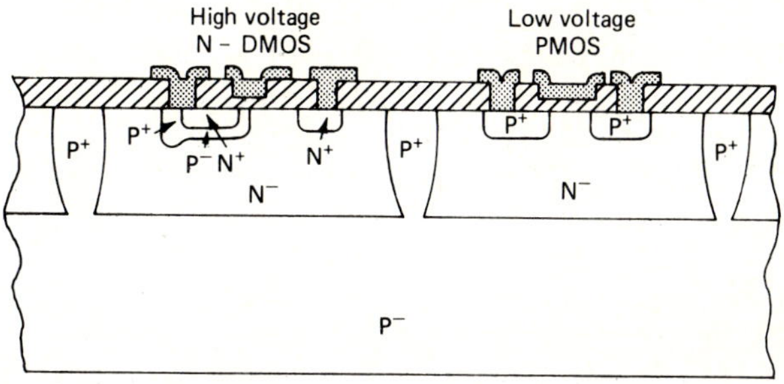

FIG. 3-48 Cross section of general-purpose high-voltage technology based upon thick epitaxial layers and LDMOS devices. Compatible logic circuits can be made from PMOS devices as illustrated [51].

Any generally useful high-voltage technology must be capable of implementing high-performance, low-voltage logic functions. This is necessary both for interfacing with other circuits and because it is very costly in terms of power and chip area to utilize high-voltage devices where they are not needed. The structure shown in Fig. 3-46 is obviously directly compatible with low-voltage logic (NMOS or DMOS) since the process is basically a conventional low-voltage one modified

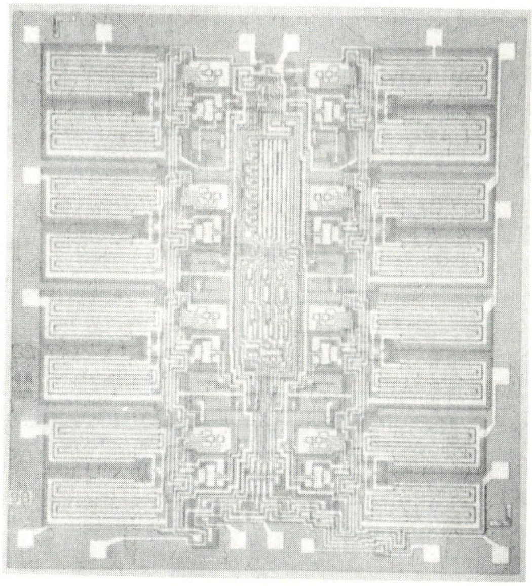

FIG. 3-49 Photomicrograph of integrated circuit fabricated with the technology of Fig. 3-48. The DMOS output devices are the large devices on the chip sides. The PMOS logic is in the center [51].

to add high-voltage outputs. The technology illustrated in Figs. 3-48 and 3-49 uses conventional PMOS devices for low-voltage logic. This is attractive because all of the logic can be placed in a single isolated N^- region. Vertical NPN and lateral PNP bipolar transistors are also compatible with the basic eight-mask process, as are N-channel DMOS devices. These latter three, however, have the major disadvantage of requiring isolated N^- regions. This is a substantial penalty in high-voltage circuits because thick epitaxial layers ($\simeq$25 to 30 μm for 200-V operation) imply large N^- regions to allow for lateral diffusion of the P^+ isolation and to support large voltages between the substrate and the epitaxial layer. Dielectric isolation techniques would clearly circumvent some of these problems.

Additional processing steps can be used to implement conventional NMOS devices. Specifically, the addition of an ion-implanted P well in an isolated N^- region and any threshold-shifting implants required have been used to fabricate high-performance CMOS logic [56]. Such a process, however, is substantially more complicated than the basic technology of Fig. 3-48, often requiring more than 11 masking steps and several critical ion implants. An example of such a technology, implementing low-voltage CMOS and high-voltage DMOS devices, is shown in Fig. 3-50 [56].

There are several problems which arise in any of these high-voltage technologies. One of these is that lightly doped regions which are necessary for high-voltage operation are inherently subject to surface inversion. In a circuit of even modest complexity, the need to route metal lines around on the surface for device interconnection will nearly always result in some situations in which the voltages on the metal lines are large enough to invert the underlying surface. The classical approach to this problem has been to use guard rings as illustrated in Fig. 3-51. In the particular example illustrated here, the metal line has

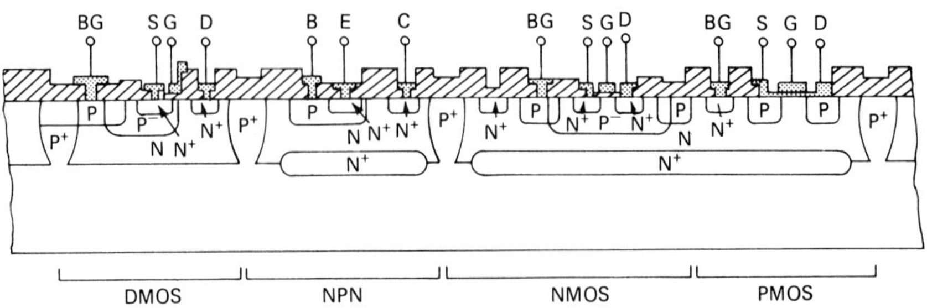

FIG. 3-50 Cross section of a high-voltage technology similar to FIG. 3-48 but modified to incorporate low-voltage CMOS logic as well as high-voltage DMOS and NPN devices [56].

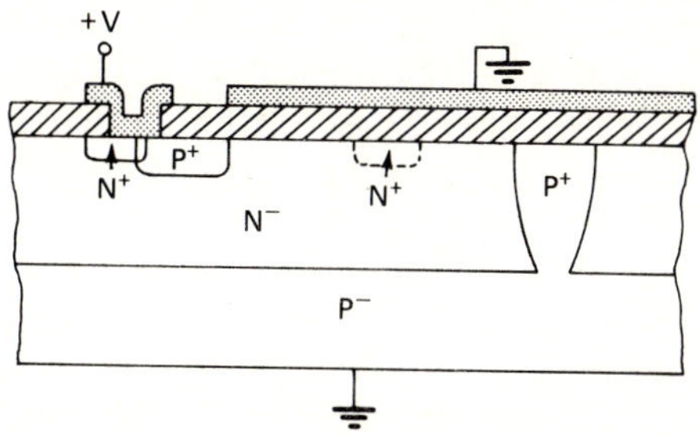

FIG. 3-51 Surface inversion problem in high-voltage ICs. The use of N^+ guard rings is often not possible.

a potential of the correct polarity to invert the underlying N^- surface, and hence to cause conduction between the P^+ region on the left and the isolation region on the right. The N^+ region in the center will prevent this inversion.

The use of these guard rings can, however, significantly degrade device voltage capability. Thus high surface fields near the edges of the N^+ guard ring can cause avalanche breakdown at a voltage which is dependent on the oxide thickness and N^+ doping profile, as described earlier and in Ref. 34. As a typical example, a 1.3-μm oxide thickness and a 2-μm-deep, heavily doped N^+ region will break down at $\simeq$130 to 150 V. This may be well below the voltage at which the circuit could otherwise operate. One solution to this problem is simply to eliminate the N^+ guard rings and simply accept the leakage currents associated with these parasitic field-oxide devices. Careful attention to layout can minimize these effects [51]. Another approach is to use a silicon-gate technology in which the polysilicon layer is used selectively as a guard ring under the aluminum interconnect lines. This is the so-called field shield technique; it can be used to minimize surface inversion problems.

Other problems which can arise in the high-voltage technologies of Fig. 3-48 or 3-50 include the need for thick gate oxides when high voltages appear on the device gate electrodes [51], and parasitic vertical PNP transistors. This latter parasitic can be minimized by incorporating buried N^+ layers in the process as is commonly done in bipolar technologies, and as is shown in Fig. 3-50.

In summary, it appears that high-voltage or high-power MOS integrated circuits may find significant commercial application in the future. Output transistors of the general type shown in Fig. 3-46 have been shown capable of reaching more than a 500-V drain breakdown, and a 1000-V capability seems possible. More general-purpose devices like the technologies of Figs. 3-48 and 3-50 will have a more limited voltage capability because of high fields across gate oxides, very thick epitaxial layers and their consequent isolation problems, and parasitic bulk and surface devices. Practical junction isolated circuits appear to be limited to about 200 V. Dielectric isolation techniques will improve this figure because all of the above problems are eliminated by this technology except the need in some cases for high gate oxide fields.

3.8 CONCLUSIONS AND FUTURE PROJECTIONS

This chapter has attempted to provide a perspective on the rapidly growing fields of discrete and integrated power MOS devices. It should be clear that MOS devices, which historically have been thought of as low-voltage logic elements for VLSI, can in fact operate at high voltages and high power levels. While it should not be expected that power MOS devices will completely replace bipolar structures, they will be used in applications which benefit from the higher input impedance and higher frequency capability they offer. Bipolar devices will retain their advantage in On resistance because they make use of minority-carrier injection and conductivity modulation, which MOS devices do not. It is to be expected that the choice of MOS or bipolar for a specific application will have to weigh the relative advantages of each type of device.

By way of overall perspective, it is interesting to note that E. O. Johnson in 1965 [57] predicted maximum limits for silicon devices based purely on material limits. He argued that a block of silicon with linear dimension L can support a voltage of $E_{cr}L$ and that carriers require a transit time of L/v_{sat}, where v_{sat} is the scattering limiting velocity of the carriers. Substituting values for silicon leads to the relationship

$$\mathrm{BV}\, f_T \leq 300\ \mathrm{V \cdot GHz} \tag{3-32}$$

where BV is the maximum voltage rating of a particular device and f_T is the common-emitter or common-source cutoff frequency (gain-bandwidth product). Modern power MOS and bipolar devices are both approaching this limit. Commercially available power MOSFETs (LDMOS devices) have achieved BV f_T products within 20 percent of this value and bipolar devices are almost as good [58]. Other materials such as GaAs provide limits which are somewhat higher (perhaps 500 V·GHz), but we can conclude that modern power devices are approaching fundamental limits in terms of some performance parameters. We can certainly still expect to see significant advances in device structures which optimize particular performance parameters, however.

REFERENCES

1. R. S. C. Cobbold, *Theory and Applications of Field-Effect Transistors*, Wiley-Interscience, New York, 1970.
2. A. S. Grove, *Physics and Technology of Semiconductor Devices*, Wiley, New York, 1967.
3. M. B. Barron, "Low-Level Currents in IGFETs," *Solid State Electronics* **15:**293 (1972).
4. J. R. Brews, "Subthreshold Behavior of Uniformly and Non-uniformly Doped Long-Channel MOSFET," *IEEE Trans. Electron Dev.* **ED-26**(9):1282 (1979).
5. R. W. Bower, H. G. Dill, K. G. Arbuchon, and S. A. Thompson, "MOS Field Effect Transistors Formed by Gate Masked Ion Implantation," *IEEE Trans. Electron Dev.* **ED-15**(10):757 (1968).
6. H. G. Dill, "A New Insulated Gate Tetrode with High Drain Breakdown Potential and Low Miller Feedback Capacitance," *IEEE Trans. Electron Dev.* **ED-15**(10):717 (1968).
7. R. A. Blanchard, P. A. Gargini, G. A. May, and R. D. Melen, "High-Voltage Simultaneous Diffusion Silicon-Gate CMOS," *IEEE J. Solid State Circuits* **SC-9**(3):103 (1974).
8. I. Yoshida, M. Kubo, and S. Ochi, "A High-Power MOSFET with a Vertical Drain Electrode and a Masked Gate Structure," *IEEE J. Solid State Circuits* **SC-11**(4):472 (1976).
9. H. J. Sigg, G. D. Vendelin, T. P. Cauge, and J. Kocsis, "DMOS Transistors for Microwave Applications," *IEEE Trans. Electron Dev.* **ED-19**(1):45 (1972).
10. V. A. K. Temple and R. P. Love, "A 600-Volt MOSFET with Near Ideal On-Resistance," *IEDM Conf. Dig.*, December 1978, p. 664.
11. A. Lidow, T. Herman, and H. W. Collins, "Power MOSFET Technology," *IEDM Conf. Dig.*, December 1979, p. 79.
12. D. B. Lee, "Anisotropic Etching of Silicon," *J. Appl. Phys.* **40:**180 (1969).
13. F. F. Fang and A. B. Fowler, "Hot Electron Effects and Saturation Velocities in Silicon Inversion Layers," *J. Appl. Phys.* **41:**1825 (1970).
14. B. E. Deal, "The Current Understanding of Charges in the Thermally Oxidized Silicon Structure," *J. Electrochem. Soc.* **121**(6):1980 (1974).
15. S. C. Sun and J. D. Plummer, "Modeling the On-Resistance of LDMOS, VDMOS, and VMOS Power Transistors," *IEEE Trans. Electron Dev.* **ED-27**(2):356 (1980).

16. R. A. Blanchard, "Optimization of Discrete High Power MOS Transistors," *Stanford Electronic Labs. TR No. 1DEZ686-2*, 1982.
17. S. C. Sun and J. D. Plummer, "Electron Mobility in Inversion and Accumulation Layers on Thermally Oxidized Silicon Surfaces," *IEEE Trans. Electron Dev.* **ED-27**(8):1487 (1980).
18. S. M. Sze, *Physics of Semiconductor Devices*, Wiley, New York, 1969.
19. S. L. Miller, "Ionization Rates for Holes and Electrons in Silicon," *Phys. Rev.* **105:**1246–1249 (1957).
20. D. P. Kennedy and R. R. O'Brien, "Avalanche Breakdown Characteristics of a Diffused P-N Junction," *IRE Trans. Electron Dev.* **ED-9:**478 (1962).
21. J. L. Moll, J. I. Su, and A. C. H. Werner, "Multiplications in Collector Junctions of Silicon N-P-N and P-N-P Transistors," *IEEE Trans. Electron Dev.* **ED-17:**420 (1970).
22. S. M. Sze and G. Gibbons, "Avalanche Breakdown Voltages of Abrupt and Linearly Graded P-N Junctions in Ge, Si, GaAs," *Appl. Phys. Lett.* **8:**111 (1966).
23. O. Leistiko, Jr., and A. S. Grove, "Breakdown Voltage of Planar Silicon Junction," *Solid State Electronics* **9:**847 (1966).
24. V. A. K. Temple and M. S. Adler, "A Substrate Etch Geometry for Near Ideal Breakdown Voltage in P-N Junction Devices," *IEEE Trans. Electron Dev.* **ED-24**(8):1077 (1977).
25. V. A. K. Temple and M. S. Adler, "The Theory and Application of a Simple Etch Contour for Near Ideal Breakdown Voltage in Plane and Planar P-N Junctions," *IEEE Trans. Electron Dev.* **ED-23**(8):950 (1976).
26. K. P. Lisiak and Josef Berger, "Optimization of Nonplanar Power MOS Transistors," *IEEE Trans. Electron Dev.* **ED-25**(10):1229 (1978).
27. F. Conti and M. Conti, "Surface Breakdown in Silicon Planar Diodes Equipped with Field Plate," *Solid-State Electronics* **15:**93 (1972).
28. L. E. Clark and D. S. Zoroglu, "Enhancement of Breakdown Properties of Overlay Annular Diodes by Field Shaping Resistive Film," *Solid-State Electronics* **19:**653 (1972).
29. T. Matrushito, T. Aoki, T. Ohtru, H. Yamoto, H. Hayaski, M. Okayano, and Y. Kawana, "High Reliability High Voltage Transistors by Use of the SIPOS Process," *IEEE Trans. Electron Dev.* **ED-23**(8):826 (1976).
30. Y. C. Kao and E. D. Wolley, "High Voltage Planar P-N Junctions," *Proc. IEEE* **55**(8):1409 (1967).
31. A. S. Grove, O. Leistiko, Jr., and W. W. Hooper, "Effect of Surface Fields on the Breakdown Voltage of Planar Silicon P-N Junctions," *IEEE Trans. Electron Dev.* **ED-14**(3):157 (1967).
32. P. P. Castrucci and J. S. Logan, "Electrode Contact of SiO_2 Passivated Planar Junctions," *IBM J. Res. Dev.*, 1964, p. 394.
33. M. S. Adler, V. A. K. Temple, A. P. Ferro, and R. C. Rustay, "Theory and

Breakdown Voltage for Planar Devices with a Single Field Limiting Ring," *IEEE Trans. Electron Dev.* **ED-24**(2):107 (1977).

34. M. J. DeClercq and J. D. Plummer, "Avalanche Breakdown in High-Voltage DMOS Devices," *IEEE Trans. Electron Dev.* **ED-23**:1 (1976).
35. C. Hu, "A Parametric Study of Power MOSFETs," *Proc. PESC*, Paper 5.5, San Diego, June 1979.
36. H. R. Camenzind, "Electronic Integrated Systems Design," Van Nostrand Reinhold, New York, 1972, p. 117.
37. M. D. Pocha, A. G. Gonzalez, and R. W. Dutton, "Threshold Voltage Controllability in Double-Diffused MOS Transistors," *IEEE Trans. Electron Dev.* **ED-21**(12):778 (1974).
38. J. D. Plummer and J. D. Meindl, "A Monolithic 200-V CMOS Analog Switch," *IEEE J. Solid-State Circuits* **SC-11**(6):809 (1976).
39. S. Ramo, J. R. Whinnery, and T. Van Duzer, Fields and Waves in Communication Electronics, Wiley, New York, 1967, p. 98.
40. M. D. Pocha and R. W. Dutton, "A Computer-Aided Design Model for High-Voltage Double Diffused MOS (DMOS) Transistors," *IEEE J. Solid-State Circuits* **SC-11**(5):718 (1976).
41. A. B. Phillips, *Transistor Engineering*, McGraw-Hill, New York, 1962, p. 216.
42. H. C. Lin, *Integrated Electronics*, Holden-Day, San Francisco, 1967, p. 110.
43. R. F. David, "Computerized Thermal Analysis of Hybrid Circuits," *ECC*, 1977, p. 324.
44. M. Chi and C. Hu, "Some Issues on Power MOSFETs," *PESC Record*, June 1982, p. 392.
45. P. L. Hower, "Bipolar vs. MOSFET: Seeing Where the Power Lies," *Electronics*, Dec. 18, 1980, p. 106.
46. R. A. Blanchard, R. H. Baker, M. Blogolji, and K. E. White, "A New High-Power MOS Transistor for Very High-Current High-Voltage Switching Applications," *Proc. Powercon 8*, 1981, N1.1.
47. J. Meador and N. Zommer, "Using Bipolar-MOSFET Combinations to Optimize the Switching Transistor Function," *Proc. Powercon 8*, 1981, F4.1.
48. J. D. Plummer and B. W. Scharf, "Insulated Gate Planar Thyristors: 1—Structure and Basic Operation," *IEEE Trans. Electron Dev.* **ED-27**(2):380 (1980).
49. B. W. Scharf and J. D. Plummer, "Insulated Gate Planar Thyristors: 2—Quantitative Modeling," *IEEE Trans. Electron Dev.* **ED-27**(2):387 (1980).
50. J. Takanyi, "Functional Integration of Power MOS and Bipolar Devices," *IEDM Conf. Dig.*, December 1980, p. 75.
51. J. D. Plummer, "Monolithic MOS High Voltage Integrated Circuits, *IEDM Conf. Dig.*, December 1980, p. 70.

52. K. Awane, K. Fujii, T. Yamano, H. Tamaki, T. Biwa, H. Hattori, and T. Fujimoto, "High-Voltage DSA-MOS Transistor for Electroluminescent Display," *ISSCC Conf. Dig.*, February 1980, p. 224.

53. I. Yoshida et al., "Device Design of an Ion Implanted High Voltage MOSFET," *Proc. 6th Conf. Solid State Dev.*, Tokyo, 1974.

54. J. A. Appels and H. M. J. Vaes, "High Voltage Thin Layer Devices (Resurf Devices)," *IEDM Conf. Dig.*, December 1979, p. 238.

55. S. Colak, B. Singer, and E. Stupp, "Lateral DMOS Power Transistor Design," *IEEE Electron Dev., Lett.*, **EDL-1**(4):51 (1980).

56. "Simplified Drivers, New Applications Spur Plasma-Panel Usage," *EDN*, Sept. 20, 1979, p. 51.

57. E. O. Johnson, "Physical Limitations of Frequency and Power Parameters of Transistors," *RCA Rev.*, 1965, p. 163.

58. S. C. Sun, "Physics and Technology of Power MOSFETs," *Stanford Electronic Labs. TR No. 1DEZ696-1*, 1982.

CHAPTER 4

Evolution of the Planar Process for PICs

F. Bertotti
SGS Microelettronica
Milan, Italy

4.1 INTRODUCTION	4.1
4.2 INTEGRATED NPN POWER TRANSISTORS	4.2
4.3 INTEGRATED PNP TRANSISTORS	4.5
4.3.1 Vertical PNP Transistors	4.6
4.3.2 Lateral PNP Transistors	4.8
4.4 CLAMPING DIODE	4.11
4.5 HIGH-VOLTAGE PROCESSES	4.13
4.6 DIGITAL STRUCTURES IN PICs	4.17
4.7 PARASITICS IN PICs	4.19
4.8 CONCLUSION	4.24

4.1 INTRODUCTION

The planar process, developed at the beginning of the 1960s, has proved to be very suitable for the development of low-cost, fairly complex circuits with low currents and with supply voltages up to 40 V.

The drive in more recent years to integrate power functions requiring higher currents and voltages (voltage regulators, audio amplifiers, etc.) has led to an improvement in integrated structures and to new process steps which have enabled these objectives to be reached without any significant increase in production costs.

This chapter describes the developments that have enabled integrated NPN and PNP transistors to reach output currents in the range of 1 to 5 A, the modifications to the basic technological process that have resulted in operating voltages of more than 200 V, and the techniques that are used to make the devices more reliable under the most severe operating conditions.

4.2 INTEGRATED NPN POWER TRANSISTORS

The planar process can produce excellent double-diffused NPN transistors that reveal only one serious limitation when compared with discrete devices: the collector contact has to be kept on the front of the chip instead of on the back (Figs. 4-1 and 4-2). Furthermore, while in discrete transistors the first N^+ substrate usually has a resistivity of 10 to 20 mΩ·cm, the average resistivity of the N^+ buried layer in integrated transistors is around 30 to 40 mΩ·cm. The dominant component of collector series resistance is the resistance of the path linking the N^+ buried layer with the collector contact, since the epitaxial layer normally has resistivity values between 1 and 10 Ω·cm.

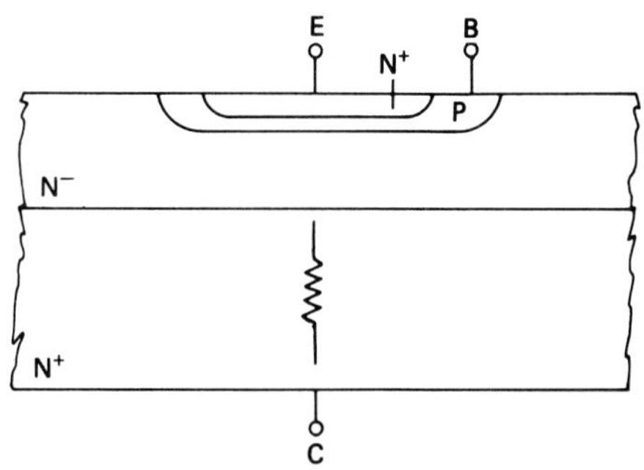

FIG. 4-1 Cross section of a discrete NPN power transistor.

Since the series resistance of the collector is given as a first approximation by R_1 plus R_2 (Fig. 4-2), the design of integrated NPN transistors operating at high currents and with low saturation voltage requires some modifications to the basic process and accurate drawing of the geometries. First of all, resistance R_1 must be minimized, and this can be done by means of a low-resistivity deep N^+ diffusion, the so-called sinker region; it is carried out after the isolation diffusion (Fig. 4-3). R_2 is minimized by modifying the process in order to lower the buried-layer resistivity. By using arsenic or antimony as dopant an average

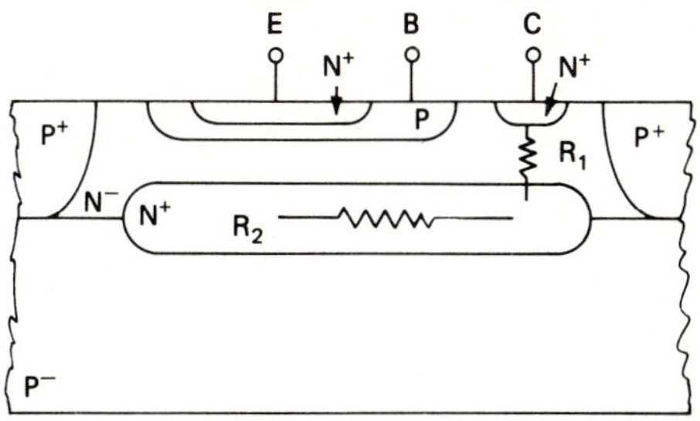

FIG. 4-2 Cross section of an integrated NPN power transistor.

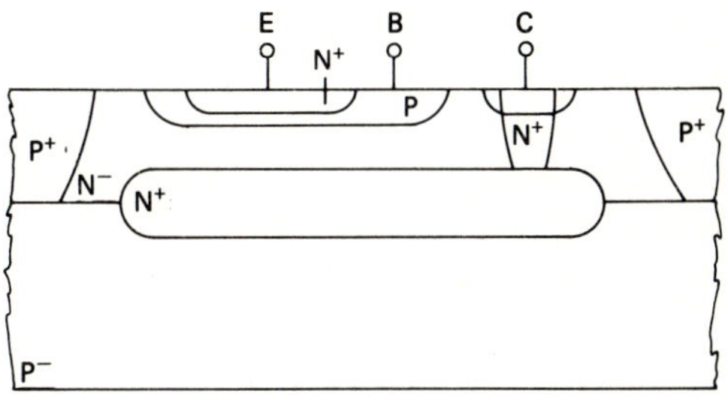

FIG. 4-3 Cross section of an integrated NPN power transistor with sinker diffusion.

resistivity of 10 to 15 mΩ·cm can be reached without compromising the quality of the epitaxial layer. From the point of view of the structure, minimization of R_2 makes it necessary to use interdigitated structures with many emitter strips (Fig. 4-4).

For high-voltage processes (see Figs. 4-5 and 4-6), very thick and resistive epitaxial layers are required; in this case one has to take into account the quasi-saturation behavior of the transistor (see Sec. 2.2.4).

By increasing only the emitter area and its efficiency, it is possible to obtain power NPN transistors operating in full saturation at high current levels. This obviously involves an increase in the area of the integrated NPN transistor as a function of the collector-to-emitter breakdown voltage BV_{CEO}, as shown in Fig. 4-7.

Slight process and device modifications make it possible to obtain collector currents up to 5 A with saturation voltages lower than 1.5 V. Compared with discrete components, the ruggedness characteristics, I_{SB} and E_{SB} (see Sec. 2.4), are the most difficult to improve. Since substantial changes to processes like epitaxial bases or double epitaxial layers are practically impossible, it is necessary to act essentially on device geometry.

Therefore to improve direct second breakdown, structures are used

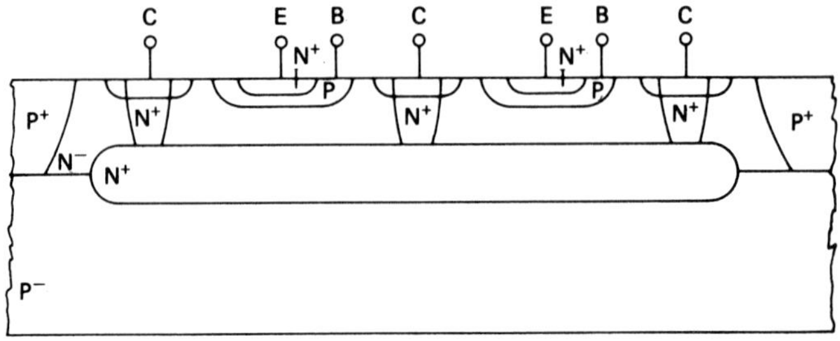

FIG. 4-4 Cross section of an interdigitated NPN power structure for low saturation voltage.

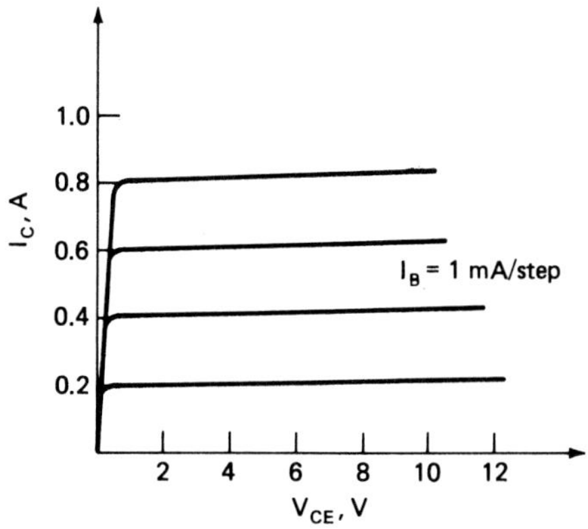

FIG. 4-5 Output characteristic of a 20-V process NPN power transistor.

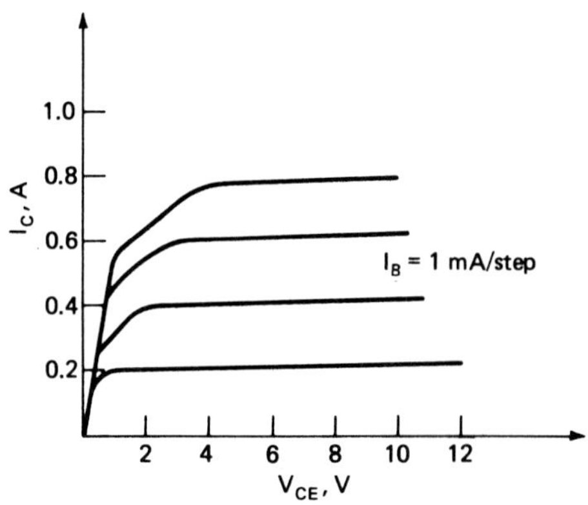

FIG. 4-6 Output characteristic of an 80-V process NPN power transistor.

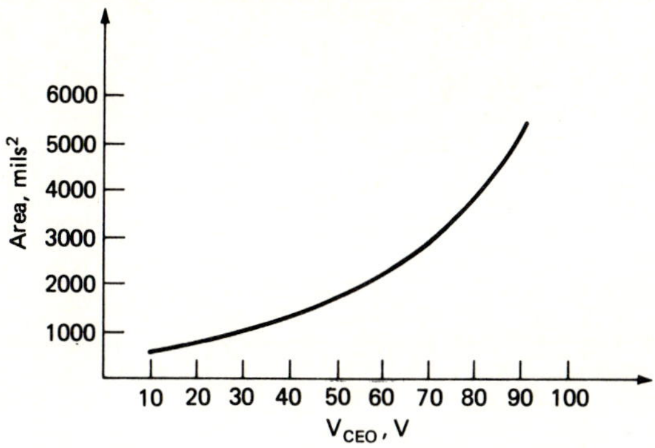

FIG. 4-7 Required area for an NPN transistor with $V_{CE,sat}$ = 1 V at I_C = 3 A, versus collector-emitter breakdown voltage.

which equalize the distribution of current in the emitter fingers [1] by using small ballast resistors; this is obtained without any additional cost in the manner illustrated in Fig. 4-8. Figures 4-9 and 4-10 show the safe operating area (SOA) and the geometry of an integrated NPN transistor using emitter ballast resistors.

4.3 INTEGRATED PNP TRANSISTORS

PNP transistors obtained with the planar process exhibit very poor performance when compared with planar-process NPN transistors [2]. This can be realized by considering at some length the two fundamental PNP structures: the vertical PNP and the lateral PNP transistor (Figs. 4-11 and 4-12).

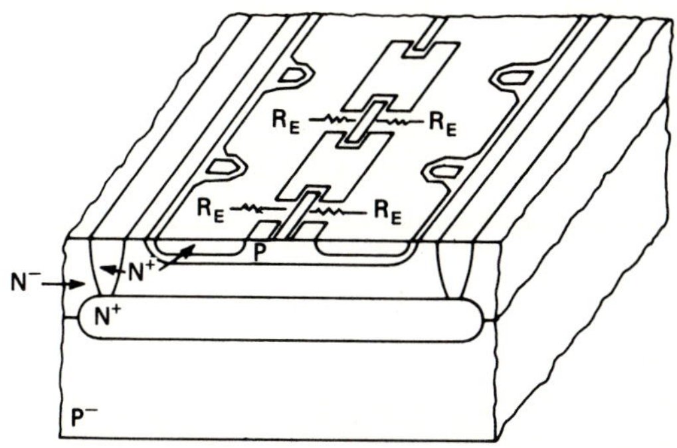

FIG. 4-8 Cross section of an integrated NPN power transistor with emitter ballast resistors.

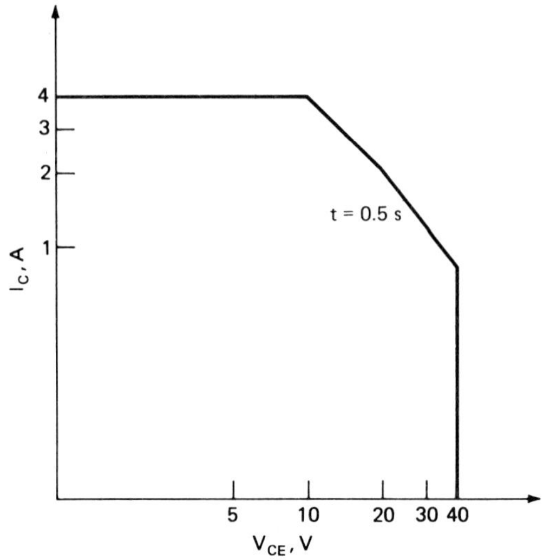

FIG. 4-9 Safe operating area for an NPN power transistor with emitter ballast resistors.

4.3.1 Vertical PNP Transistors

In a vertical PNP transistor the current flows vertically from the emitter, which has been realized using the base NPN diffusion, toward the substrate (collector); the base region lies in the epitaxial layer. The base contact is realized on a region obtained simultaneously with the NPN emitter diffusion; this region is usually overlapped with the PNP emitter diffusion area to reduce the base resistance and prevent carrier injection toward the surface; this avoids low-current h_{FE} degradation.

The collector-to-emitter breakdown voltage (BV_{CEO}) of the vertical PNP transistor is limited by the punch-through mechanism controlled by the net thickness and by the resistivity of the epitaxial layer between emitter and substrate.

The cutoff frequency f_T of the vertical PNP is determined essentially by the homogeneous-base transit-time ($\tau_B = W_B^2/2D_{pB}$); it usually ranges from a few megahertz up to 30 MHz, depending on the epitaxial-layer characteristics.

The most severe limitation upon the use of the vertical PNP transistor is that the collector is grounded so that in power integrated circuits it can be used only as an emitter follower. The grounded collector also means that the use of vertical PNP transistors requires the collector to be isolated; this, too, makes widespread use impracticable and imposes the use of lateral structures.

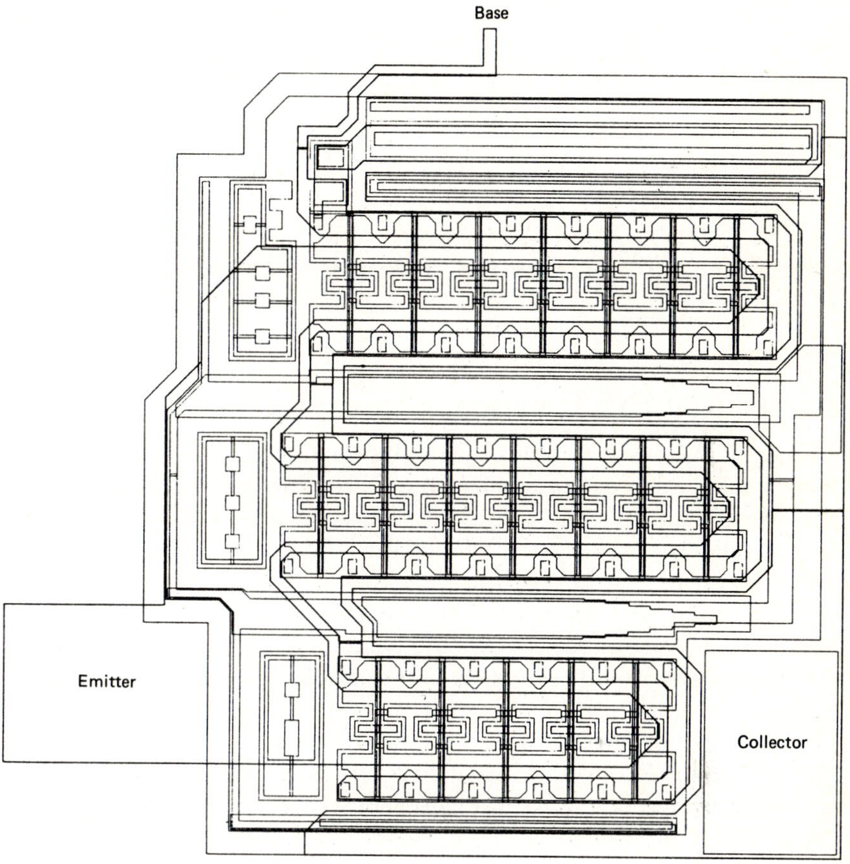

FIG. 4-10 Layout of the NPN whose SOA is shown in Fig. 4-9.

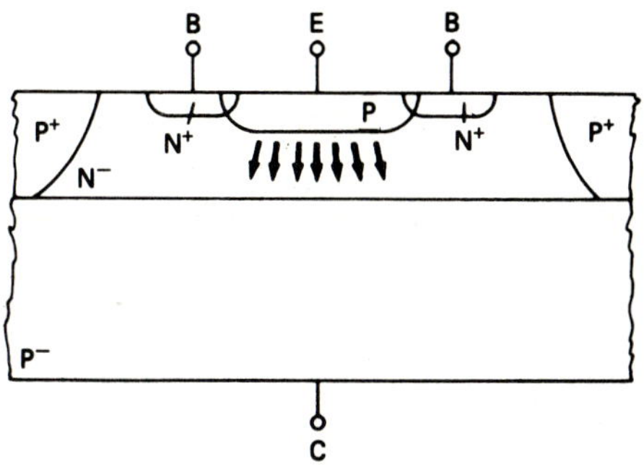

FIG. 4-11 Cross section of an integrated vertical PNP transistor.

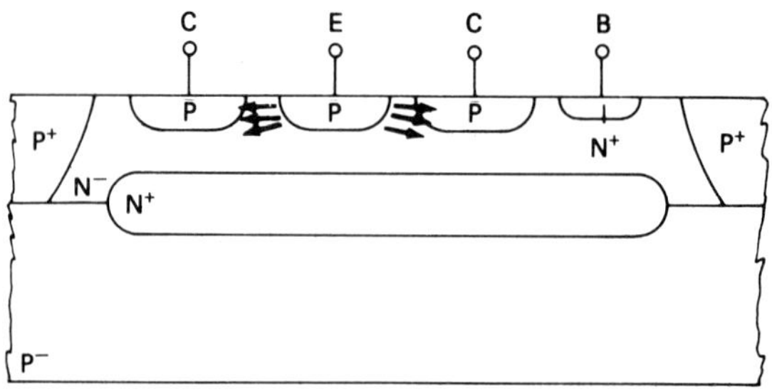

FIG. 4-12 Cross section of an integrated lateral PNP transistor.

4.3.2 Lateral PNP Transistors

The lateral PNP (Fig. 4-12) is realized using the NPN base diffusion for emitter and collector regions and the NPN emitter diffusion to realize the base contact. In this case the buried layer used for the NPN integrated transistor is inserted to reduce the efficiency of the unavoidable parasitic vertical PNP transistor.

The lateral PNP is punch-through-limited according to the effective channel width and exhibits a cutoff frequency f_T which is related to the minority carrier transit time across the channel and to storage effects resulting from parasitic vertical PNP transistor behavior. The cutoff frequency f_T of the lateral PNP is therefore lower than that of the vertical PNP and ranges from a few megahertz up to 15 MHz, depending on the required collector-to-emitter breakdown voltage.

The lateral PNP is used mainly in conjunction with a power NPN (Fig. 4-13), resulting in a structure known as a synthetic PNP; it presents a current gain equivalent to the product of the current gains of the two transistors. Attention has therefore been focused on optimizing the performance of the lateral PNP.

Since the lateral PNP transistor operates mainly on that part of the emitter periphery which faces the collecting region, while the current injected vertically is lost in the substrate, the perimeter-to-area ratio of

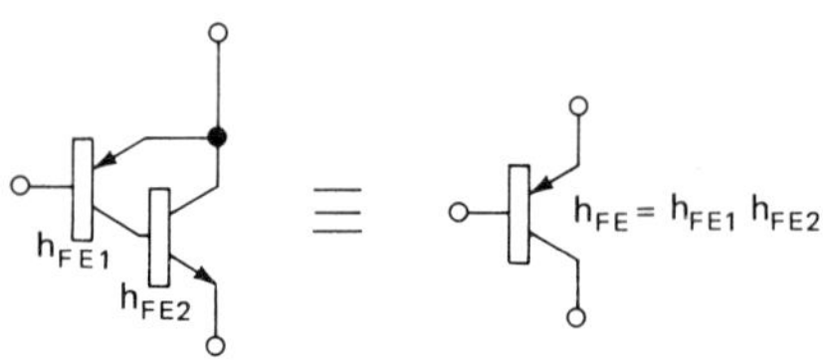

FIG. 4-13 Composite integrated PNP.

the emitter must be optimized. Whether the largest current gains are obtained with the circular structure used in the minimum-size transistor (1 disk) or in the multiple configuration depends on the operating current levels. Figures 4-14 and 4-15 show two PNPs, with 1 disk and with 18 disks, respectively, and their current gains as a function of collector current.

To limit emitter current crowding in the multidisk structures, wide base contacts are used as well as sinker diffusion in order to lower the base resistance. The current density of lateral PNPs is very low compared with that of the NPN transistors. This is due to two factors: modulation of base resistivity at high currents and poor emitter efficiency (the emitter sheet resistivity is in the range 100 to 200 Ω/□).

To improve emitter efficiency and the perimeter-to-area ratio, a P^+ diffusion has been added to the standard planar process. It is realized with a sheet resistance of 10 to 15 Ω/□ and is deeper than the base diffusion, as is shown in Fig. 4-16. This results in a much higher peak current-gain and current handling capability than exhibited by the standard PNP transistor.

Figure 4-17 shows the current gain as a function of collector current for an 18-disk structure, similar to the one shown in Fig. 4-15. As can be noted the considerable improvement in the h_{FE} makes it possible to obtain PNPs which can be used up to currents of around 100 mA (h_{FE} = 10). Structures with a greater number of disks make it possible to use lateral PNPs that can handle currents up to 0.5 A while still occupying a fairly small silicon area.

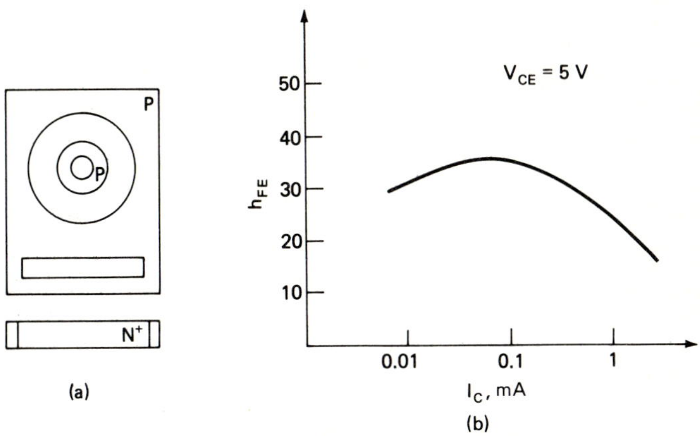

FIG. 4-14 Lateral PNP transistor: (*a*) minimum-size geometry (1 disk); (*b*) current gain vs. collector current.

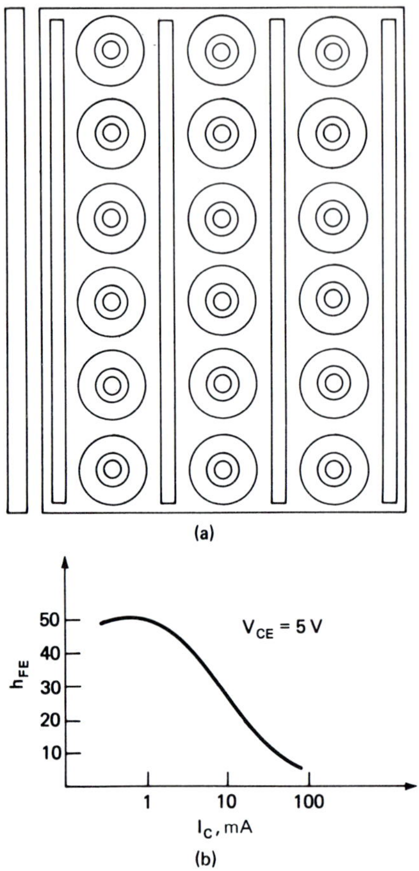

FIG. 4-15 Multidisk lateral PNP transistor: (*a*) geometry (18-disk); (*b*) current gain vs. collector current.

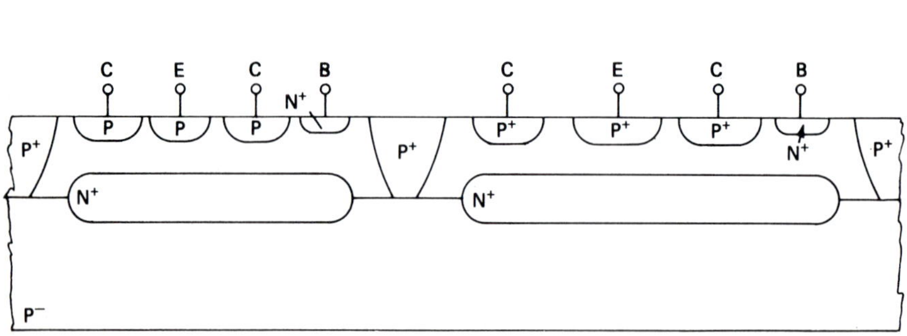

FIG. 4-16 Comparison between two lateral PNP structures (base and P^+).

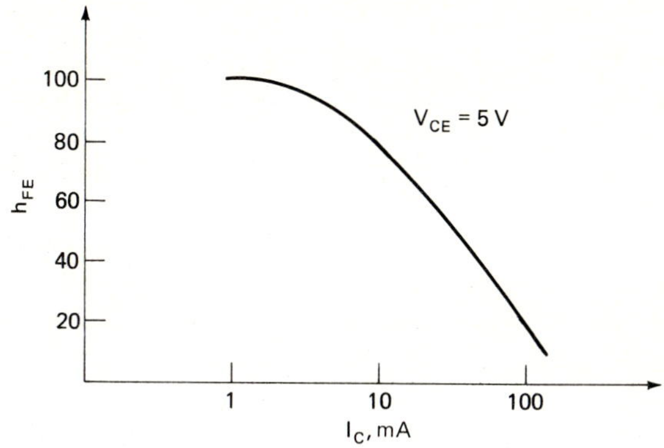

FIG. 4-17 P^+ structure (see Fig. 4-16) current gain vs. collector current.

4.4 CLAMPING DIODE

Power integrated circuits often have to drive inductive loads such as loudspeakers or solenoids. The switching of high currents in these loads gives rise to voltages which can exceed the supply voltage or drop below the ground voltage.

The negative overvoltages are clamped automatically by the collector-to-substrate diode associated with the output power transistor; however, there are some limitations, which will be discussed in Sec. 4.7. The positive overvoltages, on the other hand, require clamping diodes not available in the standard planar process.

The only diode having a high reverse breakdown voltage is in fact the base-collector diode of the NPN transistor (Fig. 4-18). However, it has a serious limitation, since a parasitic PNP transistor is associated with it, the substrate working as a collector. The current gain of this PNP transistor can be approximately 1, causing the injection of strong currents into the substrate, which, in turn, may cause malfunctioning of the circuit (e.g., the substrate might inject into a low-potential pocket of the epitaxial layer), as well as considerable dissipated power on the diode itself.

It has been necessary to develop new structures [3] for clamping diodes in order to avoid these kinds of problems. A diode structure which greatly reduces the substrate current without additional process steps is shown in Fig. 4-19. The N^+ region diffused inside the anode is realized with the emitter diffusion of the NPN and reduces the efficiency of the parasitic vertical PNP. The P ring, obtained with the base diffusion of the NPN, acts as the collector of a lateral PNP, collecting

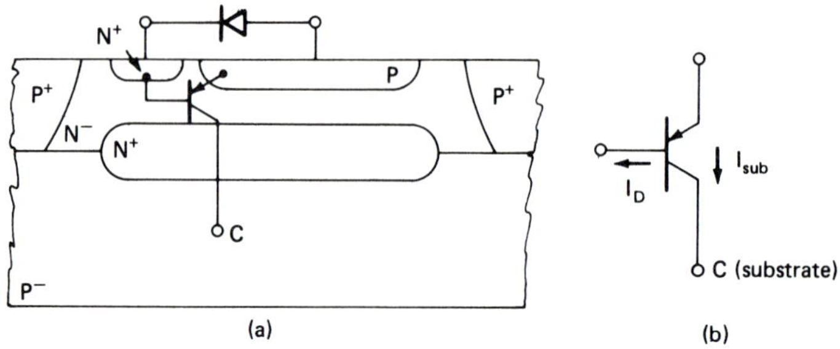

FIG. 4-18 Parasitic PNP transistor associated with a base-collector diode: (*a*) cross section; (*b*) equivalent model.

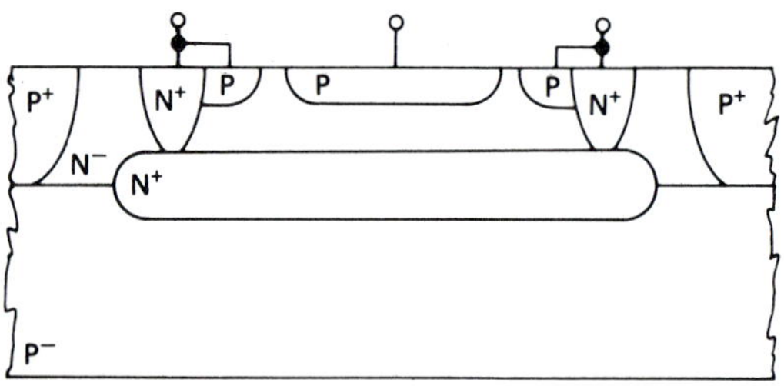

FIG. 4-19 Improved structure of an integrated diode.

the carriers injected laterally and preventing them from reaching the substrate.

Figure 4-20 shows the reduction that can be obtained in the current gain of the parasitic transistor; it goes from a value of 100, for low current, to a factor greater than 10 at currents of around 1 A. The diodes whose characteristics are illustrated here have an area of approximately 200 mil^2 and a reverse breakdown voltage of 100 V.

A different type of structure can be obtained by introducing other process steps which provide "top and bottom" isolation (see Sec. 4.5). This structure is shown in Fig. 4-21. The cathode of the diode is obtained via a P^+ diffusion, carried out at the same time as the one used for isolation from the bottom, while the anode consists of the N^+ of the emitter diffusion of the NPN.

The P^+ diffusion of the isolation from the top is used to collect the cathode of the diode on the surface, while the sinker to the P^+ region short-circuits the base and the emitter of the PNP transistor formed by the P^+ (emitter) diffusion above the buried layer (base) and the substrate (collector); thus it is possible to avoid the possibility of activating this transistor, which would lead to a leakage current toward the substrate.

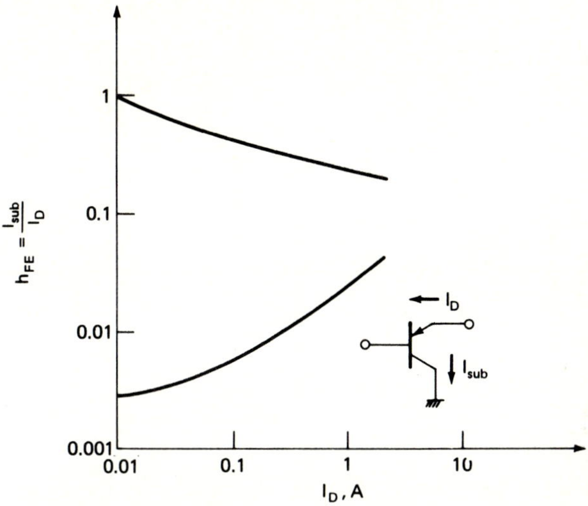

FIG. 4-20 Comparison of parasitic behavior of standard and improved integrated diodes.

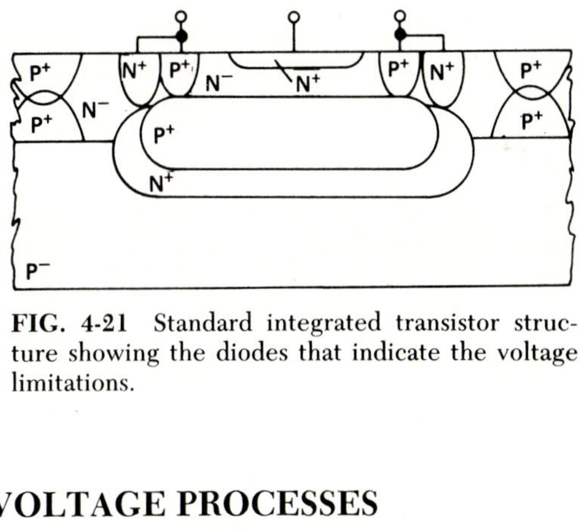

FIG. 4-21 Standard integrated transistor structure showing the diodes that indicate the voltage limitations.

4.5 HIGH-VOLTAGE PROCESSES

If we consider the structure of an integrated NPN transistor (Fig. 4-22), the collector-base breakdown voltage BV_{CBO} is defined by the breakdown voltage of diodes D_1 and D_2. The net epitaxial thickness W_1 and its resistivity define the breakdown voltage of D_1 according to the reach-through mechanism, while the breakdown voltage of D_2 depends on the radius of curvature of the base diffusion.

The collector-to-substrate breakdown voltage, on the other hand, is limited by the breakdown voltage of diodes D_3 and D_4.

The radius of curvature of the P^+-N junction (isolation-epitaxial) of

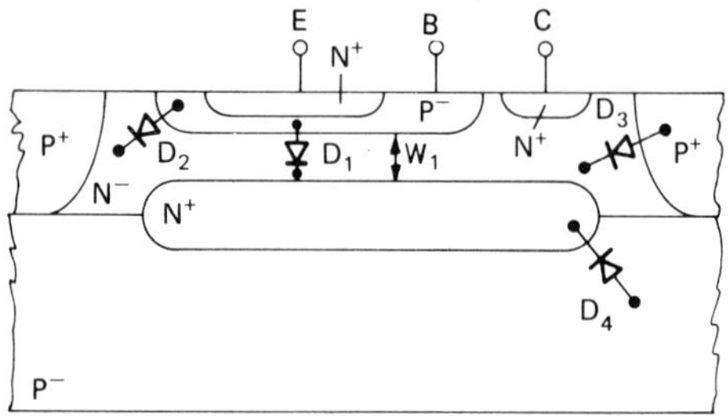

FIG. 4-22 Voltage limitation on standard integrated planar structure.

D_3 is usually greater than that of the N^+-P^- junction (buried layer–substrate) of D_4. Therefore in order to increase the collector-to-substrate breakdown voltage it is necessary to increase the resistivity of the substrate. Using the standard planar process it is not difficult to increase the collector-isolation breakdown voltage, while for the collector-base breakdown voltage the net thickness W_1 and the resistivity of the epitaxial layer must be increased. Naturally, this leads to an increase in the total epitaxial thickness and to increased device dimensions.

In fact, as the epitaxial thickness increases, there is an increase in the lateral diffusion of the isolation, which, together with the greater tolerances required to withstand higher voltage, increases the dimensions of an NPN transistor; this is illustrated in Fig. 4-23 for two processes, at 20 and 80 V of BV_{CEO}. Figure 4-24 shows the area of an NPN transistor with minimum geometry as a function of the collector-to-emitter breakdown voltage of the process.

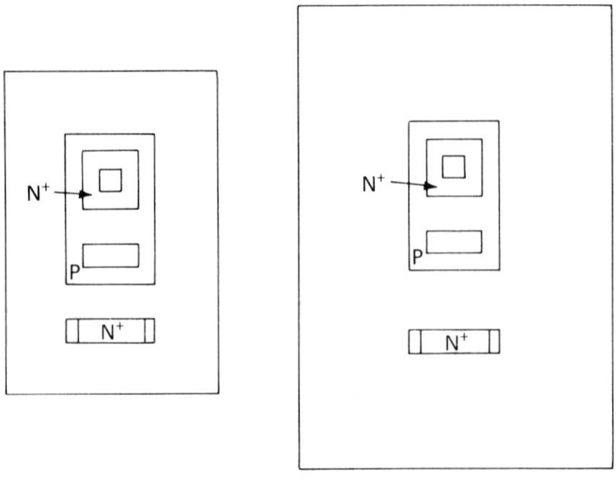

FIG. 4-23 Minimum-size NPN transistor: comparison between 20-V and 80-V processes.

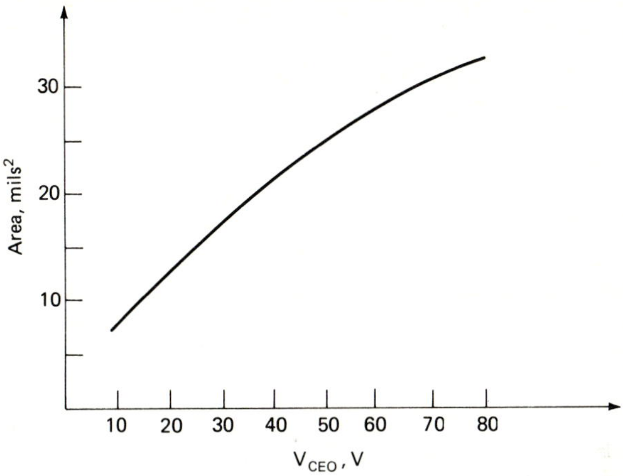

FIG. 4-24 Minimum NPN geometry: area vs. emitter-collector breakdown voltage.

A high-voltage process for small devices therefore requires that thickness W_1, needed for a given BV_{CEO}, be obtained starting from the lowest possible total epitaxial thickness. This is made possible by limiting the outdiffusion of the buried layer during the steps following epitaxial growth. At present, if we exclude the processes with dielectric isolation as too expensive, the most economical process for limiting the dimensions of the components is the so-called top-bottom isolation process [4]. The flowchart of the process is shown in Fig. 4-25. The isolation and sinker obtained from top and bottom enable the diffusion time and therefore the outdiffusion of the buried layer to be reduced considerably. Furthermore, lateral diffusion of the isolation is about half that of the standard top isolation, allowing further reduction of component dimensions.

It may be noted that only the initial phase of the process is different from the standard procedure. After the isolation has been realized the operations are the same as in the conventional process.

Using the top-bottom isolation process, a BV_{CEO} of more than 100 V can be obtained with NPN current gains from 100 to 250 using an epitaxial thickness of around 18 μm. There is no great difficulty in increasing the BV_{CEO} to 150 V with an epitaxial thickness of 25 μm.

The increased operating voltages in processes for PICs require a special metallization mask to avoid breakdown due to the junction curvature. The technique used to eliminate curvature-limited breakdown is the same as that usually adopted in discrete NPN components, i.e., overlapping of the base-collector junction with the emitter or base metallization (the field-plate technique shown in Fig. 4-26).

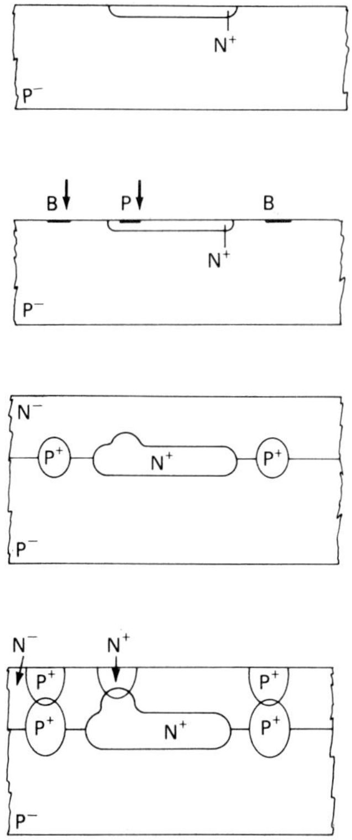

FIG. 4-25 Process sequence for top-bottom-isolated circuits.

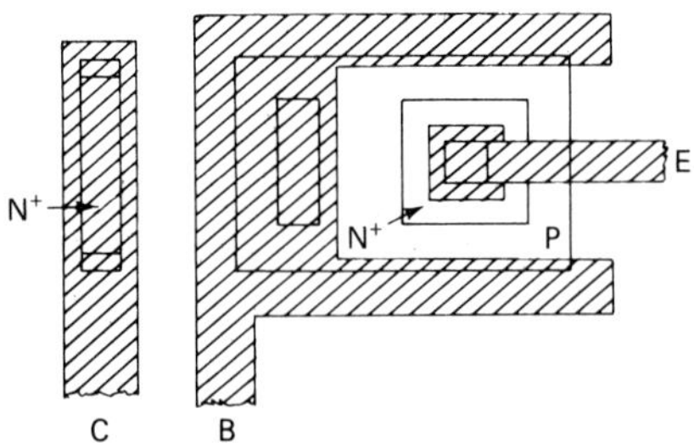

FIG. 4-26 High-voltage NPN transistor with metal field plate.

In the case of integrated circuits the problem is complicated by the fact that the base cannot be completely overlapped. Two spaces that are left for the emitter metal are defined by masking tolerances. Since these spaces may lower the base-collector breakdown voltage, a technique which has proved very efficient is the use of a polysilicon field plate shorted to the base and merged in the oxide (Fig. 4-27).

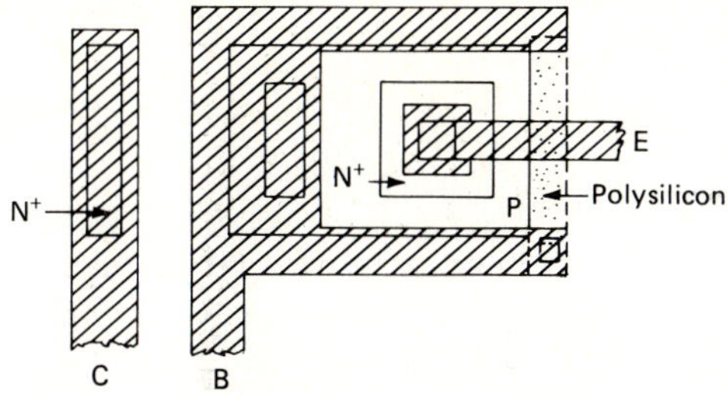

FIG. 4-27 High-voltage NPN transistor with metal and polysilicon field plate.

The same polysilicon field plate is used to protect the isolation junction from being crossed by metal held at very high potential (Fig. 4-28). In the absence of the field plate, the increased electric field strength due to curvature effects at the isolation junction crossed by metallization held at high potential, in the presence of positive mobile charges, would cause a degradation of the collector-isolation breakdown voltage.

Field protection of the collector-isolation surface junction results in breakdown voltages higher than 200 V with excellent long-term stability [5].

4.6 DIGITAL STRUCTURES IN PICs

PICs are ever increasing in complexity and often demand the implementation of analog and digital functions on the same chip. The achieved density is generally very poor when using conventional logic configurations such as RTL, TTL, and so on.

The structure which is naturally compatible with PIC technologies and also exhibits a very high function density is, undoubtedly, the so-

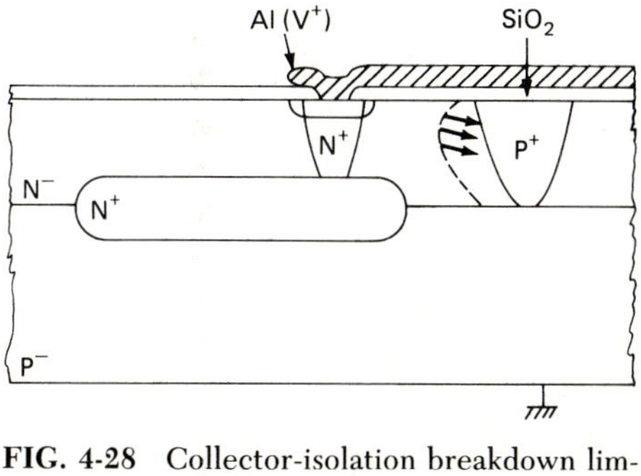

FIG. 4-28 Collector-isolation breakdown limitation on standard structure.

called integrated injection logic (I^2L). Figure 4-29 shows the basic circuit configuration of an I^2L NAND gate, together with the integrated structure cross section.

The compactness of the structure results from a physical "merging" (which justifies the acronym MTL—merged transistor logic—chosen by IBM researchers) between a lateral PNP transistor (the so-called injector) and a multicollector NPN transistor operating in the upward mode. Thus no isolation region is required, as all the NPN emitters and PNP bases share a common region. I^2L logic is of the "wired-AND" type with a fan-out given by the number of collectors belonging to the NPN transistor.

Although early developments were devoted to pure digital applications (with low operating voltage), suggesting I^2L as the bipolar alternative to MOS logic both for density and low-power operation, I^2L structures are now well-established in mixed analog-digital application because of their compatibility with analog circuit processes and thus with PICs.

It is well known that an upside-down-operating NPN transistor exhibits very poor performance for both static and dynamic behavior. This results from the reduced emitter efficiency and from a lowering of the base transport factor caused by a retarding electric field seen by the upward-injected minority carriers [6]. The effective current gain introduced in dc analysis of an I^2L gate may be very low or in some

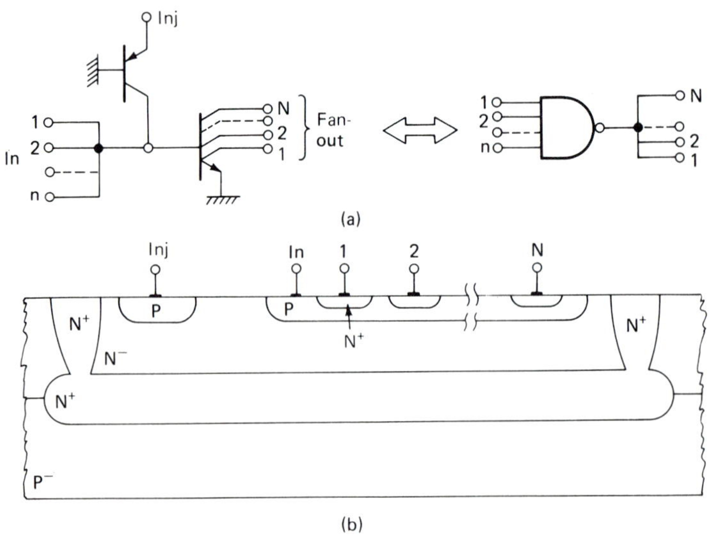

FIG. 4-29 (*a*) Schematic diagram and (*b*) cross section of an I^2L NAND gate.

cases less than unity. However, gain must be greater than unity for I^2L inverter operation.

I^2L minimum delay time (τ_{min}) is directly related to epitaxial-layer characteristics. This can be deduced from a first-order analysis, which shows that the relationship of epitaxial-layer parameters to I^2L performance can be stated as follows [7]:

$$\tau_{\min} = \frac{W_{\text{epi}}}{S_{\text{N}-\text{N}+}} \tag{4-1}$$

where W_{epi} is the base-depth to buried-layer distance and S_{N-N+} is the epitaxy to buried-layer surface recombination velocity for the holes injected downward into the emitter.

It can be easily understood that the collector-to-base breakdown voltage of the NPN transistor operating in the downward mode and I^2L minimum delay time are closely related [7].

A detailed analysis shows that the degree of process optimization can be expressed by a figure of merit given by BV_{CBO}/τ_{min}. This relationship shows the trade-off between PIC voltage requirements and expected I^2L switching performance.

PIC processes are not particularly suitable for this kind of application but, nevertheless, slight process modifications result in satisfactory I^2L structures. These process modifications require the reduction of the epitaxial-layer resistivity in the regions destined for the I^2L structures. This is obtained by ion implantation, and requires only one extra masking step. The reduction of resistivity improves emitter efficiency, lowers the base Gummel number, and improves minority-carrier transit time. I^2L performance (particularly in minimum delay time and effective current gain) becomes less sensitive to epitaxial parameter spread. A BV_{CBO} of 100 V with τ_{min} = 100 ns represents the present state of the art in the PIC field.

4.7 PARASITICS IN PICs

The parastics in PICs are either bulk parasitics (transistors) or surface parasitics (MOS).

The bulk parasitics are typically of the SCR type; they can be eliminated easily by avoiding PNP and NPN transistors in the same pocket or by proper biasing of the pockets containing transistor and resistors.

The only unavoidable bulk parasitic is the one occurring with inductive loads, which force an NPN collector (typically that of an output power transistor) below ground level. This switches on a lateral NPN, where emitter, base, and collector are formed, respectively, by the collector of the power transistor, the substrate, and the adjacent epitaxial zones (Fig. 4-30). The current gain of this transistor can vary from 0.5 to 1, and the current injected into the adjacent epitaxial region, which acts as a collector, can easily fire a parasitic SCR with the possibility of latch-up [3].

The problem can be solved by careful placement of the adjacent components or by using a structure of the type shown in Fig. 4-31. The structure realized by means of the isolation, buried-layer, and sinker diffusions is equivalent to an NPN with the base shorted to the collector

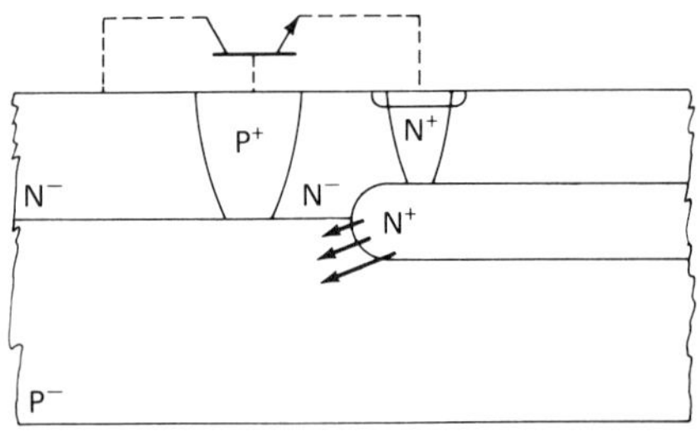

FIG. 4-30 Cross section of a parasitic NPN transistor.

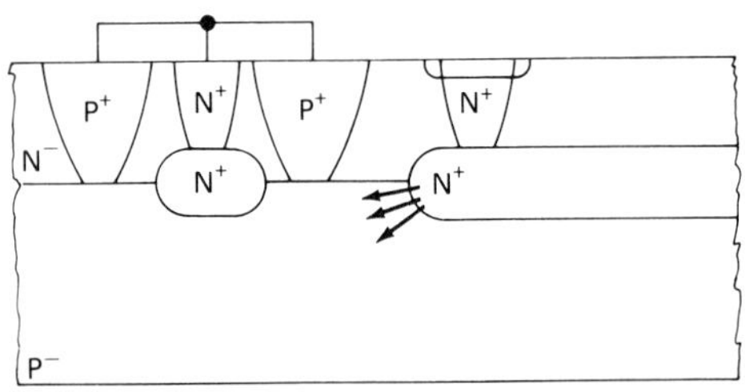

FIG. 4-31 Structure devoted to reduction of parasitic NPN effect.

and collects most of the current injected by the parasitic NPN; therefore its current gain toward the adjacent region can drop to a value of less than 0.01.

The second type of parasitic in PICs is due to the fact that each structure formed by two P diffusions crossed by a low-potential metallization (low with respect to the epitaxial region) can form a P-channel MOS transistor, where the gate oxide is the thick oxide grown on the epitaxial layer (Fig. 4-32). The threshold voltage of this MOS transistor

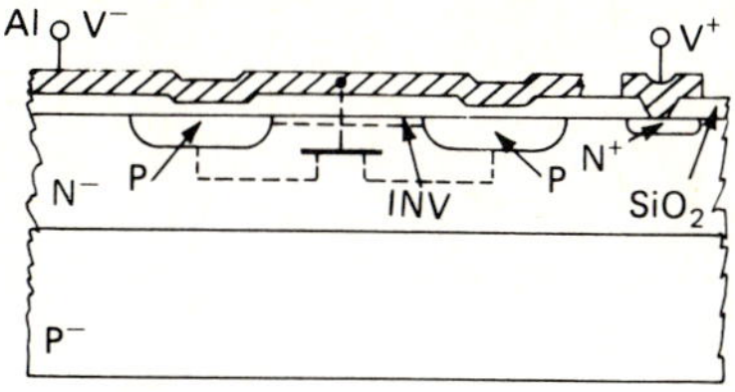

FIG. 4-32 Cross section of a parasitic MOS transistor.

depends on the oxide thickness, on the resistivity of the epitaxial layer, and on the final thermal treatment of the process, which influences the Q_{SS}. A typical value for this threshold voltage ranges from 20 to 40 V—according to the orientation of the starting substrate, whether $\langle 100 \rangle$ or $\langle 111 \rangle$, since $Q_{SS\langle 100 \rangle} = \frac{1}{3} Q_{SS\langle 111 \rangle}$—as given by the equation [8]:

$$V_{T0} = -\frac{Q_{SS}}{C_{ox}} - \frac{Q_{SD\max}}{C_{ox}} + \phi_{MS} + 2\phi_F \tag{4-2}$$

where V_{T0} = threshold voltage
Q_{SS} = fixed positive charge density per unit area at silicon–silicon dioxide interface
$Q_{SD\max}$ = maximum value of the charge density per unit area
C_{ox} = dielectric capacitance per unit area across the insulating gate oxide
ϕ_{MS} = metal-semiconductor work function difference
ϕ_{Fn} = Fermi potential on n-type semiconductor

If the source of the parasitic MOS transistor is reverse-biased with respect to the epitaxial layer with a voltage V_S, the threshold voltage is increased and becomes

$$V_T(V_S) = -\frac{Q_{SS}}{C_{ox}} - \frac{Q_{SD\max}}{C_{ox}} \left(\frac{V_S - 2\phi_{Fn}}{-2\phi_{Fn}} \right)^{1/2} + \phi_{MS} + 2\phi_{Fn} \tag{4-3}$$

Taking into account that in order to identify a parasitic MOS transistor it is more convenient to refer the gate voltage to the epitaxial potential, Eq. 4-2 becomes

$$V_T(V_S) = -\frac{Q_{SS}}{C_{ox}} - \frac{Q_{SD\max}}{C_{ox}} \left(\frac{V_S - 2\phi_{Fn}}{-2\phi_{Fn}} \right)^{1/2} - V_S + \phi_{MS} + 2\phi_{Fn} \tag{4-4}$$

which, as a first approximation, gives:

$$V_T(V_S) = V_{T0} - V_S \tag{4-5}$$

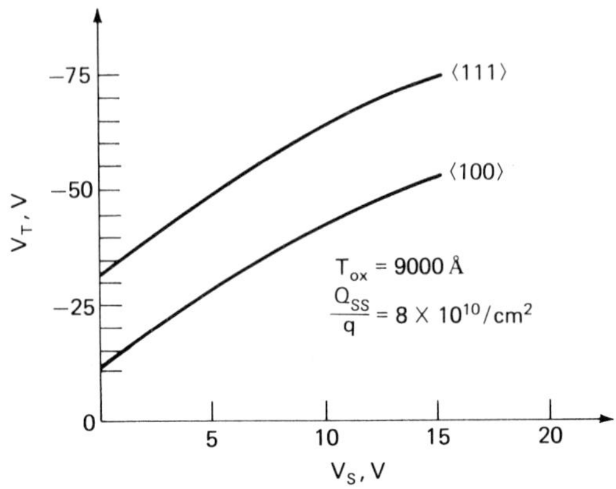

FIG. 4-33 Body effect on threshold voltage for two starting-material crystal orientations.

The threshold voltage dependence on voltage V_S is shown in Fig. 4-33, which is for an NPN PIC process with BV_{CEO} = 40 V.

The most efficient way to stop these parasitic MOS transistors is to introduce an N^+ diffusion between the potential source and drain regions, to be realized together with the emitter of the NPN, which acts as a channel stopper (Fig. 4-34).

Since the same MOS parasitic configuration may occur between a P-type resistor or the collector of a lateral PNP transistor or the base of an NPN transistor and the isolation, the channel stopper technique is widely used in PIC layouts.

For example, Fig. 4-35 shows a lateral PNP transistor in which the parasitic MOS between collector and isolation is eliminated with a channel stopper which entirely surrounds the device. In this case the

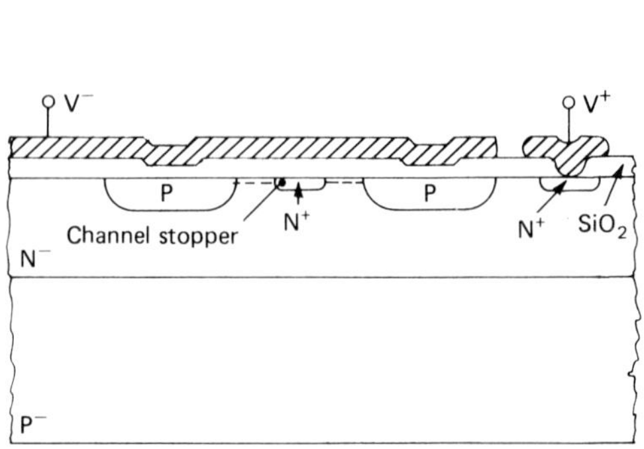

FIG. 4-34 Channel stopper technique to avoid parasitic MOS transistors.

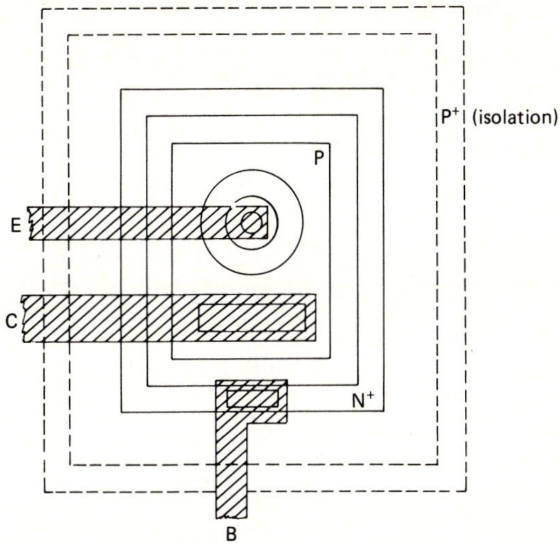

FIG. 4-35 Lateral PNP equipped with channel stopper.

N^+ diffusion of the channel stopper is also used for the base contact of the PNP transistor.

By using this technique all the parasitic MOS structures can be eliminated at time zero. As is known [9], however, with aging the conduction on the outer surface of the oxide causes an extension of the metal to a larger area (Fig. 4-36).

A typical PIC structure in which this situation can be found is the lateral PNP transistor. The collector metallization, normally at low potential with respect to the base and the emitter, can extend over the base channel, causing a parasitic MOS transistor to appear between

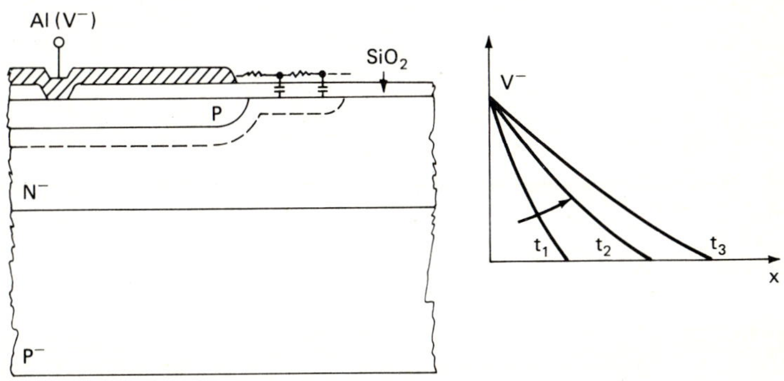

FIG. 4-36 Electrical equivalent circuit and voltage distribution of ion surface migration.

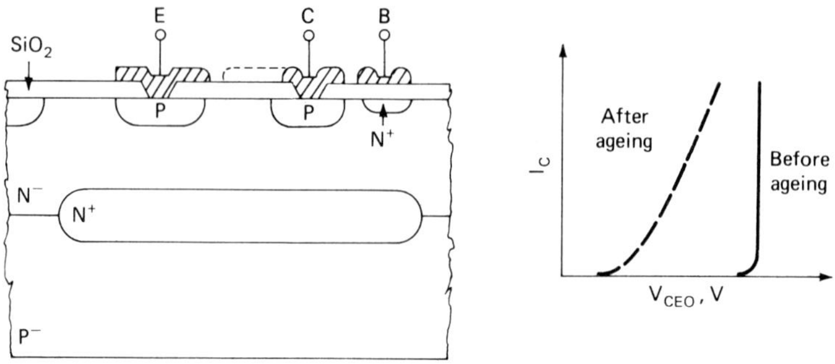

FIG. 4-37 Qualitative diagram showing V_{CEO} degradation due to surface effects.

emitter and collector; this transistor deteriorates the breakdown characteristics (Fig. 4-37) of the lateral PNP.

The most effective protection in this case consists of extending the emitter metallization to cover the base channel, thus preventing the collector metallization from "virtually" expanding over the same region.

The same technique is used in similar situations between the isolation diffusion and the high-potential P diffusion to substitute for or combine with the N^+ channel stopper technique.

The protection technique just described allows PICs to operate at voltages greater than 200 V with excellent long-term stability.

4.8 CONCLUSION

Market demands for PICs capable of ever-higher current and voltages will force a continuous evolution of the planar process. From a device point of view, isolated-collector vertical PNP transistors (see Chap. 5) will become widely used, and power NPN transistors will be characterized by an improved ruggedness as regards both forward and reverse second breakdown. In the very-high-voltage-process field, top-bottom isolation techniques give reason to expect that PICs will shatter the 400-V barrier. That means that in the mid-1980s the maturity of very-high-voltage processes at the production level will allow designing of PICs that are directly connected to the main supply.

REFERENCES

1. B. Murari, Power Integrated Circuit: Problems, Tradeoffs and Solutions, *IEEE J. Solid State Circuits*, Vol. SC-13, No. 3, June 1978.

2. D. J. Hamilton and W. G. Howard, *Basic Integrated Circuit Engineering*, McGraw-Hill, New York, 1975.

3. F. Bertotti and S. D'Arrigo, New Voltage Regulators Integrate More Functions, Have Wider Application Field, *IEEE Trans. Consumer Electronics*, Vol. CE-24, February 1978.

4. F. Bertotti, P. Erratico, and B. Murari, Future and Limits of Power I.C.'s, *ESSCIRC '80, 6th European Solid State Circuit Conf.*, Sept. 22–25, 1980.

5. F. Bertotti, V. Borghesi, and B. Murari, Video Stage I.C. Implemented with New Rugged Isolation Technology, *IEEE Trans. Consumer Electronics*, Vol. CE-27, No. 3, August 1981.

6. F. M. Klaassen, Device Physics of Integrated Injection Logic, *IEEE Trans. Electron Devices*, Vol. ED-22, March 1975.

7. R. D. Davies, "Poly I^2L: Integrated Circuit Technology for Compatible Analog and Digital Signal Processing, *Stanford Electronic Lab. Rep. SEL-78-008*, Stanford, Calif.

8. P. Richman, *MOS Field-Effect Transistors and Integrated Circuits*, Wiley, New York, 1973.

9. A. S. Grove, *Physics and Technology of Semiconductor Devices*, Wiley, New York, 1967.

CHAPTER 5

Improved Transistor Structures in Power Integrated Circuits

T. Kikkawa
NEC Corporation
Kawasaki, Japan

5.1 EPITAXIAL-BASE VERTICAL PNP TRANSISTOR 5.2
5.1.1 Device Structure and Fabrication 5.2
5.1.2 Electrical Characteristics 5.6
5.2 TRIPLE-DIFFUSED PNP TRANSISTOR 5.8
5.2.1 Device Structure and Fabrication 5.8
5.2.2 Electrical Characteristics 5.13
5.3 DIELECTRICALLY ISOLATED COMPLEMENTARY TRANSISTORS 5.15
5.3.1 Device Structure and Fabrication 5.15
5.3.2 Electrical Characteristics 5.17

NPN transistors are widely used in various analog circuits. However, PNP transistors have been trailing behind, because of the difficulty in fabricating PNP transistors complementary to NPN transistors in monolithic integrated circuits. Power circuit functions also require PNP transistors complementary to NPN transistors because of the push-pull output section of the amplifiers. A number of monolithic PNP transistor structures have been developed which are compatible with the standard NPN bipolar process technology [1–11]. The structures, fabrications, and characteristics for the improved PNP transistors as part of a complementary pair will be reviewed in this chapter.

5.1 EPITAXIAL-BASE VERTICAL PNP TRANSISTOR

Lateral PNP transistors are widely used in power integrated circuits as well as small-signal analog integrated circuits [1, 2]. The lateral PNP transistor has an advantage in that it requires no masking and diffusion steps beyond the standard NPN bipolar process. However, electrical characteristics, such as current gain, maximum collector current, and cutoff frequency, are degraded by the device structure.

The epitaxial-base vertical PNP transistor overcomes the performance limitations of lateral PNP transistors [3, 4]. The structure, fabrication, and characteristics of the epitaxial-base vertical PNP transistor as a part of a complementary pair will be described in this section.

5.1.1 Device Structure and Fabrication

Epitaxial-Base Vertical PNP Transistor. The fabrication process is illustrated in Fig. 5-1. The diffusion steps of the vertical PNP transistor are as follows. A heavily doped N-type buried layer for the NPN collector is formed in the P-type substrate (Fig. 5-1*a*) as shown in Fig. 5-1*b*. A P-type buried layer is formed in the heavily doped N-type buried layer for the PNP collector, and in the substrate for the P-type isolation barrier as shown in Fig. 5-1*c*. An N-type epitaxial layer is grown on the substrate as shown in Fig. 5-1*d*. P-type regions are formed from the epitaxial layer surface for the PNP collector wall and isolation as shown in Fig. 5-1*e*. P-type regions are formed for the NPN base and PNP emitter as shown in Fig. 5-1*f*. Heavily doped N-type regions are formed for the emitter of the NPN transistor and the base contact of the PNP transistor as shown in Fig. 5-1g.

Top views of a vertical PNP transistor and an NPN transistor are shown in Fig. 5-2*a* and *b*, respectively.

As described above, the vertical PNP transistor is fabricated by the standard NPN bipolar process with one additional step involving P-type buried collector diffusion into the heavily doped N-type buried layer. Since the vertical PNP transistor is completely isolated from the substrate, it provides a floating collector configuration.

The collector island of the vertical PNP transistor is formed by a redistribution diffusion from the P-type buried layer source at the epitaxial layer–substrate interface within a heavily doped N-type buried layer. This takes place because the diffusion coefficient of boron, the P-type buried layer dopant, is larger than that of antimony or arsenic, the N-type buried layer dopant [12, 13]. The redistribution of buried layers occurs mainly during the drive-in cycle for isolation. The doping pro-

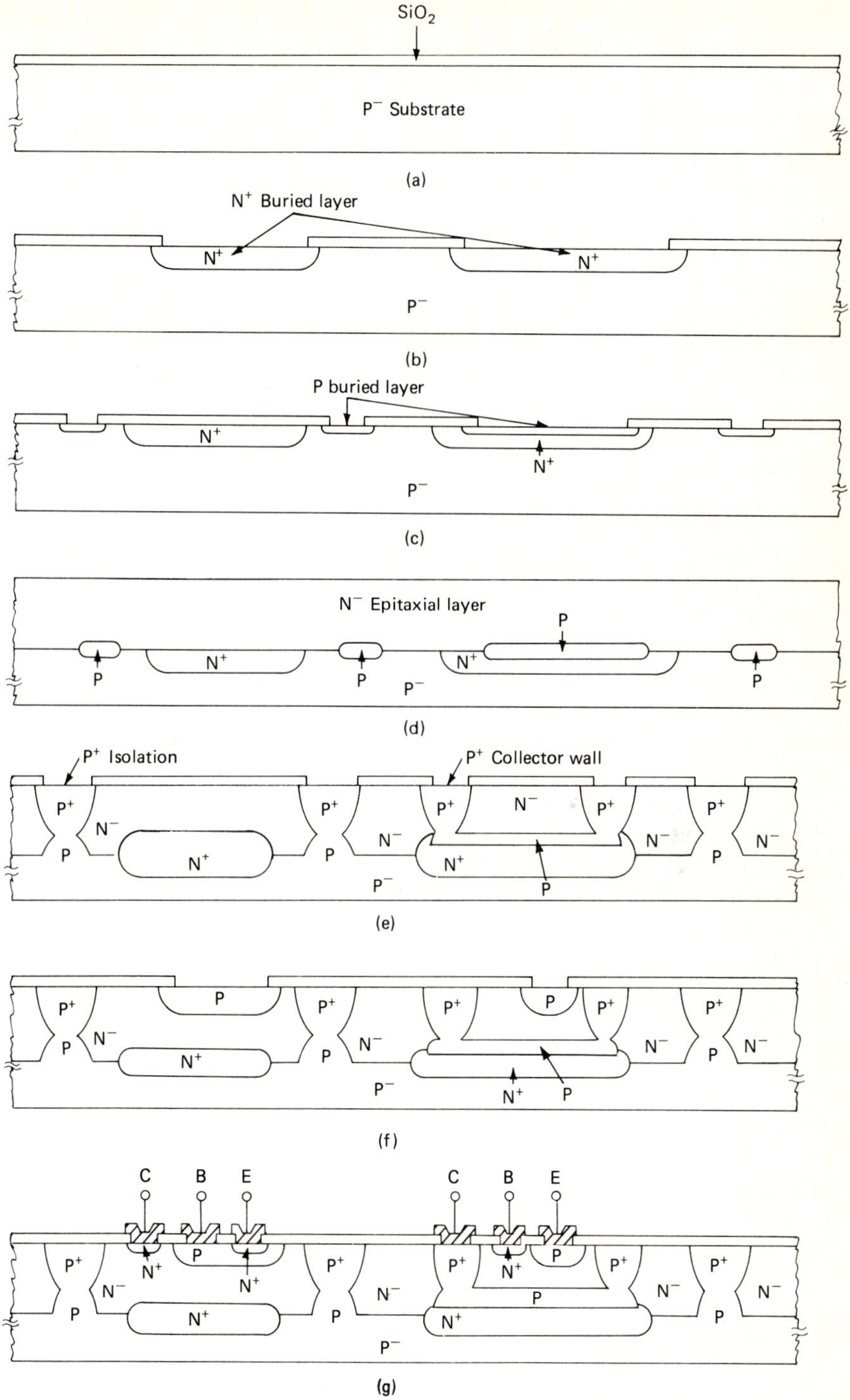

FIG. 5-1 Process sequence for the fabrication of an epitaxial-base vertical PNP transistor and an NPN transistor.

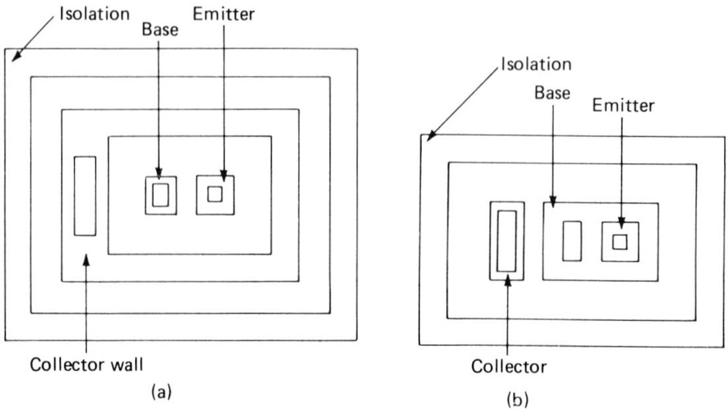

FIG. 5-2 Top views of (*a*) epitaxial-base vertical PNP transistor and (*b*) NPN transistor.

files of the vertical PNP and NPN transistors are shown in Fig. 5-3*a* and *b*, respectively.

The base width of the vertical PNP transistor is determined by the following expression:

$$W_B = X_{\mathrm{epi}} - X_{jE} - d \tag{5-1}$$

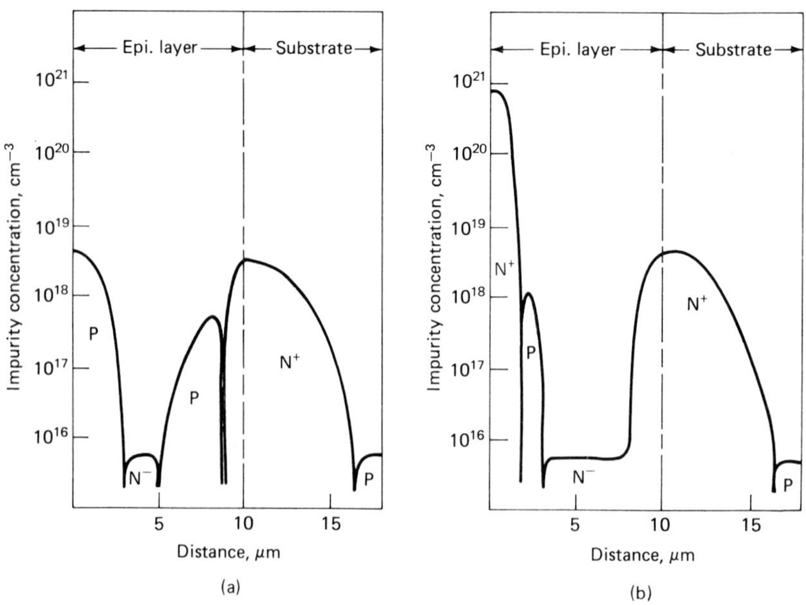

FIG. 5-3 Impurity profiles for (*a*) epitaxial-base vertical PNP and (*b*) NPN transistors.

where W_B = base width of vertical PNP transistor
X_{epi} = epitaxial layer thickness
X_{jE} = PNP emitter junction depth
d = distance to upper boundary of the P-type buried layer from the epitaxial–substrate interface

From Eq. 5-1, the base width of the vertical PNP transistor is determined mainly by the redistribution of the P-type buried layer and by the epitaxial layer thickness. The distance d represents the following expression, which assumes a gaussian distribution and is shown in Fig. 5-4 [8, 12]:

$$d = \left(4Dt \ln \frac{Q}{2N_{epi}\sqrt{\pi Dt}}\right)^{1/2} \tag{5-2}$$

where D = diffusion coefficient for boron in silicon $\quad$ cm^2/s
Q = total quantity of boron $\quad$ cm^{-2}
t = diffusion time $\quad$ s
N_{epi} = donor concentration in the epitaxial layer $\quad$ cm^{-3}

Q is estimated by measuring the sheet resistivity and junction depth of a layer. From Irvin's curves [14], C_0 is found. Then, Q is calculated from [8, 12]

$$Q = \sqrt{\pi D_0 t_0}\, C_0 \tag{5-3}$$

where D_0 and t_0 are the diffusion coefficient and the time at the temperature of the buried-layer diffusion in the substrate, respectively.

The base width of the vertical PNP transistor is smaller than that of the lateral PNP transistor. This is because the base width, which is the distance between the bottom of the emitter junction and the top of the collector buried layer, does not depend on the process limitations of photolithography and on the masking tolerances, but only on the redistribution of the collector buried layer and on the epitaxial layer thickness.

Modified Epitaxial-Base Vertical PNP Transistor. The depletion layer at the collector-base junction expands into the base toward the emitter junction since the impurity concentration of the base region is lower than that of the collector region. Punch-through appears when the depletion layer reaches the emitter junction. When high-voltage PNP transistors are required, a very thick epitaxial layer is needed to avoid the punch-through condition. However, this will cause an increase in the size of the device.

A modified vertical PNP transistor structure for high-voltage appli-

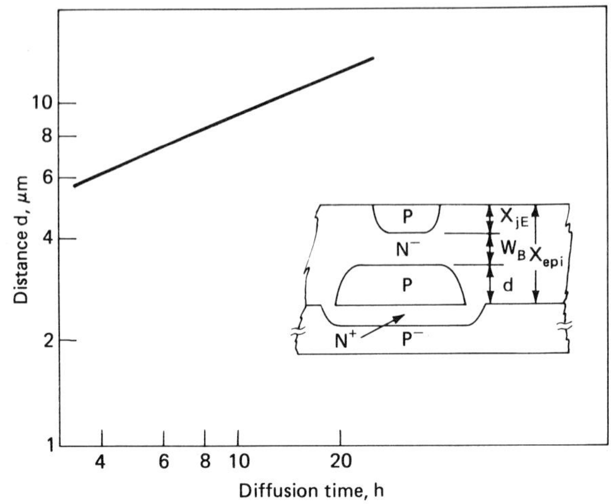

FIG. 5-4 Measured values of the distance *d* of the upper boundary of the P-type buried layer as a function of the diffusion time at 1200°C.

cations has been proposed, as shown in Fig. 5-5. It has the heavily doped N-type ion-implanted base region around the emitter to suppress the inward expansion of the depletion layer from the collector-base junction. Consequently, punch-through does not occur. The breakdown voltage for this structure is determined by the avalanche breakdown characteristic. As a result, a vertical PNP transistor with a high breakdown voltage is obtained.

5.1.2 Electrical Characteristics

The emitter current injected into the epitaxial layer, which is either a lateral or a vertical component, is all collected by the collector wall and the buried layer. Therefore, there is no current loss, and the effective base width of the vertical PNP transistor is smaller than that of the lat-

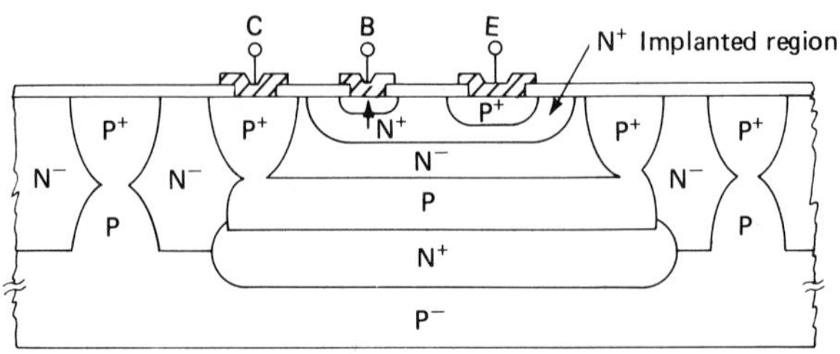

FIG. 5-5 Cross-sectional view of a modified vertical PNP transistor.

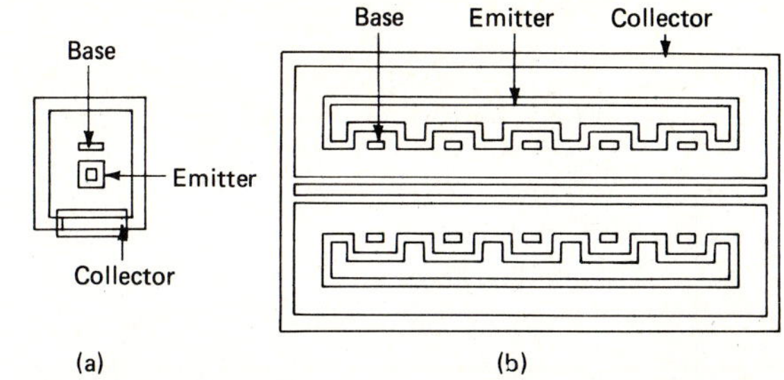

FIG. 5-6 Top views of (*a*) small-signal and (*b*) high-current epitaxial-base vertical PNP transistors.

eral PNP transistors. Consequently, common-emitter current gain and frequency response for the vertical PNP transistor can be improved in comparison with those for the lateral PNP transistor.

Since the bottom junction area of the emitter for the vertical PNP transistor is active, a large emitter periphery and area can be in use. Therefore, the vertical PNP transistor is capable of handling a greater amount of current without degrading the current gain. Top views of a small-signal transistor and a high-current transistor are shown in Fig. 5-6*a* and *b*, respectively. The corresponding common-emitter current gain as a function of collector current is shown in Fig. 5-7. The current capability is still smaller than that of an NPN transistor with comparable geometry because of the low doping of the epitaxial base of the vertical PNP transistor.

The gain-bandwidth product f_T versus collector current is shown in

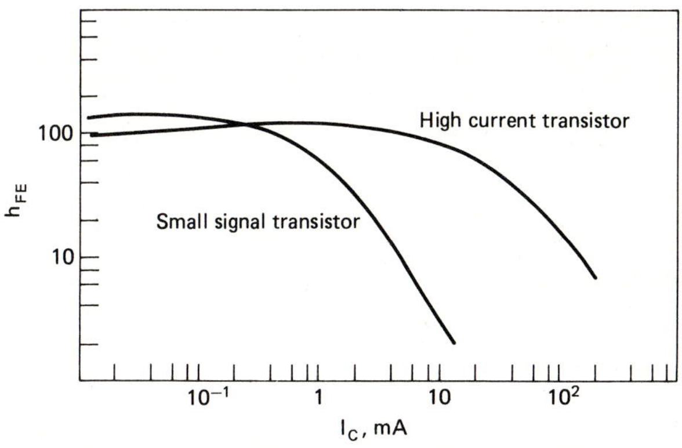

FIG. 5-7 Common-emitter current gain as a function of collector current for a high-current and a small-signal vertical PNP transistor.

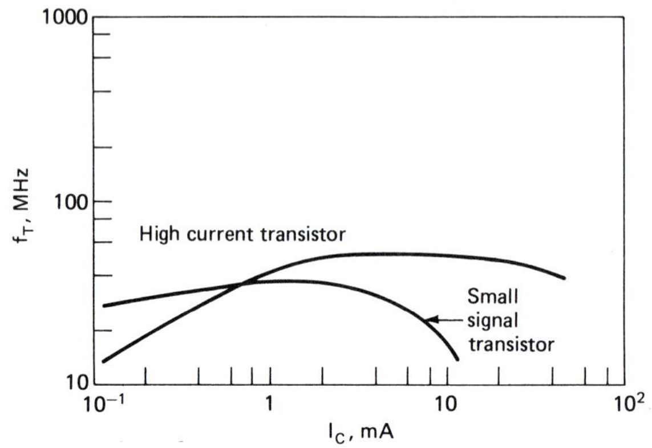

FIG. 5-8 Gain-bandwidth product as a function of collector current for a high-current and a small-signal vertical PNP transistor.

Fig. 5-8. The poor frequency response is due to the large base width and volume in comparison with pertinent NPN transistor values.

Collector saturation resistance R_{SC} for the vertical PNP transistor is still larger than that for the NPN transistor, because the P-type collector buried layer has a lower impurity concentration than that of the underlying N-type buried layer as shown in Fig. 5-3*a*.

5.2 TRIPLE-DIFFUSED PNP TRANSISTOR

In the design of high-performance analog circuits, especially for complementary power integrated circuits, the performance characteristics for lateral PNP transistors and conventional vertical PNP transistors are not sufficiently good to assure obtaining high current capability, small collector series resistance, and good frequency response. To attain these characteristics a fully complementary transistor structure, which is composed of triple-diffused PNP transistors and double-diffused NPN transistors, has been developed for analog integrated circuits [9–11]. The structure, fabrication, and characteristics of the triple-diffused PNP transistor as a part of a complementary pair will be described in this section.

5.2.1 Device Structure and Fabrication

Key drawbacks of conventional PNP transistors are the inherently low frequency response due to large base width, high saturation voltage due to high collector series resistance, and low current capability due to high base resistivity. A triple-diffused PNP transistor structure which has the same impurity profile as the double-diffused NPN transistor is needed to obtain small base width and low base resistivity. The heavily doped P-type buried layer, which is isolated from the substrate, is also needed to obtain low collector series resistance.

The fabrication processes of the complementary transistor pair are illustrated in Fig. 5-9. Impurity profiles of the PNP and NPN transistors are shown in Fig. 5-10*a* and *b*, respectively.

The lightly doped N-type buried layer under the PNP transistor is formed in the P-type substrate as shown in Fig. 5-9*a*. The heavily doped N-type buried layer under the NPN transistor is formed in the substrate as shown in Fig. 5-9*b*. The heavily doped P-type buried layer is formed in the N-type buried layer and the substrate, as shown in Fig. 5-9*c*. The N-type epitaxial layer is grown on the P-type substrate surface as shown in Fig. 5-9*d*. The lightly doped P-type collector region of the PNP transistor is formed from the epitaxial layer surface. A heavily doped P-type collector wall and isolation are formed as shown in Fig. 5-9*f*. The base of the PNP transistor is formed from the epitaxial layer surface, as shown in Fig. 5-9*g*. The base of the NPN transistor, the emitter of the PNP transistor, and the emitter of the NPN transistor are formed as shown in Fig. 5-9*h*. Finally, metals are deposited, as shown in Fig. 5-9*i*.

The PNP transistor is isolated from the P-type substrate by inserting

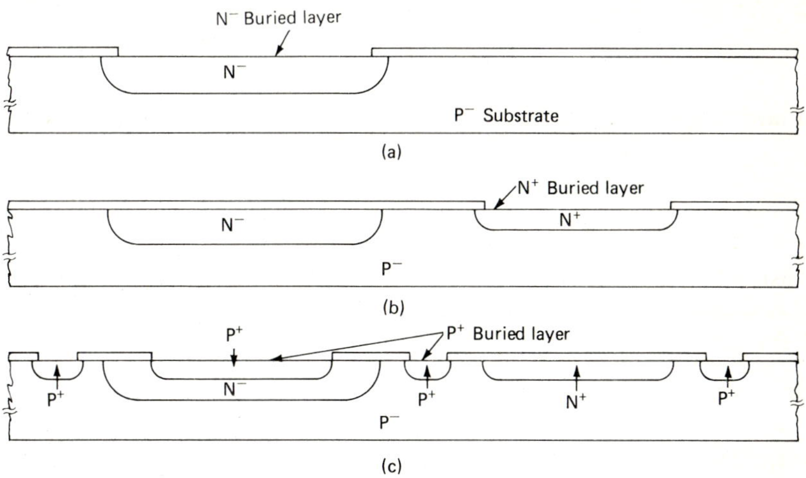

FIG. 5-9 Process sequence for the fabrication of a triple-diffused PNP and a double-diffused NPN transistor.

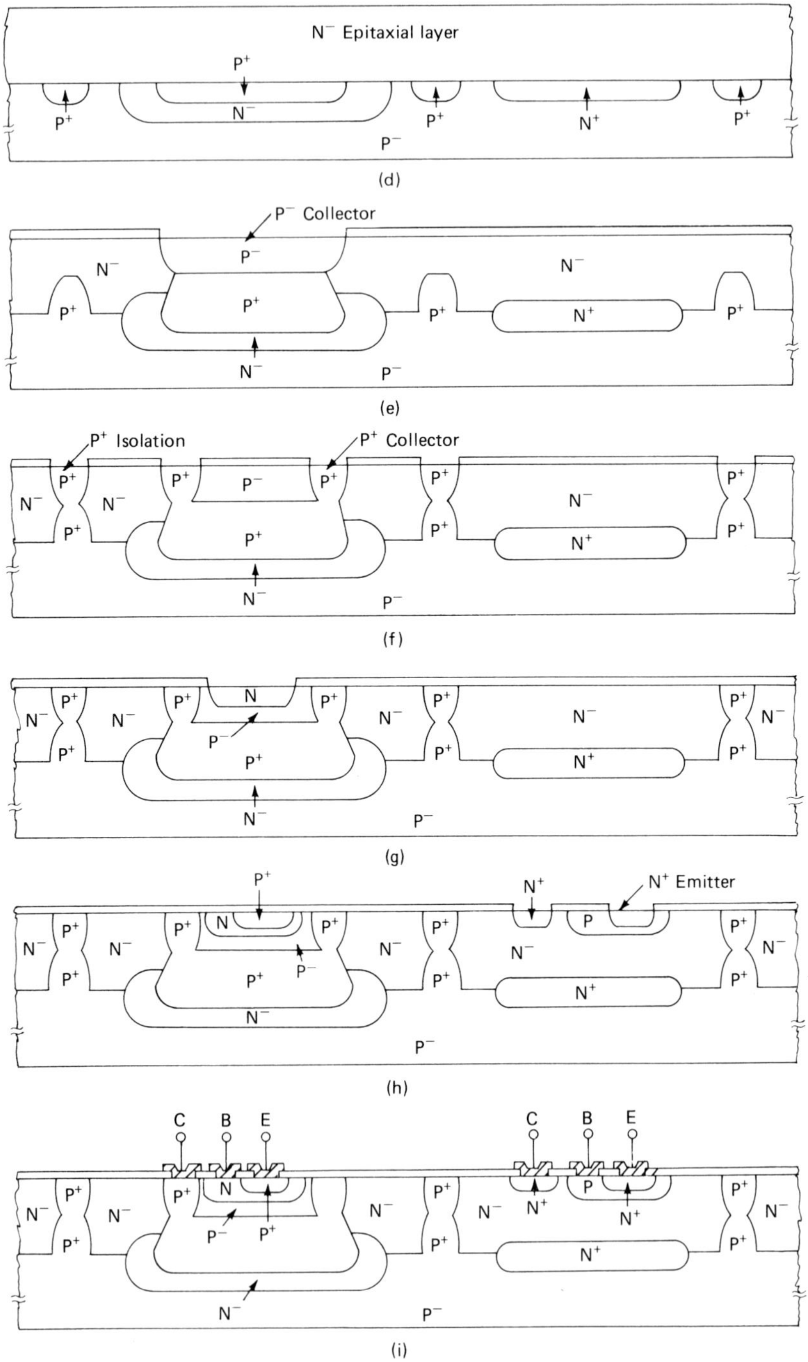

FIG. 5-9 *Continued*

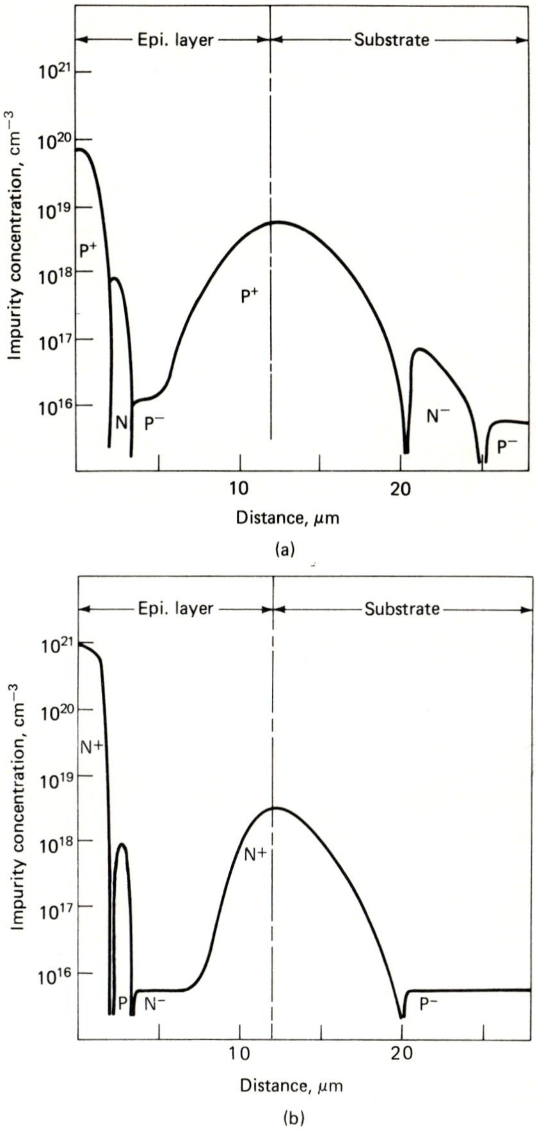

FIG. 5-10 Impurity profiles for (*a*) triple-diffused PNP and (*b*) double-diffused NPN transistors. *(After Kikkawa et al.* [9]. *Copyright © 1980 IEEE.)*

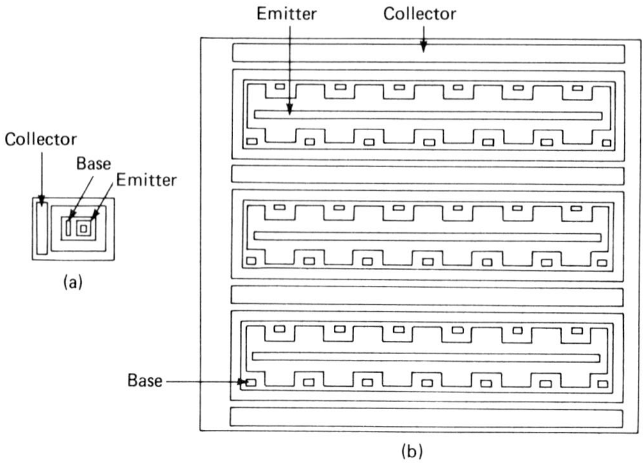

FIG. 5-11 Top views of (*a*) small-signal and (*b*) high-current triple-diffused PNP transistors.

an N-type buried layer between the P-type collector buried layer and the substrate. The P-type collector buried layer is heavily doped in order to reduce the collector series resistance of the PNP transistor. On the other hand, the N-type buried layer beneath the P-type collector buried layer is lightly doped in order to avoid an increase in the collector series resistance due to impurity compensation.

In addition, the N-type buried layer under the NPN transistor is heavily doped to reduce the collector series resistance in the NPN transistor, as shown in Fig. 5-10*b*.

The base region of the PNP transistor is diffused from the surface to control current gain precisely and is doped rather heavily to prevent conductivity modulation at high current levels. Finally, the high-concentration emitter region is diffused.

The NPN transistor is fabricated by the conventional double-diffusion process.

In the entire process, the PNP diffusion steps are interleaved with the standard NPN diffusion steps without changing the NPN impurity profile, and the optimum PNP impurity profile can be obtained independently.

For high current capability, an interdigitated or fish-bone emitter is necessary to obtain long emitter periphery to counteract current crowding.

Top views of small-signal and high-current transistors are shown in Fig. 5-11*a* and *b*, respectively.

5.2.2 Electrical Characteristics

The PNP transistor shows fully complementary characteristics to the NPN transistors in regard to current gain, current handling capability, frequency response, and saturation voltage. It provides 5 to 10 times better performance in $I_{C(\max)}$, f_T, and $V_{CE(\text{sat})}$ in comparison with lateral or conventional vertical PNP transistors.

Since the PNP base region consists of a higher impurity concentration than that of the collector region in the vicinity of the collector-base junction, most of the depletion layer at the collector-base junction

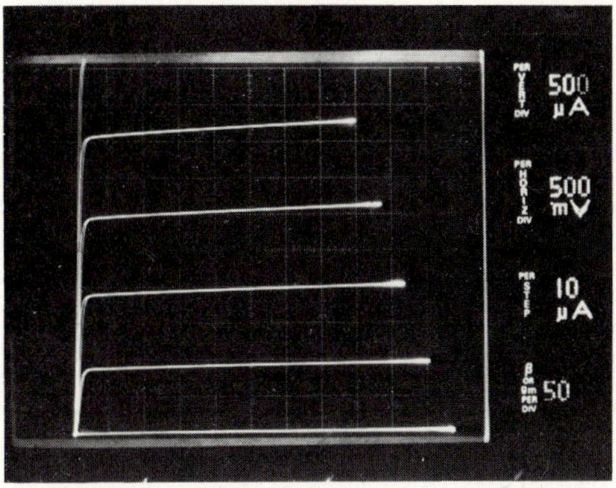

(*a*)

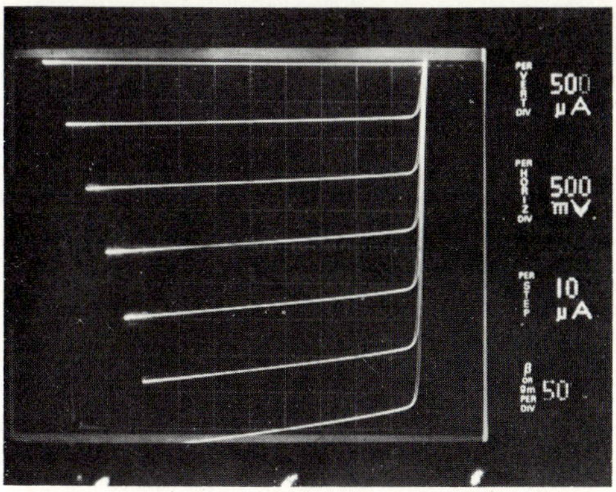

(*b*)

FIG. 5-12 I_C-V_{CE} characteristics for (*a*) double-diffused NPN and (*b*) triple-diffused PNP transistors. *(After Kikkawa et al.* [9]. *Copyright © 1980 IEEE.)*

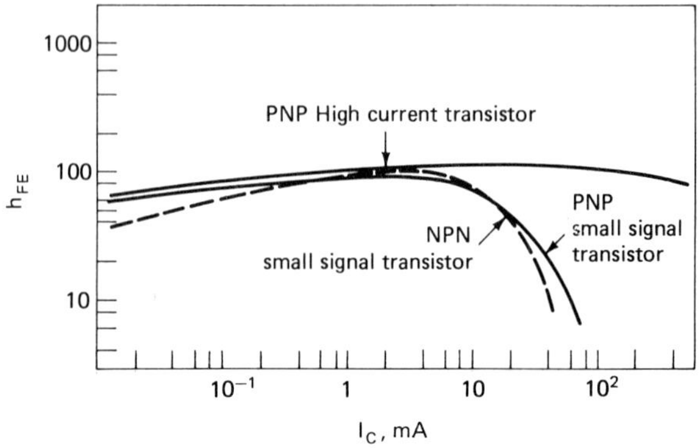

FIG. 5-13 Common-emitter current gain as a function of collector current for a high-current and a small-signal transistor.

expands into the collector region. Therefore, base width modulation of the PNP transistor resulting from changes in collector-emitter voltage is small in comparison with conventional vertical PNP transistors.

I_C-V_{CE} characteristics of the complementary transistor pair are shown in Fig. 5-12. Common-emitter current gain h_{FE} as a function of collector current for fully complementary transistors is shown in Fig. 5-13. The PNP transistor current gain can be adjusted over the same range as the NPN transistor current gain by changing the PNP transistor base implantation dose.

A collector series resistance of less than 2 Ω is achieved in the power transistor by introducing the heavily doped collector buried layer. The

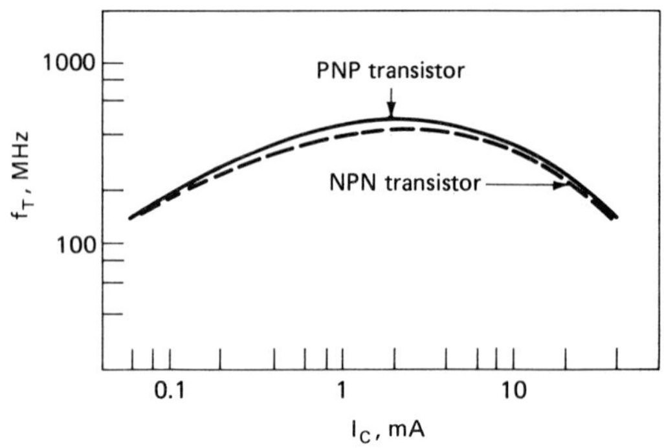

FIG. 5-14 Gain-bandwidth product as a function of collector current for a triple-diffused PNP and a double-diffused NPN transistor. *(After Kikkawa et al. [9]. Copyright © 1980 IEEE.)*

frequency response is improved by the minimizing of the base width and volume that results from the triple-diffused structure of the PNP transistor. The measured gain-bandwidth product f_T as a function of collector current for a complementary transistor pair is shown in Fig. 5-14. An f_T of more than 400 MHz is obtained for the small-signal PNP transistor.

5.3 DIELECTRICALLY ISOLATED COMPLEMENTARY TRANSISTORS

A high-voltage complementary process is necessary for power integrated circuits as well as other analog integrated circuits. The dielectrically isolated complementary transistor process that creates dielectrically isolated P-type and N-type islands on the same chip, makes it possible to obtain both the high voltage rating and the complementary transistors [15–18]. The structure, fabrication, and characteristics of the dielectrically isolated complementary transistors will be described in this section.

5.3.1 Device Structure and Fabrication

Dielectric Isolation. The PN junction isolation technique has some drawbacks. The most significant one for high-voltage integrated circuits is the low breakdown voltage of the back-to-back diodes providing the isolation. Others are the parasitic junction capacitance and the parasitic four-layer switching effect associated with the extra junctions.

Conventional dielectric isolation techniques, such as that used in fabricating epitaxial passivated integrated circuits [19], overcome the voltage limitations. However, they are not able to fabricate complementary transistors, because all the islands produced by these techniques have the same conductivity type. However, the *complementary* epitaxial passivated integrated circuit technique [17, 18] provides both the high voltage capability and the complementary transistors.

The process is illustrated as follows in Fig. 5-15 [18], the reference letters denoting the corresponding parts of the figure. (*a*) A P-type (100) silicon substrate is anisotropically etched using a silicon dioxide mask, and P islands are formed. (*b*) After washing out the SiO_2 mask, a film of Si_3N_4 is deposited on the substrate and is selectively etched off to make the P islands bare. P-type impurities are doped onto the P island surfaces. (*c*) After the P island surfaces are oxidized, the Si_3N_4 film is removed. (*d*) An N-type silicon film is deposited under epitaxial growth conditions. The deposited silicon film has three types of

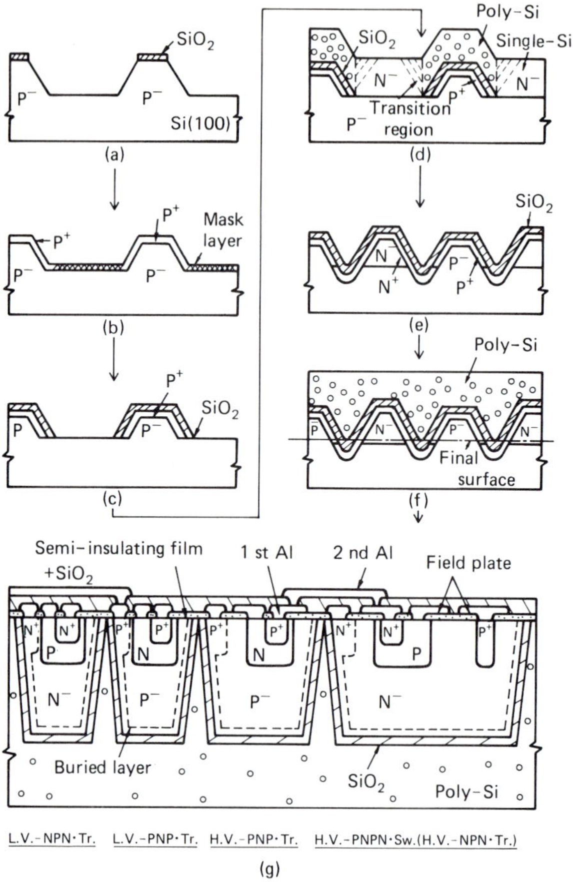

FIG. 5-15 Process sequence for the fabrication of complementary dielectric isolated islands. *(After Sakurai et al.* [18]. *Copyright © 1983 IEEE.)*

regions: single-crystalline, polycrystalline, and transition regions. The transition and the polycrystalline regions are etched off by the use of an anisotropic etching technique. The single-crystalline N islands are formed as a result. (*e*) After N-type impurities are doped onto the N island surfaces, a thick SiO_2 layer is grown. (*f*) A thick polycrystalline film is deposited as a new supporting substrate. The unnecessary single-crystalline substrate is polished off to isolate the N islands and P islands. (*g*) Low- and high-voltage transistors are formed on these islands.

High-Voltage Transistor Using a Shallow Junction. A high-voltage transistor with shallow junction is needed to obtain good frequency

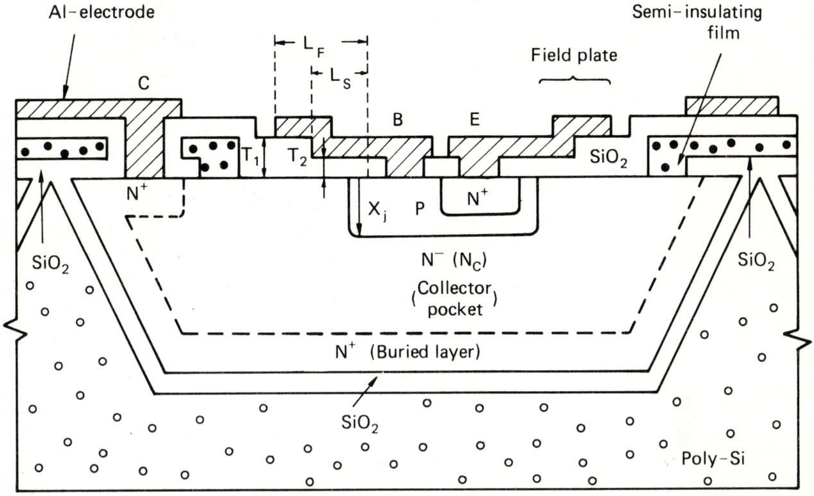

FIG. 5-16 Field plate with nonuniform oxide thickness for high-voltage transistors. *(After Sakurai et al. [18]. Copyright © 1983 IEEE.)*

response and small device size for integrated circuits. A shallow-junction transistor using the field plate, nonuniform oxide [20], and semi-insulating polycrystalline silicon [21] has been developed. A field-plated transistor with thin oxide near the junction and thick oxide near the field plate edge is formed to reduce the electric field near the junction and field plate edge, respectively, as shown in Fig. 5-16 [18].

In order to reduce the electric field in curved regions of the junction, a smaller oxide thickness near the junction, T_2 and larger field plate length, L_F and distance to the oxide step, L_S are favorable. The electric field in regions beneath the field plate edge decreases with increasing oxide thickness at the field plate edge, T_1. In the regions beneath the oxide steps, the electric field decreases with decreasing $(T_1 - T_2)$. The electric field decreases with decreasing the impurity concentration of the island, N_C.

Furthermore, the semi-insulating polycrystalline silicon film is used to prevent the breakdown at the surface near the heavily doped region.

5.3.2 Electrical Characteristics

The complementary epitaxial passivated integrated circuit technique makes it possible to form vertical double-diffused transistors in complementary islands and obtain a high breakdown voltage of more than 350 V in spite of using shallow junctions with a depth of less than 2 μm.

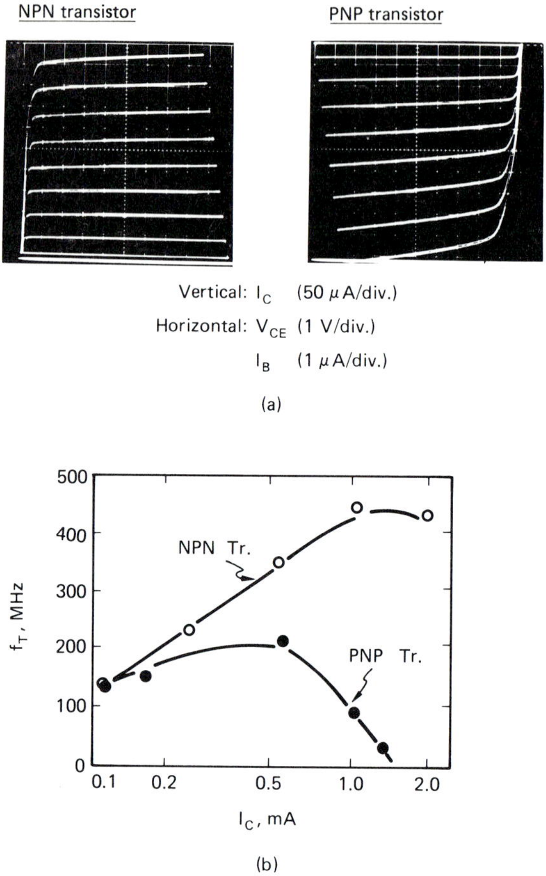

FIG. 5-17 Complementary characteristics for dielectrically isolated complementary transistors. (*a*) I_C-V_{CE} characteristics. (*b*) Gain-bandwidth product as a function of collector current. *(After Sakurai et al. [18]. Copyright © 1983 IEEE.)*

I_C-V_{CE} characteristics and the gain-bandwidth product as a function of collector current for a complementary transistor pair are shown in Fig. 5-17*a* and *b*, respectively. The f_T of 200 MHz and 450 MHz are obtained for PNP and NPN transistors, respectively.

REFERENCES

1. H. C. Lin, T. B. Tan, G. Y. Chang, B. van der Leest, and N. Formigoni, Lateral Complementary Transistor Structure for the Simultaneous Fabrication of Functional Blocks, *Proc. IEEE*, **52:**1491 (1964).

2. B. Murari, Power Integrated Circuits: Problems, Tradeoffs, and Solutions, *IEEE J. Solid-State Circuits,* **SC-13:**307 (1978).
3. S. C. Su and J. D. Meindl, A New Complementary Bipolar Transistor Structure, *IEEE J. Solid-State Circuits,* **SC-7:**351 (1972).
4. K. Ryono and I. Sasaki, A New Vertical PNP Transistor for Integrated Circuits, *Tech. Dig. 1971 IEEE IEDM,* p. 154.
5. R. A. Stehlin, The Realization of 1.0 Volt Multivibrators Using Complementary Integrated Circuits, *IEEE J. Solid-State Circuits,* **SC-4:**284 (1969).
6. R. A. Stehlin and G. W. Niemann, Complementary Transistor-Transistor Logic—An Approach to High Speed Micropower Logic, *IEEE J. Solid-State Circuits,* **SC-7:**153 (1972).
7. P. C. Davis, S. F. Moyer, and V. R. Saari, High Slew Rate Monolithic Operational Amplifier Using Compatible Complementary P-N-P's, *IEEE J. Solid-State Circuits,* **SC-9:**340 (1974).
8. R. G. Donald, Complementary Transistors in Integrated Circuits, *Solid-State Electronics,* **13:**815 (1970).
9. T. Kikkawa, T. Suganuma, K. Tanaka, and T. Hara, A New Complementary Transistor Structure for Analog Integrated Circuits, *Tech. Dig. 1980 IEEE IEDM,* p. 65.
10. D. W. Aull, D. A. Spires, P. C. Davis, and S. F. Moyer, A High Voltage IC for a Transformerless Trunk and Subscriber Line Interface, *IEEE J. Solid-State Circuits,* **SC-16:**261 (1981).
11. J. H. Green, S. F. Moyer, and J. J. Nahas, A Channel Unit Signal Controller for Shared Codec D-Type Channel Banks, *IEEE J. Solid-State Circuits,* **SC-16:**341 (1981).
12. A. S. Grove, *Physics and Technology of Semiconductor Devices,* Wiley, New York, 1967.
13. C. S. Fuller and J. A. Ditzenberger, Diffusion of Donor and Acceptor Elements in Silicon, *J. Appl. Phys.,* **27:**544 (1966).
14. J. C. Irvin, Resistivity of Bulk Silicon and of Diffused Layers in Silicon, *Bell Syst. Tech. J.,* **41:**387 (1962).
15. A. B. Grebene, *Analog Integrated Circuits Design,* Van Nostrand Reinhold, New York, 1972.
16. J. D. Beasorn, A Process for Simultaneous Fabrication of Vertical NPN and PNP's, Nch, and Pch MOS Devices, *Tech. Dig. 1973 IEEE IEDM,* p. 41.
17. T. Sakurai and K. Kato, A New Complementary Dielectric Isolation Process for High-Voltage Devices, *IEEE Trans. Electron Devices,* ***ED-28:**1199 (1981).*
18. T. Sakurai, T. Ohno, K. Kato, Y. Inabe, and T. Hayashi, A Dielectrically Isolated Complementary Bipolar Technique for Analog/Digital Compatible LSI's, *IEEE Trans. Electron Devices,* **ED-30:**1278 (1983).
19. R. M. Warner, Jr., and J. N. Fordemwalt, *Integrated Circuits,* McGraw-Hill, New York, 1965, p. 166.

20. F. Conti and M. Conti, Surface Breakdown in Silicon Planar Diodes Equipped with Field Plate, *Solid-State Electronics*, **15**:93 (1972).

21. T. Matsushita, T. Aoki, T. Ohtsu, H. Yamoto, H. Hayashi, M. Okayama, and Y. Kawana, Highly Reliable High-Voltage Transistors by Use of the SIPOS Process, *IEEE Trans. Electron Devices*, **ED-23**:826 (1976).

CHAPTER 6

Technological Limitations and Protection Techniques in Power ICs

C. Cini
SGS Microelettronica
Milan, Italy

6.1 INTRODUCTION 6.2
6.1.1 Current Limitations 6.3
6.1.2 Limitation of Maximum Current I_{max} in AC and DC Amplifiers 6.5
6.1.3 Limitation of I_{max} in Switching Amplifiers 6.7
6.2 LIMITATION OF POWER DISSIPATION P_{tot} 6.9
6.2.1 DC Amplifier 6.9
6.2.2 AC Amplifier 6.11
6.2.3 Switching Amplifier 6.12
6.2.4 Thermal Resistance 6.14
6.2.5 Thermal Capacitance 6.18
6.2.6 Circuits for Limitation of $T_{j\ max}$ 6.22
6.3 SECOND BREAKDOWN 6.24
6.4 VOLTAGE LIMITING 6.26
6.4.1 Breakdown Voltage 6.28
6.4.2 Overvoltage Protection Circuits 6.29
6.4.3 Grounded-Base Circuits 6.30

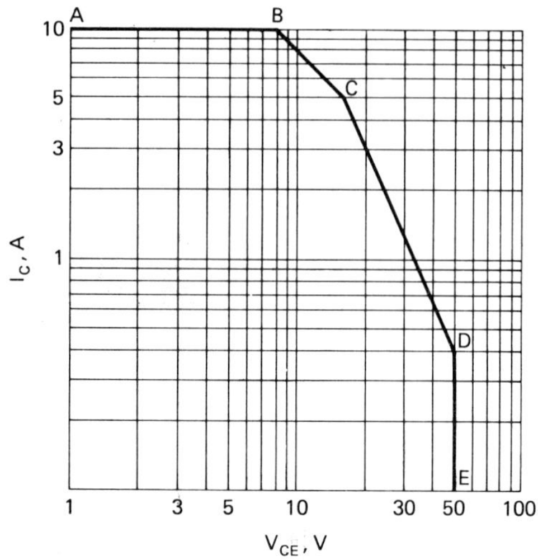

FIG. 6-1 Power transistor safe operating area (SOA).

6.1 INTRODUCTION

Power integrated circuits can supply tens of watts for low-frequency amplifiers and voltage regulators and hundreds of watts for switching amplifiers. Technological progress and circuit solutions have made it possible to increase the current, voltage, and second breakdown limits, while specially designed packages guarantee heat dissipation from the junctions of the power devices toward the ambient. At present, the limitations of power integrated transistors are inhibiting further developments in this direction for a variety of reasons. These limitations are discussed in this chapter, and some possible solutions are outlined.

Figure 6-1 shows the I_C versus V_{CE} plot of a typical power transistor. It is well known experimentally that there is a safe operating area (SOA) inside which all the transistor's operating points must lie (Chap. 2). The limitations which determine this curve can be classified according to the four distinct parts of the polygon which outlines the SOA [1,2], as described in the following subsections. The area enclosed by the curve is the minimum SOA, because it indicates the worst-case dc conditions. When the transistor operates only for short periods, this area can be greatly expanded, since the SOA is an inverse function of the applied energy.

In Secs. 6.2, 6.3, and 6.4 are described respectively the limitations on power dissipation, second breakdown, and voltage in power ICs.

6.1.1 Current Limitations

The limitation upon current is indicated by segment *AB* in Fig. 6-1. I_C must not exceed this value; otherwise the metal connections may be damaged as a result of the Joule effect or electromigration phenomena.

The current density of the different metals used in the construction of the IC has become a standard, codified with values below 10^6 A/cm^2 for aluminum interconnections and 2×10^5 A/cm^2 for 25-μm gold bonding wires.

First consider the metallization connecting the collectors and that

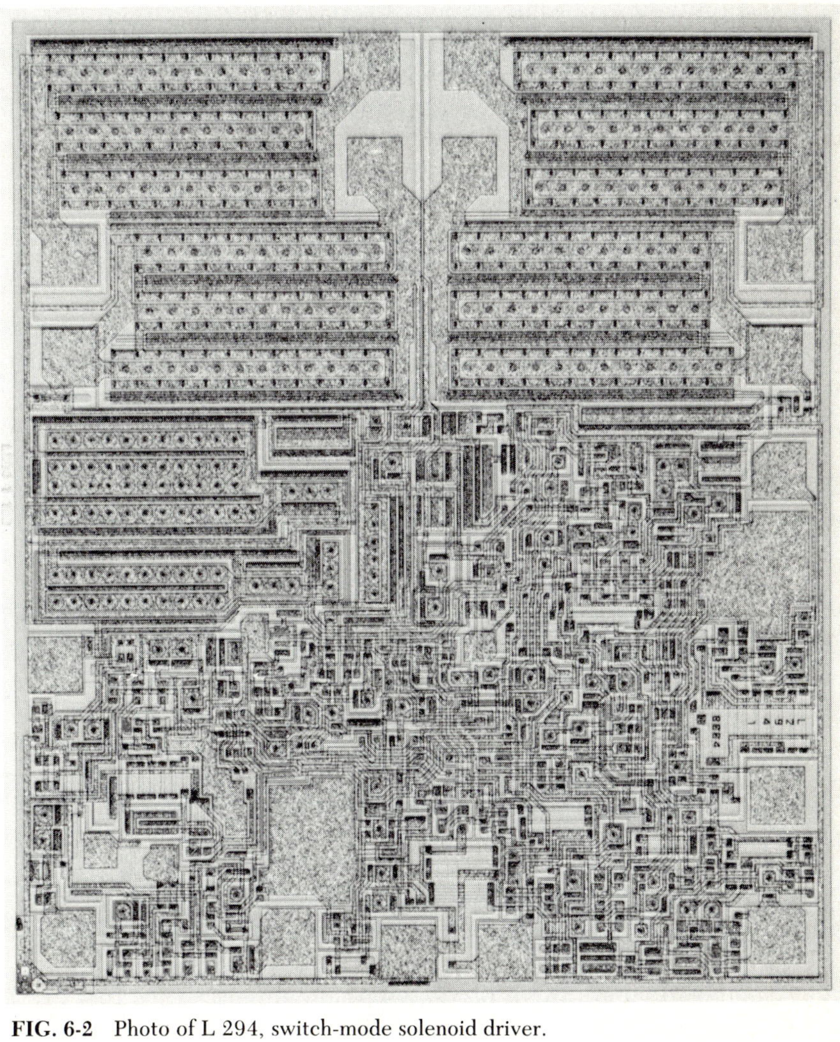

FIG. 6-2 Photo of L 294, switch-mode solenoid driver.

connecting the emitters of the power device. In order to guarantee a mean time between failure (MTBF) of 10^4 h at a temperature of 150°C [3, 4], the metallization must be fairly wide; however this is not a serious limitation, especially with metal thicknesses of 3 μm or more.

Next consider the bonding wires. In multipin ICs the bonding wires are usually gold and have a maximum diameter of 50 μm. In packages such as Versawatt and Pentawatt, which have few pins, aluminum wires with a greater diameter can be used. The 50-μm gold wire can carry a maximum dc current of 2.5 A. When higher currents are required several wires must be connected in parallel.

For a power transistor with $V_{CEO} \geq 50$ V and $I_C \geq 5$ A, a process with current density of 5 mA/mil^2 of emitter area for $h_{FE} = \frac{1}{2}h_{FE\ \text{peak}}$ can be used; this is an index for defining the current-carrying capacity of an IC process. To obtain $I_C = 5$ A, an emitter area of 1000 mil^2 is required; if other requirements, such as the V_{sat}, have to be satisfied and a uniform activation is required, this will lead to transistors with at least six emitter strips and seven collector strips. This transistor structure, which occupies about 4000 mil^2, does not present any problem for the current density on the metal interconnections, since a large emitter area is necessary. Figure 6-2 shows a device where the above mentioned power transistors are used.

To produce the same power device with $V_{CEO} \geq 20$ V, in which case the current density is 20 mA/mil^2, an emitter area a quarter of the size of the area used in the previous transistor is sufficient.

The current density J in the metal, which must be less than or equal to 10^6 A/cm^2, imposes a minimum surface requirement upon the power device apart from that determined by the emitter current density requirements. Therefore, for a high-voltage process ($V_{CEO} \geq 50$ V), device size is a function of the emitter area, while for low-voltage processes ($V_{CEO} \geq 20$ V) it is linked to the current density in the metal.

The power transistor shown in Fig. 6-2 has two pads on the collector and two on the emitter for two bonding wires. The double pads offer two advantages. First, and very important, they enable better activation of the transistor; and second, they permit simple electrical testing to ensure that both bonding wires, necessary to carry the higher current, have been connected.

The bonding wires needed to connect chip and lead are gold wires. The maximum diameter allowed is 50 μm. Larger wires are not advisable since the pressure that needs to be exerted on the silicon to obtain a good bond can cause microcracks in the crystal.

The advantage of gold wires for connecting the chip and leads, in addition to reliability, is that they allow the use of automatic thermosonic bonders, which cannot be used with aluminum wires. It is preferable for economic reasons to use small-diameter wires. Since most of

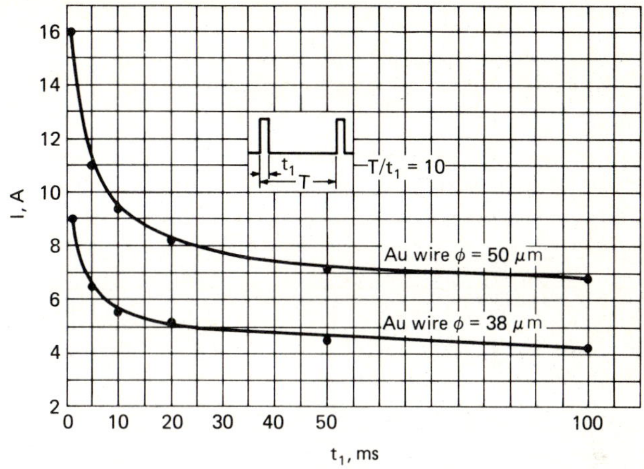

FIG. 6-3 Melting current for gold bonding wires pulse width.

the pads are usually signal connections and all the bonding wires of an IC must be of the same diameter, it is therefore more economical to use multiple pads and bonds for the power section.

The curves in Fig. 6-3 give the melting current of 50-μm and 38-μm gold wires encapsulated in epoxy resin in a power package as a function of pulse width. As can be seen the bonding wire can carry currents almost an order of magnitude greater than the dc value for short periods.

Similar considerations can be made for the metallizations. Melting of the wire and of the metal occurs as a result of the Joule effect, but for short periods the thermal capacity of the metal is great enough to withstand a much greater power. The current peak is limited even for short periods by the transistors, which are dimensioned to guarantee the maximum current.

6.1.2 Limitation of Maximum Current I_{max} in AC and DC Amplifiers

The maximum current which flows in power transistor T_1 (Fig. 6-4) is measured by a resistor R_S, in series with the collector and formed by the interconnection metal (Chap. 9, Sec. 8). The voltage measured goes to a differential amplifier with threshold. When this threshold is reached the current is sinked from the base of the power transistor T_1 by transistor T_3. In dc amplifiers (power op-amp) both power transistors, T_1 and T_2, are protected.

Referring to Fig. 6-4, the value of I_{max} is linked to the value of R_S and to the matching of transistors T_4 and T_5 and of the resistors R_3 and R, which define the level of the current I_1. Matching of the resistors can

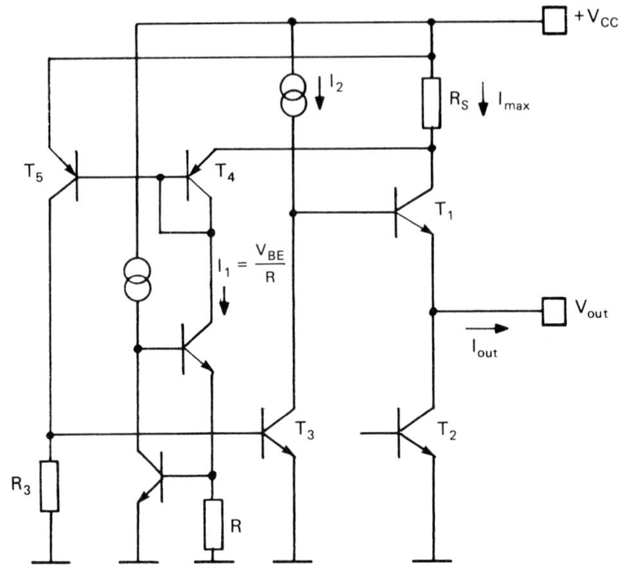

FIG. 6-4 Real circuit for defining the maximum current.

be within 1 percent, while the spread of the R_S value, which is linked to the thickness of the metal, to the resistivity, and to masking, is much greater.

Variations in I_{max} can be expected to be less than ± 20 percent and are determined by the spread of R_S and by mismatching of the transistors T_4 and T_5 and of the resistances R_3 and R. From Fig. 6-4 we can say

$$V_{BE4} + R_S I_{max} - V_{BE5} = 0 \tag{6-1}$$

Developing the equation and introducing the current density per unit emitter area, we get:

$$I_{max} = \frac{1}{R_S}\frac{kT}{q}\ln\left(\frac{I_{T5}}{I_{T4}}\frac{A_4}{A_5}\right) \tag{6-2}$$

where A_4 = area of emitter of T_4
A_5 = area of emitter of T_5

Furthermore, if

$$I_{T4} = I_1 = \frac{V_{BE}}{R}$$

and

$$I_{T5} = \frac{V_{BE}}{R_3}$$

then substituting in Eq. 6-2 gives

$$I_{max} = \frac{1}{R_S}\frac{kT}{q}\ln\left(\frac{A_4}{A_5}\frac{R}{R_3}\right) \tag{6-3}$$

From Eq. 6-3 it can be seen that, since resistor R_S is in series with the collector of the power transistor, it must have a low value to avoid reducing the output dynamic range and at the same time increasing the dissipated power. For these reasons R_S is realized with the metal which connects the collector of the power transistor to the supply pad. The consequence of this choice is that its values can be neither too low, for reasons of precision, nor too high, for the reason already mentioned and because it would require excessive dimensions. In practice the value of this resistor is in the range 10 to 50 mΩ. Greater flexibility is obtained in the value of the resistors R and R_3 and in the emitter area of transistors T_4 and T_5.

We can add that, if T_4 is driven by another current generator which is a function of the V_{CE} voltage of the power transistor T_1, the output current I_{out} is equal to that supplied by T_1 and is an inverse function of its V_{CE} voltage. This function is realized to exploit better the voltage-current characteristics of the power transistors without leaving their SOA.

Differentiating I_{max} with respect to the temperature T, we find that the function is decreasing because the variation in R_S over temperature is .43%/°C, while the variation of kT/q over temperature is .33%/°C; therefore $\partial I_{max}/\partial T = -.1\%/°C$.

6.1.3 Limitation of I_{max} in Switching Amplifiers

Figure 6-5 shows a switching amplifier with a current limitation circuit. I_{max} is measured at the terminals of R_S (Fig. 6-5) and monitored by the offset comparator, where the input transistors are T_1 and T_2. The com-

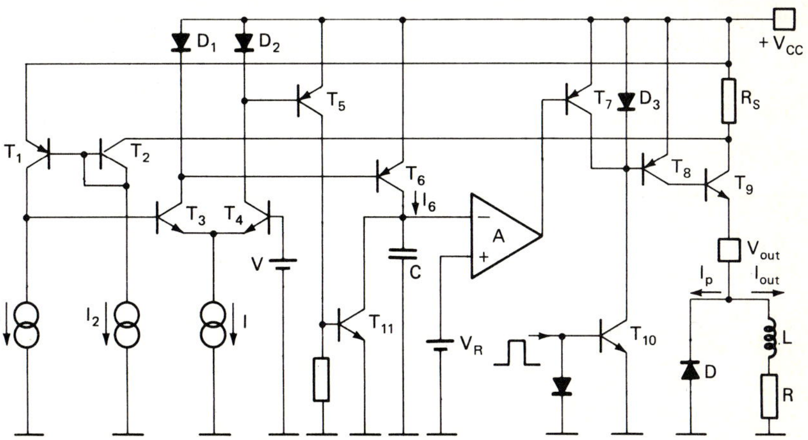

FIG. 6-5 Switching amplifiers with a current limitation circuit.

parator remains inactive until the threshold, which corresponds to the maximum current value required for the power device, is reached.

From Eq. 6-2 and referring to the circuit of Fig. 6-5, we obtain:

$$I_{max}R_S = \frac{kT}{q} \ln \left(\frac{A_2}{A_1} \frac{I_1}{I_2} \right) \tag{6-4}$$

Choosing the ratio of emitter area to the current of transistors T_1 and T_2, since kT/q is known, we can find from Eq. 6-4 the value of the threshold voltage for which the current-limiting circuit intervenes.

When the current flowing in R_S reaches the threshold voltage $V_{thr} = I_{max}R_S$, T_3 turns on and consequently T_6 begins to charge capacitor C; during this phase T_{11} is off. The functioning of T_{11} is necessary because it prevents C from accumulating the charge supplied by T_6 in the preceding cycle.

Power transistor T_9, which drives an inductive load, is driven alternately on and off at a fixed frequency by the input signal at the base of T_{10}.

Let us examine the operation starting from the phase in which T_9 is off; the voltage at the terminals of inductor L reverses polarity, diode D conducts, and the energy previously stored in the inductor is released. When transistor T_9 is on, the diode is reverse-biased, and the current that the transistor must supply is that necessary to charge the inductor (I_{out}) plus the peak current (I_p) necessary to turn off the diode.

The value of the current of T_9 in this phase is $I_{T9} = I_{out} + I_p$, where I_p can have an amplitude and duration very variable, even with the same value of I_{out}, because it is a function of the type of diode used (typically the t_f time is less than 0.5 μs and the amplitude $I_p = 0.3I_{out}$). Since the current in the transistor is much higher than that in the load, it can trigger the current limiter (otherwise the threshold value would have to be so high as to be ineffective for protection).

It is therefore necessary to delay the intervention of the limiting circuit. The delay time must be sufficient to allow the current I_p to turn off diode D, but it cannot be too long, or it will not be effective in avoiding forward second breakdown of the power device.

If the current flowing in T_9 exceeds I_{max} for a time longer than the delay time t_{delay} (e.g., when there is an output short circuit), capacitor C will be charged to voltage V_R, and at this point comparator A will change status, driving T_7, which turns off T_8 and T_9.

The delay time, defined by the following equation, is a function of the current in T_6, the value of C, and the voltage V_R:

$$t_{delay} = \frac{CV_R}{I_6} \tag{6-5}$$

6.2 LIMITATION OF POWER DISSIPATION P_{tot}

The dissipated power is linked to the maximum temperature T_d that the junction can tolerate. Figure 6-1 shows this limitation in the *BC* stretch. The temperature is proportional to $P_{tot} = I_C V_{CE}$ and has a slope of -1 on a log-log scale.

The power available at the output of an amplifier is a function of the type of amplifier (ac, dc, or switching) and especially of the power that can be dissipated.

In low-frequency ac amplifiers and dc amplifiers the output power is limited not by the electrical characteristics of the components, but by the power that can be dissipated, which is a reverse function of the (poor) thermal conductivity of silicon. As of the mid-1980s the maximum power that can be dissipated in operating conditions is 20 W.

In switching amplifiers the situation is much better because it is not the dissipated power that limits the output power, but the characteristics of the power transistor, such as its V_{CEO}, switching time, and immunity to direct and reverse second breakdown.

In the high-voltage high-current process, 400 W can be obtained on unipolar loads and 200 W on bipolar loads. Devices made according to this process have an installed power of 1000 W, given by the voltage-current product. To obtain this output power we need a technology capable of integrating high-voltage high-current power devices, a multilead power package with low thermal resistance, and high-efficiency amplifiers.

For a supposed steady state with dissipated power P_{tot} we can find the power delivered to the load by the different kinds of amplifiers: dc, ac, and switching. For the examples which follow we assume an ambient temperature T_a of 70°C, a junction temperature T_j of 150°C, a total (package + heat sink) thermal resistance $R_{th,ja}$ of 4°C/W. In this case we can find from Eq. 6-34 P_{tot}, which is equal to 20 W.

6.2.1 DC Amplifier

The transfer efficiency, defined as the ratio between the power on the load and the power supplied by the supply, increases linearly as a function of the output voltage. Its limits are fixed mainly by the losses due to saturation of the output transistors [5].

For the application of Fig. 6-6, the maximum dissipated power is obtained when the output voltage is

$$V_{out} = \frac{V_{CC}}{2} \tag{6-6}$$

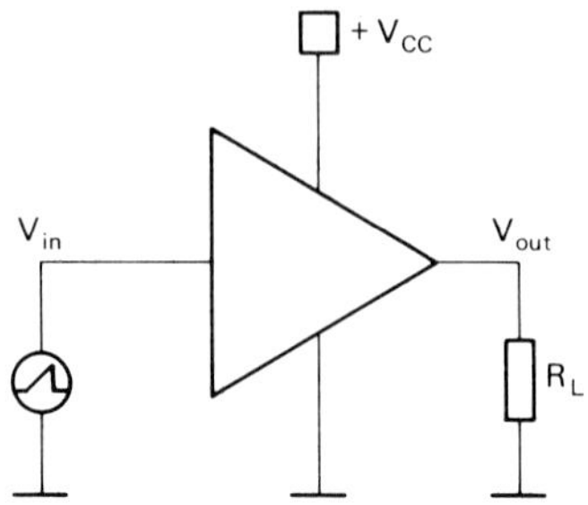

FIG. 6-6 Schematic of a dc amplifier.

From this we get

$$P_{\text{tot,max}} = \frac{V_{CC}^2}{4R_L} \tag{6-7}$$

If saturation losses were zero ($V_{CE\ \text{sat}} = 0$), the maximum voltage on the load would be

$$V_{\text{out}} = V_{CC} \tag{6-8}$$

and the power would be

$$P_{\text{out,max}} = \frac{V_{CC}^2}{R_L} \tag{6-9}$$

therefore the ratio $P_{\text{out,max}}/P_{\text{tot,max}}$ would be 4, and

$$P_{\text{out,max}} = 4P_{\text{tot,max}} \tag{6-10}$$

where $P_{\text{out,max}}$ and $P_{\text{tot,max}}$ are respectively the maximum output power and the maximum dissipated power.

In practice there are limitations that must be taken into consideration. First, the saturation voltage causes the loss of about 10 percent of $P_{\text{out,max}}$. Another important limitation is the supply voltage, which has a precision with respect to V_{out} of at least ± 10 percent, due to tolerances and poor load regulation.

In calculating the dissipated power $P_{\text{tot,max}}$, we have to find the worst case, which corresponds to $V_{CC\ \text{max}}$; while for the output power P_{out}, it will be the opposite, corresponding to $V_{CC\ \text{min}}$.

The voltage difference from $V_{CC\ \text{max}}$ to $V_{CC\ \text{min}}$ is 20 percent, while the power on the load will be reduced, due to this effect, by 40 percent (the power is a function of V_{CC}^2); to this we must add the loss due to $V_{CE\ \text{sat}}$ ($\approx$ 5 percent, or 10 percent of $P_{\text{out,max}}$).

The useful power on the load will therefore be reduced by 50 percent with respect to $P_{out,max}$:

$$P_{out} = 0.5P_{out,max} \tag{6-11}$$

We can therefore conclude from the previous example that in a dc amplifier the power that can be supplied to the load is twice the maximum dissipated power:

$$P_{out} = 2P_{tot} = 40\text{ W} \tag{6-12}$$

6.2.2 AC Amplifier

Figure 6-7 shows an ac amplifier; the transfer efficiency of this amplifier is worse than that of the dc amplifier in Sec. 6.2.1.

The maximum dissipated power (Chap. 9) corresponds to an output voltage of

$$V_{out\ pp} = \frac{2V_{CC}}{\pi} \tag{6-13}$$

Therefore

$$P_{tot,max} = \frac{(V_{out\ pp}/2\sqrt{2})^2}{R_L} = \frac{V_{CC}^2}{R_L}\frac{1}{2\pi^2} \tag{6-14}$$

The maximum output power, with no losses due to saturation of the power devices (i.e., with $V_{out} = V_{CC}$), will be

$$P_{out,max} = \left(\frac{V_{CC}}{2\sqrt{2}}\right)^2 \frac{1}{R_L} = \frac{V_{CC}^2}{8R_L} \tag{6-15}$$

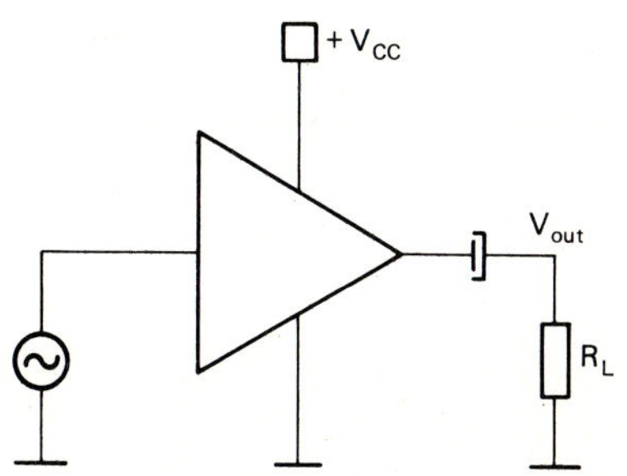

FIG. 6-7 Schematic of an ac amplifier.

The ratio between $P_{out,max}$ and $P_{tot,max}$ is

$$\frac{P_{out,max}}{P_{tot,max}} = \frac{\pi^2}{4} \tag{6-16}$$

or

$$P_{out,max} = \frac{\pi^2}{4} P_{tot,max} \tag{6-17}$$

In this case, too, considering what we said before about losses due to $V_{CE\ sat}$ and to the variation in supply voltage, P_{out} is 50 percent of the $P_{out,max}$:

$$P_{out} = 0.5P_{out,max} \tag{6-18}$$

By substituting Eq. 6-18 into Eq. 6-17, we obtain

$$P_{out} = \frac{\pi^2}{4} \times 0.5P_{tot,max} \cong 1.23P_{tot,max} \tag{6-19}$$

We can say, from Eq. 6-19, that the output power in ac amplifiers is only 23 percent higher than the power dissipated. Fortunately, in audio applications the power to be supplied to the load is usually very low and it is only for short periods that a much higher power (musical power) will be required.

From the results obtained and from these considerations we can show:

1. The $P_{tot,max}$, indicated in Eq. 6-19, is the power that can be dissipated in the steady-state condition. In our example $P_{tot,max}$ is equal to 20 W and $P_{out} \cong 25$ W.
2. The power which can be dissipated for brief periods, i.e., before the junction temperature T_j exceeds the normal operational limit, is much higher.
3. As a consequence of this, the output power, being a function of the dissipated power, is for short periods greater than the steady-state power. Thanks to this, it is possible to obtain, even with integrated circuits, useful powers of several tens of watts in the load.

6.2.3 Switching Amplifier

In switching amplifiers the transfer efficiency is very high; ideally, with saturation voltage equal to zero, there is no dissipated power (Fig. 6-

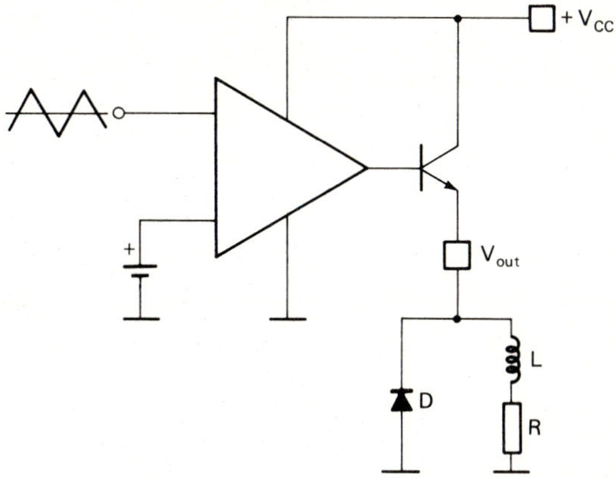

FIG. 6-8 Schematic of a switching amplifier.

8). The transistors which transfer power from the supply to the load operate alternatively on and off, while the current is maintained constant by the inductive load (motor, solenoid, electromagnet, etc.)

The efficiency is affected not only by V_{sat}, but also by switching losses due to the time taken by the power devices to switch completely on or off, to the current in the diode needed to switch it off, and to the diode's direct voltage V_D.

We will attempt to evaluate the theoretical efficiency of the output stage of these amplifiers; from this, and from the power dissipated by the package in which the IC is assembled, it is possible to find the power transferred to the load.

The total power loss (i.e., dissipated power) is given by the sum of the power dissipated in saturation and the power dissipated during switching. In saturation, if $t_{on} = t_{off}$ (50 percent duty cycle), we have

$$P_{tot,sat} = \frac{V_{sat} + V_D}{2} I_{out} \tag{6-20}$$

In switching, what determines the dissipation is the rise and fall times t_r and t_f (see Chap. 2, Sec. 2.5) of the transistor in comparison with the switching frequency f:

$$P_{tot,sw} = (t_r + t_f) f I_{out} V_{CC} \tag{6-21}$$

The total power dissipated is

$$P_{tot} = P_{tot,sat} + P_{tot,sw} = I_{out}\left[\frac{V_{sat} + V_D}{2} + (t_r + t_f) f V_{CC}\right] \tag{6-22}$$

The output power P_{out} is given by the load current I_{out}, multiplied by the output voltage $V_{out} = V_{CC} - V_{sat}$:

$$P_{out} = I_{out}\,(V_{CC} - V_{sat}) \qquad (6\text{-}23)$$

The efficiency η, using Eqs. 6-22 and 6-23, will be

$$\eta = \frac{P_{out}}{P_{out} + P_{tot}} = \frac{V_{CC} - V_{sat}}{V_{CC} - V_{sat} + (V_{sat} + V_D)/2 + (t_r + t_f)fV_{CC}} \qquad (6\text{-}24)$$

Giving numerical values for Eq. 6-24, we find, for $V_{CC} = 50$ V, $I_{out} = 4$ A, $V_{sat} = 2$ V, $t_f + t_r = 0.2\ \mu$s, and $f = 100$ kHz, that the efficiency η is 95 percent.

From Eq. 6-24 we can obtain the output power P_{out}:

$$P_{out} = \frac{\eta}{1 - \eta}\,P_{tot} = 19P_{tot} \qquad (6\text{-}25)$$

We can conclude that in switching amplifiers, since the efficiency is very high ($\eta = 95$ percent), it is possible to transfer to the load 19 times the power which can be dissipated, or $19P_{tot}$. When $P_{tot} = 20$ W, we therefore obtain a P_{out} of 380 W.

6.2.4 Thermal Resistance

The electrical energy *VIt*, applied to the junction of the power transistor is transformed into thermal energy. To prevent excessive temperature increase the heat must be transferred from the junction to the package and then to the ambient.

This transfer occurs through the thermal resistance. Therefore the lower this resistance, the greater will be the power that can be dissipated at the same junction temperature. Fourier's law of heat conduction, assuming unidirectional heat flow, gives the equation

$$P_{tot} = \sigma_{th}\,A\,\frac{dT}{dx} \qquad (6\text{-}26)$$

Figures 6-9 and 6-10 show a simplified structure of the heat flow path. The *heat source* is at the upper surface, while the lower surface is at ambient temperature; the heat will flow from the higher to the lower temperature. In the equations in this subsection, the dissipated power P represents the heat source, dT is the differential temperature between the source (T_j) and the ambient (T_a), dx is the differential length along the heat flow path, σ_{th} is the thermal conductivity of the material, and A is the cross-sectional area of the thermal path.

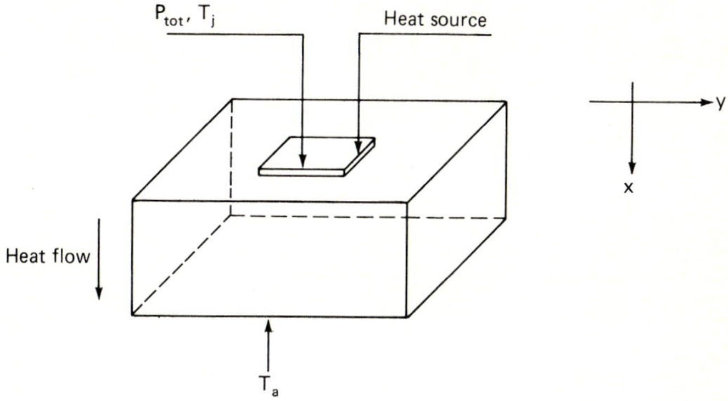

FIG. 6-9 Heat flow path.

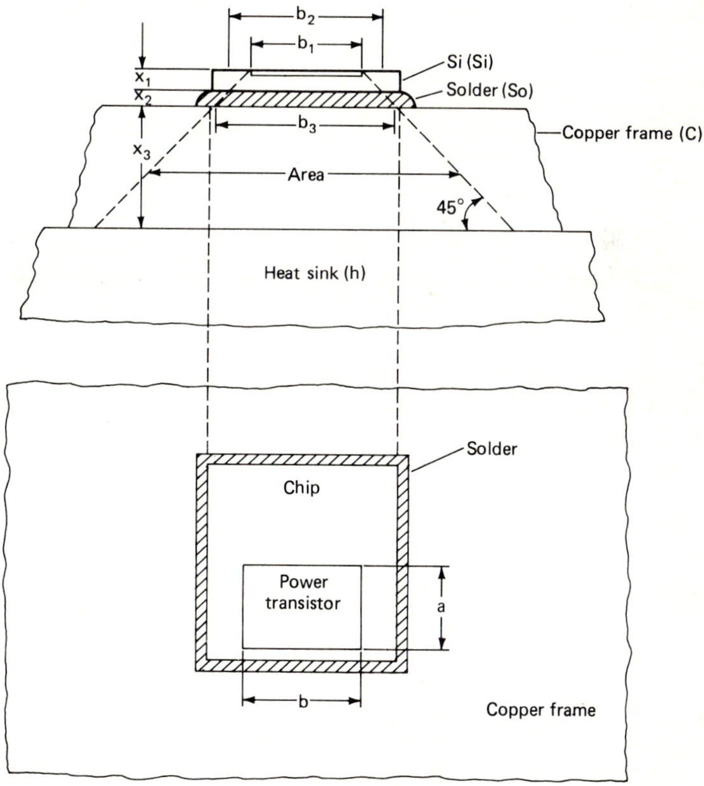

FIG. 6-10 (*a*) Top view and (*b*) cross section of a PIC with case and heat sink.

Integrating Eq. 6-26, we obtain

$$P_{tot} \frac{x}{\sigma_{th} A} = \Delta T \tag{6-27}$$

where the term $x/\sigma_{th} A = R_{th}$ is the thermal resistance. The following equation relates the temperature gradient ΔT to the dissipated power P and the thermal resistance R_{th}:

$$\Delta T = R_{th} P_{tot} \tag{6-28}$$

The model considered in Fig. 6-10 consists of a silicon chip soldered with soft solder onto a copper support (the package), which in turn is normally attached to a heat sink, to facilitate the exchange of heat between package and ambient.

To reach a simple equation to calculate the thermal resistance R_{th} of the various materials, we make the following assumptions:

1. The heat is propagated from the inside to the outside at an angle of 45° (Fig. 6-10).
2. Propagation is only by conduction (convection and irradiation are not considered, as they are less important by orders of magnitude).
3. The heat source is rectangular, with sides a and b and with uniform temperature.

The thermal resistance will be lower if the area of the heat source is increased, if the length of the heat path inside the package is shorter, or if the thermal conductivity of the metals in the heat flow path is higher.

Silicon is a given element; we can either increase its surface, in which case cost problems will be encountered, or we can reduce the thickness; however, for 4-in or 5-in wafers this creates handling problems due to fragility.

The solder must act as a cushion between the silicon and the package. Because of its more favorable thermal expansion coefficient, soft solder must be used; however its influence on the thermal resistance will be almost negligible.

The package plays a fundamental role in determining the thermal resistance value. Consequently, the use of high-conductivity metals, such as copper, seems to be indispensable [6–8].

The thermal resistance of a truncated pyramid will be

$$R_{th} = \frac{1}{\sigma_{th}} \int_0^x \frac{dx}{A} \tag{6-29}$$

For a rectangular heat source with sides a and b, neglecting the variation of σ_{th} with temperature we obtain:

$$R_{th} = \frac{1}{2\sigma_{th}(b_1 - a_1)} \ln\left(\frac{b_1}{a_1} \frac{2x + a_1}{2x + b_1}\right) \qquad (6\text{-}30)$$

With reference to Fig. 6-10 we have

$$\begin{aligned} R_{th,\,jc} &= \frac{1}{2\sigma_{th,Si}(b_1 - a_1)} \ln\left(\frac{b_1}{a_1} \frac{2x_1 + a_1}{2x_1 + b_1}\right) \\ &+ \frac{1}{2\sigma_{th,So}(b_2 - a_2)} \ln\left(\frac{b_2}{a_2} \frac{2x_2 + a_2}{2x_2 + b_2}\right) \\ &+ \frac{1}{2\sigma_{th,c}(b_3 - a_3)} \ln\left(\frac{b_3}{a_3} \frac{2x_3 + a_3}{2x_3 + b_3}\right) \end{aligned} \qquad (6\text{-}31)$$

The three terms of Eq. 6-31 indicate the thermal resistance of the silicon (Si), the solder (So), and the case (c) respectively; the sum of the three terms represents the thermal resistance between the silicon die (or junction) and the external package of the device (or case), $R_{th,\,jc}$.

As an example we will calculate the thermal resistance of a device with an active area of 2000 mil^2 (Fig. 6-10) and the dimensions indicated below:

$$\begin{aligned} b_1 &= 1600\ \mu m & a_1 &= 810\ \mu m \\ b_2 &= 2200\ \mu m & a_2 &= 1410\ \mu m \\ b_3 &= 2230\ \mu m & a_3 &= 1440\ \mu m \\ x_1 &= 300\ \mu m & \sigma_{th,Si} &= 1.5\ W/(cm \cdot {}^\circ C) \\ x_2 &= 15\ \mu m & \sigma_{th,So} &= 0.36\ W/(cm \cdot {}^\circ C) \\ x_3 &= 1500\ \mu m & \sigma_{th,c} &= 4\ W/(cm \cdot . {}^\circ C) \end{aligned}$$

The approximate values given in Fig. 6-11 are then found:

$$R_{th,\,jc} = R_{th,Si} + R_{th,So} + R_{th,c} = 1.5^\circ C/W \qquad (6\text{-}32)$$

The power that can be dissipated with an infinite heat sink—i.e., with thermal resistance from case to ambient equal to zero ($R_{th,ca}$ = 0)—and T_j = 150°C and T_a = 25°C is

$$P_{tot} = \frac{T_j - T_c}{R_{th,\,jc}} = \frac{150 - 25}{1.5} = 83\ W \qquad (6\text{-}33)$$

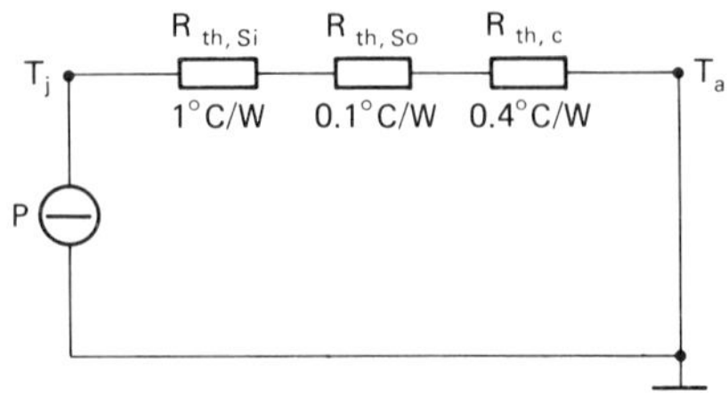

FIG. 6-11 Thermal equivalent resistance, junction to ambient.

This power, however, is reduced drastically if more realistic assumptions are made, e.g., an external heat sink with $R_{\text{th},ca} = 2.5°\text{C/W}$ (as opposed to $R_{\text{th},ca} = 0$) and an ambient temperature $T_a = 70°\text{C}$ (as opposed to 25°C). With these new assumptions we obtain

$$P_{\text{tot}} = \frac{T_j - T_a}{R_{\text{th},jc} + R_{\text{th},ca}} = \frac{80}{1.5 + 2.5} = 20 \text{ W} \tag{6-34}$$

It should be noted that the case–heat sink thermal resistance $R_{\text{th},ch}$ is considered zero. The substrate of an integrated circuit is soldered to the case.

6.2.5 Thermal Capacitance

A very high dissipated power cannot safely be handled in steady-state fashion, but thermal capacitance can cause the dissipation intervals to be short enough that the device will not be harmed [9–10].

The thermal circuit shown in Fig. 6-11 is considered to be in steady state when there is a continuous source of heat. As heat is first applied, however, the temperature gradient $\Delta T = T_j - T_a$ will increase as shown in Fig. 6-12 from t_0 to t_1. This shows that there is a finite time required

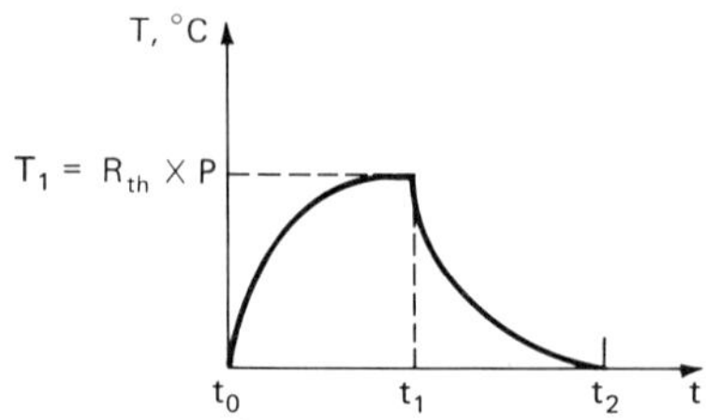

FIG. 6-12 Transient junction temperature.

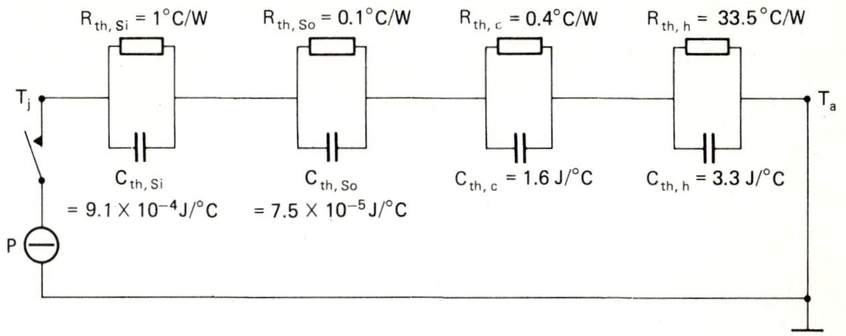

FIG. 6-13 Thermal equivalent circuit, junction to ambient.

to reach steady-state thermal gradients. The rate of differential temperature increase is identical to that of voltage buildup in an *RC* network.

The thermal capacitance value is needed to determine the rate at which equilibrium is achieved. If the power is cut off at t_1, then from t_1 to t_2 the junction temperature returns to its original value.

This kind of device behavior, in which the thermal circuit does not reach equilibrium, is ideal for many applications of switching amplifiers in which they are used to drive a motor or to open or close electrically operated valves. In these cases the operation time is on the order of milliseconds and is followed by long periods of inactivity. It is therefore important to know the thermal resistance and the thermal capacitance, since, depending on the application, the external heat sink may not be required, although the power dissipated momentarily is very high.

The curve in Fig. 6-12 is similar to that of the charge-discharge cycle of an *RC* network; to represent the equivalent circuit, the thermal circuit of Fig. 6-11 is modified by adding to each thermal resistor its thermal capacitance, which is a function of the metal and of its dimensions.

Figure 6-13 shows the thermal circuit of a power package (Multiwatt®) without external heat sink. The thermal capacitance of each element is given by

$$C_{\mathrm{th}} = VP_SC_S \qquad \mathrm{J/°C} \tag{6-35}$$

where V = volume, P_S = specific weight, and C_S = specific heat are referred to different materials such as silicon, solder, and copper.

The junction temperature T_j is a function of the ambient temperature T_a, of the power P_{tot}, and of the thermal impedance Z_{th}:

$$T_j = T_a + P_{\mathrm{tot}}Z_{\mathrm{th}} \tag{6-36}$$

where

$$Z_{\text{th}} = Z_{\text{th,Si}} + Z_{\text{th,So}} + Z_{\text{th},c} + Z_{\text{th},h} \tag{6-37}$$

We can then write

$$\begin{aligned} T_j = T_a + P_{\text{tot}}[R_{\text{th,Si}}(1 - e^{-t/\tau_{\text{Si}}}) \\ + R_{\text{th,So}}(1 - e^{-t/\tau_{\text{So}}}) + R_{\text{th},c}(1 - e^{-t/\tau_c}) + R_{\text{th},h}(1 - e^{-t/\tau_h})] \end{aligned} \tag{6-38}$$

In general, with K cells we can say that the junction temperature variation ΔT_j is

$$\Delta T_j = \Delta P \sum_{n=1}^{K} R_{\text{th},n}(1 - e^{-t/\tau_n}) \tag{6-39}$$

where τ is the product of the thermal resistance and the thermal capacitance:

$$\tau = R_{\text{th}} C_{\text{th}} \tag{6-40}$$

The greater the value of τ, the more slowly the equilibrium value is reached.

For the model in Fig. 6-13 we have

$$\tau_{\text{So}} < \tau_{\text{Si}} < \tau_c < \tau_h$$

With good approximation we can consider that the time constants of the silicon and of the solder are independent of their size. The thermal resistance is an inverse function of the silicon area in which the power is dissipated, while the thermal capacitance increases with the same silicon area, so that the product is constant. For silicon we have

$$R_{\text{th,Si}} = \frac{1}{A} \alpha \tag{6-41}$$

$$C_{\text{th,Si}} = A\beta \tag{6-42}$$

and therefore

$$\tau = R_{\text{th,Si}} C_{\text{th,Si}} = \alpha\beta \tag{6-43}$$

where α and β are constants.

The situation is different for the package and the heat sink because the metals used are not always the same and their geometries vary in form and dimensions, with corresponding variation in thermal parameters.

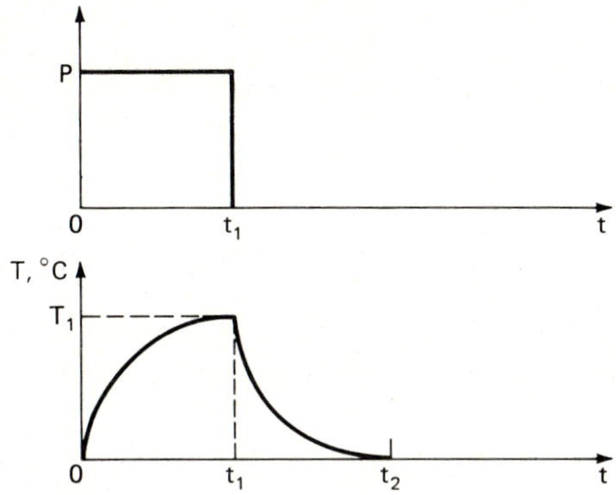

FIG. 6-14 Rectangular power pulse and temperature variation.

Let us suppose we have a rectangular power pulse (Fig. 6-14) with duration t_1 and a single cell $R_{th}C_{th}$. In this case the temperature increase is

$$\Delta T = \Delta P_{tot} R_{th}(1 - e^{-t/\tau}) \tag{6-44}$$

for $0 \leq t \leq t_1$, while it will be

$$\Delta T = \Delta P_{tot} R_{th}[e^{-(t-t_1)/\tau}(1 - e^{-t/\tau})] \tag{6-45}$$

for $t \geq t_1$.

Let us suppose that we want to know the value of T_j when a power pulse of 50 W is applied for a time $t_1 = 100$ ms to a chip assembled in a Multiwatt® power package without external heat sink. Knowing that $\tau_{Si} = 0.91 \times 10^{-3}$ s and $\tau_{So} = 7.5 \times 10^{-6}$ s, we can say that in time t_1

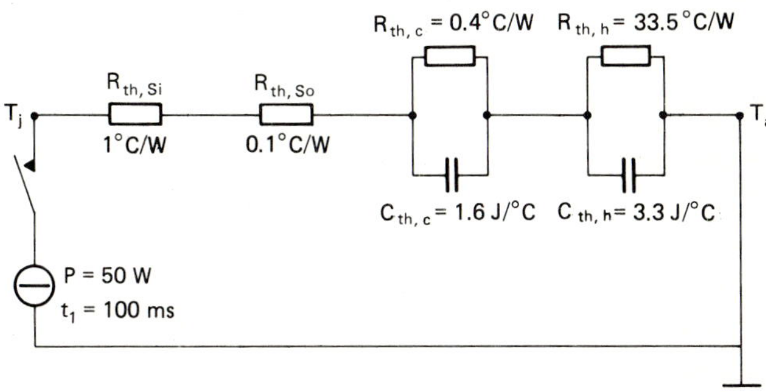

FIG. 6-15 Thermal equivalent circuit, junction to ambient, with a power pulse of $t_1 = 100$ ms.

these elements reach thermal equilibrium; the network that expresses this condition is indicated in Fig. 6-15.

From Eq. (6-38) we get

$$\begin{aligned} T_j &= T_a + P_{\text{tot}}[(R_{\text{th,Si}} + R_{\text{th,So}}) + R_{\text{th},c}(1 - e^{-t/\tau P}) + R_{\text{th},h}(1 - e^{-t/\tau h})] \\ &= T_a + 50(1.1 + 0.06 + 0.03) = T_a + 60°\text{C} \end{aligned} \tag{6-46}$$

Therefore the junction temperature T_j increases by 60°C.

We can conclude that in steady-state amplifiers the power is limited first by the dissipation capacity of the silicon and then by that of the package.

In ac amplifiers a power of 25 W (at clipping) can be transferred to the load since 20 W can be dissipated; in amplifiers where $t_{\text{on}} \ll t_{\text{off}}$ and t_{on} is in tens of milliseconds (e.g., in switching amplifiers), 50 W can be dissipated without external heat sink, while up to 1000 VA can be supplied to the load.

6.2.6 Circuits for Limitation of $T_{j\ \max}$

The junction temperature must not exceed $T_{j\ \max}$. If P_{tot} exceeds the values allowed, or if the heat sink is not sufficient, the temperature may go beyond $T_{j\ \max}$. To avoid this, thermal sensors have been introduced to measure the temperature and reduce P_{tot} or, in switching amplifiers, to switch off the device and, after the temperature has fallen, to switch it on again.

a. Thermal Protection with T_j Constant

With reference to Fig. 6-16, T_3 and T_4 are physically close to power devices T_1 and T_2; when T_3 and T_4 switch from Off to On, the effect of their temperature reduces the current in power transistor T_1. The circuit reduces P_{tot} and P_{out} to maintain T_j at a constant level, and the device remains in operating condition, but with greatly reduced output power. At a temperature of 25°C the voltage at point A should be $V_A = V_{BE3} + V_{BE4} = 1.2$ V to absorb the current I_G.

The thermal drift of $V_{BE3} + V_{BE4}$ has a value of $2\partial V_{BE}/\partial T = -4$ mV/°C, while the temperature step is

$$\Delta T = T_{j\ \max} - T_a = 125°\text{C} \tag{6-47}$$

and hence

$$\Delta V_{BE} = \Delta T \frac{2\partial V_{BE}}{\partial T} = 500 \text{ mV} \tag{6-48}$$

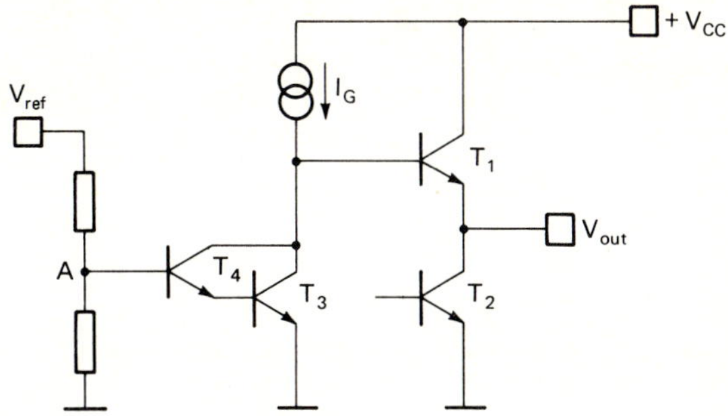

FIG. 6-16 Thermal shutdown circuit.

Therefore to make the protection intervene at $T_j = 150°C$, the voltage at point A must be

$$V_A = (V_{BE3} + V_{BE4})_{25°C} - \Delta V_{BE} = 700 \text{ mV} \tag{6-49}$$

The current variation ΔI_C for a temperature variation ΔT_j in the operating region of the transistor is found from the equation

$$\frac{\Delta I_C}{\Delta T_j} = \frac{\Delta I_{out}}{\Delta T_j} \tag{6-50}$$

The output power for this current variation will vary quickly (experimental results show about 35%/°C). Only a few degrees of junction temperature variation will lead to a drastic reduction in P_{out} and P_{tot}.

b. Thermal Protection with Hysteresis

In switching amplifiers a slight reduction of the driving current cannot be tolerated because the power transistors would no longer be able to saturate during the On phase; consequently the transistors would leave the safe operating area (SOA) and break down. The hysteresis thermal protection circuit solves this problem. Figure 6-17 shows this circuit.

Transistor T_5, the thermal sensor, is biased by a voltage divider at a voltage V_A ($V_A \approx 400$ mV) so that it is cut off at junction temperatures below 150°C. When the junction temperature reaches 150°C, the voltage V_B is:

$$V_B = V_{ref} - I_5 R_3 < V_C \tag{6-51}$$

In this condition the comparator, with the input transistors T_6 and T_7, switches and drives transistor T_4, which cuts off the power transistor T_1.

Transistor T_6 has two collectors because it performs two functions. The first collector drives the base of T_4; the second provides the hys-

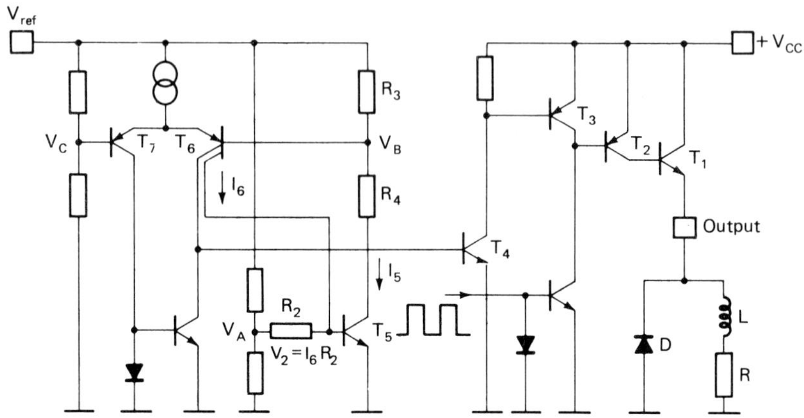

FIG. 6-17 Thermal shutdown circuit with hysteresis.

teresis voltage V_2, which is given by the product of the current I_6 and the resistance R_2, or $V_2 = I_6R_2$.

The voltage V_2 keeps T_5 on even if the output transistor is cut off and the junction temperature T_j falls below the 150°C threshold level. During this cooling phase, when the temperature change ΔT_j is greater than or equal to the hysteresis voltage V_2 divided by the temperature variation of the V_{BE} of T_5 ($\Delta V_{BE}/\Delta T$), the condition $V_C < V_B$ is reached and the comparator changes state, the protection circuit turns off, and the output stage begins to operate again.

The thermal hysteresis is:

$$\Delta T_j \geq \frac{V_2}{\Delta V_{BE}/\Delta T} = \frac{I_6R_2}{2\ \text{mV/°C}} \tag{6-52}$$

6.3 SECOND BREAKDOWN

In the plot shown in Fig. 6-1 segment *CD* represents the phenomenon of second breakdown [11]; the slope of this segment is steeper than -1, and therefore it indicates a further limitation with respect to P_{tot} (Chap. 2, Sec. 2.4). The direct second breakdown is a function of:

1. Duration of the power pulse. Direct second breakdown increases as the pulse time decreases.
2. Thickness and resistivity of the base. Direct second breakdown increases as the thickness increases and as the resistivity decreases.
3. Transistor geometry.

4. V_{CE}. This voltage, reverse-biasing the base-collector junction, causes a reduction in the thickness of the base.

When the load is inductive and the power transistor works in the switch mode, conditions can arise in which the working collector-emitter voltage V_{CE} is roughly equal to the supply voltage V_{CC}; in this case the current the power transistor must supply at switch-on is that of the load plus that required to switch off the recirculation diode D (Fig. 6-5) or $I_C = I_{out} + I_D$; during this phase—switching from Off to On—the voltage and current of the transistor are at a maximum and the transistor can reach a forward second breakdown condition (I_{SB}).

During the On-Off phase we still have all the V_{CC} at the terminals of the transistor, with current $I_C = I_{out}$ and therefore lower than in the previous case (now there is no I_D), but with fall time longer than rise time ($t_f > t_r$). To reduce t_f, current is removed from the base, and the power device in this case is stressed by the reverse second breakdown (E_{SB}), which in terms of energy is much less than the direct second breakdown.

To improve the direct second breakdown I_{SB}, structures using ballast resistors distributed in series to the emitter and the base have been proposed [12, 13]. The best structures for reverse second breakdown E_{SB} are those with high interdigitation, with a large emitter area and low R_{bb}.

Figures 6-18 and 6-19 show the I_{SB} and E_{SB} curves of the power device shown in Fig. 6-2. The process used for this power transistor has

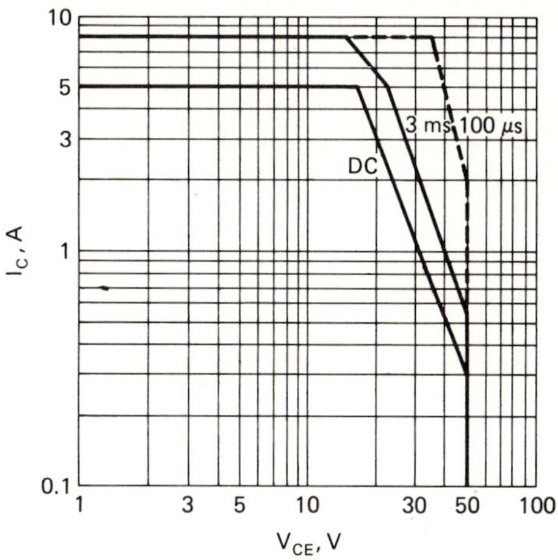

FIG. 6-18 Forward safe operating area for an NPN power transistor of the integrated circuit shown in FIG. 6-2.

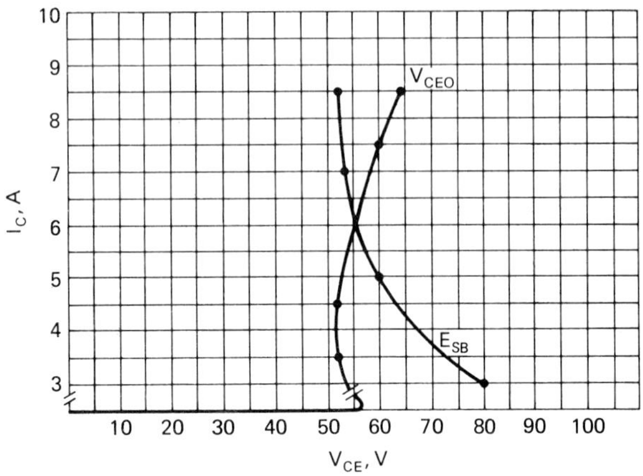

FIG. 6-19 V_{CEO} and E_{SB} limitation of a power transistor.

$BV_{CEO} > 50$ V, and as can be seen, there is a good trade-off between I_{SB} and E_{SB}.

6.4 VOLTAGE LIMITING

The maximum voltage which can be applied to a device must not exceed the breakdown voltage of the junction, as indicated in Fig. 6-1 by the segment *DE*.

The limits of segment *DE* lie between the breakdown voltage of the collector-emitter junction with open-circuited base, BV_{CEO}, which is the minimum, and the breakdown voltage of the N (collector) junction with the substrate or isolation, BV_{CS}, which is the highest value. The ratio of these values (BV_{CS}/BV_{CEO}) is in the range 1.5 to 3.

In some applications, e.g., automotive, the nominal operating voltage is within well-known limits (i.e., 6 to 16 V), but for brief periods ($t = 100$ ms) it may accidentally reach levels from 3 to 10 times the nominal value. For these applications it is preferable that the IC be capable of withstanding the overvoltages rather than that external devices be added for transient-voltage protection. There is, however, no universally agreed-upon standard.

For signal processing and control circuits, where the current consumption is low (on the order of 100 mA or less), it is preferable to use the more costly but necessary solution of adopting special voltage regulators to guarantee correct operation. These regulators, fabricated using high-voltage processes, are capable of guaranteeing a nominal output voltage even in the presence of overvoltages at the input. In power circuits this approach is impractical because the currents are very high (2 to 6 A) and it is too costly and inefficient to filter the overvoltages with electronic systems.

The use of high-voltage processes is also impractical because the silicon area of a power IC is proportional to the square of the voltage, and therefore the cost would be prohibitive.

Power devices are connected directly to the power supply (battery) and are fabricated with low-voltage processes. Functions are added to measure the supply voltage; when this exceeds a preset value, the circuit is inhibited.

When the circuit is inhibited, it is placed in a condition that permits it to withstand the voltage transient without damage; normal operation is restored when the supply voltage V_{CC} returns within given limits. The use of a control circuit operating at low voltage, e.g., 5 V, and always active, with a power circuit turned off and left in standby during the transient, has become a de facto standard in this sector.

In power ICs for industrial applications where the nominal voltage is less than or equal to 50 V (and consequently where there are no overvoltages), different solutions have been adopted. These devices perform interface functions between microprocessors and power transducers (i.e., motors, solenoids, or electromagnets). They contain a small (in terms of silicon area) section for signal processing functions, digital or analog, which operates at a low voltage. The high-voltage section is the most important and consists of the final power stages.

The division of the functions in one of these interfaces is organized as shown in Fig. 6-20. Output power transistors, drivers, some protection circuits, and the voltage regulator are supplied by V_{CC}, whereas the logic and analog control circuits, the reference voltage, and the thermal

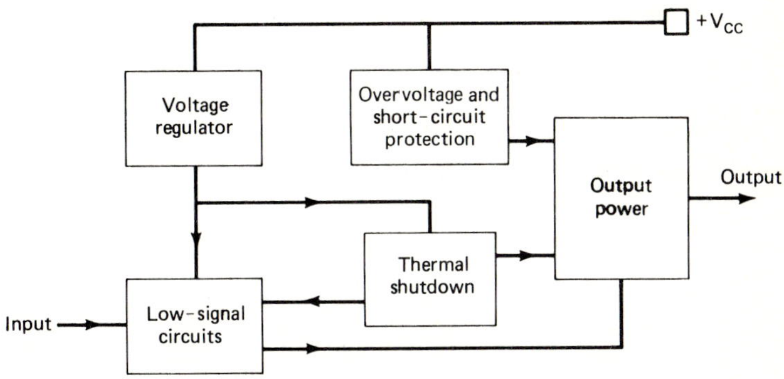

FIG. 6-20 Block division of a power device.

and short-circuit protection are supplied by an internal voltage regulator which supplies a voltage that can vary from 3 to 10 V.

The choice of a single IC containing both signal processing circuits working at low voltage and the larger (in area) power section working at high voltage makes the IC more interesting than a two-chip solution because it is simpler to use, more compact, and more economical.

6.4.1 Breakdown Voltage

IC transistors, like discrete transistors, have breakdown voltages which, for a given process, are a function of the junction considered.

The first limit, since it is the lowest, is BV_{CEO}, which indicates the junction breakdown voltage of the collector-emitter junction with the base open. The BV_{CEO} is a fundamental parameter used for the characterization of a process.

The second limit encountered is the collector-base breakdown voltage with emitter open-circuited, BV_{CBO}. This value is greater than the BV_{CEO}; the two are linked by the expression:

$$BV_{CBO} = BV_{CEO} \sqrt[n]{h_{FE}} \tag{6-53}$$

where n is a function of the technology. For linear ICs $n = 5$.

The third limit, greater than BV_{CBO}, is the breakdown voltage of the collector toward the substrate or insulation, BV_{CS}.

A breakdown voltage between BV_{CEO} and BV_{CBO} is BV_{CER}, which varies as a function of R, where R is a resistor connected between base and emitter. When $R = 0$, we have BV_{CES} (instead of BV_{CER}) equal to BV_{CBO}, while with $R = \infty$, we get BV_{CEO} for BV_{CER}.

Figure 6-21 shows the different breakdown voltages in an IC.

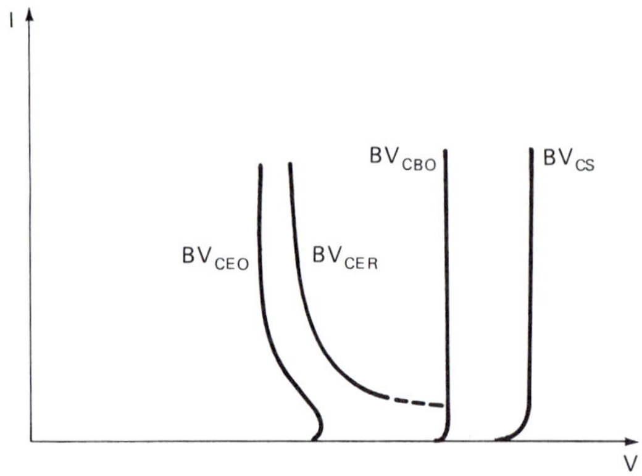

FIG. 6-21 Different breakdown voltages and their relations in an IC transistor.

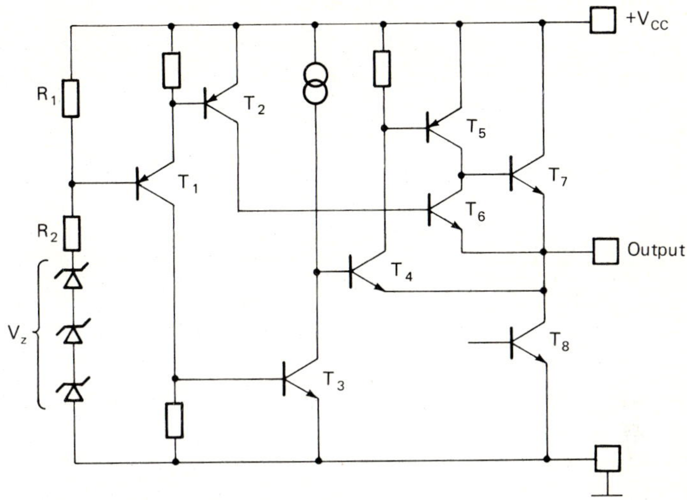

FIG. 6-22 Overvoltage protection circuit.

6.4.2 Overvoltage Protection Circuits

When the supply voltage V_{CC} exceeds the transistor limits imposed by the process, it is preferable to set it in a nonoperational state during the overvoltage to prevent damage to the IC. In this phase the transistors exposed to the high voltage are connected with the base grounded so that they can withstand higher voltages. Figure 6-22 shows a circuit which implements this function.

In operating conditions the zener diodes are not switched on and T_1, T_2, and T_3 are off. When V_{CC} exceeds the value

$$V_{CC1} = V_Z + V_{BE}\left(\frac{R_2}{R_1} + 1\right) \tag{6-54}$$

T_1 and T_3 switch on and the final stage is turned off.

When V_{CC} reaches the value

$$V_{CC2} = V_Z + 2V_{BE}\left(\frac{R_2}{R_1} + 1\right) \tag{6-55}$$

transistor T_2 switches on and makes T_6 saturate, thus short-circuiting the base to the emitter of the power device T_7. In these conditions the circuit is not operating, but it can withstand a maximum voltage V_{CC} equal to the BV_{CES} of transistor T_7. The voltage BV_{CES} is approximately equal to BV_{CBO} because the base is short-circuited to the emitter.

There are two intervention thresholds to guarantee first switch-off of

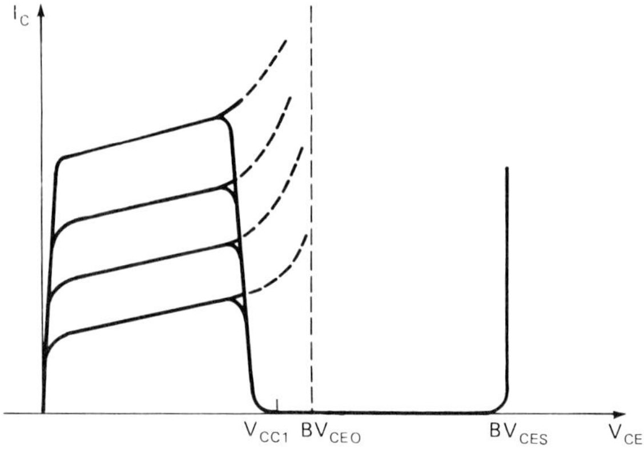

FIG. 6-23 Behavior of the power transistor T_7 of the circuit shown in FIG. 6-22.

the driver and then closing of the base-emitter junction of the power device T_7. The behavior of the power device is shown in Fig. 6-23.

6.4.3 Grounded-Base Circuits

In the circuits shown in this section solutions are indicated in which the transistors operate in V_{CER} rather than V_{CEO} conditions. V_{CER} working conditions prevail when a resistor (or, better, an equivalent low impedance) is connected between base and emitter. If the value of the resistor R connected between the base and emitter is such that the $V_{CE\ \mathrm{on}}$ of the transistor is not obtained, i.e., if the condition $I_C/h_{FE} < V_{BE\ \mathrm{on}}/R$ occurs, the transistors operate in a V_{CER} condition which is almost equal to V_{CBO}. The same results are obtained when the transistor is driven by a voltage generator. In this condition the transistor works with grounded base.

Using these techniques, the devices operate up to the breakdown voltage $V_{CER} = V_{CBO}$; this is a considerable advantage in terms of applications (Chap. 2, Sec. 2.1.5). For example, in the low-voltage process it means operating up to 40 V, because $BV_{CEO} \geq 20$ V, while $BV_{CBO} \geq 40$ V.

A typical diagram used is the Wilson current generator (T_1, T_2, and T_3) shown in Fig. 6-24. Transistor T_2 is operating with grounded base because the impedance at point A is low and equal to $Z_A = 2kT/qI_1$; T_2 is driven by T_3 and its collector voltage can reach, in this application, BV_{CBO} [14].

A solution which makes it possible to supply the IC at the maximum possible voltage, represented by the breakdown voltage of the N junction (the collector of an NPN or the base of a PNP) to the substrate, BV_{CS}, is shown in Fig. 6-24. The collector voltage of T_2 can reach as a

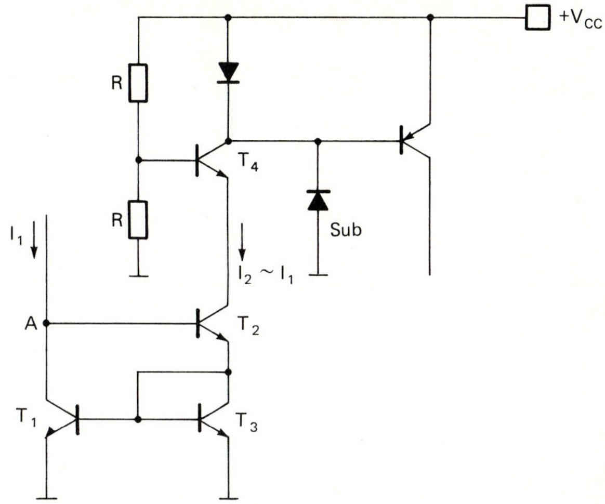

FIG. 6-24 Circuit with operating voltage equal to or less than $2V_{CBO}$.

maximum value BV_{CBO}, which is defined by the resistive divider network R and is available on the emitter of T_4. Transistor T_4, biased at $V_{CC}/2$, works in BV_{CER} conditions.

The maximum V_{CC} voltage, if there were no other limits, would be given by the maximum voltage of T_2 plus that of T_4, which is

$$V_{CC\,\max} = BV_{CBO\,T_2} + BV_{CER\,T_4} = 2BV_{CBO} \tag{6-56}$$

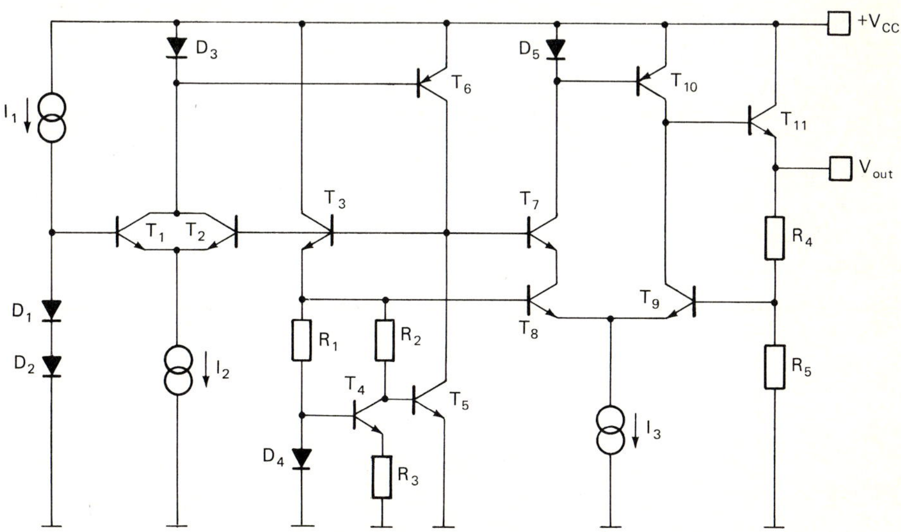

FIG. 6-25 Circuit with operating voltage equal to or less than V_{CBO}.

Following this technique it would be possible to reach very high voltages, but at this point the breakdown voltage of the junction to substrate, BV_{CS}, intervenes, a characteristic of every process, and this determines the maximum value of V_{CC}.

Normally we can say that in IC processes the voltage $BV_{CS} < 2BV_{CBO}$.

Another application using this technique (grounded-base configuration) is the voltage regulator shown in Fig. 6-25.

REFERENCES

1. W. M. Dorn, W. T. Hetterscheid, C. W. J. V. A. Pol, and W. Smulders, Power Transistor Limitations and Ratings, *Philips Electronic Applications Bull.*, Vol. 32, No. 4, 1974.
2. B. Murari, Power Integrated Circuits: Problems, Trade-Offs and Solutions, *IEEE J. Solid State Circuits*, Vol. SC-13, No. 3, June 1978.
3. J. R. Black, Electromigration: A Brief Survey and Some Recent Results. *IEEE Trans. Electron Devices*, Vol. ED-16, No. 4, April 1969.
4. J. R. Black, Mass Transport of Aluminum by Exchange with Conducting Electrons, *IEEE Proc. Ann. Symp. Reliability Physics*, Cat. 7-15 C 58, November 1967.
5. F. Bertotti, P. Erratico, and B. Murari, Future and Limits of Power ICs, *ESSCIRC '80, 6th Eur. Solid State Circuit Conf.*, September 22–25, 1980.
6. L. H. Holway, Jr., and M. G. Adlerstein, Approximate Formulas for the Thermal Resistance of Impatt Diodes Compared with Computer Calculations, *IEEE Trans. Electron Devices*, February 1977.
7. C. Mitchell and H. M. Berg, Thermal Studies of a Plastic Dual-in-Line Package, *IEEE Trans. Components, Hybrids, and Mfg. Technol.*, Vol. CHMT-2, No.4, December 1979.
8. F. F. Oettinger, D. L. Blackburn, and S. Rubin, Thermal Characterization of Power Transistors, *IEEE Trans. Electron Devices*, Vol. ED-23, No. 8, August 1976.
9. M. L. Buller, Thermal Transients in Electronic Packages, *IEEE Trans. Components, Hybrids, and Mfg. Technol.*, Vol. CHMT-3, No. 4, December 1980.
10. D. L. Blackburn and F. F. Oettinger, Transient Thermal Response Measurements of Power Transistors, *IEEE Int. Conv. Record*, Power Electronic Specialist Conference, 1974.
11. R. M. Scarlett and W. Shockley, Secondary Breakdown and Hot Spots in Power Transistors, *IEEE Int. Conv. Record*, Pt. 3, March 1963.
12. D. Stolnitz, Experimental Demonstration and Theory of a Corrective to Second Breakdown in Si Power Transistors, *IEEE Trans. Electron Devices*, Vol. ED-13, Nos. 8 and 9, August and September 1966.

13. R. J. Nienhuis, Second Breakdown in the Forward and Reverse Base Current Region, *IEEE Trans. Electron Devices,* Vol. ED-13, Nos. 8 and 9, August and September 1966.

14. P. R. Gray and R. G. Meyer, *Analysis and Design of Analog Integrated Circuits,* Wiley, New York, 1977.

CHAPTER 7

Packaging of PICs

C. Rovere
SGS Microelettronica
Milan, Italy

7.1 INTRODUCTION TO PACKAGE DESIGN 7.1
7.2 FEATURES OF THE PIC TO BE "PACKAGED" 7.3
7.3 PACKAGE THERMAL DESIGN 7.4
7.3.1 Thermal Resistance 7.5
7.3.2 Calculation of $R_{th,jc}$ 7.5
7.3.3 Measurement of $R_{th,jc}$ 7.5
7.3.4 Junction Temperature (T_j) Measurement 7.8
7.3.5 Case Temperature (T_c) Measurement 7.8
7.3.6 Measurement of $R_{th,ca}$ 7.9
7.4 PACKAGE MECHANICAL DESIGN 7.9
7.4.1 Thermoelastic Stress Measurement 7.13
7.4.2 Mechanical Design of Leads 7.15
7.5 PACKAGE ELECTRICAL DESIGN 7.15
7.6 CONCLUSION 7.16

7.1 INTRODUCTION TO PACKAGE DESIGN

The package designer for power integrated circuits has to make compatible with a printed circuit (PC) board a silicon chip requiring electrical connections, heat dissipation, and environmental protection. A package for a PIC (see Fig. 7-1) consists of:

- Silicon chip or die
- Heat spreader for heat transmission

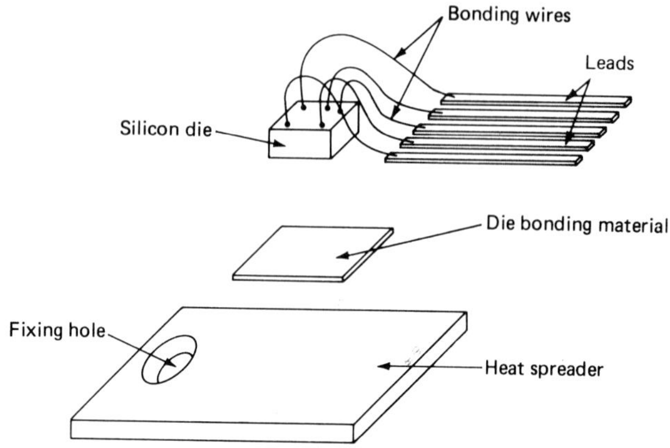

FIG. 7-1 Main parts of a PIC package.

- Die bonding material to connect heat spreader and die
- Lead frame for electrical connection to a PC board
- Wire for connecting die and leads
- Encapsulant or cap for environmental protection

The factors limiting package design and choice of material are as follows:

- The die temperature must not exceed 150°C.
- The wire must be able to carry several amperes.
- The package must be able to resist mechanical stress like that encountered during assembly by users.
- The package must meet reliability targets fixed by military standards with reference to:

  Thermal shock between −65 and 150°C

  Thermal cycles between −65 and 150°C

  Thermal fatigue (from On-Off cycles)

  High-temperature storage

  High-moisture, high-temperature storage (85°C at 85% relative humidity)

In this chapter we will examine some fundamental aspects that a package designer must take into account to design a PIC package.

7.2 FEATURES OF THE PIC TO BE "PACKAGED"

The silicon chip that is the "heart" of a PIC is a small die with a surface area of 5000 to 20,000 mil^2 and a thickness of 8 to 12 mil. As can be seen in Fig. 7-2, package construction must address the following requirements to a silicon die:

- Back-side bondable
- Bonding pad for wiring
- Front-side surface passivation

Surface passivation (P. Vapox or silicon nitride) has the purpose of sealing the junctions against ionic contamination and protecting the interconnecting metallization from mechanical damage. The *bonding pads* to be connected to the lead frame must be compatible with the wiring techniques and must be dimensioned to allow high assembly yield. *Back-side finishing* of the wafer has three purposes:

1. Thinning of the silicon chip to improve heat transfer
2. Providing a metallized surface for an ohmic contact between the silicon body and the heat spreader
3. Providing a metallized surface solderable with the die bonding technique employed

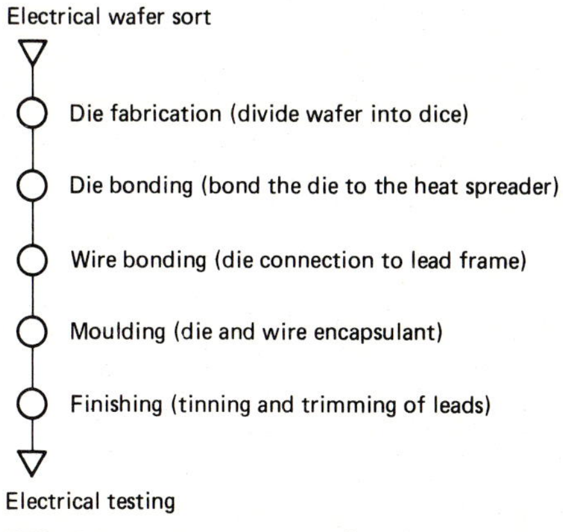

FIG. 7-2 Main operations flowchart for plastic PIC packaging.

7.3 PACKAGE THERMAL DESIGN

For a packaged semiconductor device, heat generated at the junction area of the powered chip causes the junction temperature T_j to rise above the ambient temperature T_a. The maximum acceptable T_j is determined by users' reliability requirements; it is well known that a 10°C increase in T_j is enough to reduce device life by a factor of 2. Maximum T_j at present is fixed at 150°C as an industry standard, and the T_a rating is typically 70°C.

Power (P_d) generated at the junction "flows" through a thermal resistance from the source (at T_j) to the ambient (at T_a), just as an electric current flows through an electrical resistance; thus thermal resistance is a physical parameter describing the ability of the material to transfer heat. If we define

$R_{th,jc}$ = thermal resistance between junction and case

$R_{th,ca}$ = thermal resistance between case and ambient

then the equivalent electrical circuit of the heat transfer is described in Fig. 7-3, and in an analogy with Ohm's law, we have:

$$T_j - T_c = P_d \times R_{th,jc} \tag{7-1a}$$

$$T_c - T_a = P_d \times R_{th,ca} \tag{7-1b}$$

$$T_j - T_a = P_d (R_{th,jc} + R_{th,ca}) \tag{7-1c}$$

From the above expressions it is evident that the package designer must minimize the contributions of $R_{th,jc}$ and $R_{th,ca}$.

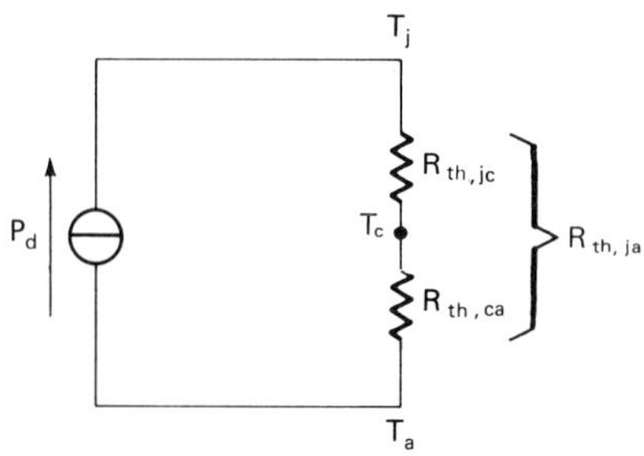

FIG. 7-3 Electrical equivalent thermal resistance network.

7.3.1 Thermal Resistance

If we analyze the package structure, we can see that $R_{th,jc}$ depends on several factors associated with the different materials used for the package construction, while $R_{th,ca}$ depends also on the user application [1–2]. Hence the equivalent electrical circuit network can be modified as in Fig. 7-4. This figure shows that the different packages have different heat transfer paths that can modify $R_{th,ca}$. The term $R_{th,jc}$ can be calculated following the truncated pyramid model for each element i, as follows,

$$R_{th,jc} = \sum_i R_{th,i} = \frac{1}{\sigma_{th,i}} \int_0^{x_i} \frac{dx}{(a_i + 2x)(b_i + 2x)} \tag{7-2}$$

for the rectangular surface with sides a_i, b_i and thickness x_i. Integrating Eq. 7-2 when $a > b$, we obtain

$$R_{th,i} = \frac{1}{2\sigma_{th,i}(a_i - b_i)} \ln \frac{a_i(b + 2x_i)}{b_i(a + 2x_i)} \tag{7-3a}$$

or when $a = b$, we obtain

$$R_{th,i} = \frac{1}{\sigma_{th,i}} \frac{x_i}{a_i^2 + 2a_i x_i} \tag{7-3b}$$

7.3.2 Calculation of $R_{th,jc}$

Table 7-1 gives some examples of $R_{th,jc}$ calculated from a single power source with sides 1600 μm × 800 μm and 300 μm thick using $\sigma_{th,jc}$ values reported in Table 7-2. From this comparison it appears that $R_{th,jc}$ can vary from 1.3 to 14.5°C/W. This indicates that the best heat transfer is obtained when we use 200-μm thick silicon, Au-Si bonding material, and OFHC copper as heat spreader. Any other material combination will be a compromise among production limitation, material cost, and really acceptable thermal resistance, as in the fourth column of Table 7-1.

7.3.3 Measurement of $R_{th,jc}$

From Eq. 7-1a we have:

$$R_{th,jc} = \frac{1}{P_d}(T_j - T_c) \tag{7-4}$$

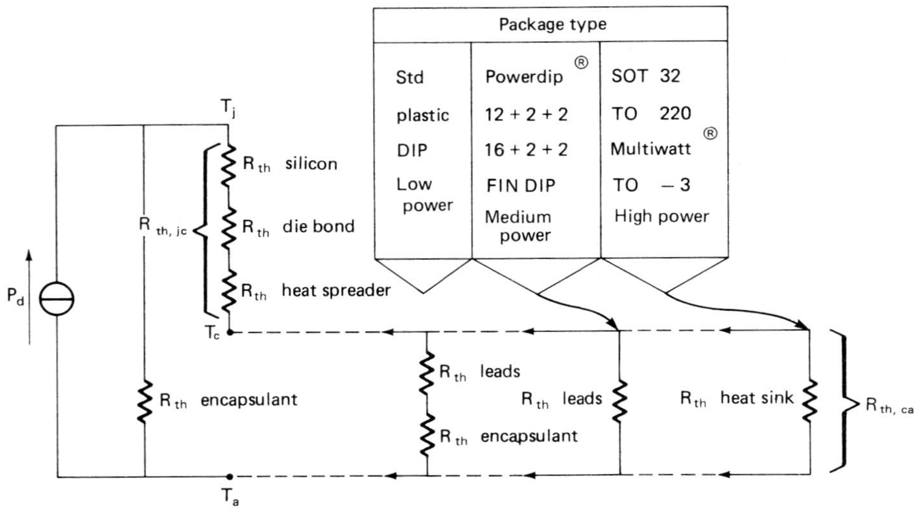

FIG. 7-4 Electrical equivalent thermal resistance network for different packages.

TABLE 7-1 Thermal Resistance Calculation Examples

Power source dimensions: 1.6 × 0.8 mm

Material	$R_{th,i}$, °C/W	$R_{th,jc}$, °C/W			
		14.5	1.8	1.3	1.9
Silicon					
Thickness 200 μm	0.9		X	X	
Thickness 300 μm	1.2	X			X
Preforms for die bonding					
Au-Si eutectic (30 μm)	0.033			X	
Au-Sn eutectic (30 μm)	0.037				
Pb-Sn alloy (30 μm)	0.26		X		X
Silver epoxy glue (50 μm)	7.65	X			
Heat spreader					
Alloy 42 (thickness 0.25 mm)	5.6	X			
Copper alloy (thickness 1.27 mm)					
OFHC	0.4			X	
OLIN C151	0.44				X
0.12 Sn OFC	0.46				
DHP	0.48				
OLIN 194	0.62		X		

TABLE 7-2 Thermal Conductivity $\sigma_{th,i}$

Material	$\sigma_{th,i}$, W/(cm·°C)
Silicon die	1.3
Die bonding material	
Pb/Sn alloy	0.35
Au/Si eutectic	2.8
Au/Si eutectic	2.5
Silver epoxy glue	0.005
Heat spreader	
Copper OFHC	3.96
Copper OLIN C151	3.61
Copper 0.12 Sn OFC	3.44
Copper DHP	3.34
Copper OLIN 194	2.58
Fe/Ni Alloy 42	0.105
Encapsulant	
Siliconic (standard)	0.005
Epoxy (standard)	0.006
Epoxy (high, thermal conductivity)	0.015

When T_j and T_c are known at a power P_d, it is possible to find $R_{th,jc}$. Power P_d is the electrical power dissipated in the device, measured in watts, while the measurements of T_j and T_c are described in Sec. 7.3.4, which follows.

7.3.4 Junction Temperature (T_j) Measurement

Several methods have been developed to measure T_j. These include:

1. Direct surface temperature reading by an infrared detector after chemical etching of a small portion of the encapsulant [3]
2. Monitoring of a temperature-dependent electrical parameter of the complete device
3. V_{BE} measurement of a power transistor similar to that of a complete PIC after an accurate calibration

Methods 2 and 3 are more commonly used because the calibration process is easier. Figure 7-5 presents a suitable set of equipment for reading V_{BE} or a temperature-dependent parameter. The procedure consists of alternating a long On power interval with a short Off cycle of the transistor under test and reading V_{BE} during the power Off interval. In this case the T_j is obtained by reading a calibration curve of $V_{BE} = f(T)$.

7.3.5 Case Temperature (T_c) Measurement

T_c is measured by positioning a thermocouple on the outside package surface directly below the chip; because this point is one of the hottest on the package surface it has, of course, the minimum value of $R_{th,jc}$.

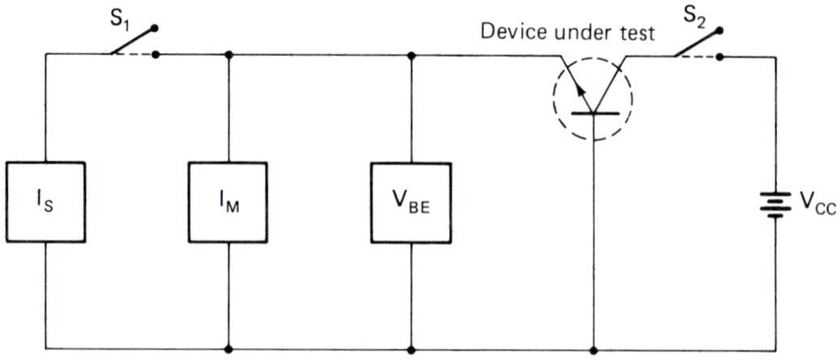

I_S = current generator for power P_d (S_1, S_2 "on")
I_M = current generator for measure V_{BE} (S_1, S_2 "off")
V_{BE} = voltage drop dependent on T_j

FIG. 7-5 Equipment for junction temperature measurement.

A second method of determining $R_{\text{th},jc}$ involves heat-sinking of the bottom of the case with a massive water-cooled copper plate to allow $R_{\text{th},ca}$ to drop to zero. Forced-air cooling can be introduced for more uniform package temperature distribution.

Table 7-3 gives the maximum values for $R_{\text{th},jc}$ and $R_{\text{th},ja}$ for more commonly used PIC packages.

7.3.6 Measurement of $R_{\text{th},ca}$

$R_{\text{th},ca}$ can easily be deduced from

$$R_{\text{th},ja} = R_{\text{th},jc} + R_{\text{th},ca} \tag{7-5}$$

when $R_{\text{th},ja}$ and $R_{\text{th},jc}$ are known. As was true of $R_{\text{th},jc}$, $R_{\text{th},ja}$ can also be measured, but its value depends to a large extent on application. From the manufacturer's point of view, it is useful to measure $R_{\text{th},ja}$ in air with natural convection because this is the maximum $R_{\text{th},ja}$ that will be encountered. The method used for the $R_{\text{th},ja}$ measurement is the same as for $R_{\text{th},jc}$, using Eqs. 7-1*c* and 7-5 when T_a and T_j are known.

From these considerations and from the data reported as an example we can conclude that the choice of material for the heat spreader is very important. In fact, many copper alloys are suitable for PIC package design, but other parameters must be considered. These include:

- Electrical conductivity
- Mechanical formability
- Softening temperature
- Thermal expansion coefficient
- Price

Table 7-4 gives the above parameters as available from several material suppliers.

7.4 PACKAGE MECHANICAL DESIGN

For a complete PIC package design we must take into account that all materials for the package construction are assembled at temperatures higher than ambient or working temperature and that these materials must withstand mechanical stress, such as lead forming, automated electrical testing, automated PC board insertion, and heat-sink connection to the heat spreader.

TABLE 7-3 Maximum Value of R_{th} for Some PIC Packages

Package type	$R_{th,jc}$, °C/W	$R_{th,ja}$, °C/W	Maximum power, W	
14- or 16-lead DIP Powerdip®* 12 + 2 + 2† 16 + 2 + 2‡	— 14 14 14	80 70 80 80	1 5 5 5	
14 or 16 leads for bar	3	40	25	
Quadruple in-line	12	70	5	
SOT-32	9	100	8	

TO-220	4	40	20	
Multiwatt®§	3	25	25	
TO-3	4	35	20	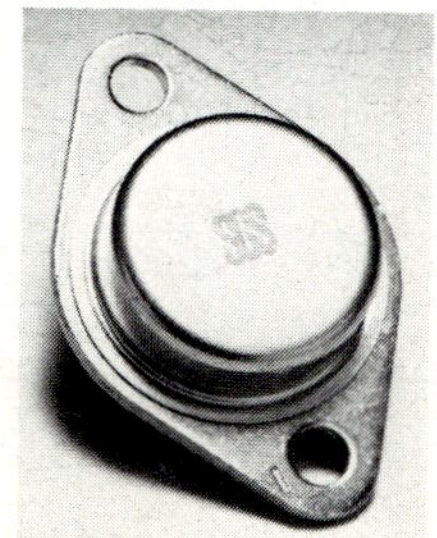

*Powerdip® is a registered trade name of SGS-ATES. The package has eight leads (leads 9 to 16) connected to the heat spreader.
†The package has four leads (4, 5 and 12, 13) connected to the heat spreader.
‡This has four leads (5, 6 and 15, 16) connected to the heat spreader.
§Multiwatt® is a registered trade name of SGS-ATES.

TABLE 7-4 Comparison of Some Copper Alloys

Supplier and type	(Generic) OFHC	Olin C194	Hussey C155	Olin C151	Trefimetaux Cupronics	Mitsui Sn-OFC
Composition, %	99.995 Cu	97.5 Cu 2.35 Fe 0.03 P 0.12 Zn	0.11 Mg 0.06 P trace Ag	0.1 Zr balance Cu	99.6 Cu 0.15 Cd balance Zn	0.15 Sn balance Cu
Electrical conductivity, % IACS	101	65	85	90	89	85
Thermal conductivity, W/(cm·°C)	3.96	2.58	3.44	3.61	3.66	3.44
Modulus of elasticity, × 10^6 psi	17.5	17	17	17	17	17
Lead bend* fatigue (cycle)	11	16	13	15–20	14	10
Softening temperature (1-h exposure)	200°C	575°C	425°C	450°C	375°C	400°C
Cost factor	102	100	111	103	102	103

*Per Mil-Std 883-B, method 2004: 0.137 × 0.010.

During the PIC assembly process two steps can produce permanent thermoelastic stress:

- Die bonding with a $\Delta T \approx 300°C$
- Transfer molding and curing of encapsulant with a $\Delta T \approx 150°C$

Thermoelastic stress can be also produced during

- Preconditioning thermal cycles (from -65 to 150°C)

and mechanical shock can be produced during

- Lead trimming and forming
- Marking

The theory for calculation of mechanical stress is derived from that used for stress prediction on slab composite material [4].

Since a realistic model for stress calculation is not readily available, the PIC package designer has to make accurate stress measurements using a practical model.

7.4.1 Thermoelastic Stress Measurement

In recent years many papers have been dedicated to mechanical stress measurement inside a package. The most common method is that employing test patterns constituted by diffused resistors sensitive to mechanical deformations induced by the packaging process and materials. These test patterns are similar to that in Fig. 7-6 and are designed

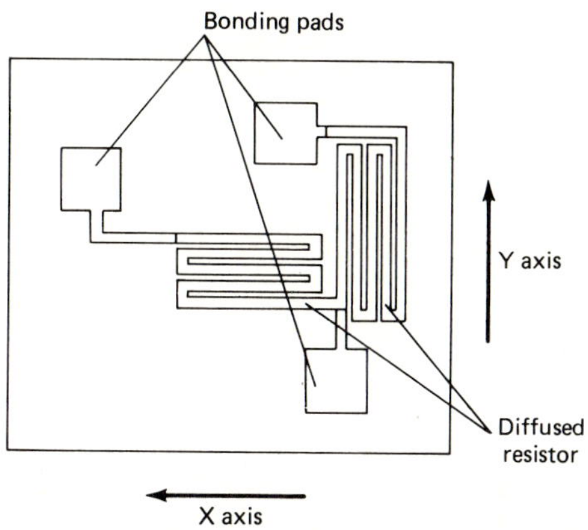

FIG. 7-6 Test pattern for internal stress evaluation.

to have the sensitive elements parallel to the *X* and *Y* axes of the package and will be built in the four corners and at the center of a silicon die used as a stress gauge. An accurate measurement of the resistor values before and after each step of the assembly process can indicate the stress level induced.

Some authors [5–7] report that a silicon chip packaged in a standard DIP is subjected to a uniform pressure mainly due to the encapsulant's modulus of elasticity and to its different thermal expansion coefficient (TEC) relative to that of the silicon. In the case of a PIC very little information is available in the literature. Several measurements indicate the relative stress level for a device mounted in a nonsymmetrical package, such as TO-220, as shown in Table 7-5.

The values on the table show the important role of molding compounds in inducing stress. Siliconic molding compounds were used at first for encapsulation of PICs; however, because of their poor reliability in moist environments and low resistance to thermal shocks, these were later replaced by epoxy compounds. Although this change improved reliability with respect to moisture and thermal cycling, it raised the level of stress on the silicon chip by a factor of 2.

As a result of this, today many molding compound suppliers are offering new "low-stress" compounds. These new compounds are obtained by modifying the filler type and the proprietary production process. The resulting molding compound has a lower thermal expansion coefficient without affecting the humidity and thermal shock resistance that had made the change from siliconic to expoxy molding desirable in the first place.

Authors [7] have also investigated molding conditions and found that by lowering the molding and postcuring temperature the stress can be reduced to about half of its "normal" value, where normal means mold-

TABLE 7-5 Stress during Assembly on TO-220 Package

Operation	Δ TEC,* ppm/°C	Relative stress level
Die bonding material†		
Silver epoxy glue	46.5	1
Pb/Sn	25.2	2
Encapsulation material		
Epoxy	21.5	5–12
Siliconic	26.5	4–6

*Referred to silicon thermal expansion coefficient (TEC).
†On copper heat spreader.

ing conditions suggested for acceptable productivity. Present low-stress compounds are able to reduce the stress by about 15 to 25 percent of that found in standard compounds.

7.4.2 Mechanical Design of Leads

In a PIC, where good thermal and electrical conductivity are required, copper alloys must be used. In comparison with the iron-nickel alloys first used for semiconductor packaging, copper alloys exhibit lower resistance to lead bend fatigue.

The bonding and the mechanical application of a package require a copper alloy that will remain rigid after thermal exposure to die bonding, wire bonding, and encapsulant postcuring. For this purpose the package designer will select the alloy with the highest thermal conductivity and a high resistance to lead bend fatigue, as stated in the highest standards [8–9]. Table 7-4, in Sec. 7.3.6, compares the composition, thermal conductivity, lead bending fatigue, and price of some of the most commonly used copper alloys.

7.5 PACKAGE ELECTRICAL DESIGN

When designing the PIC package, it is necessary to select the type of wire to be used for connecting the die to the lead frame and the alloy to be used for die bonding. The choice of die bonding alloy is dependent on considerations from previous sections: thermal resistance must be minimized; stress must be reduced by using soft alloys capable of reducing the mismatch between the silicon chip and the heat spreader material; and the cost of the alloy must be minimized by avoiding the use of precious metals.

Thermosonic gold ball bonding is the most commonly used wire bonding technology. The sizes of wire normally available are 0.7, 1.0, 1.5, and 2.0 mil. The most commonly used wires for PICs are 1.5 and 2.0 mil; larger gold wires cannot be used, for price and bonding reasons. In the case of high currents, the wires can be doubled.

In the space of only a few years, especially since the rapid increase in the price of gold, many suppliers have developed low-cost wires, such as palladium, palladium-silver, and gold-silver. Palladium wire has been successfully used only for low-current devices, on account of its conductivity, which is 4 times lower than that of gold. Copper wire is at an advanced stage of development for applications requiring high current.

Table 7-6 compares the main properties of the wires mentioned in this section.

TABLE 7-6 Comparison of Gold, Palladium, and Copper Wire

Variable	Au	Pd	Au–50% Ag	Ag–10% Pd	Cu
Modulus of elasticity, kN/mm^2	78	121	90	92	85
Thermal conductivity, W/(m·K)	312	75	75	120	200
Electrical conductivity, % IACS	78	17.4	16.3	26.5	95
Price (relative)	100	30	60	20	18

7.6 CONCLUSION

In this chapter on package design for PICs, the primary role of the thermal aspects in power applications has been outlined. We have seen plastic packages that are more widely used on account of their low price and high degree of automation. This type of package is in continuous evolution in an effort to meet reliability targets fixed by military standards stated for hermetic packages as TO-3, and to keep abreast of "quality competition" among semiconductor producers.

The next few years will see considerable development in automation to further reduce working costs, and improvements in material performance to produce plastic packages for PICs similar or better than hermetic ones.

REFERENCES

1. C. Mitchell and H. M. Berg, Thermal Studies of Plastic Dual-in-Line Package, *IEEE Trans. Components, Hybrids, and Manufacturing Technol.*, Vol. CHMT-2, No. 4, pp. 500–511, December 1979.
2. K. E. Manchester and D. W. Bird, Thermal Resistance: A Reliability Consideration, *IEEE Trans. Components, Hybrids, and Manufacturing Technol.*, Vol. CHMT-3, No. 4, pp. 580–587, December 1980.
3. C. A. Lidback, Scanning I.R. Microscopy Techniques for Semiconductor Thermal Analysis, *Proc. 17th Ann. Conf. Reliability Physics*, pp. 183–189, 1979.
4. R. Zeyfang, Stress and Strains in a Plate Bonded to a Substrate, *J. Appl. Phys.*, Vol. 42, p. 1182, 1971.
5. W. H. Schroen and J. L. Spencer, Reliability Test and Stress in Plastic Integrated Circuits, *Proc. 19th Ann. Conf. Reliability Physics*, pp. 81–87, 1981.
6. J. L. Spencer and W. H. Schroen, New Quantitative Measurement of IC Stress Introduced by Plastic Package, *Proc. 19th Ann. Conf. Reliability Physics*, pp. 74–80, 1981.

7. R. J. Usell, Jr., and S. A. Smiley, Experimental and Mathematical Determination of Mechanical Strains within Plastic IC Packages and Their Effect on Devices during Environmetal Tests, *Proc. 19th Ann. Conf. Reliability Physics*, pp. 65–73, 1981.

8. J. R. Howell, Reliability Study of Plastic Encapsulated Copper Lead Frame Epoxy Die-Attach Packaging System, *Proc. 19th. Annual Reliability Physics*, pp. 104–110, 1981.

9. S. P. Zarlingo and R. J. Scott, Lead Frame Materials for Packaging Semiconductors, *1st Ann. Int. Packaging Soc. Conf.*, 1981.

CHAPTER 8

Power Integrated Circuit Reliability

F. R. Borri
G. D'Espinosa
SGS Microelettronica
Milan, Italy

8.1 INTEGRATED CIRCUIT RELIABILITY 8.3
8.1.1 Some Considerations about Power Integrated Circuit Stresses 8.4
8.1.2 Basic Design Points for Reliable Power Integrated Circuits 8.6
8.2 RESISTANCE OF PLASTIC PACKAGES TO HUMIDITY 8.6
8.2.1 Phenomena Caused by Humidity 8.7
8.2.2 Correlation between Various Laboratory Tests and Models 8.8
8.2.3 Correlation between Laboratory Tests and the Field 8.10
8.2.4 Humidity Test Elements Not Yet Fully Explained 8.11
8.3 QUALITY OF ENCAPSULATING RESINS 8.12
8.3.1 Polymerization Kinetics and Influence of Degree of Cure 8.14
8.3.2 Thermal Stability of Encapsulating Resins 8.18
8.3.3 Linear Thermal Expansion Coefficient and Glass Transition Temperature 8.21
8.3.4 Determination of Ion Content in Encapsulating Resins 8.24

8.3.5 Physical and Electrical Characteristics of Encapsulating Resins 8.24
8.4 THE PROBLEM OF Au-Al INTERMETALS IN POWER ICs 8.27
8.4.1 Formation Kinetics of Au-Al Intermetals 8.28
8.4.2 Problems of Reliability in Au-Al Bonds 8.31
8.4.3 Bonding Considerations 8.32
8.4.4 Evaluation of Bond Reliability 8.34
8.5 ELECTROMIGRATION 8.36
8.5.1 Electromigration Theory 8.38
8.5.2 Electromigration in Thin Films 8.39
8.5.3 Electromigration in Integrated Circuits 8.44
8.5.4 Effects of Metal Impurities on Electromigration 8.46
8.6 PARASITIC MOS 8.47
8.6.1 Lateral PNP Transistors: Structures and Electrical Characterization 8.47
8.6.2 Results of Life Tests on PNP Structures in Plastic Packages 8.50
8.6.3 Theory of the Inversion Phenomenon 8.53
8.7 THERMOMECHANICAL EFFECTS ON THE DIE AND ON Al METALLIZATION 8.55
8.7.1 Principal Materials Used in Power ICs 8.56
8.7.2 Stresses on the Silicon Die 8.58
8.7.3 Effects of Temperature Excursion on the Heat Sink–Solder–Die System 8.60
8.7.4 Effects of Temperature Excursion on the Metallization 8.61
8.8 MAIN RELIABILITY TESTS 8.64
8.8.1 Thermal Cycles 8.64
8.8.2 Thermal Shock 8.65
8.8.3 Life Test 8.66
8.8.4 Humidity Test (Moisture Resistance) 8.72
8.8.5 Other Tests on PICs 8.74

The reliability of power ICs, like that of all other integrated circuits, depends on three main factors:

- Design
- Manufacturing
- Application

We shall examine the various aspects of these factors, giving special attention to those most characteristic of power devices which today are almost all produced in plastic packages.

8.1 INTEGRATED CIRCUIT RELIABILITY

Evaluating the reliability of an IC involves the study of numerous factors. The subject is extremely vast, and here we shall mention only a few aspects.

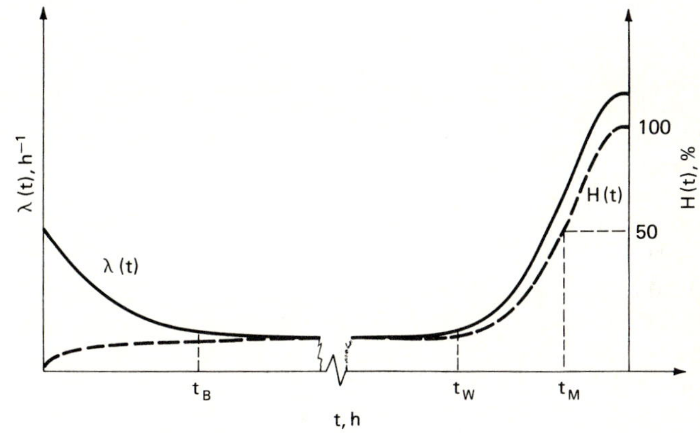

FIG. 8-1 Failure rate λ and cumulative failed units H vs. lifetime *t*.

Figure 8-1, the well-known "bathtub" curve, which links the instantaneous failure rate to the device lifetime, indicates the following reliability elements which the user needs in order to use a family of devices in the correct application and to forecast the reliability of the equipment or system:

- The early failure rate, or *infant mortality*
- The time t_B at which infant mortality can be considered exhausted
- The failure rate in the *useful life*
- The start of *wear-out* t_W
- The *median life* (the time t_M at which 50 percent of devices fail)

The failure rate and times t_B, t_W, and t_M depend on the type and intensity of stress both in the laboratory and in the actual application.

The *main activities in the reliability field* are:

- Study and determination of the *stresses* which the ICs encounter in applications.
- Study and definition of *laboratory tests* to be used to check reliability. The laboratory tests consider one stress or a few simultaneous stresses, as opposed to the large variety of stresses encountered by

the device during operation in the field. Numerous standardized tests are required to research the effect of each stress on the device. These can be divided into standard tests (used for checks) and accelerated tests (which aim to give results in a short time through acceleration of the test stresses).

- Study of *physical laws* governing the various failure mechanisms, and determination for each failure mechanism of its dependence on a particular stress.
- Development of more and more sophisticated analysis techniques to find the *cause of failure* of devices that fail during testing or while in use.
- Study and determination of *screening techniques* and *preconditioning* to remove infant mortality before the product is used.
- Theoretical studies of *general laws* governing reliability (reliability models).
- Study of the best systems for *collecting* and *interpreting the data* coming from laboratory tests and from the field (data banks and statistical analysis for a correct interpretation of the results).
- Transfer to production people the *reliability knowledge* acquired during the design of devices and processes, designing at the same time suitable reliability checks during the production process.
- Transfer of acquired reliability knowledge to the designer of the IC application in order to *forecast* and to *optimize reliability.*

Since semiconductor technology is continually evolving toward more and more sophisticated and complex devices, obviously the problems involved in studying the reliability of these devices are also more and more complex. The study process involves the continuous exchange of information among reliability, design, and application experts.

It is impossible to give all the knowledge acquired in the field of IC reliability in one chapter only, so we shall consider a few of the most important elements for power ICs.

8.1.1 Some Considerations about Power Integrated Circuit Stresses

The five major stresses an integrated circuit can encounter in its life are:

- Current
- Voltage
- Temperature

- Mechanical stresses
- Moisture

Complex electronic systems, which include many active and passive components, cannot be more reliable than the cumulative reliability of all of their components taken separately. From the reliability point of view, power integrated circuits (PICs) used in such systems can be the weakest point, for two main reasons:

1. Although the dangers represented by current, voltage, and temperature are widely recognized, the importance of *high mechanical stresses* and of *moisture* for PICs is often underestimated. The concern about high mechanical stresses and moisture can be easily justified considering that:
    a. A PIC must be firmly connected to an external heat sink. Inadequate planarity of the surfaces in close mechanical contact (the surface of the package and that of the heat sink) and excessive torque used to apply mounting screws, etc., can generate mechanical stresses that are not present in other integrated circuits.
    b. Mechanical stresses are also generated internally by different thermal expansion of the silicon and of the header of the package. The greater the temperature excursions, the higher the mechanical stresses. Power integrated circuits normally have the highest temperature excursion of a system because they dissipate more power internally than other active components.
    c. The sensitivity to moisture of power integrated circuits becomes obvious if we consider that a PIC is mounted in a package that exhibits large interface areas between the resin and the tabs used to decrease the junction-to-case thermal resistance. These interface areas generally show a preferential path for moisture to reach the silicon chip; hence moisture may penetrate deep into a PIC package, more quickly than into other integrated circuits.
2. Power integrated circuits are, by definition, the *lowest-impedance point* of a complex system and the most exposed to the external environment. These circuits have to handle power with the highest possible efficiency, i.e., with the lowest power dissipation; therefore their internal impedance must be the lowest possible. Any form of overload (voltage or load impedance) is immediately converted into heavy current flow that, in some cases, can have catastrophic consequences. Since the PICs are the power actuators of a system,

they are very often directly connected to external electromechanical components (loudspeakers, solenoids, motors, etc.). The probability of mishandling due to wrong connections between PICs and external components is higher than that for the other active components present in a system.

8.1.2 Basic Design Points for Reliable Power Integrated Circuits

A reliable power integrated circuit is a complex structure controlling electrical power to a specified degree of accuracy within a determined time period. Such a structure can be divided into many components:

- The *active regions*, i.e., the PN junctions where signal processing and high voltage and current handling are performed
- The *metallization* that interconnects the active regions
- The *passivation layer* that, among other things, protects the active regions from the influence of external impurities
- The *soldering to the heat sink*, which removes the heat generated by power dissipation caused by the simultaneous presence of high voltage and high current
- The *bonding wires* that connect the chip to the external circuitry
- The *package*, which protects the chip and the bonding wires against the dangers of the outside world and provides a low-resistance path for heat conduction

A designer of PICs must consider all these components to obtain a reliable device, because reliability has to be designed in from the very beginning.

In the following, some optimizations for the design of the many components of a PIC are described. The design criteria for the package are discussed in Chap. 7.

8.2 RESISTANCE OF PLASTIC PACKAGES TO HUMIDITY

Today a large percentage of PICs are produced in plastic packages. It is important to determine that these devices can resist the most serious temperature and humidity conditions encountered during the life of the device. Accelerated laboratory tests are normally used for this purpose:

1. *Pressure cooker:* 121/100 (121°C, 100% relative humidity, or RH) at 2.08 atm. This test can be carried out with and without applied bias.
2. *85/85* (85°C, 85% RH). This test can be carried out with and without applied bias. It is named TH (temperature humidity) and THB (temperature humidity bias), respectively. The present trend is toward testing with bias, even though it is more costly and causes more complex interpretation problems.

The laboratory tests have a dual aim. If we know the acceleration factors between one test and another and between the conditions in the laboratory and those in the field, it is possible to evaluate a certain useful device life if certain laboratory tests are passed.

Laboratory tests are also used to compare different constructive solutions or products from different suppliers, even if the acceleration factors of each test are not known exactly.

8.2.1 Phenomena Caused by Humidity

When environmental humidity reaches the die, after a certain time it can cause *corrosion of the aluminum.* Corrosion phenomena are very complex.

Corrosion may be *galvanic* or *electrolytic.*

Galvanic Corrosion. Galvanic corrosion requires two metals and an electrolyte. The corrosion processes are complicated by the fact that the metals are usually protected by oxide films, which are themselves attacked by impurities, such as the Cl^- ion, which start the reaction.

If we consider Au wires on Al metallizations, humidity penetration, dragging in chlorine impurities (Cl^-), attacks the surface oxide of the Al, uncovering the metallic Al. In this case an Al-Au galvanic cell has been formed, where the Au acts as a cathode and the Al as an anode, with +1.50 V and −1.62 V, respectively, in the electrochemical series of the elements.

The following reactions will therefore occur:

$$Al \rightarrow Al^{3+} + 3e \qquad \text{(anodic reaction on the Al)} \qquad (8\text{-}1)$$

$$3H^+ + 3e \rightarrow \tfrac{3}{2}H_2 \qquad \text{(cathodic reaction on the Au)} \qquad (8\text{-}2)$$

$$Al + 3H^+ \rightarrow Al^{3+} + \tfrac{3}{2}H_2 \qquad \text{(total reaction)} \qquad (8\text{-}3)$$

In the anode zone there may be a further reaction:

$$Al^{3+} + 3H_2O \rightarrow Al(OH)_3 + 3H^+ \qquad (8\text{-}4)$$

Therefore in the anode zone the metal dissolves (corrodes). The Cl^- does not participate directly in the reaction, but its presence (or that of other aggressive ions) with that of the H_2O is indispensable for the reaction to occur.

Electrolytic Corrosion. On the other hand, electrolytic corrosion occurs when there is a cell consisting of two metallizations (even of the same type of metal—here it is Al) but with externally applied bias. There too the presence of impurities sparks off the reactions:

$$Al + 3Cl^- \rightarrow Al^{3+} + 3Cl^- + 3e \qquad \text{(anodic reaction)} \tag{8-5}$$

The ionized Al is transported to the cathode, where we have

$$Al^{3+} + 3e \rightarrow Al \qquad \text{(cathodic reaction)} \tag{8-6}$$

But the Al is not able to deposit in these conditions, and in the presence of humidity the following reaction occurs:

$$2Al + 3H_2O \rightarrow Al_2O_3 + 3H_2 \tag{8-7}$$

Corrosion appears as an interruption (open circuit) in the Al or in the bonds preceded at times by the degradation of the electrical characteristics of the device (e.g., increased leakage current).

The corrosion phenomenon is therefore accelerated by the impurities carried by the H_2O when it crosses the resin and laps against the metal surfaces of the frame and by the voltage applied to the device (electrolytic corrosion). The phenomenon is delayed by passivation of the die and by increasing the thickness of the Al metallizations.

Humidity tests are therefore used to evaluate:

- Plastic-frame adhesion and possible package cracks
- Permeability of plastic to water and corrosive atmospheric pollutants
- Plastic, die attach, and frame-plating purity (ionic contamination)
- Passivation quality (condensation occurs mainly in passivation cracks)
- Design characteristics (i.e., aluminum thickness, quality, and morphology; internal slug and frame geometric design; passivation type; phosphorous content for P. Vapox).

8.2.2 Correlation between Various Laboratory Tests and Models

Various *reliability models* were presented in the 1960s to find the correlation between the various tests and conditions. These included:

- Pressure-vapor model #[1]
- $(RH)^2$ model #[2]
- Eyring model #[3]
- Peck model #[4]
- Reich-Hakim model #[5]

In recent years detailed studies have been made to find out which of these models corresponds most closely to the experimental results.

SGS experience has shown that the *Reich-Hakim* [5] is the best model. This model is based on an empirical law that links the median life exponentially to a number which is the sum of the test temperature expressed in degrees Celsius and the test relative humidity expressed as a percent:

$$t_M = e^{A+B(T+\mathrm{RH})} \tag{8-8}$$

where A and B are two constants to be determined experimentally.

Figure 8-2 shows the difference in acceleration factor between one test type and the other, with bias (15 V) and without, and numerous

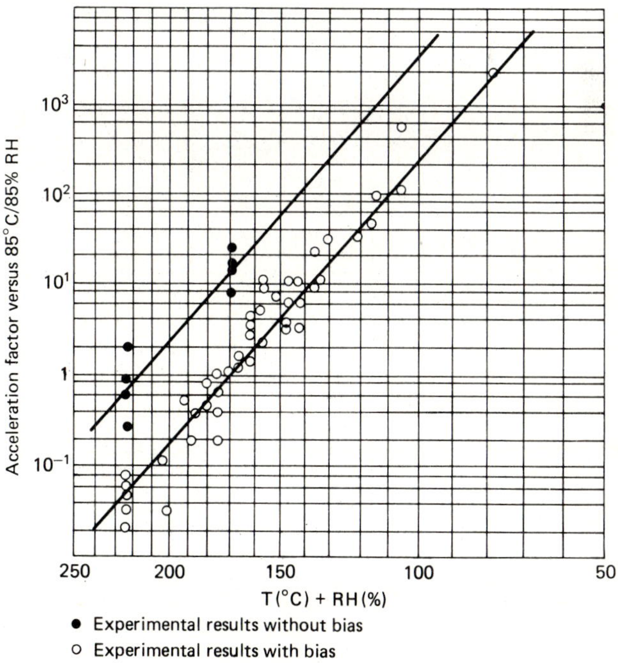

FIG. 8-2 Acceleration factors of various humidity tests compared, vs. 85°C/85% RH, with and without bias.

experimental results confirming the validity of the Reich-Hakim model. Figure 8-2 was prepared by giving a value of 1 to the median life obtained with 85°C/85% RH bias.

The Reich-Hakim model, which did not originally contain the applied voltage, is used at SGS both for applied-voltage tests and for tests without applied voltage.

8.2.3 Correlation between Laboratory Tests and the Field

When the integrated circuit operates in equipment it dissipates power, which causes an increase in heating of the die with respect to the surrounding environment. Figure 8-3 enables us to calculate how the rel-

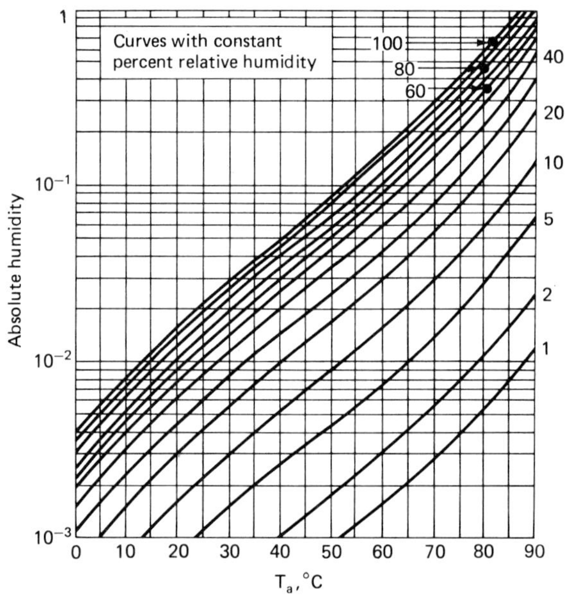

FIG. 8-3 Relative humidity vs. absolute humidity and ambient temperature.

ative humidity varies on the die as a function of the increase in temperature.

The following example will help to explain the process. If we assume that:

- The ambient temperature of the equipment is 30°C
- The relative humidity of the environment is 80 percent

- The On time of the equipment is 25 percent (21,500 h in 10 years)
- The Off time of the equipment is 75 percent (65,500 h in 10 years)
- ΔT_j (when equipment is on) is 20°C
- The useful life required by the equipment is 10 years

we can make the following calculations:

a. *Dormant Conditions*

- Ambient conditions: 30°C/80% RH (no bias)
- Laboratory test conditions: 85°C/85% RH (no bias)

Considering the Reich-Hakim diagram of Fig. 8-2, the ratio between the two conditions (acceleration factor) is (85/85)/(30/80) → 100; that is, 655 h of 85/85 testing without bias is equivalent to 65,500 h of field testing without bias.

b. *Operating Conditions*

If the die temperature is increased by 20°C during operation of the equipment, the ambient conditions of the die are modified as follows:

$$T_j = 30° \text{ (amb)} + 20° \ (\Delta T_j) = 50°\text{C} \tag{8-9}$$

A relative humidity of 80 percent at 30°C becomes 30 percent at 50°C (see Fig. 8-3). For a die temperature increase of 5°C during the test, 85/85 becomes 90°C and 70% RH (see Fig. 8-3). Considering the Reich-Hakim diagram, the ratio between the two conditions (acceleration factor) is

$$\frac{90/70}{30/50} \rightarrow 750 \tag{8-10}$$

that is, around 30 h of 85/85 testing with bias is equivalent to 21,500 h of operation in the field. Therefore it would be necessary to carry out an 85/85 laboratory test without bias for 655 h plus a 30-h 85/85 test with bias. Considering an average acceleration factor of 13 for the tests with and without bias, we could carry out 80 h of 85/85 tests with bias or 1045 h of 85/85 tests without bias to guarantee the same results.

8.2.4 Humidity Test Elements Not Yet Fully Explained

The effect of corrosion on devices immersed in humid environments for long periods without bias followed by short intervals of voltage and then applied power has still to be evaluated completely. Laboratory tests with On-Off cycles of the applied voltage are being proposed for this purpose [7]. In any case, up to now these tests have not been standardized because the number of experiments has been insufficient to show a correlation with the various applications.

Finally, the Reich-Hakim law is no longer valid when the humidity falls below certain values. Various experts maintain that below about 30 percent relative humidity the corrosion phenomena become negligible and therefore of little practical interest.

8.3 QUALITY OF ENCAPSULATING RESINS

Semiconductor packages can be of two types: *plastic* and *hermetic*. In addition to providing a support for the die, connecting it to the external leads, and dissipating power, the package must also protect the integrated circuit or transistor from external agents, in particular humidity.

In devices in *plastic packages* the predominant failure mode is an open circuit or intermittent contact, which can be attributed to mechanical stress caused by the different expansion coefficients of the package material and the metal parts. Furthermore, if these devices are exposed to a humid environment for long times, corrosion of the internal metallizations becomes a big risk for device reliability. The *hermetic package*, on the other hand, is less resistant to mechanical stresses such as vibrations and shocks, since it does not form a compact block between the package and the internal connections. The mechanical and/or thermal stresses can also compromise the hermeticity.

Despite these limitations the hermetic package is normally preferred for military and space applications and in general for all applications requiring high intrinsic reliability of the component.

We do not intend to discuss the reliability of hermetic power packages here since a vast amount of literature already exists on these packages, which were the first to be used. Today the majority of power integrated devices are supplied in plastic packages, which are much more economical than the hermetic ones.

The study of the quality aspects of encapsulating resins has therefore become more and more important in recent years. More and more sophisticated criteria and methods have been used to investigate their chemical and physical characteristics with regard to the aspects that may influence the performance of discrete and integrated components.

A complete examination of these criteria and methods would go beyond the limits of this book. We shall therefore only discuss the most significant methods of evaluating the reliability of an encapsulating resin.

Two main types of resin are used in the fabrication of semiconductor devices:

1. Silicon resins
2. Epoxy resins

Silicon resins, which were widely used in the past, are today mainly used in the production of discrete power components and in general for devices to be used in high-frequency or high-voltage applications, where good dielectric characteristics are necessary.

The weak point of the present silicon resins is their poor adhesion to the metal surfaces of the frame, due to the formation of water during polymerization. Humidity and other polluting elements present in the external environment can therefore easily reach the die through the frame-resin interface.

Epoxy resins are almost always used in the production of linear and digital integrated circuits in plastic packages. In addition to being easily workable these resins offer high purity, good mechanical resistance and thermal conductivity, a low thermal expansion coefficient up to 150°C, excellent shrinkage after polymerization (0.4 to 0.6 percent), and limited costs.

The evaluation of an encapsulating resin requires not only a wide chemical and physical characterization but also a complete series of reliability tests.

The chemical and physical characterization tests include:

1. Determination of *polymerization kinetics,* including study of the effect of postcure conditions on the physical and chemical characteristics of the resin and of the relative consequences on the reliability of the device
2. Determination of *thermal stability* and of thermal decomposition mode in inert atmosphere and in air
3. Determination of *workability* characteristics of the compound
4. Determination of *ion content*
5. Determination of *electrical characteristics*

Reliability tests, on the other hand, must be carried out on significant test patterns and test vehicles to check the compatibility of the fabrication process and of the dice passivation with the characteristics of the encapsulating resin.

In the following paragraphs the fundamental aspects of each point will be examined and their importance evaluated.

8.3.1 Polymerization Kinetics and Influence of Degree of Cure

The basic component of the epoxy compounds used to seal integrated circuits in plastic packages via transfer molding is a thermo setting resin in the form of a powder which, once thermally activated, undergoes a highly exothermic polymerization reaction. This chemical reaction, which involves setting up crossed links between the various molecules, gives rise to a complex polymer which is unmeltable, insoluble, and stable at normal operating temperatures. The number of links formed is expressed as a percentage of the total and is known as the *degree of cure*.

The polymerization reaction develops mainly during the first few minutes after activation, but it is completed only after several hours of storage at suitable temperatures (postcure time).

A high degree of cure provides considerable advantages from the point of view of the final product characteristics (such as dielectric, physical, chemical, and mechanical properties) that affect the quality and reliability of the device.

The polymerization kinetics and the degree of cure can be measured with the help of a *Differential Scanning Calorimeter* (DSC), which operates on the principle of maintaining the sample of material examined in an environment with strictly controlled temperature and continuously monitoring the heat exhange between sample and environment.

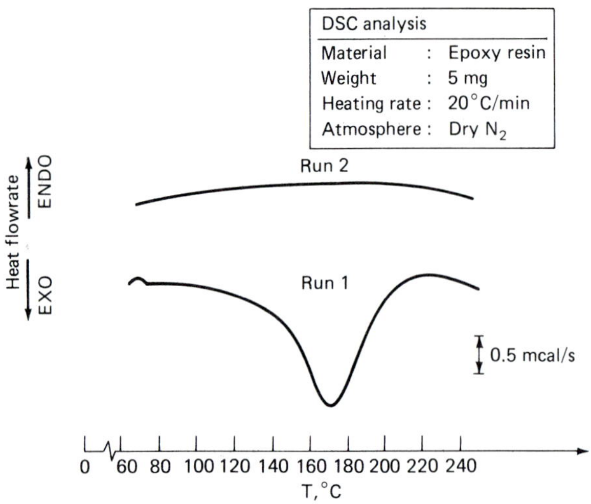

FIG. 8-4 DSC analysis of completely unpolymerized (Run 1) and of completely polymerized (Run 2) epoxy resin (y axis: arbitrary origin).

The measurements shown in Figs. 8-4 to 8-10 were made using a thermal analysis system including a DSC. In the model used here, any change of thermal energy (whether endothermic or exothermic) in the test sample, with respect to a reference, is automatically compensated by the electrical supply in order to keep the sample at the set temperature even during a transition or a reaction. The compensation power is supplied as an output of the instrument directly in millicalories per second and is shown on the chart recorder.

Figure 8-4 shows the typical behavior of an epoxy resin subjected to DSC analysis.

The lower trace (Run 1) is the calorimetric response obtained during scanning of a completely unpolymerized material and carried out in an atmosphere of N_2 to prevent oxidation at high temperatures. Note the large exothermic peak due to the development of the polymerization reaction. The width and profile of the curve depend not only on the base resin but also on the various additives used in the fabrication of the compound (catalysts, flame retardant, fluidizers, mold release agents, etc.).

Since in practice there is an almost perfect repeatability of the curves produced by samples belonging to different lots of the same compound, the DSC analysis on the raw material is an excellent method of identifying and checking the compound. Note, for example, in Figs. 8-5 to 8-8 the small difference in the curves obtained from identical scans on two resin samples of two different suppliers.

In both compounds, before the exothermic peak appears, a small

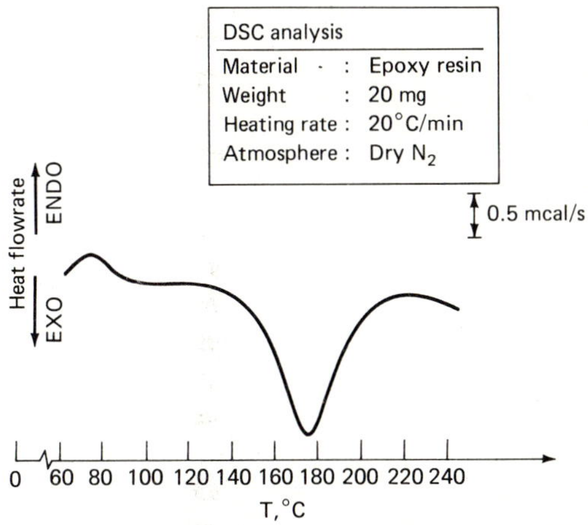

FIG. 8-5 DSC analysis of completely unpolymerized epoxy resin: supplier A, lot A (y axis: arbitrary origin).

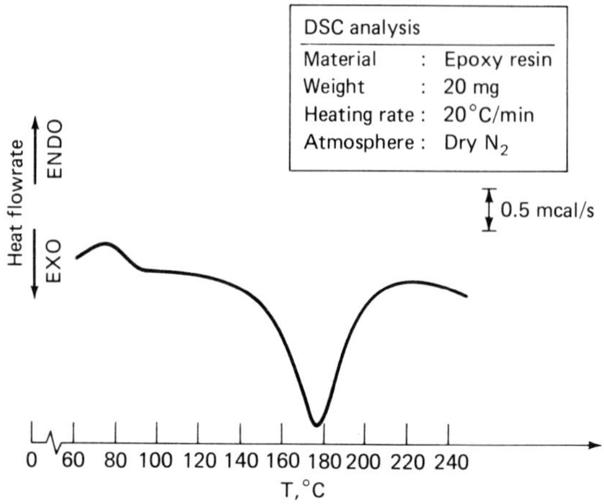

FIG. 8-6 DSC analysis of completely unpolymerized epoxy resin: supplier A, lot B (y axis: arbitrary origin).

endothermic peak is produced due to liquidization of the additive used as a fluidizer and for removal from molds.

To show the curve of the endothermic peaks of various types of resin coming from different suppliers, the scans made in the temperature range of interest (60 to 110°C) are given in Fig. 8-9. The presence of natural compounds (wax) is detected by endothermic peaks showing melting at temperatures between 60 and 100°C. Note the absence of endothermic peaks in the silicon resin due to the lack of release agents, and in the epoxy resin of supplier C due to the use of a completely artificial release agent.

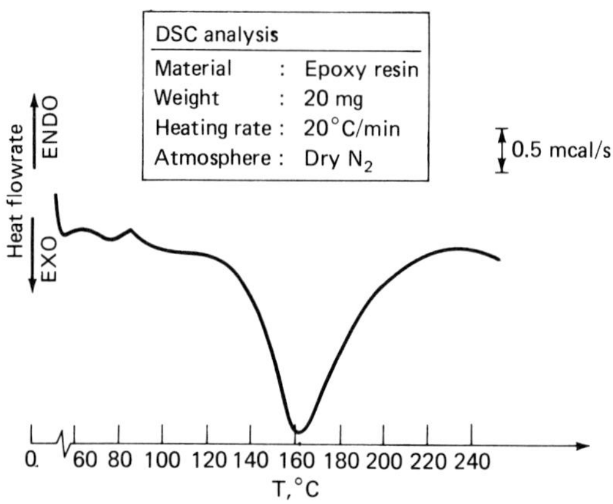

FIG. 8-7 DSC analysis of completely unpolymerized epoxy resin: supplier B, lot A (y axis: arbitrary origin).

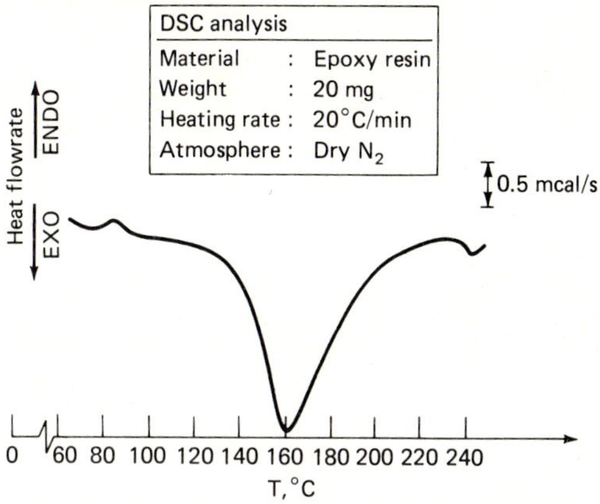

FIG. 8-8 DSC analysis of completely unpolymerized epoxy resin: supplier B, lot B (y axis: arbitrary origin).

As regards the exothermic peak in the epoxy compound of supplier A (Fig. 8-5), it begins around 120°C, reaches a maximum at around 175°C, and is exhausted around 225°C. In that of supplier B (Fig. 8-7), on the other hand, it begins around 100°C, reaches a maximum around 162°C, and is exhausted around 240°C.

Note how the curves produced by samples of different lots of the same resin are almost exactly the same.

The upper trace (Run 2) in Fig. 8-4 is the response obtained from

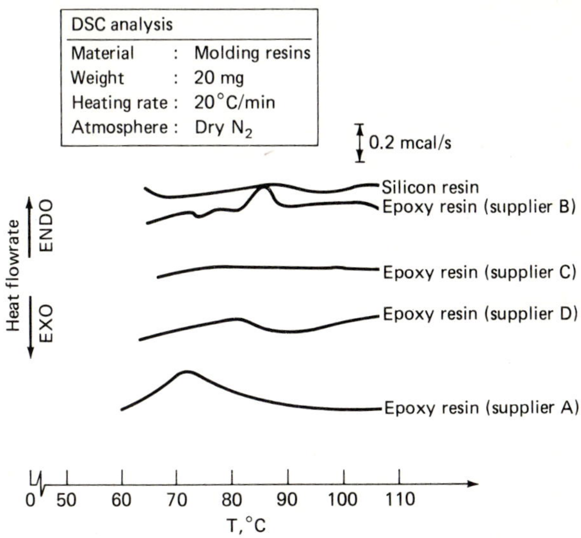

FIG. 8-9 DSC analysis of various resin types.

another scan made in the same conditions and on the same sample as Run 1. The absence of an exothermic peak indicates that the material is completely polymerized. This trace is not rectilinear since the specific heat of the resin varies as the temperature varies.

If the two curves Run 1 and Run 2 are superimposed, it is possible to integrate in time the area subtended from the exothermic polymerization peak, whose value, expressed in suitable units, provides the total amount of heat emitted by the sample during the reaction.

A sample subjected to scanning when already partly polymerized produces an intermediate response between the two curves (Run 1 and Run 2) of Fig. 8-4, and in this case the area subtended is directly proportional to completion at 100 percent of the degree of polymerization. The area subtended from the peak is in fact proportional to the number of residual links still available in the material being examined. It is thus possible to determine the *degree of polymerization* of a material by comparing the areas of the peaks developed by the sample with those produced by an identical quantity of material completely unpolymerized.

Furthermore, since the curves for different lots of the same resin are almost identical, the chemical age of the resin can be determined both when the material is received and after long periods of storage. An old lot of resin in fact can give rise to material which is difficult to work and which after polymerization is full of vacancies and permeable by humidity in the environment.

Figure 8-10 shows a DSC analysis carried out on samples of epoxy resin isothermically polymerized (the rise-and-fall speed to reach operating and ambient temperatures is 320°C/min) for 10 s, 90 s, and 1 h. The degree-of-cure values obtained were, respectively, 49, 86.7, 98.7 percent.

We can immediately note the exponential nature of the growth in time of the parameter in question. Later we shall see how it affects the fundamental physical characteristics and therefore how it affects the reliability of the finished IC.

8.3.2 Thermal Stability of Encapsulating Resins

The thermal stability of molding compounds is a characteristic which is continually gaining importance in the field of electronic components. In fact, although the polymers normally used today are stable up to temperatures higher (200 to 220°C) than the maximum operating temperatures of the device, the need to subject devices in plastic packages to accelerated tests or to higher and higher temperatures is becoming increasingly important.

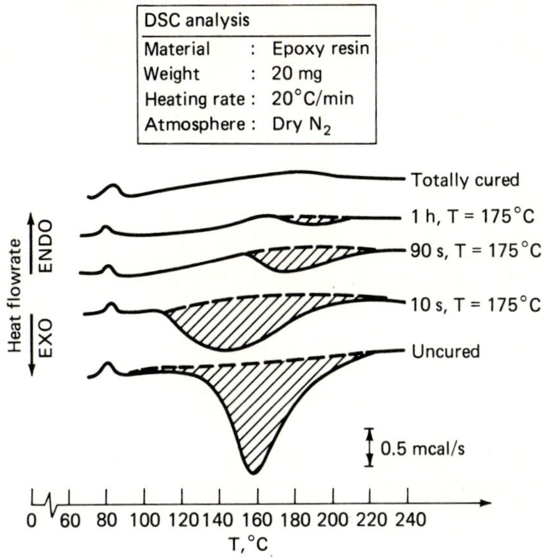

FIG. 8-10 DSC analysis of uncured, totally cured, and partially cured samples of epoxy resin. Dashed areas are proportional to degree of cure.

Thermogravimetrical techniques are used to determine the thermal stability of the encapsulating resins. These involve recording the loss (or gain) in weight of a material as the temperature varies. The *thermogravimetrical profile* obtained is a function of the conditions used for the analysis (type of atmosphere, heating rate, etc.). The width and position of the various stretches of the curve provide another element of diversification, different from those seen in the DSC analysis, within the single types of resin. Use of the oxidizing atmosphere, in particular, enables determining what type of compound is being examined. In fact the final inorganic residue (consisting only of a filler for the epoxy resins and of filler plus the silicon content of the resin for silicon resins) is:

55 to 65 percent for the *epoxy anhydride cured*

65 to 75 percent for the *epoxy Novolac cured*

90 to 95 percent for the *silicon*

In an oxidizing atmosphere the efficiency of the flame retardants can also be evaluated. The system used for the thermogravimetrical measurements mentioned here allows the analysis in static or flowing inert or active gas up to 1000°C, ± 2°C, and with a sensitivity of 0.1 μg.

Obviously it is not possible to evaluate the difference between mate-

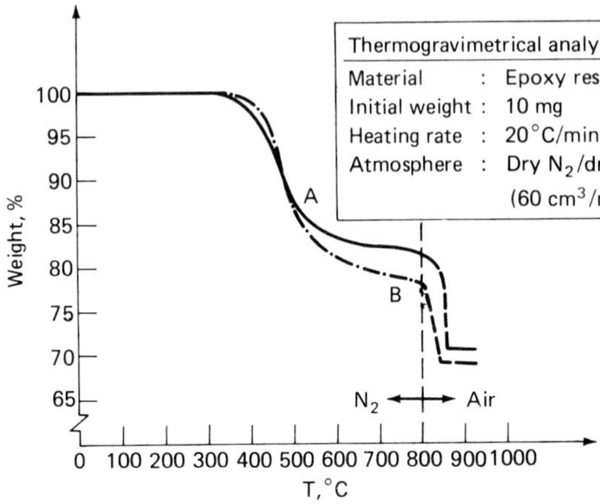

FIG. 8-11 Thermogravimetrical analysis of two different epoxy resins.

rials with different degrees of cure using the same method, since, if kept at a high temperature, the polymerization progresses rapidly, attenuating the differences until they disappear.

Figure 8-11 shows the typical decomposition profiles (in an atmosphere of inert N_2 up to 800°C and then in air) of two different epoxy resins. In the curves we can detect a first phase of initial decomposition (at about 400°C) of the base resin which forms the compound, a second intermediate phase of cracking and progressive volatilization of the polymer, and a final phase (at 800°C) of combustion in air of the polymer residue and of the coloring pigments. The final rectilinear stretch identifies the percentage of inorganic material present.

Figure 8-12 shows the decomposition in air of the same resins as in Figure 8-11 for two different scanning speeds. The resin of supplier B is not flame-retarded.

The flame retardant agent has a significant effect on the decomposition rate, noticeable especially in conditions of maximum temperature rise time (320°C/min), which are most similar to the process of combustion alone. These are the only conditions that tend to indicate any difference in stability between materials with different degrees of cure, since they simulate the case in which the material suddenly finds itself beyond the decomposition temperature without having spent a long time at a temperature which would complete the polymerization.

Experiments have shown that the various degrees of cure are more or less equivalent as regards decomposition with temperature. The profiles for unpolymerized materials in powder, on the other hand, are quite different.

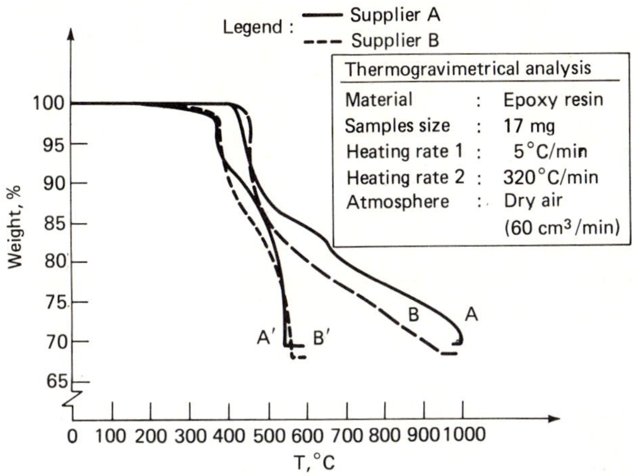

FIG. 8-12 Thermogravimetrical analysis of two different epoxy resins for two different scanning speed rates (sample B is not flame-retarded).

8.3.3 Linear Thermal Expansion Coefficient and Glass Transition Temperature

One of the most important characteristics of plastic compounds used to seal electronic devices is the linear thermal expansion coefficient (CTE). This parameter affects the reliability of the device especially in severe operating conditions as regards thermal stresses (shocks, cycles, thermal fatigue). Its value is affected by any changes in the composition or structure of the basic epoxy resin (e.g., degree of cure).

To determine and study the linear thermal expansion coefficient and the characteristics which influence it, a dilatometric system was used. This sytem allows operation in the temperature range -170 to $+325°C$ and the appreciation of variations in length up to 0.1 μm.

Figure 8-13 shows the typical result of temperature scanning on a sample of epoxy compound. There is an evident increase in the expansion rate when the temperature at which the glass transition of the material occurs is exceeded.

The *glass transition temperature* T_g is universally defined as the intersection point between two almost linear sections of the dilatometric curve. The value of T_g is strictly correlated to the degree of polymerization of the compound. The expansion coefficients before and after the transition are calculated from the slopes of the two segments of the curve in Fig. 8-13 according to the following formulas:

$$\gamma_1 = \frac{\Delta L_1}{\Delta T_1 \cdot L_0} \tag{8-11}$$

$$\gamma_2 = \frac{\Delta L_2}{\Delta T_2 \cdot L_0} \tag{8-12}$$

where γ_1 = thermal expansion coefficient of resin in temperature range below T_g

$\Delta T_1 = T_1 - T_0$, temperature variation (in the range below T_g) used in a resin test to determine γ_1

$\Delta L_1 = L_1 - L_0$, variation in length of specimen due to temperature variation ΔT_1

L_0 = length of specimen at temperature T_0

γ_2 = thermal expansion coefficient of resin in temperature range above T_g

$\Delta T_2 = T_2 - T_0$, temperature variation (in the range above T_g) used in a resin test to determine γ_2

$\Delta L_2 = L_2 - L_0$, variation in length of specimen due to temperature variation ΔT_2

A series of measurements carried out on various lots of resin from two different suppliers made it possible to evaluate the variation in the thermal expansion coefficient and in T_g as a function of the degree of cure.

Figure 8-14 shows the average percentage variation of the thermal expansion coefficients for temperatures above and below the glass transition temperature as the cure time varies. On Fig. 8-14*a* and *b*, the variations of CTE are referred to a cure time of 16 h at T_a = 175°C. The variation of T_g as the cure time varies is shown in Fig. 8-15.

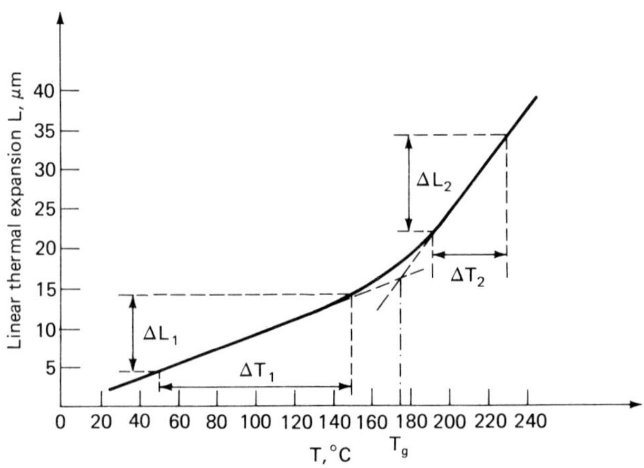

FIG. 8-13 Typical curve of epoxy resin linear thermal expansion vs. temperature. T_g is the glass transition temperature.

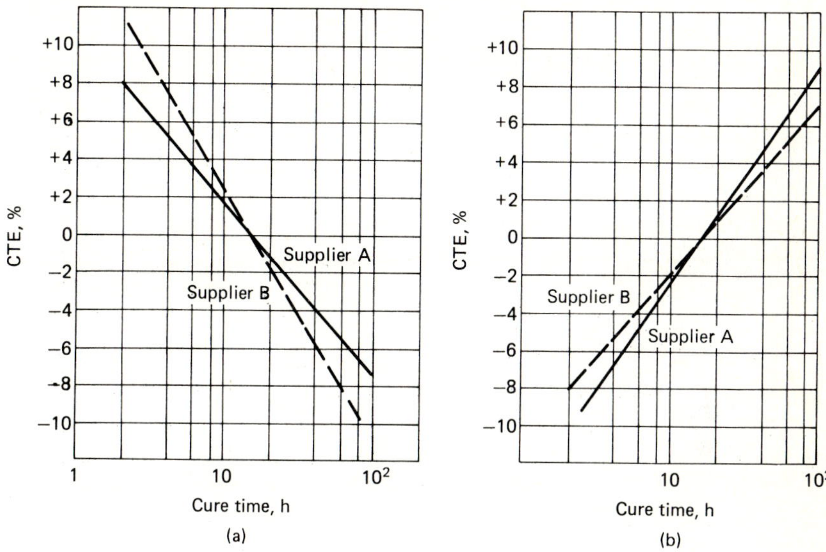

FIG. 8-14 Average percentage variation of the CTE vs. cure time: (a) with $T < T_g$; (b) with $T > T_g$.

Note in the three graphs of Figs. 8-14 and 8-15 the progressive modification of the parameters as the degree of cure proceeds. In particular the thermal expansion coefficient tends to decrease and T_g to increase as the degree of cure grows. Both these elements improve the reliability of the device (IC or transistor).

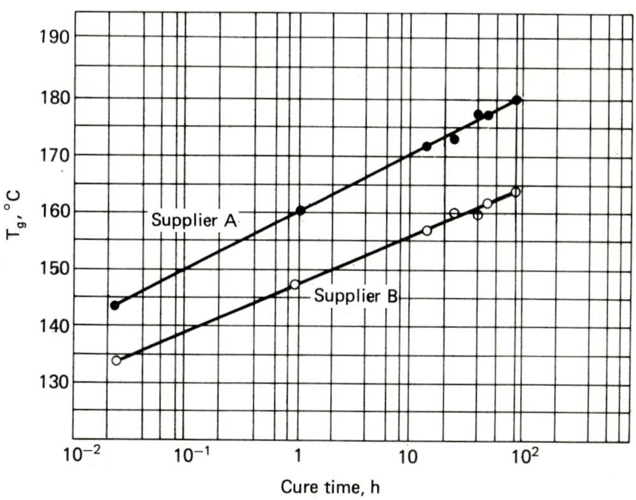

FIG. 8-15 Typical variation of T_g vs. cure time.

8.3.4 Determination of Ion Content in Encapsulating Resins

Chemical analysis of plastic compounds for transfer molding aims to quantify the content of hydrolizable ions, which are potentially dangerous because they contaminate the surface of the die. Table 8-1 shows the contents of the principal substances, parts of which are chemically linked to molecular chains of the polymers forming the resin. This table gives the results of a series of analyses carried out according to the specifications of the British Post Office publication M219C [33] on silicon and Novolac epoxy resins.

Although these analyses may be significant for comparing first-generation plastic compounds, they are no longer sufficient for a similar comparison of different compounds which the technological progress of recent years has brought to high degrees of purity. Their validity must therefore be confined to the comparison of different lots of the same material [8].

The presence of hydrolizable ions and in particular their mobility has a significant effect on the stability in time of some integrated structures. This mobility can be evaluated through thermally simulated discharging [9].

8.3.5 Physical and Electrical Characteristics of Encapsulating Resins

Table 8-2 shows the main physical and electrical characteristics of some silicon and epoxy encapsulating resins published in the data sheets of the respective manufacturers. Note the asymmetry of the thermal expansion coefficients and the release of postcure shrinkage of the silicon resins (the first two resins). These two factors are the reason why the silicon resins show low resistance to thermal stress while the silicon-epoxy and all epoxy resins in general provide good performance under these stresses.

It should also be remembered that the deterioration of the electrical characteristics, as either the temperature or the humidity permeation increases [10,11,12], is worse in epoxy resins than in silicon resins. These factors have had a negative influence on the introduction of epoxy resins into discrete power components, which require good electrical performance at high temperatures and voltages. In power integrated circuits, on the other hand, epoxy resins are widely used because of the lower operating temperatures required.

The introduction of epoxy resins has also been encouraged by the fact that manufacturers have managed to obtain nonflammable com-

TABLE 8-1 Analytical Data of Various Silicon and Epoxy Resins

		Type							
Analytical data	Units	A*	B*	C†	D†	E†	F†	G†	H†
Filler content	%	88‡	90‡	72	69	68	75	70	67
Cl content	ppm	358	320	264	226	285	338	346	345
Na content	ppm	982	1100	20	16	20	21	19	24
K content	ppm	228	471	13	7	22	19	31	9
Ca content	ppm	28,400	24,200	65	53	91	110	63	195
Fe content	ppm	573	384	104	63	256	253	311	215
Sb content	%	absent	absent	0.44	0.99	absent	absent	1.28	0.66
N content	%	absent	absent	0.55	0.42	0.45	0.44	0.59	0.44
Water extract conductivity§	μS/cm	122	120	105	85	116	105	121	72
Water extract pH§		7.7	7.2	5.3	5.8	5.2	5.2	5.4	5.5
Na leach§	μg/mL	1.5	0.4	0.5	0.1	0.3	0.4	0.6	0.1
K leach§	μg/mL	0.2	0.2	0.6	0.1	1	1.3	0.2	0.7
Cl leach§	μg/mL	4.6	1.5	16.7	7.9	20	11.5	28.5	11.7

*Silicon compound.
†Epoxy Novolac compound.
‡Filler plus silicon resin residue.
§In accordance with M219C [33].

TABLE 8-2 Data Sheet Characteristics of Cured Silicon and Epoxy Resins

Cured material characteristics	Units	KMC-10	DC306	EM4700	P410B/SG	MG 17 F	P415SG	P410C
Glass transition temperature T_g	°C	absent	absent	165	165	175	160	165
CTE	10^{-6} °C^{-1}	24* (49)	22* (40)	25	25	25	24	29
Linear shrinkage	%	0.4† (0.25)	0.5† (0.38)	0.6	0.6	0.6	0.5	0.6
Specific gravity	g/cm^3	1.9	1.9	2	1.7	1.8	1.8	2.1
Thermal conductivity	10^{-4} cal/(s·cm·°C)	12	15	30	15	17	15	35
Flexural strength	10^{-3} lb/in^2	9	9	18	20	18	20	20
Dielectric constant		3.9	3.6	4	3.7	3.8	3.8	4.2
Dissipation factor	10^{-3}	1.8	1.6	6	4	3	3	4
Dielectric strength	V/mil	380	320	450	350	103	350	350
Arc resistance	s	250	290	180	190	190	190	190
Volume resistivity	10^{15} Ω/cm	2	2.8	1	1.5	5	1.5	1.5
Surface resistivity	10^{15} Ω	absent	absent	absent	1	5	1	1
Flammability‡		VO§	VO§	VO§	absent	VO§	VO§	VO¶

*In the same direction as molding flow; the number in parentheses is the value normal to molding flow.
†After molding; the number in parentheses is the value after postcure.
‡In accordance with UL 94 [34].
§1/16-in specimen.
¶1/8-in specimen.

pounds (like the silicon resins) without deteriorating the purity characteristics.

Reliability Tests. The evaluation, selection, and control of encapsulating resins are completed with reliability tests of the products. The principal ones are:

1. *HTRB tests* to evaluate the effects of contaminating ions on the devices (see Sec. 8.8.3*c*)
2. *Humidity tests* to evaluate the permeability of the resins and control the action on the dice (see Sec. 8.8.4)
3. *Thermal tests* (cycles and shocks) to evaluate the thermomechanical compatibility of the various components (wires, die, frame with resin) (see Secs. 8.8.1 and 8.8.2)

8.4 THE PROBLEM OF Au-Al INTERMETALS IN POWER ICs

One of the biggest problems faced by the semiconductor industry has been that of understanding the reason for failure of the bonds connecting the Au wires to the Al pad and trying to prevent it; the size of the problem is due to the important effect this type of failure has on the life of most devices in plastic packages. Much effort has therefore gone into the study of the failure modes, of the Au-Al intermetal phases present during bonding, and of how these evolve in time following stress at variable or constant temperature.

A comprehensive analysis has been made of the *phase diagram* of the Au-Al binary system, the *influence of contaminants, optimal bonding conditions* (time, temperature, pressure), and the *Al metallization thickness.* Various conclusions have been reached, at times contradictory because of the different methods used to prepare the sample and to carry out the tests. Some studies were made using thin Au and Al films in close contact, others produced the Au-Al bond by bonding Au or Al wires on metals deposited on various types of substrate (for example, SiO_2), and still others created the bond through electrical soldering of the Au and Al wires.

Each of these methods has its limitations. In the case of superimposed thin films the limiting factor is the dominating role of the surface effects compared with the volume effects. In the case of the bonding of wires to metals, the limitations are not only the different techniques used in the bonding operation, but also the practical impossibility of repeating (even by using the same equipment) exact evaluations at different times and obtaining the same results because of the numerous variables involved.

Finally, in the case of electrical soldering of Au or Al wires (even though they contain Si impurities to simulate the composition of the metals of the IC), the system is limited by the insufficient likeness to the practical case of ICs.

Nevertheless, many of the results obtained correspond to those found in practice, and they explain phenomena occurring in ICs after the application of electrical and/or thermal stresses.

8.4.1 Formation Kinetics of Au-Al Intermetals

When soldering Au and Al, five intermetallic compounds may be formed ($AuAl_2$, AuAl, Au_2Al, Au_5Al_2, and Au_4Al) in accordance with the diagram of the Au-Al binary system phases (see Fig. 8-16). Each of the

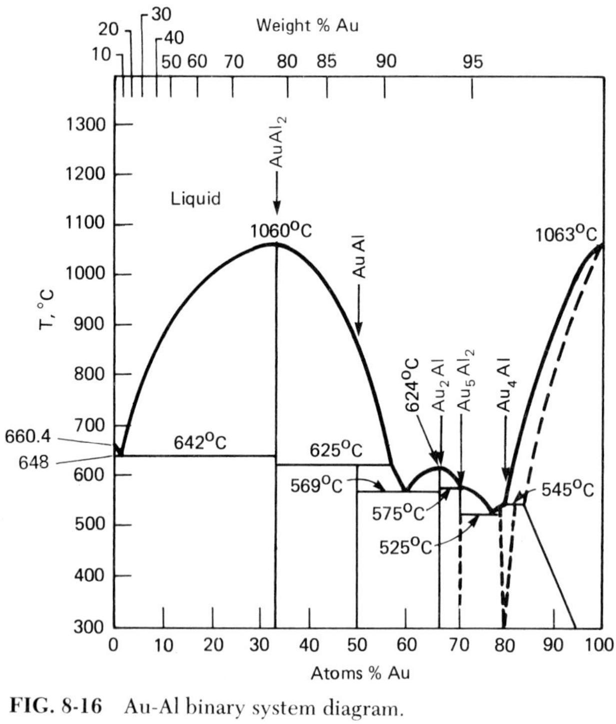

FIG. 8-16 Au-Al binary system diagram.

five compounds can be obtained by subjecting thin Au and Al films, superimposed and deposited in stoichiometrically exact quantities, to the effects of time and temperature.

In the practical case of power ICs, however, the thick Al metallizations ($> 1\ \mu m$) and the quantity of Au in the ball at the point of contact make it difficult to establish which phases will form during bonding and in what sequence they will evolve in time.

The phenomenon of interdiffusion of the two metals in these circumstances is controlled by the parabolic law, which is valid for a given temperature:

$$X = (D \times t)^{1/2} \qquad \text{cm} \tag{8-13}$$

where X = thickness of compound layer formed
D = rate constant, cm^2/s
t = time, s

The rate constant D is a function of the temperature and is expressed as follows:

$$D = D_0 e^{-E_A/kT} \qquad cm^2/s \tag{8-14}$$

where D_0 = constant, cm^2/s
E_A = activation energy, eV
T = temperature of bond, K
k = Boltzmann constant, eV/K

Equation (8-14) is valid (in the case of Au-Al bonds) in the temperature range between 200 and 460°C [13] and works on the assumption that the diffusion phenomenon is continuously supplied by the presence of both materials.

In this assumption, all authors today accept that

$$D_0 = 5.2 \times 10^{-11}\ cm^2/s \tag{8-15}$$

which represents the value of the constant relative to the growth speed of Au_5Al_2, the predominant compound during the formation of the Au-Al bonds, whose growth rate is up to 3 orders of magnitude greater than that of the other four intermetals.

Nevertheless, if the Al or Au were to become exhausted, the diffusion would continue, although very slowly, with the formation of richer compounds in the residual metal. The formation of $AuAl_2$ and Au_4Al would thus be accentuated. Their D_0 constants are a few orders of magnitude smaller than that of the Au_5Al_2 intermetal.

The value of the activation energy E_A varies considerably from one author to another. Table 8-3 gives some values found experimentally and the temperature range for which the values are valid.

Various experiments have been carried out on the Au-Al joints obtained by subjecting them to high-temperature storage, both by researchers [13 to 17], and by SGS laboratories, to establish the for-

TABLE 8-3 Different Activation Energies E_A Experimentally Found by Several Authors

Author	E_A value, eV	Operative range, °C
J. A. Cunningham [14]	1	200–300
G. K. Chen [15]	1.07	175–275
M. Kashiwabara and S. Hattori [16]*	0.67–0.78	150–350
E. Philofsky [13]	0.69	200–460

*The E_A value depends on the modality of treatment of the sample.

mation sequence of the five intermetal compounds and to understand the relative mechanisms.

The first compound formed was the intermetal Au_5Al_2. It must be underlined, however, that there are no precise reasons for its being formed first, and we cannot be certain that this will always happen in the many different conditions in which the bond may be generated.

The formation of the intermetal Au_5Al_2 as the first compound could, at first sight, seem reasonable, since from the diagram of the Au-Al phases (Fig. 8-16) we can see that Au_5Al_2 is a compound whose composition is similar to that of the low–melting point eutectic. This hypothesis, however, is contradicted by the fact that the Au_4Al compound has a composition even nearer to that of the same eutectic.

A second hypothesis is that on contact the five intermetal compounds form simultaneously, and when the two metals are in contact the appearance of one compound in place of another depends only on its growth in time.

If this hypothesis explains the presence of Au_5Al_2 as the first compound (because it has the fastest growth rate), it does not explain the experimentally found variation in the sequences of the intermetal phases according to the method of formation of the bond subjected to tests [17]. Nevertheless, if we accept the hypothesis of a volume interdiffusion phenomenon, the conclusions drawn by E. Philofsky [13], following tests conducted on bonds obtained by soldering Au wires to Al wires and stored at various temperatures, are today commonly accepted.

Philofsky found that at 400°C two intermetals, Au_5Al_2 and $AuAl_2$ (one very rich in Au and the other richer in Al), are generated immediately. These are followed after 3 min by the formation of Au_2Al and after 1.30 h by Au_4Al—the richest in Au—and by AuAl. The last forms in the body of the intermetal $AuAl_2$ due to a gradual increase in Au.

This is also demonstrated by the simultaneous reduction in the quantity of Au_2Al present in the bond. The five phases form in the bond, remaining separate, and tend to arrange themselves according to an order determined by the stoichiometry of the compound.

In the bond, in fact, there is a progression from the phases richest in Al, near the Au_2Al, to the phases richest in Au at the other end.

Philofsky [13] has also studied and determined the behavior of the bonds when the Al contains 5 percent of Si. This situation simulates the practical case of a semiconductor, where the Si can always be assumed to be present both on the surface of the Al metals and in its bulk.

In this case the growth rate of the compounds AuAl, Au_2Al, Au_5Al_2, and Au_4Al does not seem to be affected significantly, while the behavior of the purple $AuAl_2$ phase, in which Philofsky found about 6 percent of the weight to be Si, was different. This led him to consider that there might be a third phase, $AuSiAl_4$.

8.4.2 Problems of Reliability in Au-Al Bonds

The Au-Al intermetal compounds were for a long time believed to be mechanically weak and therefore the main cause of failure of bonds of Au wires to Al pads. Furthermore, since in many failed bonds the purple-colored $AuAl_2$ intermetal was found to be present, the purple plague was taken to be the primary cause. Further investigations [13] showed that the five intermetal compounds formed by Au and Al (even when the latter contains Si), although porous and typically fragile, are mechanically more robust than the original metals. If free of vacancies, they have a resistance to traction higher than the 18,000 lb/in^2 of the Au and have a Vickers hardness about 5 times greater than that of each of the two materials taken individually.

Deterioration and then failure of an Au-Al bond appear first with a gradual increase in the contact electrical resistance, followed by opening of the bond even when it is subjected only to a moderate mechanical stress.

The increase in the contact electrical resistance is a result of the formation of the intermetal phases. In fact, while the electrical resistivity of the Au is $\rho = 2.3 \times 10^{-6}\ \Omega\cdot$cm and that of the Al is $\rho = 3.2 \times 10^{-6}\ \Omega\cdot$cm, the electrical resistivity of the intermetals is always greater than that of the original metals. Therefore, for example, the resistivity ρ of the $AuAl_2$ phase is $7.9 \times 10^{-6}\ \Omega\cdot$cm, and it reaches 25.5×10^{-6} and $37.5\ \Omega\cdot$cm for the richer Au phases such as Au_5Al_2 and Au_4Al, respectively [18].

The later definitive failure of the bond, when it becomes detached, is due to the formation of Kirkendall vacancies in the bond as a result of the faster diffusion of Au in Al than of Al in Au. These vacancies have been found to form mainly in the richest Au zones, the Au_5Al_2 phase, where it meets the Au_4Al and/or the Au. Due to the effects of time and temperature, these vacancies tend to extend until they form almost continuous zones. Although the presence of these zones does not necessarily lead to detachment of the bond, the zones do contribute to increasing its contact electrical resistance, and only the minimum mechanical stress (shocks, vibrations) is needed for the bond to become detached.

The vacancies can form under the bond, as in the case described, or around it. Vacancies around the bond are prevalent when the Al is very thin ($< 0.5\ \mu m$). As the phenomenon evolves, it leads to a gradual increase in the electrical resistance of the bond so that, in the final phase, the bond appears to be mechanically good and electrically open.

This failure mode appears mainly in integrated circuits in which the Al pad lies on SiO_2 (except for rare cases), while obviously it cannot appear in cases where the metal is in direct contact with the Si (e.g., transistors and diodes).

In the final bond degradation phase, if the vacancies are under the bond and the bond is exposed to high temperature and then returns to ambient temperature, this may cause the bond to open definitively.

The five intermetal compounds of the Al and Au have *different lattice constants* and *different thermal expansion coefficients* from one another and from those of the original materials. For example, the crystallographic constants of Au and Al are 4.078 and 4.048 Å, respectively, while those of $AuAl_2$ and Au_4Al are 6 and 7 Å, respectively [18]. The thermal expansion coefficient of Al is $23 \times 10^{-6}\ °C^{-1}$, that of Au is $14.2 \times 10^{-6}\ °C^{-1}$, while that of $AuAl_2$ is $9.4 \times 10^{-6}\ °C^{-1}$.

These crystallographic differences and different thermal expansion coefficients of the intermetal compounds are the cause of their intrinsic weakness when subjected to thermal shocks.

E. Philofsky [13] has found that the cooling of bonds rich in Au_5Al_2 (with Au_5Al_2 rich in vacancies) from 400°C to ambient temperature probably caused the formation of the cracks found by him along the zone where the vacancies had formed previously.

The $AuAl_2$ also tends to form Kirkendall vacancies when exposed to temperature of more than 400°C. While in Au_5Al_2 this tendency derives from the faster diffusion of Au toward the other intermetals than that of Al, in $AuAl_2$ (the rich phase of Al) it derives from the fact that Al leaves this phase faster than Au enters it. The phenomena which occur are, nevertheless, similar to those described for the case of Au_5Al_2.

8.4.3 Bonding Considerations

We have seen the negative effects for the reliability of an Au-Al bond when Kirkendall vacancies form and evolve in it. These vacancies are generated due to the combined effect of temperature and time even in the bond obtained in optimum conditions. It must be emphasized, however, that at normal operating temperatures ($< 200°C$) the time required for formation of vacancies is longer than the estimated useful life of the device.

In practice vacancies may also form for other reasons. When the bonding operation is carried out by using thermocompression, the Au and Al are heated to activate the mutual diffusion of the two materials. Since the diffusion of Au in Al occurs at a different speed from diffusion of Al in Au, Kirkendall vacancies may again form as the bond is created, unless the bonding conditions are optimal. Furthermore, a layer of contaminating ions is often present on the pads.

Deformation of the parts, induced by the bonding operation, breaks this contaminating layer at several points and tends to disperse it, thus encouraging different bonding zones. In operation the vacancies which have thus formed tend to extend until the bond breaks. The degradation phenomenon is accelerated by current focusing and localized overheating in the areas where the bond was made.

Attempts to eliminate this problem have led to bonding techniques (despite the difficulties in regulating the equipment) in which the Au ball is still heated, but a vibration is applied at a tangent to the Al pad (e.g., thermosonic bonding). In ICs the Al of the pad is normally deposited on SiO_2, and in power devices it can be up to 3 μm thick. If the bonding operation is not carried out in optimal conditions, this thickness may be reduced considerably.

In operation, due to the effect of the temperature, the Al of the bonding area interferes with the Au of the wire, changing into AuAl intermetal compounds. When the Al of the pad begins to become exhausted, it starts to separate from the underlying SiO_2. Figures 8-17 to 8-19 illustrate the phenomenon described.

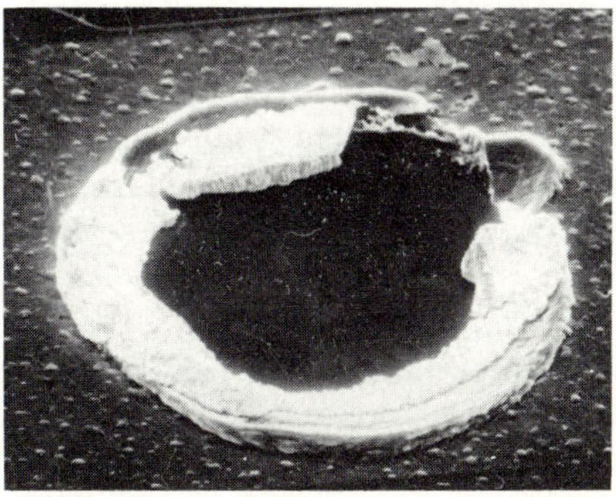

FIG. 8-17 Al pad of an IC after high-temperature storage.

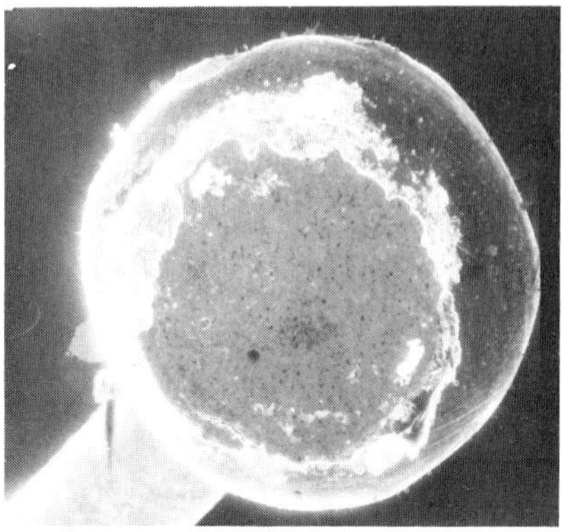

FIG. 8-18 Bonding zone of the Au wire detached from the pad in Fig. 8-17.

In fact Fig. 8-17 shows the Al pad of an IC after storage at 250°C for 64 h. Note the Al around the bond zone and the clear zone rich in intermetal compounds which extend in depth to the underlying SiO_2. The mechanical action exercised to detach the bond wire encourages lifting of the intermetal compounds of the Au-Al bond present in the peripheral zone of the bond, which at the center shows the SiO_2 completely uncovered.

Figure 8-18 shows the bonding zone of the Au wire detached from the pad in Fig. 8-17. Figure 8-19 shows the bonding zone of the Au wire from a different angle. Note the profile of the part where this wire, uncovered by the intermetal compounds, was in direct contact with the SiO_2 of the pad.

8.4.4 Evaluation of Bond Reliability

To establish optimal bonding conditions we must act on several variables, the main ones being:

- Die temperature
- Wire temperature
- Pressure of wire on pad
- Frequency and energy of vibration on wire

The operating conditions listed in turn depend on other variables such as the type of Al of the metallization (grain, thickness, etc.) and the type of wire.

Setting up a bonding system is therefore very complex, involving quality and reliability considerations. *Quality evaluations* cover the

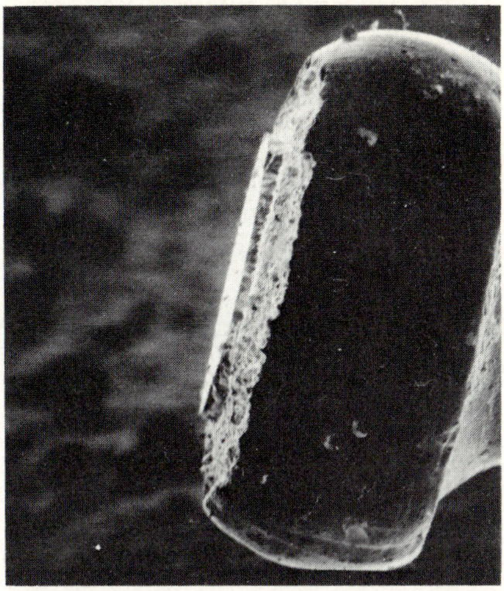

FIG. 8-19 Another view of the bonding zone of the Au wire detached from the pad in Fig. 8-17.

form and dimensions of the bond, visual analysis, and mechanical resistance of the bond to the pull test and/or to shearing stress.

Reliability evaluations are more complex and consist mainly of:

1. *Storage tests* on open parts and on parts after molding, carried out at high temperature in air and in an inert atmosphere. After the test the bonding zones are examined carefully, both inside the Au wire bond zone and in the Au wire–Al pad interface zones after detachment of the wire from pad.
2. *Thermal cycles* and *thermal shocks* on open parts and on parts after molding followed by examination of the bond zone as described in item 1. All these tests are carried out to compare the operating conditions and choose those which guarantee the best median life for the bond.

Figures 8-20 and 8-21 show experimental examples [19, 20] of the resistance to shearing stress and the degradation of the electrical resistance in thermovibrated bonds of Au wire to Al pad in PICs, respectively, after storage tests of 1000 h at a temperature of $T_A = 150°C$, 1000 thermal cycles, and 1000 thermal shocks between extremes of temperature $T_A = -65°C$ and $T_A = +150°C$. Bonding of Au wires to Al pads was carried out under a pressure of 100 g and by subjecting the Au wire to 400 Hz vibration at a tangent to the Al pad for 20 ms.

Three different temperature conditions were also used:

Condition	Au wire temperature,°C	Al pad temperature,°C
A	180	290
B	180	260
C	200	260

The devices were subjected to the tests after molding in epoxy resin and a postcure time of 16 h at T_A = 175°C. Checks were made after careful removal of the resin from virgin samples and from samples taken during tests carried out at various intermediate steps. Note how after completion of the tests solution C still shows excellent mechanical resistance and sufficiently low electrical resistance (< 3 mΩ).

8.5 ELECTROMIGRATION

In its widest definition electromigration is that phenomenon of mass transport caused by the application of an electrical field to a conductor.

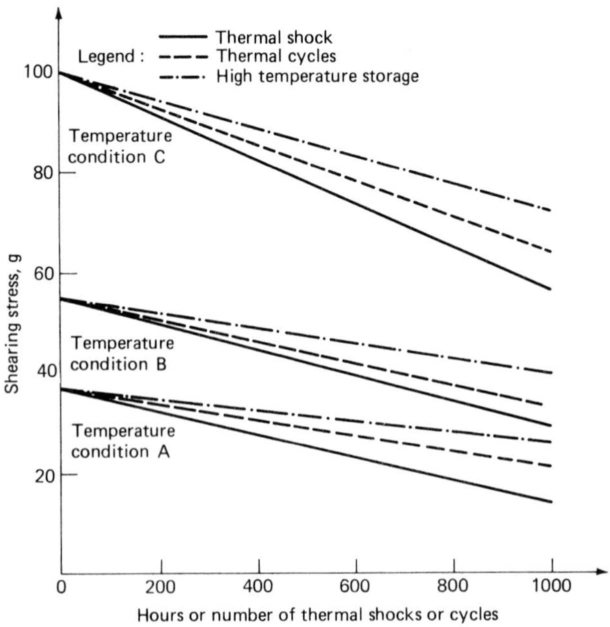

FIG. 8-20 Resistance to shearing stress in thermovibrated bonds after, respectively, high-temperature storage, thermal cycles, and thermal shocks.

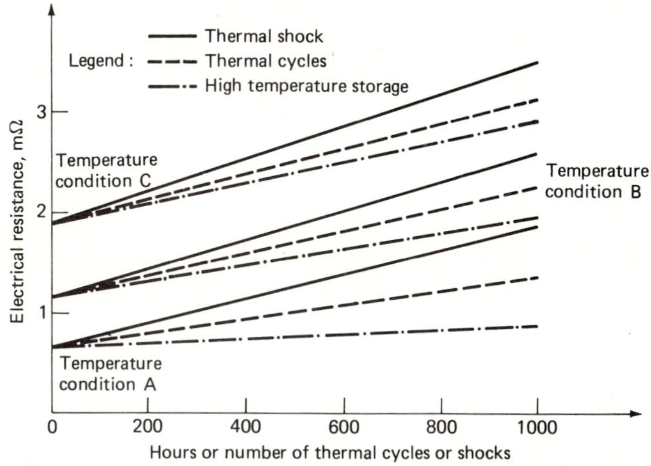

FIG. 8-21 Electrical resistance of thermovibrated bonds after, respectively, high-temperature storage, thermal cycles, and thermal shocks.

If the other conditions remain unchanged, the higher the resulting current level the faster the phenomenon of degeneration of the conductor. This phenomenon has been known for many years, but it was not studied in detail until the late 1960s, when it was discovered to be a failure mechanism in semiconductor devices.

Today it has acquired a new importance for two reasons:

1. Integrated circuits are affected by higher current levels due to the ever-increasing power levels that can be controlled by them.
2. The new generation of LSI and VLSI circuits, which have been made possible due to the development of new diffusion processes, has set the trend to reduce dimensions.

Although recent studies on Au metallizations do exist, the phenomenon has been studied in particular for Al, both as regards deposition methods (electron beam, sputtering) and the influence of final passivation of the die (with which the Al finds itself in contact). In recent years experiments on Al metallizations with other doping elements such as Si and Cu have been conducted to, for example, improve the useful life.

With reference to the studies and experiments carried out, today we are not only in a position to explain the physical phenomenon but we can also extrapolate design rules for dimensioning metallizations with the required degree of reliability.

There are still areas to be explored, however, and many studies are published on the subject.

8.5.1 Electromigration Theory

The most commonly accepted theory, to explain the phenomenon of mass transport in a conductor subjected to an electrical field **E**, is that this field exercises an electrostatic force $\mathbf{F}_1$ on each atom reduced to the ionic state by the thermal effect (Fig. 8-22). Due to the effect of this field, the metal ions tend to move from their original position to an empty one (vacancy) created within the crystalline lattice of the conductor.

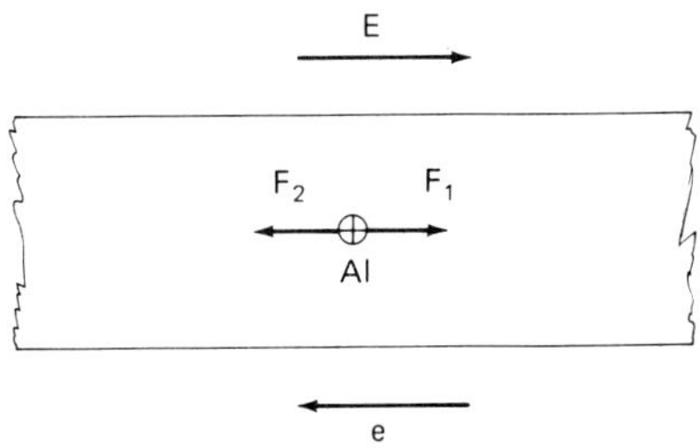

FIG. 8-22 Al metallization subject to electrical field **E**. Arrow e indicates the direction of electron flow (opposite to that of the current flow).

The electron flow in the conductor exercises a force $\mathbf{F}_2$ against the force $\mathbf{F}_1$ due to the effect of **E**, causing a momentum exchange due to collisions between the electrons and each of the atoms in the crystal.

Force $\mathbf{F}_1$ goes in the direction opposite to the electron flow; the main component of force $\mathbf{F}_2$ is in the same direction as the electron flow. Force $\mathbf{F}_1$ is expressed by

$$\mathbf{F}_1 = nq\mathbf{E} \tag{8-16}$$

where q = electron charge
n = ion valence

Force $\mathbf{F}_2$ can be expressed by

$$\mathbf{F}_2 = Aq\mathbf{E} \tag{8-17}$$

where A is a constant that depends on the resistivity of the conductor and on the density of the ions moving in the material. If $\mathbf{F}_1 > \mathbf{F}_2$, the metal ions move in the direction opposite to the electron flow.

If $\mathbf{F}_1 < \mathbf{F}_2$, the ion movement is in the same direction as the electron flow, and there is a tendency of the metal to accumulate toward the most positive zones of the electrical field. This is the case most commonly encountered in practice. The resulting force **F**, to which each atom is subjected, is given by the difference between Eqs. (8-16) and (8-17):

$$\mathbf{F} = \mathbf{F}_1 - \mathbf{F}_2 = (n - A)q\mathbf{E} = A'q\rho\mathbf{J} \tag{8-18}$$

where $\mathbf{E} = \rho\mathbf{J}$, with ρ indicating the resistivity of the conductor and **J** the current density crossing it, and $A' = n - A$.

Force **F** will determine the diffusion of the atom of the lattice with the velocity given by Einstein's ratio

$$\mathrm{v} = \frac{D\mathbf{F}}{kT} \tag{8-19}$$

where k = Boltzmann constant
T = absolute temperature
D = diffusion constant

D can be expressed by

$$D = D_0 e^{-E_A/kT} \tag{8-20}$$

where D_0 = constant
E_A = activation energy of the phenomenon

By substituting the expressions D and **F** in Eq. (8-19), we get the general expression valid for the diffusion speed in single-metal crystals:

$$\mathrm{v} = (A'q\rho\mathbf{J})\left(\frac{D_0}{kT} e^{-E_A/kT}\right) \tag{8-21}$$

in which the first quantity in parentheses expresses the force exerted on each atom by the resulting force **F**, while the second quantity in parentheses correlates mobility of the atoms and lattice temperature.

The main parameters affecting the electromigration phenomenon are, therefore, those characterizing the conductor material as well as the temperature and current density.

8.5.2 Electromigration in Thin Films

Equation (8-21) is generally valid for single-crystal conductors with uniform temperature distribution.

In the case of thin films, consisting of several adjacent monocrystals, although the physical phenomenon remains unchanged and therefore on the whole so does the law which governs it, the mechanisms for transport of material vary considerably.

H. U. Schreiber [21] distinguishes five different causes:

1. *Bulk electromigration* with activation energy $E_A = 1.4$ eV. This energy was measured by using different techniques such as nuclear relaxation, radioactive isotope diffusion of Al, and resistivity recovery experiments after quenching.
2. *Grain boundary electromigration* with $E_A = 0.4$ to 0.5 ± 0.04 eV.
3. *Vacancy motion from grain boundaries into the bulk* with $E_A \approx 0.62$ eV.
4. *Motion of vacation from defects in the bulk* with $E_A > 0.62$ eV.
5. *Surface electromigration* with $E_A = 0.28$ eV due to movement of metal ions toward positions in the crystal lattice which are empty because of broken surface links.

The importance of each of these factors varies as the structural characteristics of the conductor film vary.

J. Black [22] distinguishes between various average dimensions of Al grains:

1. *Small* (1.2 μm): with activation energy $E_A = 0.48$ eV.
2. *Large* (8 μm) and *well-ordered:* $E_A = 0.84$ eV.
3. *Large* (8 μm) and *protected by an SiO_2 film:* $E_A = 1.2$ eV, which is near the value of 1.4 eV found for bulk electromigration.

Surface passivation with dielectrics would thus slow down the surface diffusion components.

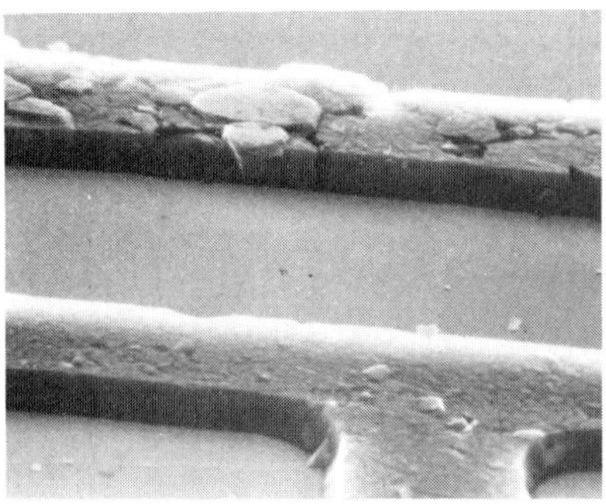

FIG. 8-23 Sample that has failed during electromigration test. One metal is integral since it has not been subjected to current.

The experience of the authors in the SGS laboratories confirms the hypothesis of the presence of a surface component in mass transport by diffusion. Al metallizations 3 μm thick and deposited by evaporation on a hot (300°C) SiO_2 substrate with small grain size ($\approx$ 1.4 μm) were used for the tests. Passivated and unpassivated devices were tested simultaneously. The passivated devices were passivated with:

- *Double layer of P. Vapox–Vapox* deposited at 400°C for 30 min
- *Silicon nitride* deposited at about 400°C for 15 min
- *PIQ*, polyamidic resin, deposited by spinning at ambient temperature followed by a postcure of 1 h at 300°C + 1 h at 200°C + 1 h at 300°C

The samples were supplied with a current generator and were monitored continuously to measure the failure time of each one.

Before the test the temperature coefficient of the resistivity of the Al film was evaluated so that its average temperature during execution of the test could be measured. The tests were conducted with current density $J = 1.5 \times 10^{-6}$ A/cm^2 and temperatures of T = 155, 205, and 265°C.

Figure 8-23, a photo of a sample which failed during tests, shows the case of two metals, one of which is integral because it has not been subjected to current, while the second shows numerous holes and accumulations of Al as a result of electromigration. Figure 8-24 is an enlargement of part of Fig. 8-23.

Figure 8-25 shows the log-normal diagram describing the curve of failures over time, at the three temperature levels indicated previously, of substrates passivated with Si_3N_4 and whose metallizations have been stressed with current density.

Figure 8-26 is obtained both from the diagram of Fig. 8-25 and from

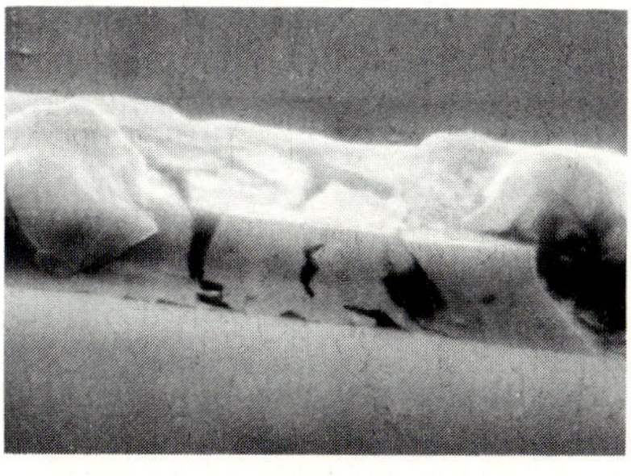

FIG. 8-24 Detail from Fig. 8-23.

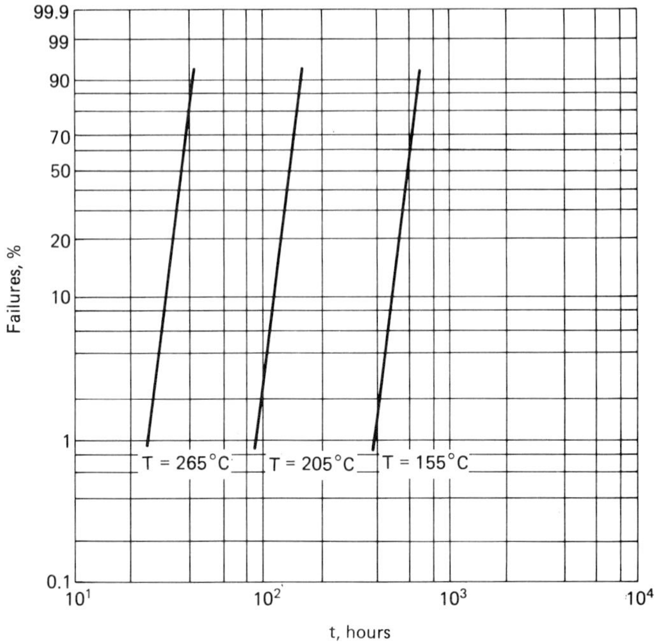

FIG. 8-25 Log-normal diagram of failure over time vs. temperature.

similar diagrams relative to unpassivated substrates and substrates passivated with PIQ or with a double layer of P. Vapox–Vapox. Figure 8-26 shows the median life (50 percent of failures) as a function of the temperature and can be used to find the activation energies E_A, given the slope of each straight line.

Analytically, to calculate the median life t_M and the activation energy E_A, the empirical law which follows was used:

$$t_M = \frac{Wt}{B} J^{-\beta} e^{E_A/kT} \tag{8-22}$$

where W = width of metallization
t = thickness of metallization
B = proportionality constant which also takes into account metal resistivity (average time between electron-ion collisions and electron-ion collision section)
β = coefficient of current density function
J = current density
T = temperature of metallization, K
k = Boltzmann constant

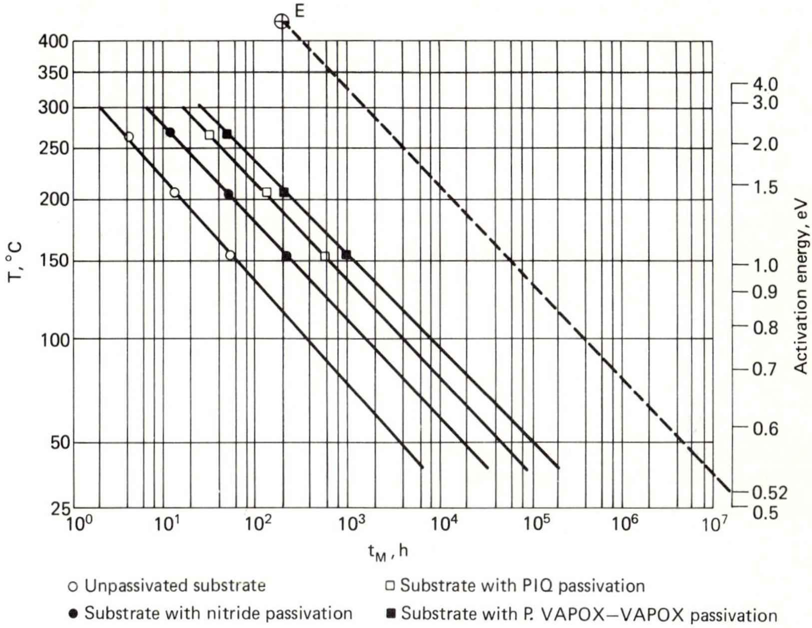

FIG. 8-26 Diagram of median life vs. metal.

For thin films with $J = 10^{-5}$ to 10^{-6} A/cm^2, β is between 1 and 3; β depends on the failure mechanism and has been found to be near 1 when structural variations in the film are the main cause of failure and near 3 when there are high thermal gradients in the film [23].

The values found for E_A are:

- Unpassivated substrate $E_A = 0.48$ eV
- Substrate with nitride passivation $E_A = 0.52$ eV
- Substrate with P. Vapox–Vapox passivation $E_A = 0.53$ eV
- Substrate with PIQ passivation $E_A = 0.52$ eV

A structural examination of the Al, unpassivated and after thermal treatment for the deposition of the passivating layers, showed no increase in the grain size. The higher values of activation energy in the passivated layers was attributed to chemical and physical interactions at the metal passivation interface, which would delay the diffusion of the Al. In the case of Al deposited by sputtering on substrates passivated with P. Vapox and grain dimensions of around 1 μm, similar tests gave $E_A = 0.54$ eV.

The current density suggested by the MIL M STD 38510D is 5×10^5 A/cm^2. In practice the density of 1×10^6 A/cm^2 is never exceeded.

8.5.3 Electromigration in Integrated Circuits

Equation (8-21) is valid in general for monocrystalline conductors with uniform temperature distribution. In power integrated circuits the metal generally used for the die metallization is Al deposited in thin film and consisting of contiguous monocrystalline grains. The temperature and current density usually vary along the metallization. The physical phenomena regulating the transport of material and therefore, in principle, the laws governing it remain unchanged.

Figure 8-27 shows an example of the temperature distribution that might be present in a power planar transistor with interdigitated structure. The hottest regions of each emitter finger (marked E) are those in which a current-crowding phenomenon occurs; the lowest-temperature zones are naturally those with the lowest current density. If the current densities involved and the temperatures are high enough to make possible the phenomenon of mass transport, the most significant possibilities are:

1. The temperature first increases and then decreases along one of the two directions of the metallization, and the electron flow e flows in the direction indicated in Fig. 8-28.

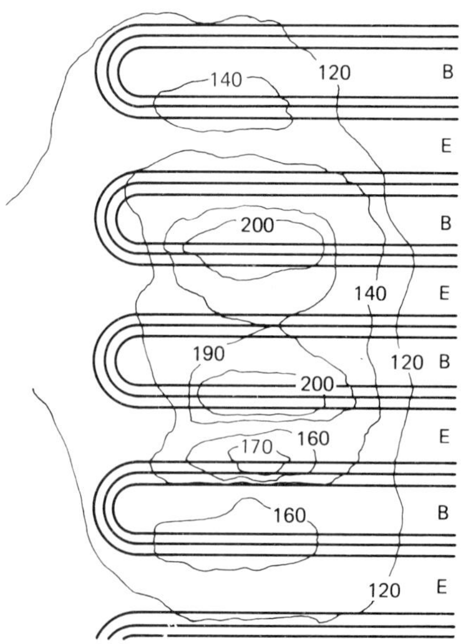

FIG. 8-27 Example of temperature distribution in a power planar transistor.

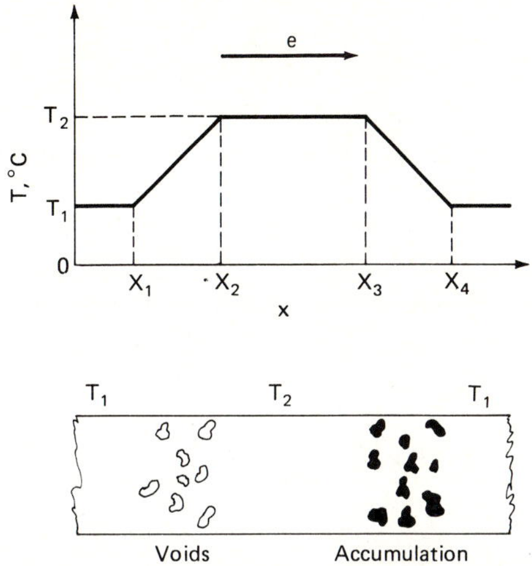

FIG. 8-28 Example of temperature distribution in Al metal and possible voids generation and Al accumulation. Arrow *e* indicates the direction of electron flow.

2. The temperature curve along the metallization is as in possibility 1, but the electrons flow in the opposite direction (Fig. 8-29).

In the hypothesis shown in Fig. 8-28, voids are generated in zone X_1–X_2 since as the temperature increases, in accordance with (8-21), the mobility of the vacancies increases and in X_2 their flow is greater than that entering X_1.

Zone X_1–X_2 is therefore a zone of accumulation of vacancies and formation of voids. Similar reasoning shows that zones X_3–X_4 (Fig. 8-28) and X_1'–X_2' (Fig. 8-29) are zones where material accumulates with the formation of hillocks and whiskers, and zone X_3'–X_4' is a zone of void formation.

The phenomenon of void formation is obviously degenerative, since the voids determine local current and temperature increases until the metallization is destroyed by fusion. To increase the current density allowed, with the same median life, several systems were used:

1. Grains of Al with the largest possible average dimensions so as to reduce the mass transport along the edges of the grains
2. Metallization with clearly defined grains and no crystallographic defects

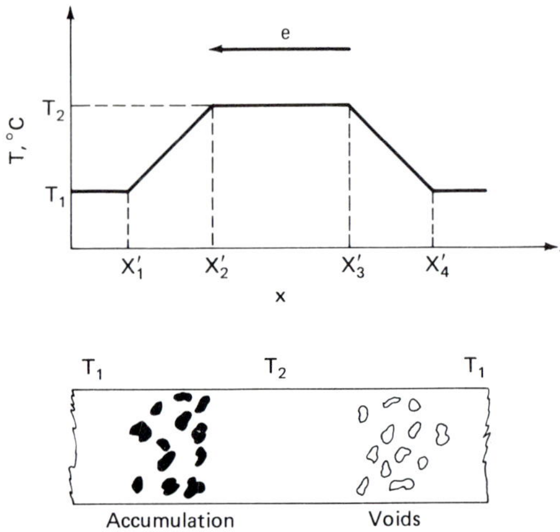

FIG. 8-29 Example of temperature distribution in Al metal and possible voids generation and Al accumulation. Arrow *e* indicates the direction of electron flow, opposite to that of Fig. 8-28.

3. Uniform sections of the film where these sections tend to thin out above the oxide steps
4. Doping Al with metal impurities (Si, Cu, Mg, Cr, Ag, Au, etc.)

8.5.4 Effects of Metal Impurities on Electromigration

To improve the performance in electromigration, metal impurities can be introduced into the Al metallization. Metals with an atomic radius equal to that of Al (such as Ag, Au) do not improve the average life of the metallization, since they settle as replacement atoms in the crystalline lattice of the Al, while metals with a different atomic radius tend to settle on the grain edges and considerably improve the average life.

Chromium and magnesium, for example, improve the average life of the films up to 2 orders of magnitude. However, these two metals are not practical, because they increase the resistivity of the Al. Silicon impurities, on the other hand, are widely used. The action of the Si encourages the formation of a lower density of structural defects than in the pure Al films.

J. Black [24] studied the behavior of these metals with small grain dimensions (1.4 μm) and found an activation energy $E_A = 0.54$ eV and different failure mechanisms for temperatures above and below 210°C.

For temperatures below 210°C, the failure mechanism is the same as that of the pure Al with small grains; i.e., there is a typical formation of voids all along the metallization until the interruption. For temperatures above 210°C, on the other hand, interdiffusion between Al and Si occurs. The Si of the die is evacuated in the zones where it is in contact with the Al, and the flow of electrons leaves the Si (cathode zone). Inversely, in the anode zones the Si tends to accumulate in the Al.

During the Si evacuation (etch pits) of the cathode zones, diffusion in the solid phase of the Al of the metal in the die occurs with the formation of short circuits. In the anode zones, on the other hand, the accumulation of Si in the metallization causes an increase in the contact resistance, with consequent overheating.

In the case of Al–4% Cu film, the average life improves up to 100 times. The Cu behaves in a similar way to the Al with regard to electromigration, but it has a higher diffusion speed. The Al film is interrupted in the evacuated zones of Cu. This means that in the initial phase the Cu is located mainly along the grain edges of the Al, stopping the diffusion. Later, when the zones have been evacuated of their impurities, the Al migration begins in accordance with the laws already explained. The activation energy is not very different from that of pure Al.

8.6 PARASITIC MOS

Linear ICs have geometrically critical structures, which can give rise to serious degradation phenomena in the presence of applied voltage and temperature. In fact these structures are sensitive to the effects of contamination by the ions present in the encapsulating resins and plating residues as well as contaminations through faulty handling or inadequate washing in the production process. Lateral PNP transistors are structures which typically suffer from failures of this type.

Here we shall examine in detail the effects of the presence of these contaminated ions and the methods of solving the resulting reliability problems.

8.6.1 Lateral PNP Transistors: Structures and Electrical Characterization

In the failure analysis carried out on some plastic PICs after the life test [27], V_{CEO} versus I_{CEO} characteristics of lateral PNPs were noted to have deteriorated considerably (Fig. 8-30). It was also noted that storage at ambient temperature $T_a = 150°C$, for a period of 6 to 8 h, was sufficient to restore these characteristics (Fig. 8-31).

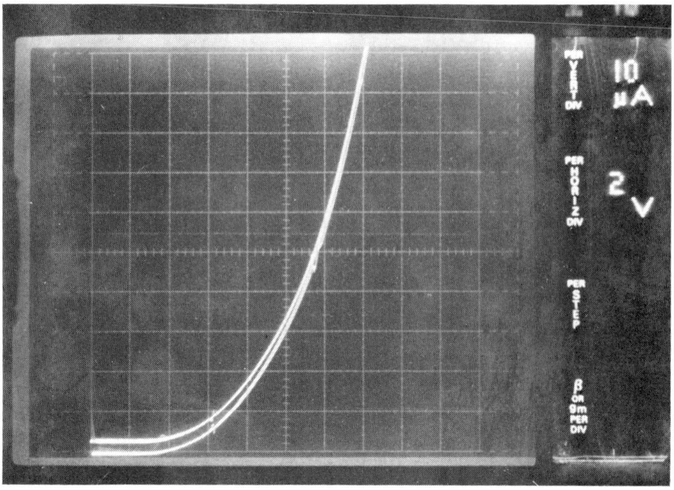

FIG. 8-30 Deteriorated V_{CEO} vs. I_{CEO} characteristic of a lateral PNP transistor.

To understand the failure modes and successive recovery in storage, special dice containing various lateral PNP structures with P. Vapox passivation and using Si ⟨100⟩ were prepared in SGS laboratories (Fig. 8-32).

Of all the tests conducted on these lateral PNP structures, we shall examine one with an uncovered channel (Fig. 8-33) and one with a

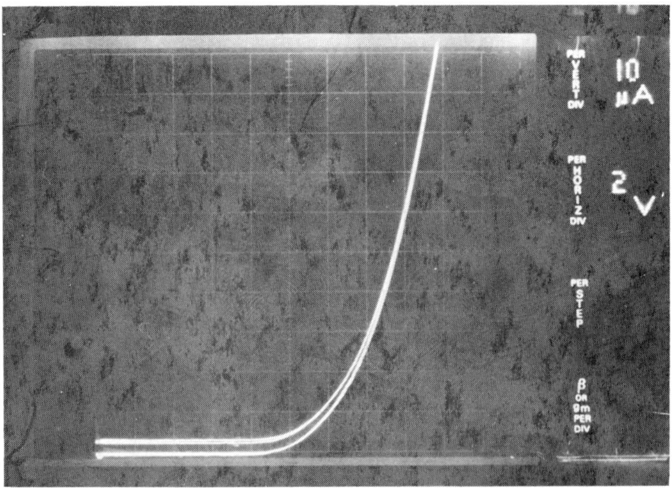

FIG. 8-31 Improvement in the V_{CEO} vs. I_{CEO} graph of Fig. 8-30 after high-temperature storage.

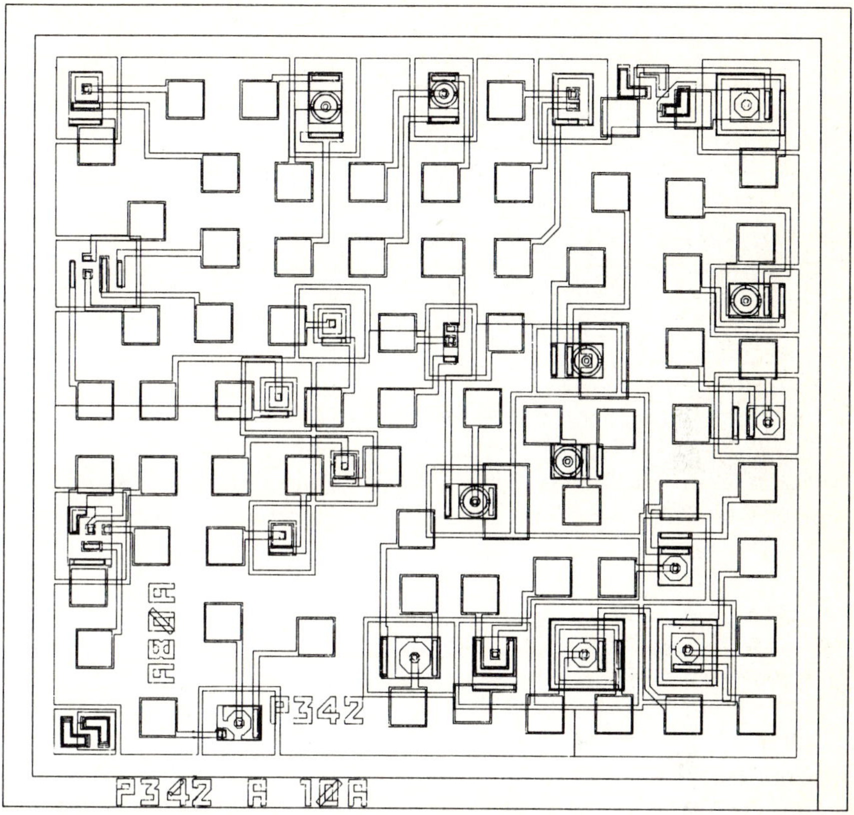

FIG. 8-32 Test pattern for lateral PNP transistor testing.

channel completely covered (Fig. 8-34) by a field plate (FP) in electrical contact with the emitter region.

These structures were tested both in metal packages and in plastic packages. The life test carried out on devices in metal packages did not give rise to the degradation phenomena shown in Fig. 8-30. The results of life tests on devices in plastic packages, however, were fundamental for improved reliability of linear integrated devices produced with conventional processes.

Initially all the fundamental electrical parameters were measured at ambient temperature.

Figure 8-35 shows the curve of the average gain h_{FE} versus the collector current I_C. It can be seen that the typical gain of PNP structures with covered channels is considerably greater than that of PNP structures with uncovered channels due to the annealing effect of the surface states present on the channels of the PNPs.

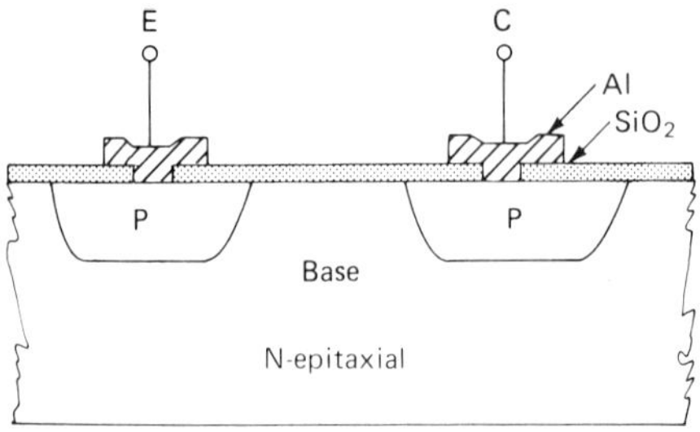

FIG. 8-33 Lateral PNP structure with uncovered channel.

Figure 8-36 shows the curve of I_{CEO} versus V_{CE} for the same structures with covered and uncovered channels.

Figure 8-37 shows the curve of I_{CES} versus V_{CE} for both structures. As can be seen, the covered channel structures of the same die have an I_{CES} considerably lower than those with uncovered channels, despite the fact that they have a higher I_{CEO}, mainly due to a gain effect.

8.6.2 Results of Life Tests on PNP Structures in Plastic Packages

Dice with the structures mentioned in the previous paragraphs were subjected to life tests with T_j increasing, V_{CEO} biased.

Figure 8-38 compares the curve of the average I_{CEO} versus time for both structures, with covered and uncovered channels. The following important points emerge from this figure:

1. The increase in I_{CEO} on PNPs with uncovered channels is considerable even in the first hours of life test, while the PNPs with covered channels show no sign of degradation.
2. In structures with covered channels the I_{CEO} doubles every 10°C as expected theoretically. This shows that when these currents

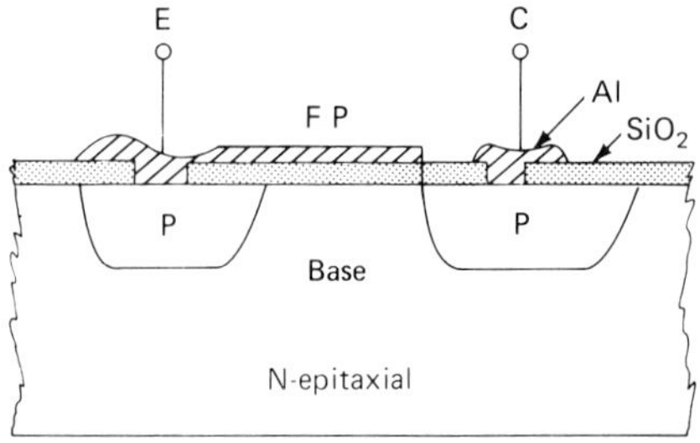

FIG. 8-34 Lateral PNP structure with base channel covered by a field plate.

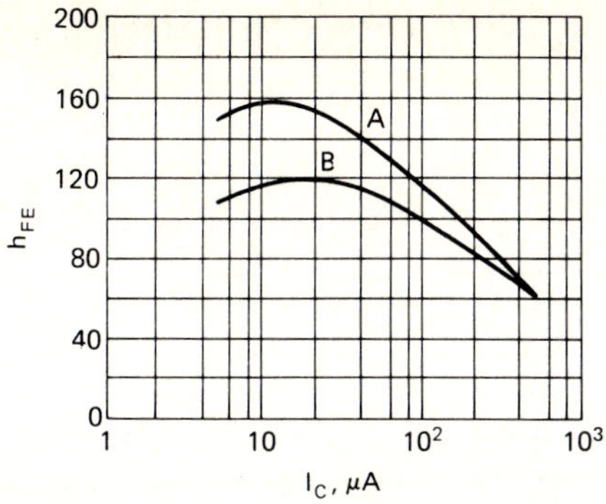

FIG. 8-35 Average gain h_{FE} vs. collector current I_C of lateral PNP transistors with covered (curve A) and uncovered (curve B) channels.

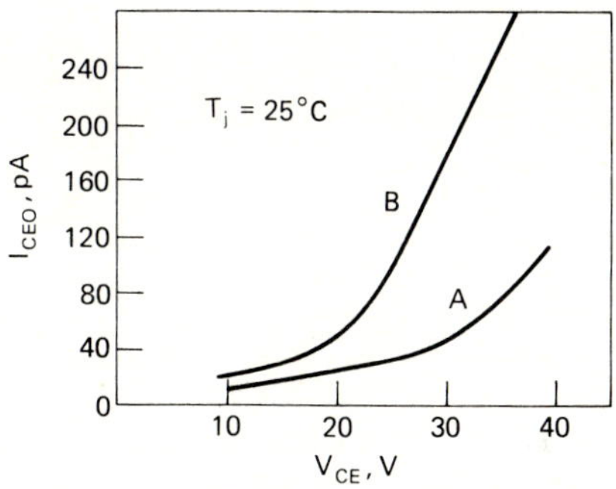

FIG. 8-36 Average value of I_{CEO} vs. V_{CE} for PNP lateral transistors with uncovered (curve A) and covered (curve B) base channels.

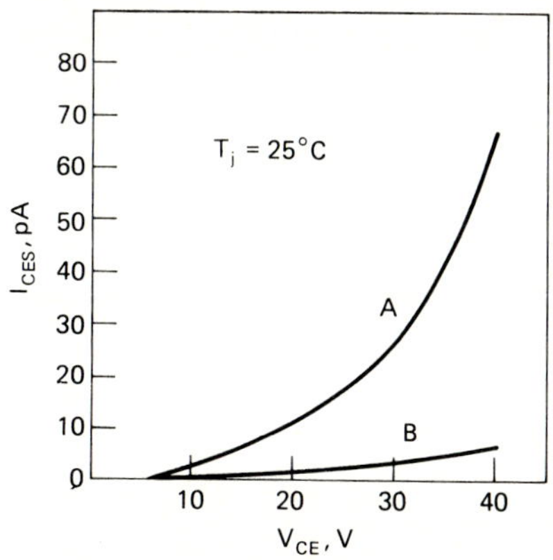

FIG. 8-37 Average value of I_{CES} vs. V_{CE} for PNP lateral transistors with uncovered (curve A) and covered (curve B) channels.

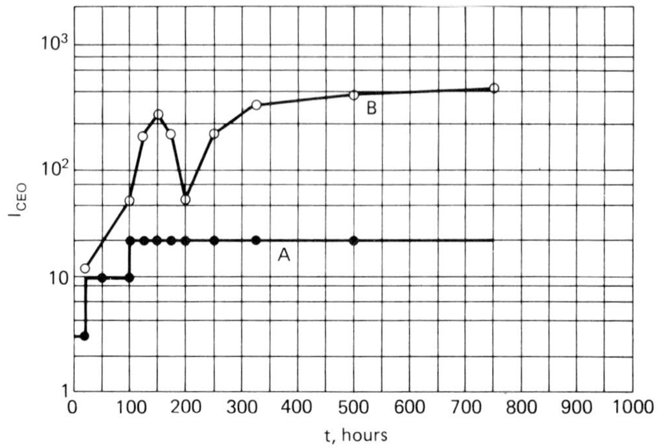

FIG. 8-38 Average I_{CEO} vs. life test hours with T_j increasing for PNP lateral transistors with covered (curve A) and uncovered (curve B) base channels. *Test conditions:* V_{CES} = 45 V, T_j = 120°C (0 to 25 h), T_j = 140°C (25 to 100 h), T_j = 150°C (100 to 750 h).

increase, the degrading effect of the encapsulating resin does not contribute to the increase in these currents.

3. As the hours of life test are increased, there is a definite recovery of the devices with uncovered channels, and the I_{CEO} decreases until it approaches the values for the structures with covered channels. As the hours of life test are further increased, however, this current tends to increase again. This phenomenon, of little practical importance, is explained in the literature on the subject [28].

The life test results given have shown the difference in behavior between the structures of lateral PNPs in plastic and metal packages. The degradation phenomena found in the plastic packages did not occur in the metal packages.

It was also seen that the epoxy resins normally used in the plastic packages are sources of ions Na^+, K^+, Cl^-. Since these resins are in direct contact with the structures examined, the degradation phenomena seen are, without doubt, attributable to the ions present in them. The presence of these ions, the die temperature, and the simultaneous effect of the polarization on the devices may create an increase of ions in zones of the resin-oxide interface, causing surface inversion of the underlying Si.

Considering the geometry of a lateral PNP (Fig. 8-33), the ions which determine surface inversion phenomena are the negative ions present in a larger quantity in encapsulating resins. From a physical point of view, the degradation phenomenon can be explained by simply considering the presence of surface conduction on the oxide, which caused extension of the negative collector potential to the channels, giving rise to the phenomenon known as *phantom gate*. This propagation, as is well known, follows the law of *complementary error functions* (erfc) [29].

Covering the base channels (Fig. 8-34) with emitter metal, which is always at the most positive potential of the PNP transistor, means that degradation phenomena will be avoided.

8.6.3 Theory of the Inversion Phenomenon

The minimum voltage required to create a surface inversion channel in silicon is known as the *threshold voltage*. The formula for threshold voltage value [30] is

$$V_{TH} = V_{FB} - \frac{Q_{SD,\max}}{C_{OX}} + 2\phi_F \tag{8-23}$$

where $Q_{SD,\max}$ = charge density associated with surface depletion region (< 0 for P-type substrates, > 0 for N-type substrates)

C_{OX} = dielectric capacitance per unit area of insulating oxide

ϕ_F = Fermi potential (> 0 for P-type substrates, < 0 for N-type substrates)

V_{FB} = flatband voltage

For MOS structures, V_{FB} is given by

$$V_{FB} = \phi_{MS} - \frac{Q_{SS}}{C_{OX}} \tag{8-24}$$

where Q_{SS} = fixed positive charge density at silicon–silicon dioxide interface (always > 0)

ϕ_{MS} = metal-semiconductor work function difference (always < 0)

Figure 8-39 shows the curve of the threshold voltage value for P-type substrates and an oxide thickness of 0.75 μm versus the acceptor concentration N_A for two values of Q_{SS}. Similarly, Fig. 8-40 shows the curve of the threshold voltage values for N-type substrates for the same oxide thickness and Q_{SS} versus the donor concentration N_D.

Let us consider two P-type resistances separated by distance L, crossed by metal of width W (Fig. 8-41). Since the N epitaxial layer is

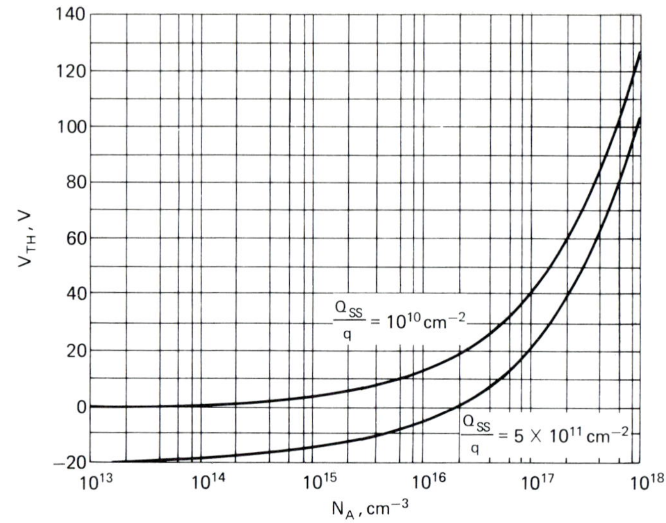

FIG. 8-39 Threshold voltage V_{TH} vs. acceptor concentration N_A.

biased at $+V_{CC}$, there will be inversion if the metal is at a voltage $V_{CC} - V_{TH}$.

It should also be noted that since the parasitic MOS is a P channel, it is equivalent to a lateral PNP transistor whose source corresponds to the emitter and consists of the resistance with a higher potential. The drain, on the other hand, corresponds to the collector and consists of

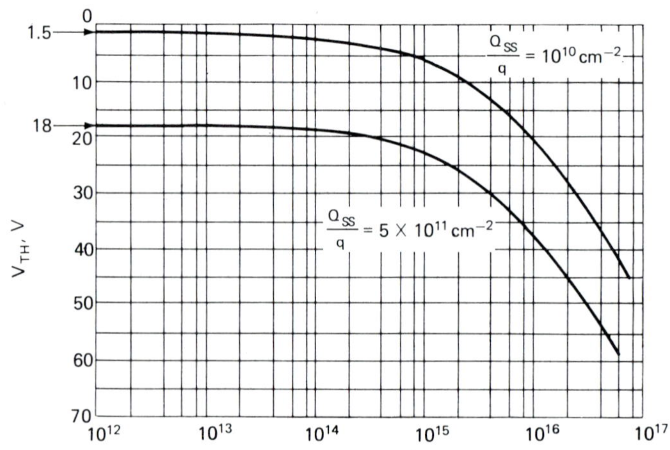

FIG. 8-40 Threshold voltage V_{TH} vs. donor concentration N_D.

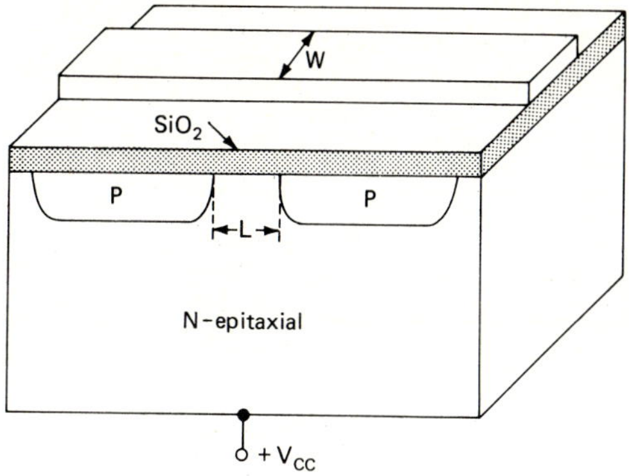

FIG. 8-41 P-type resistance diffused in N epitaxial layer.

the resistance connected to a potential lower than that of the first resistance.

If V_S is the episource voltage, Eq. (8-23) becomes

$$V_{TH} \approx V_{FB} - \frac{Q_{SD,\max}}{C_{OX}} \left(\frac{-2\phi_{FN} + V_S}{-2\phi_{FN}} \right)^{1/2} + 2\phi_{FN} - V_S \qquad (8\text{-}25)$$

V_{FB} is always less than 0.

In conclusion, as regards the technological process it is necessary to increase the oxide thicknesses or find passivation processes capable of minimizing the C_{OX} capacitances of the structures. An increase of the Q_{SS} in the oxides on traditional structures, using N-type epitaxial layers, is useful to increase the inversion threshold voltages but gives generally a poorer noise performance.

Particular care must be taken in identifying those geometrical details which, as in the case of the two resistances examined, may give rise to parasitic phenomena. In the case of lateral PNPs these phenomena can be eliminated by covering the base channels.

8.7 THERMOMECHANICAL EFFECTS ON THE DIE AND ON Al METALLIZATION

The different thermal expansion coefficients (CTEs) of the materials forming the power IC and the polymerization shrinkage of the encapsulating resin are sources of mechanical stress which appear during assembly of the IC. These physical characteristics of the materials become particularly important when the device is heated or cooled.

In fact temperature excursions may result in degradation and then failure of the IC due to thermal fatigue of the solder connecting the die to the heat sink, cracking of the die, recrystallization of the aluminum metallization, and breaking of the bonds or internal connection wires.

Here we shall examine the main characteristics of the soldering alloys (preforms) and of the materials used for the heat sink; we shall also describe the stress present on or transmitted to the Si die and the failure modes that may occur in the cyclic temperature variations.

8.7.1 Principal Materials Used in Power ICs

a. Heat Sink Materials

The basic material used for the heat sink of all power integrated devices is Cu OFHC (Oxygen Free High Conductivity) or one of its alloys. The most widely used Cu alloys (e.g., Olin 194* and Cuprofor†, whose main characteristics are given in Table 8-4) contain additives which aim to

TABLE 8-4 Important Characteristics of Various Materials Used in PICs

Materials	CTE, ppm/°C	Melting temperature, °C	Thermal conductivity, W/(cm·K)
Au–20% Sn	15.93	280	2.50
Au–6% Si	13.7	370	2.84
Cu-OFHC	16.8	1083	3.96
Au	14.2	1063	3.15
Pb–5% Sn	28.7	312	0.35
Si	2.3	1410	1.30
Silver epoxy	30–50	–	$5–20 \times 10^{-3}$
Alloy 42	5.01	1425	0.105
Kovar	5.86	1450	0.167
Olin 194	16.3	1089	2.58
Cuprofor	17	1080	3.17

give the Cu the desired permanent hardness characteristics, while modifying as little as possible the thermal conductivity and electrical characteristics. In fact following the thermal treatment to which it is subjected during die attach, molding, and polymerization of the

*Nominal weight composition: Cu = 85.67 percent, Fe = 2.3 percent, P = 0.03 percent, Zn = 12 percent.

†Nominal weight composition: Cu = 99.71 percent, Co = 0.22 percent, P = 0.07 percent.

encapsulating resin, the Cu OFHC tends to lose its original hardness, becoming softer (even 50 degrees Vickers from 120 to 140 initially) and less resistant to bending stresses.

The fact that manufacturers prefer materials with specific mechanical characteristics stems from precise reliability requirements. If we

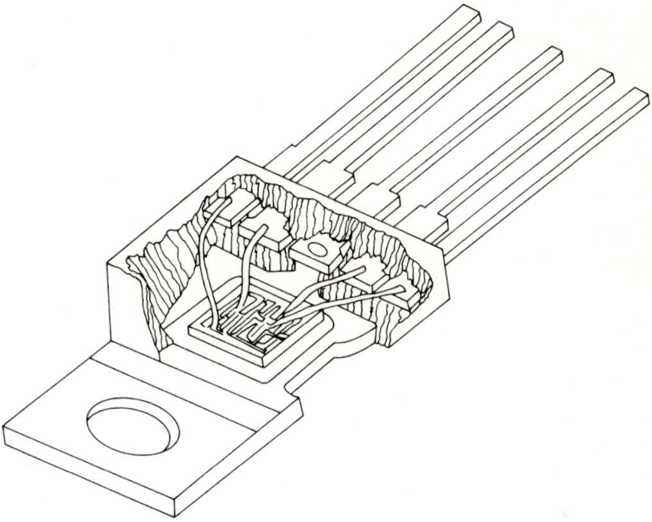

FIG. 8-42 Internal view of a Pentawatt device.

examine the heat sink of the Pentawatt package in Fig. 8-42, we can see that the polymerization shrinkage of the resin causes stress on the heat sink, and if the heat sink material is not very resistant to bending stresses, the heat sink will bend, causing cracking of the die. The use of hard metals which are not very flexible, however, has other drawbacks which are often just as serious. In fact stress on the outside can easily be transferred to the inside of the device, reaching the die and sometimes cracking it. These stresses can, for example, derive from fixing the device heat sink to an external heat sink which is not perfectly flat. We therefore try to use materials with suitable mechanical characteristics and to design heat sinks in such a way that the transfer of stresses to the die is minimized.

b. Types of Solder (Preform) for Fixing a Die to the Heat Sink

In power integrated circuits the die is soldered to the heat sink with materials which have a low melting point ($\leq$ 400°C). These can be divided into three groups: *hard solders*, *soft solders*, and *epoxy resins* (glue–die attach).

- Historically, *hard solders* are die attach alloys which do not fail because of thermal fatigue and which have a melting point under 400°C. The most commonly used hard solder alloys in ICs are gold eutectics with low melting points such as Au–3% Si, Au–12% Ge, and Au–20% Sn.
- *Soft solders* are binary or ternary alloys, usually with a high content of Pb. They belong to the category of soft solder alloys (such as Pb/Sn, Pb/Sn/Ag, Pb/Sn/In) with a melting point of around 300°C.

Hard solders are characterized by high levels of permanent stress transferred to the Si chip during die attach and caused by the different thermal expansion coefficients of the Si and Cu of the heat sink. Soft solders, on the other hand, are typically low-cost alloys capable of absorbing stresses on the chip during die attach.

- The mechanical behavior of *epoxy resins* with conductive materials (mainly Au and Ag) used as solder *(glue die attach)* is similar to that of the soft solders.

8.7.2 Stresses on the Silicon Die

a. Hard Solder

To get an idea of the stresses on a die let us examine the die–solder–heat sink structure (Fig. 8-43) in nonstress situations, i.e., when the solder has melted and the system is at temperature T. Figure 8-44 shows cooling of the system including a hard solder. When the temperature decreases by ΔT, the length of L of each of the three layers considered insulated from the other two decreases to $L(1-\alpha\,|\Delta T|\,)$, where

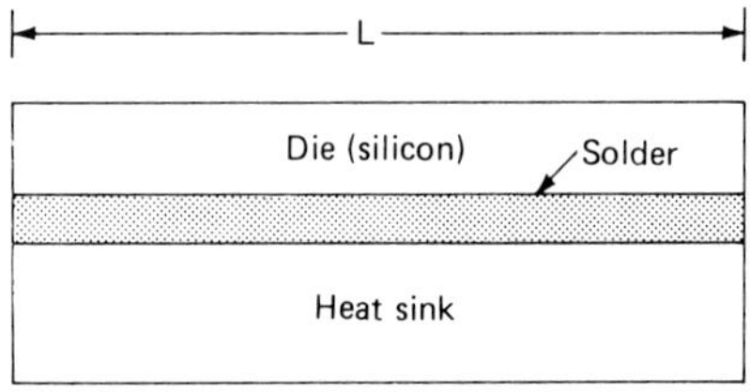

FIG. 8-43 Die–solder–heat sink system at temperature T, when the solder is melted. The mutually induced stresses are nil.

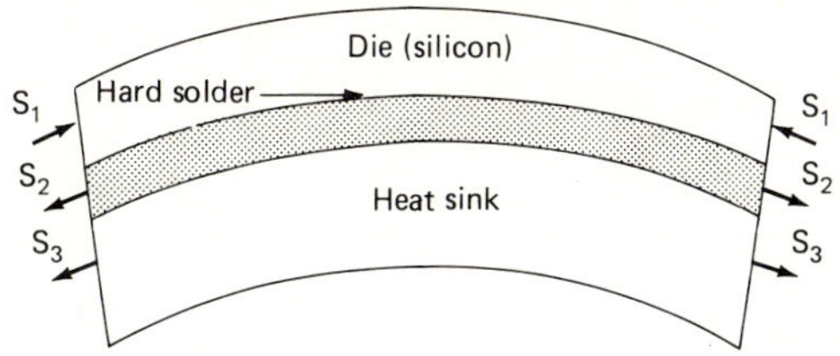

FIG. 8-44 Die–hard solder–heat sink system after solidification of the preform. $\mathbf{S}_1$, $\mathbf{S}_2$, and $\mathbf{S}_3$ indicate the direction and stresses existing on each of the three parts forming the system.

α is the thermal expansion coefficient of the individual layer. Since the layers are tightly linked to one another, each depends on the presence of the other two. Tension is thus created inside the system, which reacts by bending. Calculation of the forces acting on each layer and of the curvature radius of the system is fairly simple, provided the deformation occurs within the limits of elastic deformation of the material.

In general it can be said that the size of the force acting on the die and the curvature radius do not depend on the thermal expansion coefficient of the solder. Furthermore, taking into account the thicknesses involved, the Si die is compressed by the heat sink, which has a higher thermal expansion coefficient (see Table 8-4). The forces in the layers increase with the difference in temperature between the melting point of the solder and the temperature of the finished device. Therefore at temperatures below zero the possibility of cracking of the Si die increases.

Immediately after die attach, the alloys normally used for hard solder do not undergo any appreciable plastic deformation when the device cools from melting temperature to ambient temperature. Therefore, after temperature-cycling variations, the hard solder does not suffer from fatigue. Plastic deformation may, however, occur in the Cu of the heat sink.

b. Soft Solder

In devices soldered with soft solder (Fig. 8-45), because of the big difference in thermal expansion coefficients between the die and the heat sink, during cooling after the die attach process, the solder undergoes plastic deformation even though the die is not stressed.

Shearing stress to which the solder is subjected does not normally cause cracking, provided the resulting deformation does not reach high values ($>$ 20 percent). In this case the solder can be damaged permanently since dangerous cracks can be generated in it.

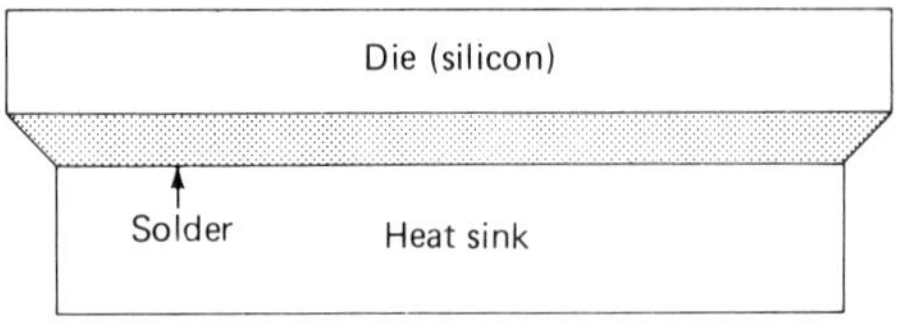

FIG. 8-45 Die–soft solder–heat sink system after solidification of the preform.

The typical defect of soft solder is the tendency to the formation of voids, both in the solder and in the soldering zones, as a result of the problems of wettability of the finishing metal on the back of the die (often Ni) or of the heat sink.

Finally, if the solder is not thick enough (lower than 10 μm), because of the different thermal expansion coefficients of the Si of the die and the Cu of the heat sink, strong residual tension occurs on the die, which in some packages (after encapsulation in epoxy resin, which causes further stress on the system due to polymerization shrinkage) may lead to cracking of the die.

8.7.3 Effects of Temperature Excursion on the Heat Sink–Solder–Die System

We have already said that the different thermal expansion coefficients of the various layers forming the heat sink–solder–die system are the cause of thermal stresses when the system is heated or cooled. Furthermore the system is in a nonstress situation when the solder is melted.

In the case of *hard solder* the die is stressed (even at operating temperatures) by bending and compression. Consequently the die, if subjected to cooling and heating cycles, may tend to flake as shown in Fig. 8-46. This failure mode may be aggravated by the encapsulating resin of the IC in plastic package. In fact the epoxy resins used for encapsulation have a shrinkage during polymerization of around 0.5 to 0.7 percent, which tends to compress the silicon die.

In the case of *soft solder*, the cyclic temperature variations normally cause both cracks in the solder, often parallel to the back of the die and near to it, and extensive voids (Fig. 8-47).

Degradation of the solder due to the formation of voids is probably accelerated by those voids already present in devices newly manufactured. These voids may be generated when the die attach temperature is not sufficiently in excess of the alloy melting temperature and is not able to remove the oxide present on the back of the die.

To reduce these drawbacks the die attach operation should be car-

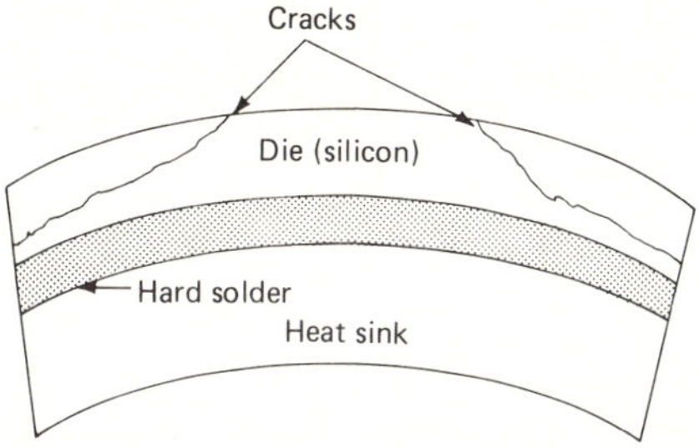

FIG. 8-46 Example of cracking, after thermal cycles, in a die attached to heat sink with hard solder.

ried out either in the furnace, where the parts are subjected to a rigorously controlled temperature profile and reducing atmosphere, or by using automatic equipment, which does an energetic die scrubbing.

Modern technological processes make it possible for power integrated devices to pass 50,000 thermal cycles in thermal fatigue tests with temperature excursions on the heat sink of 70°C without any appreciable degradation of the device.

8.7.4 Effects of Temperature Excursion on the Metallization

Figure 8-48 shows the initial conditions of an Al metallization, deposited by evaporation, of an SGS integrated circuit [31]. Figure 8-49 illustrates part of the metallization which shows up the grain size [31].

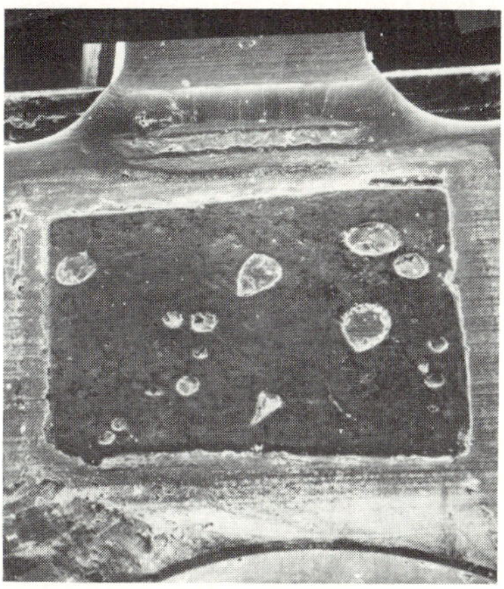

FIG. 8-47 Degradation of soft solder due to voids formed during thermal cycles.

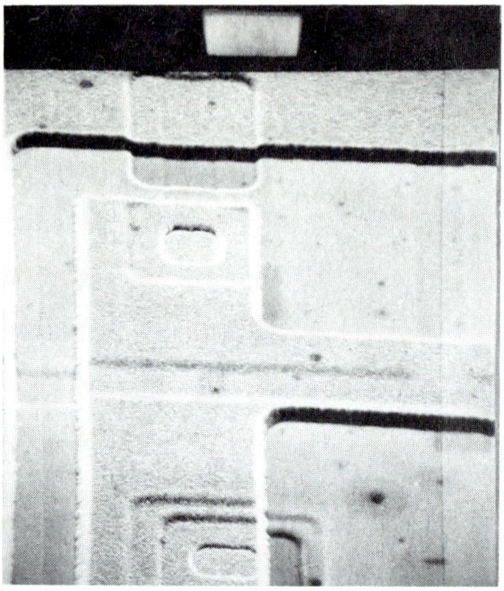

FIG. 8-48 Al metallization of a PIC.

Figure 8-50 shows the condition of the metallization after 11,500 thermal cycles between −65 and +150°C. Numerous hillocks can be seen [31]. Figure 8-51 is a close-up of part of Fig. 8-50, showing an example of recrystallization in the cubic system with centered faces found in the Al metal [31].

The hillocks seen in Fig. 8-50 are formed because of recrystallization of the Al due to the difference in thermal expansion coefficients of the Al ($\approx 24 \times 10^{-6}$ °C^{-}1), the underlying Si ($\approx 2.3 \times 10^{-6}$ °C^{-1}), and

FIG. 8-49 Detail of Fig. 8-48.

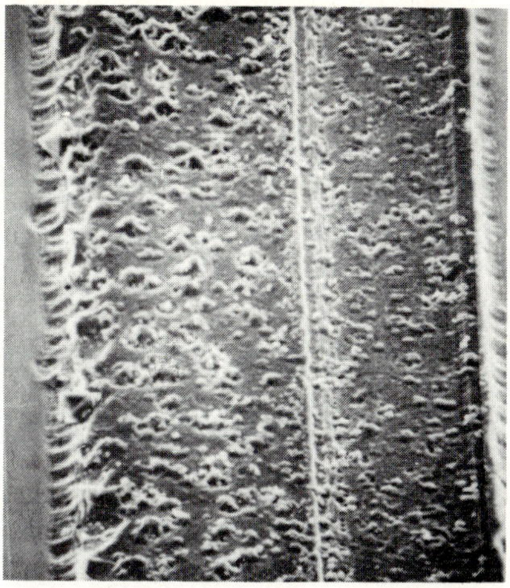

FIG. 8-50 Metallization of a PIC after 11,500 thermal cycles.

the SiO_2 (0.5×10^{-6} °C^{-1}). The stress generated as a result of any temperature excursion causes movement of the edge of the grains and, consequently, recrystallization of the metal. This phenomenon is greatly slowed down by the presence of metal impurities (Si, Cu, Sn, Mn) that tend to settle along the edges of the grain of Al, blocking the interdiffusion responsible for the abnormal grain growth and therefore of the hillocks.

As the dimensions of the original grains increase, the number of hillocks and their dimensions tend to decrease, while with the same grain

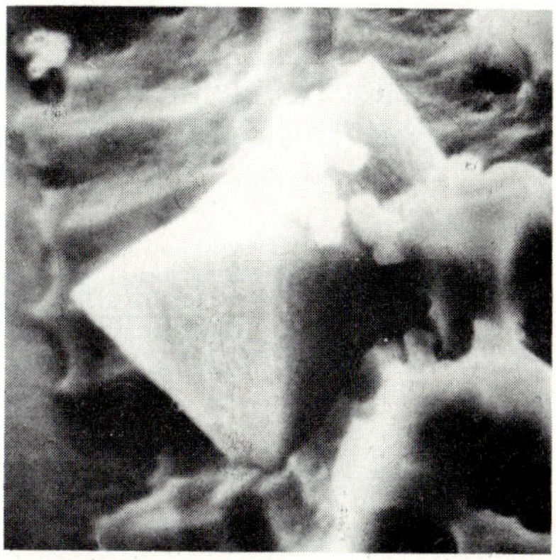

FIG. 8-51 Example of Al recrystallization.

size, as the temperature increases so does the formation of hillocks increase.

The formation of hillocks has numerous dangerous effects:

1. If present in the phase preceding the Au wire bonding operation, hillocks tend to undermine the quality of the contact and its mechanical resistance.
2. The hillocks tend to break the surface passivation layers of the dice, thus encouraging corrosion in the presence of humidity.
3. The recrystallization phenomenon causes an increase in the overall sheet resistance of the Al film. Since this increase is not uniform over the whole metallization, when current is applied, there are areas with high current density, and consequently electromigration of the Al may occur in those areas.

8.8 MAIN RELIABILITY TESTS

Completing this chapter on the reliability of PICs, we shall give a brief glance at the main laboratory tests carried out to check the reliability of integrated circuits.

To get repetitive and comparable results for tests carried out in different laboratories, it is advisable to carry out these tests in standardized conditions. The standards most commonly used are MIL (United States) and IEC (International Electrotechnical Commission).

8.8.1 Thermal Cycles

This test involves exposing the devices alternately to the highest and lowest temperatures at which they will be used, in each case for periods long enough to bring the whole device to a uniform temperature (usually 10 min minimum). The changeover from one temperature to the other occurs gradually after a short period at ambient temperature (usually 5 min maximum).

The test is carried out in air to show up failure mechanisms linked to the temperature excursions which may be encountered in equipment not operating continuously in a low-temperature environment or during transport. Variations in the dimensions of the materials forming the device can cause permanent alterations to their operating characteristics, fracture of the die, breakage of internal interconnecting wires, etc.

The thermal cycling test, as we shall see later, provides additional information to complete that provided by the thermal fatigue test with applied power.

The behavior of PICs during the thermal cycles is closely linked to the type of resin used. In fact the epoxy resins used at present have a polymerization shrinkage of 0.4 to 0.6 percent, which means that at 25°C the whole frame-chip-wire system is kept in compression. If the device is brought to a high temperature (e.g., 150°C), the compression of internal connection wires tends to be reduced, provided the glass transition temperature T_g of the resin is not exceeded ($T_g \geq 155°C$), beyond which the CTE (coefficient of thermal expansion) of the resin becomes high, introducing traction stresses in the wires. Today's devices, therefore, have a high resistance to thermal cycles at high temperature, provided they do not exceed the T_g.

Moreover, the thermal cycling test may cause cracks in the intermetal phases present at the Au wire–Al pad bonds, due to the different CTEs of Al, Au, and their intermetal phases. These cracks may appear mainly in the $AuAl_2$ and Au_5Al_2 phases and in the zones where these intermetals meet areas with a higher Au content.

In practice modern PICs do not reach wear-out due to material fatigue before a number of thermal cycles (−65 to +150°C) that is of an order of magnitude of some thousands of cycles. This is proof of the quality of the design.

These or similar thermal cycles can be used as a screening to eliminate incorrect bonding of the internal wires or cracks and weakening of the silicon chip. For this purpose 5 or 10 cycles is sufficient to remove most of the production rejects (freak population).

In the past, when silicon resins with practically nonexistent low shrinkage were used, the high-temperature cycles brought the wires into traction with resulting fatigue that appeared at a number of cycles about an order of magnitude lower than that of present epoxy resins.

8.8.2 Thermal Shock

Like the thermal cycling test, this test involves exposing the devices alternately to the highest and lowest possible temperatures at which they could be used, in each case for periods long enough to bring the whole device to a uniform temperature. While thermal cycling is carried out in air, the thermal shock test involves transferring the devices from one container to another, each containing liquid at the desired temperature.

The time for transfer from one bath to another must be as short as possible. While in thermal cycling the temperature varies gradually, in thermal shocks the temperature changes very quickly.

Thermal shocks are more rapid and more expensive than thermal cycles. The cost is mainly due to the use of special liquids for the high and low temperatures; on the other hand, by using liquids instead of air, less time is required to bring the devices being tested to uniform temperatures (typically less than 5 min instead of 10 min minimum).

The failure modes in the two tests are fairly similar. Thermal shocks cause more stress, especially for the Au-Al bonds, due to the rapid temperature changes. Thermal cycles, because of their duration, may introduce some failures, typical of storage tests caused by the movement of contaminating impurities.

In any case wear-out normally occurs at a lower number of shocks than of cycles.

8.8.3 Life Test

A life test aims to determine the "failure rate" of a semiconductor device and to prove its reliability under specified conditions.

The life test has to last for a period of time sufficient to show up the characteristic phenomenon of early failures and to explore part of the device's useful life, pointing out any traces of early wear-out. Moreover, intermediate checks, performed during the test phase, give precise information about the parameter variations in time.

When the life test aims to establish the device limitations, the test conditions have to be as stressful as possible in terms of input, output, load, bias, and junction temperature.

The life test duration is normally 1000 h, except when its object is to ascertain the device's conformity with a specified failure rate or its wear-out.

The test end parameter checks have to be concluded within 96 h of the life test end and must include the most important parameter and functional characteristic checks to determine catastrophic or degradation rejects and any shift from the initial value.

In the case of tests with applied voltage, the tested devices have to be cooled to ambient temperature before the voltage switch-off. The checks to be performed at ambient temperature have to be concluded before any further heating of the device to impede the surface rediffusion of contaminating ions that have migrated to critical areas of the device due to the combined effect of temperature and voltage applied.

Life test conditions are strictly correlated to the possible failure mechanism to be pointed out.

The following tests are analyzed under this profile:

- High-temperature storage test
- Operating life test

- High-temperature reverse bias (HTRB) test
- Thermal fatigue test

a. *High-Temperature Storage Test*

This test is carried out at high temperature without voltage applied to the device. In the case of plastic packages the temperature is limited to 150°C since the glass transition temperature of the encapsulating resin must not be exceeded.

The storage test is justified by the fact that much equipment, especially military, may be stored for 10 years and then required to operate even for very short periods.

This test usually has failure rates lower than the normal life test with voltage applied (generally 2 orders of magnitude).

It is sometimes useful to carry out the test by measuring the parameter drift in order to increase the test sensitivity and detect trouble due to incomplete stabilization.

b. *Operating Life Test*

This test is carried out to detect PIC failure rates and early-failure phenomena. It is normally carried out at the maximum junction temperature, reached with the combination of high ambient temperature and the power dissipated by the devices being tested.

The minimum test time is 1000 h, with two intermediate measures at 168 and 500 h. When operating with a 150°C junction temperature, normally the early failures stop before 168 h; 80 h is usually sufficient to eliminate 90 percent of the freak population.

In the case of devices with internal thermal shutdown (external short-circuit protection), the test T_j must be limited from 5 to 10°C under the temperature of thermal protection intervention, and the test time should be consequently prolonged in accordance with the time-temperature regression curve calculated with a 1-eV slope. The following table gives the test time as a function of T_j.

T_j, °C	150	145	140	135	130	125
Test time, h	1000	1400	1900	2700	3800	5400

When the test junction temperature is reduced, the early-failures period is prolonged. The early-failure duration can be estimated by using the regression curve, with an activation energy of 0.44 eV instead of 1 eV, since the early-failure mechanisms are less accelerated by the temperature than the useful-life ones. The following table gives the early-failure time at different T_j.

T_j, °C	150	145	140	135	130	125
Early-failure time, h	80	90	105	120	135	160

An operating life test is often performed in PICs at an ambient temperature of 25°C to maximize electrical stress (current, voltage, and then power) applied to the device, thus simulating as much as possible stress in the field. As an example, Figs. 8-52 and 8-53 show, respectively, the dynamic life test circuit and the dynamic life test board for SGS TDA 2006.

The vast experience accumulated through the years has demonstrated that if these tests give less than 1 percent of failures, no field

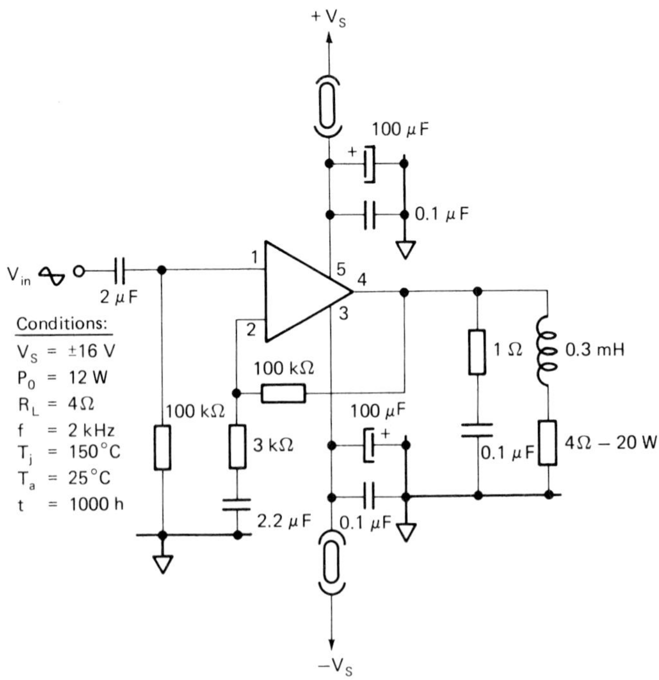

FIG. 8-52 Dynamic life test circuit for SGS TDA 2006 (10-W audio power amplifier).

FIG. 8-53 Dynamic life test board for SGS TDA 2006.

trouble is found where failure rates of around 10 FITs are reached (1 FIT = 1×10^{-9} h^{-1}) in normal operating conditions during the useful device life.

c. *High-Temperature Reverse Bias (HTRB) Test*

This test is complementary to the operating life test. Devices are generally subjected to the maximum junction temperature with the maximum number of reverse-biased junctions and to the maximum reverse dc voltages of the tested device.

In this test the high temperature serves to increase the mobility of contaminant charges, while the reverse bias brings these charges into the junction areas, which may give rise to parameter drift.

While the HTRB test traditionally lasts 1000 h, the failure mechanisms detected by it are usually evident before 168 h, when operating with a T_j of 125 to 150°C.

d. *Thermal Fatigue Test*

The term *thermal fatigue test* is commonly used to describe a series of thermomechanical stresses induced in a semiconductor device, causing cyclic self-heating up to a specified maximum junction temperature (but not greater than T_j max) and then cooling it to near ambient temperature. The test aims to evaluate the ability of the device to withstand thermomechanical stresses caused by temperature variations which occur every time the device changes from a standby condition to that of dissipating power.

This test has the following parameters:

1. Power P_d dissipated by the device that heats it from a temperature near ambient up to the required one
2. Time t_{On} at which power P_d is applied
3. Time t_{Off}, during which, without power applied, the device cools to the desired temperature
4. Number of cycles N, each with duration $t_{On} + t_{Off}$

Generally for the thermal fatigue test the case temperature variations (ΔT_C) are specified because they are directly measurable by a thermocouple; the ΔT_j junction temperature variations do not appear. The two numbers are easily correlated when the stationary internal thermal impedance value, commonly known as the *thermal resistance* $R_{TH,jc}$, is known.

The internal thermal impedance in a semiconductor device, instant after instant, links the temperature difference between the heating junction and a reference point (e.g., the case), referred to the power unit flowing in the junction.

The junction-case $R_{TH,jc}(t)$ transient thermal impedance is expressed by the relation

$$R_{TH,jc}(t) = \frac{T_j(t) - T_C(t)}{P_d} \tag{8-26}$$

where $T_j(t)$ = instantaneous junction temperature
$T_c(t)$ = case reference point instantaneous temperature
P_d = dissipated power

In stationary conditions, the above expression becomes independent of time and detects the device's internal thermal resistance $R_{TH,jc}$, which may be written

$$T_{j,\max} = T_{C,\max} + R_{TH,jc}P_d \tag{8-27}$$

$$T_{j,\min} = T_{C,\min} \tag{8-28}$$

and due to the difference between Eqs. (8-27) and (8-28),

$$\Delta T_j = \Delta T_C + R_{TH,jc}P_d \tag{8-29}$$

where ΔT_C is connected to ΔT_j.

Following are a few considerations about t_{On} and t_{Off} times. In normal conditions, equipment runs at a steady thermal condition for periods ranging from a few minutes up to a few hours. In this period stability conditions are reached by the internal ambient temperature of the apparatus and by the semiconductor devices, including all the other internal elements, dissipating or not.

To compare thermal fatigue test results with field operation conditions, both the device ambient temperature and the junction temperature should be varied simultaneously.

Variations should also be made in the device cooling times (related to the apparatus cooling methods when in standby) and the heating times related to the presence (or absence) of a heat sink outside the device, to the dissipated power, and to the position of other elements, dissipating or not.

Since it is not possible to consider all these variants, it is preferable to make the test conditions more severe whenever possible. So the thermal fatigue test is performed with the device in free air or supported by an external heat sink with high thermal resistivity and low thermal capacitance. Then the power necessary to reach the required case (or junction) temperature within time t_{On}, generally between 1 and 3 minutes, is applied. The device is then cooled with forced air or cold water, flowing inside the heat sink.

In the forced air case the t_{Off} times are rarely longer than 90 s; in the case of water cooling the time is 20 to 30 s. In the case of fast cooling with water, tests carried out at Carnegie-Mellon University on transistors using hometaxial technology in a TO-3 case (2N 3055) have shown that Pb-Sn solder wear-out occurs at a number of cycles from 1.5 to 3 times lower than on devices cooled in free air [32]. Presumably, slow cooling allows a better settling of the Pb-Sn solder than fast cooling. However, when the time for cooling exceeds 40 to 50 s, the result is no longer influenced.

To accumulate a high cycle number (10×10^3 to 20×10^3) in a short period, t_{On} and t_{Off} times must be kept reasonably low. But the test with low t_{ON} time is not acceptable for plastic package wires.

In fact, plastic packages have interconnection wires encapsulated in plastic, and the t_{On} power application time, 2 to 3 min, is not sufficient to reach the system's steady thermal condition, that, for some plastic packages puts the interconnection wires in traction with the risk of generating intermittency phenomena. Temperature cycling and thermal shock tests make it possible to stress the frame-wire-die system and check that the resin never places it under traction. The temperature cycling test is therefore necessary to complete the thermal fatigue test for plastic case power devices.

We still have to clarify which power has to be dissipated by the test

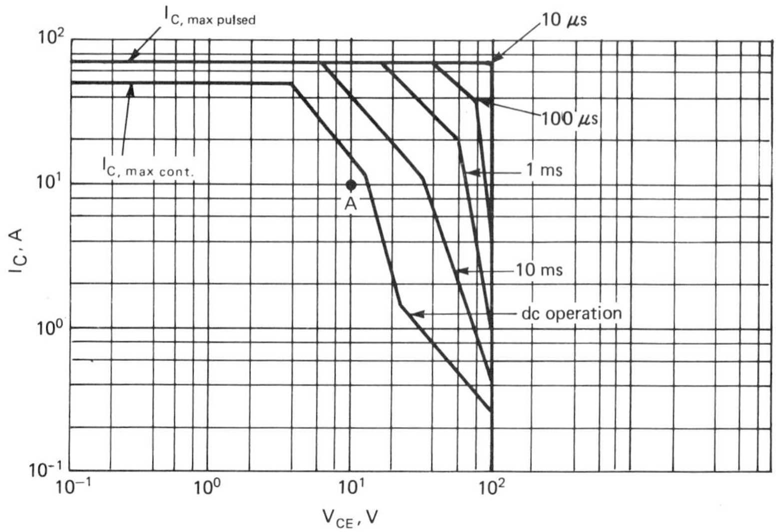

FIG. 8-54 Safe operating area (SOA) for output transistors of SGS TDA 2002 (8-W car radio audio amplifier).

device. Obviously, the greater the specific power (W/cm^2) the more severe the thermal fatigue applied to the device, since for a given thermal resistance the junction heat sink thermal difference is greater. A suitable value must be chosen for power P_d applied to the test device.

In the case of a power transistor, V_{CE} and I_C values, when possible, must allow the transistor to work near the point A of SOA (safe operating area) where the constant dissipation line and the secondary breakdown line meet (Fig. 8-54). This criteria is always possible for transistors but not always for PICs. It is therefore advisable to choose the most stressful work conditions for final power transistors, because a deterioration in soft solder always corresponds to an increase in the thermal impedance of the test device with a consequent junction temperature increase during the t_{On} period, leading to a decrease in the SOA of the test device.

8.8.4 Humidity Test (Moisture Resistance)

This test is fundamental to determine a PIC's resistance in highly humid environments.

a. 85°C/85%

This test is performed in a climatic chamber, with strict control of temperature (85°C) and relative humidity (85 percent). The test may have voltage applied (THB, or temperature humidity bias) or not (TH). Although the test has not yet been standardized with international rules, it is widely used by users and manufacturers because of its repeatability and greater speed than other standardized tests (IEC 68-2-38 test Z/AD, T_A = 25 to 65°C, RH = 80 to 96%, 10 cycles of 24 h; IEC 68-2-3 test Ca, T_A = 40°C, RH = 93%, 56 days without bias).

Test repeatability requires some fundamental precautions:

1. The climatic chamber must be kept clean; the chamber must have a stainless steel inside and no possibility of water dripping into the parts being tested; deionized water has to be used to generate the humidity because the impurities contained by water accelerate the test corrosion effect.
2. A clear drying procedure before intermediate or test-end measurements is essential to prevent surface water from altering the electric parameter measurements. It is necessary to wait 30 min, at least, before performing the electrical measurements which must be made within 24 h after the test end.
3. Tests with voltage applied require a minimum of dissipated power since even slight internal overheating drastically alters the test effects [6].

Today's PICs resist thousands of hours in 85/85 tests and about 1000 h in 85/85 bias tests. For this reason the test is mainly performed with bias, even if this requires the use of expensive boards that need frequent checks and maintenance interventions.

Humidity results are well-represented by log-normal distributions.

b. Pressure Pot

Semiconductor device resistance to humidity is influenced by many product variables, such as impurities inside the plastic package, passivation characteristics and integrity, encapsulating resin adherence to metal surfaces, cracks or voids in the resin, and differences from lot to lot of resin caused by contaminating impurities. On the other hand, the 85/85 bias test performed on every production lot would be expensive and give answers with too much of a time lag to represent an efficient production feedback.

For these reasons, another extremely accelerated test, called the *pressure pot* or *pressure cooker*, has been set up: it exploits the pressure as an element of further acceleration to obtain quicker answers than with the 85/85 bias test.

The pressure pot test is performed in an autoclave containing saturated wet vapor at 121°C and at 2.08 atm pressure. The test is generally carried out without applied voltage since the acceleration is sufficient, even without polarization. The pressure pot test requires numerous precautions, including perfect cleanliness (the autoclave inside and the device carrier baskets must be made of stainless steel) and the use of deionized water; it is necessary to avoid testing devices encapsulated with different resins at the same time to prevent reciprocal contamination; it is necessary to remove the residual air from the autoclave in pressure to get 100 percent relative humidity; perfect device drying procedures are necessary before the test devices pass to measurements.

SGS has standardized the device drying phase to 30 min at 80°C. Other procedures require a drying time between 1 and 48 h at ambient temperature.

The intermediate measurement number may also influence the average device life since corrosion is more evident when voltage is applied for the device electrical measurements.

This test may be further accelerated by increasing the test pressure with subsequent temperature increase. However, the interpretation of such a result must be done very carefully, since the more the tests are accelerated the more the test results differ, so correlations become less precise.

In conclusion, the pressure pot test gives fast, relative, fairly effective indications, but it must be completed with the more rigorous 85/85 bias test, which introduces the applied voltage effects and is repeatable and more easily correlated to the field results.

8.8.5 Other Tests on PICs

The preceding paragraphs are about the main tests performed to characterize PIC reliability. Other tests to be performed to get a complete characterization of these devices or to check other quality aspects, referring either to the international or to the national standards for a complete description of test methods, are as follows:

- *Lead bending test:* Checks the material strength and the correct work procedure during the various production phases.
- *Solderability:* Measures the lead solderability during soldering processes in the PCB (Printed Circuit Board).
- *Soldering heat resistance:* Verifies that the soldering process of devices in PCBs does not damage their electrical characteristics.
- *Logo permanence:* Not a reliability test, but it is important to ensure that the device logo is not erased during the various handling phases.

- *Resistance to electrostatic discharge:* Checks that the PIC parts with a high impedance (especially inputs) are not damaged during handling or the various processes.
- *Thermal resistance:* A very important measure to verify that the PIC has been designed and built to easily discharge the high power dissipated inside.
- *Salt atmosphere:* Simulates the sea atmosphere conditions. Even if it cannot really be correlated with actual sea conditions, it is used as a comparison test to check the resistance of the PIC pins to corrosion.

Finally, *mechanical vibration, shocks,* and *centrifuge tests,* which are very important for hermetic devices, have no practical interest for integrated circuits in plastic packages since they do not provide any damage.

REFERENCES

1. J. L. Flood, Reliability Aspects of Plastic Encapsulated Circuits, *Reliab. Phys.*, 1972.
2. R. W. Lawson, The Accelerated Testing of Plastic Encapsulated Semiconductor Components, *Reliab. Phys.*, 1974.
3. S. Glasstone, K. J. Laidler, and H. Eyring, *The Theory of Rate Processes*, McGraw-Hill, New York, 1941.
4. D. S. Peck and C. H. Zierdt, The Reliability of Semiconductor Devices in the Bell System, *Proc. IEEE*, February 1974.
5. B. Reich and E. B. Hakim, Temperature-Humidity Acceleration Factors for Plastic Transistors and Integrated Circuits, *Solid State Technol.*, September 1972.
6. G. Boccaletti, F. R. Borri, G. D'Espinosa, and G. Fioravanti, Le Prove Accelerate, *Affidabilità e Fisica dei Meccanismi di Guasto dei Semiconduttori,* Seminario Scientifico Tecnico di Lecce, 3º corso, Castromarina-Lecce, Sept. 22–27, 1980.
7. T. Ajiki, M. Sugimoto, H. Higuchi, and S. Kumada, A New Cyclic Biased THB Test for Power Dissipating IC's, *Reliab. Phys.*, 1979.
8. J. C. Harrison, Control of the Encapsulation Material as an Aid to Long Term Reliability in Plastic Encapsulated Semiconductor Components (PEDs), *Microelectron. Reliab.*, **16** (1977).
9. J. B. Woodard, Thermally Stimulated Discharge Effects in Polymer Encapsulants, *Hysol Tech. Symp.*, 1979.
10. G. Jabbour, Differences between Fused and Crystalline Silicon as Fillers for Semiconductor Molding Compounds, *Hysol Tech. Symp.*, 1979.

11. G. Jabbour, Comparison of Phenolic Novolac and Anhydride Cured Semiconductor Molding Compounds, *Hysol Tech. Symp.*, 1979.
12. W. L. Hunter and R. J. Zettech, Dielectric Properties of Encapsulants under Temperature and Humidity Conditions, *Semicond. Int.*, April 1979.
13. E. Philofsky, Intermetallic Formation in Gold-Aluminum Systems, *Solid State Electron.*, **13** (1970).
14. J. A. Cunningham, Expanded Contacts and Interconnections to Silicon Monolithic Integrated Circuits, *Solid State Electron.*, **8** (1965).
15. G. K. Chen, On the Physics of Purple-Plague Formation and the Observation of Purple Plague in Ultrasonically Joined Gold-Aluminum Bonds, *IEEE Trans. Parts, Mater. Packag.*, **PMP-3** (1967).
16. M. Kashiwabara and S. Hattori, Formations of Al-Au Intermetallic Compounds and Resistance, Increase of Ultrasonic Al Wire Bonding, *Rev. Electr. Commun. Lab.*, **17** (1969).
17. G. Majini, C. Nobili, G. Ottaviani, M. Costato, and E. Galli, Gold-Aluminum Thin-Film Interactions and Compound Formation, *J. Appl. Phys.*, **52** (1981).
18. A. P. Kemeny, G. Kalmar, and V. Stefaniay, Results of a $160 \cdot 10^6$ Device-Hour Reliability Assessment and Failure Analysis of TTLSSI Integrated Circuits; Part 2, Survey of Dominant IC Failure Mechanisms and Analysis of Failure Causes, *Microelectron. Reliab.*, **14** (1975).
19. L. Lonzi, Gold Wire Bonding on Aluminum Pad Reliability, *ESSDERC 1982,* **Sez. 87,** Munich, 1982.
20. L. Lonzi, SGS-ATES Internal Technical Report CQA/35 (1982), unpublished.
21. H. U. Schreiber, Activation Energies for the Different Electromigration Mechanisms in Aluminum, *Solid State Electron.*, **24** (1981).
22. J. Black, Electromigration—A Brief Survey and Some Recent Results, *IEEE Trans. Electron Devices*, **ED-16** (1969).
23. Kwaku A. Danso and L. Tullas, Thin Film Metallization Studies and Device Lifetime Prediction Using Al-Si and Al-Cu-Si Conductor Test Bars, *Microelectron. Reliab.*, **21** (1981).
24. J. Black, Electromigration of Al-Si Alloy Films, *Reliab. Phys.*, 1978.
25. G. J. Van Gurp, Electromigration in Al Films Containing Si, *Appl. Phys.*, **15** (1971).
26. A. J. Learn, Electromigration Effects in Aluminum Alloy Metallization, *J. Electron. Mater.*, **3** (1974).
27. G. D'Espinosa and F. Terenziani, Les Effets de Impuretés dans les Resines époxidiques sur la Fiabilité des Circuits Integrés bipolaires, *Colloq. Int. Fiabilité Maintenabilité*, Paris, 1979.
28. H. C. Potter and D. R. Reber, A Study of Surface Charge Induced Inver-

sion Failure of Junction Isolated Monolithic Silicon IC's, *Reliab. Phys.*, 1976.

29. A. S. Grove, *Physics and Technology of Semiconductor Devices*, Wiley, New York, 1967.

30. P. Richman, *MOS Field Effect Transistors and Integrated Circuits*, Wiley, New York, 1973.

31. G. D'Espinosa and L. Lonzi, L'Affidabilità nei Circuiti Integrati di Potenza, *Elettronica Oggi*, **11** (1983).

32. N. D. Zommer, D. L. Feucht, and R. W. Heckel, Reliability and Thermal Impedance Studies in Soft Soldered Power Transistors, *IEEE Trans. Electron Devices*, **ED-23** (1976).

33. *Type Approval and Quality Assurance of Thermosetting Moulding or Casting Material for the Solid Encapsulation of Silicon Integrated Circuits*, British Post Office, M219C.

34. *Tests for Flammability of Plastic Materials for Parts in Devices and Appliances*, UL 94, Underwriters' Laboratories, Inc., 207 E. Ohio St., Chicago, IL, 60611.

CHAPTER 9

Audio Amplifiers

S. Palara
SGS Microelettronica
Milan, Italy

9.1 AUDIO AMPLIFIER EFFICIENCY: THE THERMAL PROBLEM 9.2
9.1.1 Direct-Coupled Resistive Load 9.3
9.1.2 Nonregulated Power Supply: R_S Effects 9.8
9.1.3 Amplifier Fidelity 9.10
9.1.4 Maximum Effective Series Resistance 9.11
9.1.5 Amplifier Economics 9.12
9.1.6 Effects of $V_{CE,sat} > 0$ of Output Transistors 9.12
9.1.7 Thermal Problem in Audio ICs in the Presence of Real Load 9.14
9.1.8 Effect of the Output Capacitor 9.16
9.1.9 Effects of an Ohmic-Inductive Load 9.22
9.2 PERFORMANCE AND LIMITS OF A LOW-DISTORTION INTEGRATED OUTPUT STAGE 9.22
9.3 OPTIMIZING VOLTAGE DROPS IN OUTPUT POWER STAGES 9.28
9.3.1 Historical Evolution in Designing Output Stages 9.28
9.3.2 The Bootstrap Solution 9.31
9.4 DISTORTION IN THE OUTPUT STAGES OF AUDIO CIRCUITS 9.35
9.4.1 h_{FE}-Difference Distortion 9.35
9.4.2 Crossover Distortion 9.39
9.5 GENERAL AMPLIFIER ORGANIZATION 9.42
9.5.1 The Input Stage: Noise and Dynamic Performance 9.42
9.5.2 Thermal Feedback Effects 9.49

9.5.3 Turn-On 9.53
9.5.4 DC Biasing and Supply Voltage Rejection 9.55
9.6 DIFFERENT SYSTEM CONFIGURATIONS 9.58
9.6.1 Mono Amplifier 9.59
9.6.2 Monolithic Stereo Amplifier 9.59
9.6.3 Bridge Configuration 9.60
9.7 REDUCTION OF EXTERNAL PARTS 9.62
9.8 PROTECTION TECHNIQUES 9.64
9.8.1 Short-Circuit Protection in Power Transistors 9.64
9.8.2 Protection Techniques in Presence of Reactive Loads 9.66
9.8.3 Delayed Short-Circuit Protection 9.68
9.8.4 Loudspeaker Protection Technique 9.72
9.8.5 Protection Circuit with Inhibit 9.74

9.1 AUDIO AMPLIFIER EFFICIENCY: THE THERMAL PROBLEM

An amplifier system is generally formed by several stages in cascade. The input stage and intermediate stages operate with small signal in class A. Their function is to amplify the input signal to a level which will allow it to drive the final stage.

The final stage, which operates mainly in class AB, must be able to supply high voltage, current, or power to a load, which in the case of audio amplifiers is a loudspeaker. Since high currents can be required to drive a loudspeaker, several problems arise due to the power dissipated inside the silicon chip used to realize the integrated amplifier.

The thermal problem is caused by the intrinsic characteristics of the integrated transistors which degenerate beyond a certain temperature and cause considerable reliability problems. Therefore, once a maximum junction temperature T_j has been fixed, the transfer of power to a loudspeaker by the audio amplifier depends on the efficiency of the amplifier.

The efficiency of an audio amplifier is defined as follows:

$$\eta = \frac{P_0}{P_d + P_0} \tag{9-1}$$

where P_0 = power dissipated on load
P_d = power dissipated inside chip

In practice η is a measure of the amplifier's capacity to convert the dc power of a supply into ac power transferred onto a load.

In the case of a class B amplifier we shall examine the two most significant cases:

Direct-coupled resistive load

Resistive load decoupled by an output capacitor in low-frequency condition

In both cases the dissipated power is the sum of the dissipated dc power on the stages preceding the final stage and bias network and of the dissipated power in the final stage. Generally only the second P_d is taken into consideration, since it is much more important.

9.1.1 Direct-Coupled Resistive Load

Let us consider the case of a resistive load coupled directly to the output of a split-supplied amplifier (Fig. 9-1). Supposing we are in sinusoidal operation, the instantaneous output P_{oi} is given by

$$P_{oi} = \frac{V_0^2}{R_L} \sin^2 \omega t \tag{9-2}$$

The average power $\overline{P}_0$ in a semiperiod is

$$\overline{P}_0 = \frac{1}{\pi} \int_0^{\pi} \frac{V_0^2}{R_L} \sin^2 \omega t \, d(\omega t) = \frac{V_0^2}{2R_L} \tag{9-3}$$

In the same semiperiod the instantaneous power dissipated on the transistor P_{di} is

$$P_{di} = (V_{CC} - V_0 \sin \omega t) \frac{V_0}{R_L} \sin \omega t \tag{9-4}$$

where V_{CC} = supply voltage.

The average dissipated power $\overline{P}_d$ is therefore

$$\overline{P}_d = \frac{1}{\pi} \int_0^{\pi} \frac{V_{CC} V_0}{R_L} \sin \omega t \, d(\omega t) - \frac{1}{\pi} \int_0^{\pi} \frac{V_0^2}{R_L} \sin^2 \omega t \, d(\omega t) \tag{9-5}$$

The first term on the right represents the power taken from the supply, while the second term is the utilized power on R_L. Therefore

$$\overline{P}_d = \frac{2 V_{CC} V_0}{\pi R_L} - \frac{V_0^2}{2R_L} \tag{9-6}$$

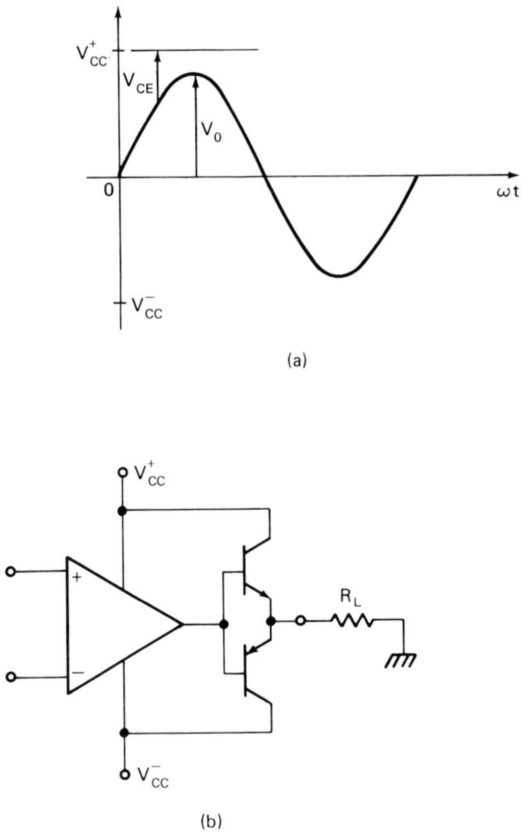

FIG. 9-1 (*a*) Output sinusoidal waveform of a (*b*) split-supplied audio power amplifier with direct-coupled load.

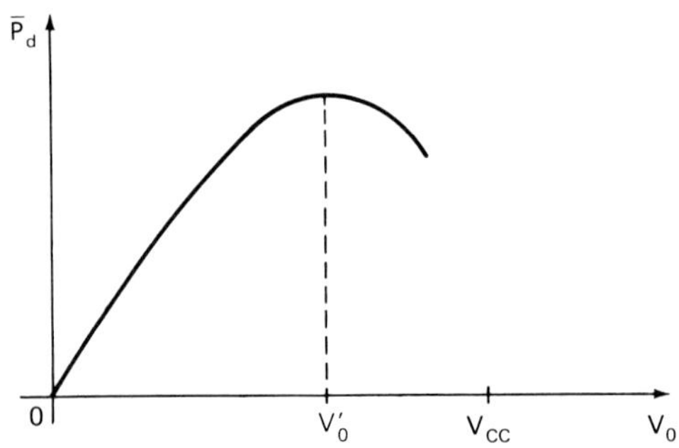

FIG. 9-2 Average dissipated power inside the silicon chip vs. output voltage swing.

By plotting this equation we get the curve in Fig. 9-2, which shows that there is a maximum dissipated power for a certain output peak voltage V_0'. This value is found by differentiating Eq. 9-6 with respect to V_0 and setting the result equal to zero:

$$\frac{\partial P_d}{\partial V_0} = \frac{2V_{CC}}{\pi R_L} - \frac{V_0}{R_L} = 0 \tag{9-7}$$

$$V_0 = \frac{2V_{CC}}{\pi} = 0.637V_{CC} \tag{9-8}$$

By substituting in Eq. 9-6 the value of V_0 found in Eq. 9-8, we get the maximum average dissipated power $\overline{P}_{d,\max}$:

$$\overline{P}_{d,\max} = \frac{2V_{CC}^2}{\pi^2 R_L} \tag{9-9}$$

This result is important because it fixes precise limits for the maximum average output power $\overline{P}_{0,\max}$, which is closely linked to the efficiency of the amplifier.

We said that efficiency η is given by

$$\eta = \frac{\overline{P}_0}{\overline{P}_s} = \frac{\pi}{4}\frac{V_0}{V_{CC}} \tag{9-10}$$

where $\overline{P}_s$ = power taken from the supply. The efficiency is best when $\overline{P}_0$ is at a maximum, that is, when

$$\overline{P}_{0,\max} = \frac{(V_{CC} - V_{CE,\text{sat}})^2}{2R_L} \tag{9-11}$$

Supposing that $V_{CE,\text{sat}}$ is negligible with respect to V_{CC}, then

$$\overline{P}_{0,\max} = \frac{V_{CC}}{2R} \tag{9-12}$$

and

$$\eta_{\max} = \frac{\pi}{4} = 0.785 \tag{9-13}$$

Instantaneous Dissipated Power. Another important parameter, one which makes it possible to evaluate the thermal behavior of an audio IC, is the instantaneous dissipated power, which can reach very high levels and can damage the device when it is operating at low frequency (20 to 100 Hz) even though the $\overline{P}_d$ is within the acceptable values.

The instantaneous power P_{di} is obviously

$$P_{di} = v_{CE} i_c \tag{9-14}$$

where

$$v_{CE} = V_{CC} - R_L i_c \tag{9-15}$$

$$P_{di} = V_{CC} i_c - R_L i_c^2 \tag{9-16}$$

By differentiating with respect to i_c and setting the result equal to zero, we obtain the value of i_c for which P_{di} is at a maximum:

$$\frac{\partial P_{di}}{\partial i_c} = V_{CC} - 2 i_c R_L = 0 \tag{9-17}$$

$$i_c = \frac{V_{CC}}{2R_L} \tag{9-18}$$

Therefore

$$P_{di,\max} = \frac{V_{CC}^2}{4R_L} \tag{9-19}$$

Once the efficiency η and dissipated power $\overline{P}_{d,\max}$ are known, the $\overline{P}_{0,\max}$ of an audio amplifier can be found after fixing a suitable junction temperature $T_{j,\max}$ and by knowing the thermal resistance $R_{\text{th},ja}$ of the device connected to a dissipator with known characteristics.

Example. Dimension the splitting supply of an ideal audio amplifier of $V_{CE,\text{sat}} = 0$ having fixed $T_{j,\max} = 150°\text{C}$, $T_{\text{amb}} = 70°\text{C}$, $R_{\text{th},ja} = 5°\text{C/W}$, and $R_L = 4\ \Omega$ by supposing that the final stage is a pure class B and ignoring the power dissipated on the signal stages.

Solution. The maximum dissipated power is

$$\overline{P}_{d,\max} = \frac{T_j - T_{\text{amb}}}{R_{\text{th},ja}} = 16\ \text{W} \tag{9-20}$$

The maximum supply voltage is found from Eq. 9-9 as

$$\pm V_{CC} = \pm \frac{\pi}{\sqrt{2}} \sqrt{\overline{P}_{d,\max} R_L} = \pm 17.77\ \text{V} \tag{9-21}$$

We can deduce that the output swing corresponding to $\overline{P}_{d,\max}$ is

$$V_0 = 11.3\ \text{V} \tag{9-22}$$

which corresponds to an output power

$$\overline{P}_0 = 16\ \text{W} \tag{9-23}$$

and the efficiency η in this case is

$$\eta = 0.5 \tag{9-24}$$

The $\overline{P}_{0,\max}$ in the ideal case of $V_{CE,\text{sat}} = 0$ for the final transistors is

$$\overline{P}_{0,\max} = \frac{V_{CC}^2}{2R_L} = 39.46 \text{ W} \tag{9-25}$$

In this case, since $\eta_{\max} = 0.785$, we get

$$\overline{P}_d = \overline{P}_0 \frac{1 - \eta_{\max}}{\eta_{\max}} = 10.8 \text{ W} \tag{9-26}$$

Therefore the maximum P_{supply} is $= 50.26$ W. The results are summarized graphically in Fig. 9-3.

The values found in the example refer to the ideal case of a stabilized supply and amplifier with the final stage at $V_{CE,\text{sat}} = 0$. These are also the conditions of greatest stress for the device from the point of view of power and junction temperature.

In practice the supplies are generally soft, and the saturation of the final transistors is not negligible. This affects the output power, in particular the dissipated power. It may therefore be useful to evaluate the dissipation in the audio IC as a function of the internal resistance R_S of

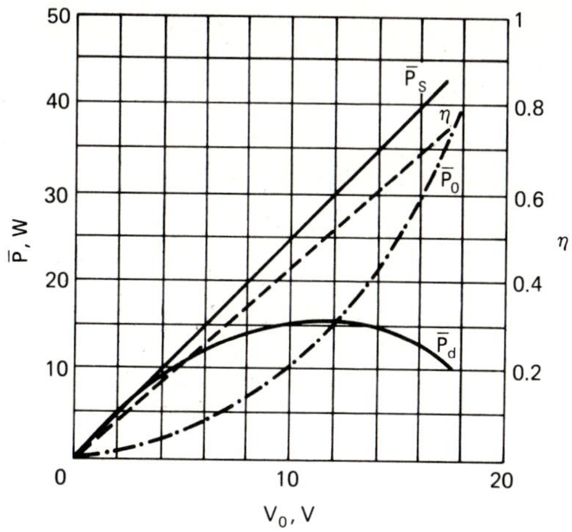

FIG. 9-3 Output power, dissipated power, and efficiency vs. output voltage swing.

the supply and as a function of the fact that the final transistors have $V_{CE,sat} > 0$.

We shall examine these two aspects separately.

9.1.2 Nonregulated Power Supply: R_S Effects

The regulation curves for a typical rectifier supply with capacitive filter show that the drop in supply voltage $\pm V_{CC}$ is roughly a linear function of the dc output current I_{dc} [1]. The supply voltage V_{CC} in relation to the no-loaded supply voltage E_0 is

$$V_{CC} = E_0 - R_S I_{dc} \tag{9-27}$$

In this case the dissipated power for each transistor in the semiperiod is

$$\overline{P}_d = \frac{1}{\pi}\left[\int_0^{\pi} E_0 i\, d(\omega t) - \int_0^{\pi} R_S i^2\, d(\omega t) - \int_0^{\pi} R_L i^2\, d(\omega t)\right] \tag{9-28}$$

where $i = (V_0/R_L)\sin \omega t$.

From Eq. 9-28 we obtain

$$\overline{P}_d = \frac{2E_0 V_0}{\pi R_L} - \frac{R_S V_0}{2R_L^2} - \frac{V_0^2}{2R_L} \tag{9-29}$$

Differentiating Eq. 9-29 with respect to V_0 and making it equal to zero, we obtain the maximum dissipated power $\overline{P}_{d,\max}$:

$$\overline{P}_{d,\max} = \frac{2E_0^2}{\pi^2(R_S + R_L)} \tag{9-30}$$

for

$$V_0 = \frac{2E_0 R_L}{\pi(R_S + R_L)} \tag{9-31}$$

The output power at clipping $\overline{P}_{0,\text{clipp}}$ is obtained when

$$V_0 = \frac{R_L E_0}{R_S + R_L} \tag{9-32}$$

and its value is

$$\overline{P}_{0,\text{clipp}} = \frac{E_0^2 R_L}{2(R_S + R_L)^2} \tag{9-33}$$

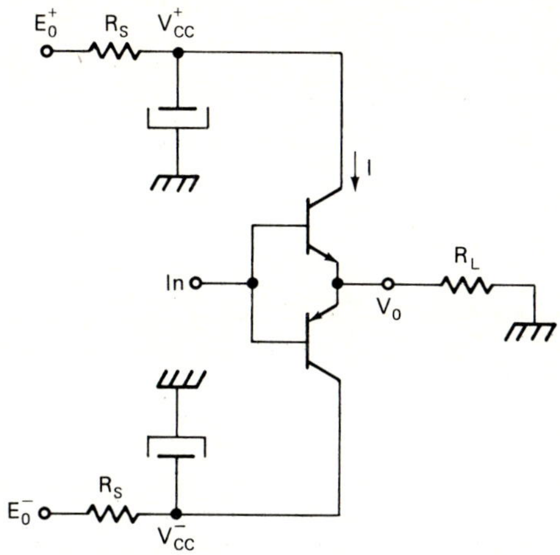

FIG. 9-4 Equivalent circuit for a split capacitive-input power supply.

The output power $\overline{P}_{0,\text{max diss.}}$ at maximum dissipated power condition, is obtained from Eq. 9-31:

$$\overline{P}_{0,\text{max diss.}} = \frac{2E_0^2 R_L}{\pi^2 (R_S + R_L)^2} \tag{9-34}$$

The output music power $P_{0,\text{music}}$ is defined as follows:

$$P_{0,\text{music}} = \frac{E_0^2}{2R_L} \tag{9-35}$$

It is interesting to see how the output power and the dissipation vary as a function of the ratio R_S/R_L. From the previous equations we obtain

$$\frac{\overline{P}_{d,\text{max}}}{P_{0,\text{music}}} = \left[\frac{\pi^2}{4}\left(1 + \frac{R_S}{R_L}\right)\right]^{-1} \tag{9-36}$$

$$\frac{\overline{P}_{0,\text{clipp}}}{P_{0,\text{music}}} = \left[1 + 2\frac{R_S}{R_L} + \left(\frac{R_S}{R_L}\right)^2\right]^{-1} \tag{9-37}$$

$$\frac{P_{0,\text{max diss.}}}{P_{0,\text{music}}} = \left\{\frac{\pi^2}{4}\left[1 + 2\frac{R_S}{R_L} + \left(\frac{R_S}{R_L}\right)^2\right]\right\}^{-1} \tag{9-38}$$

These equations are expressed graphically in Fig. 9-5, which shows that for high values of ratio R_S/R_L the dissipated power decreases considerably if compared with the music power supplied by the amplifier.

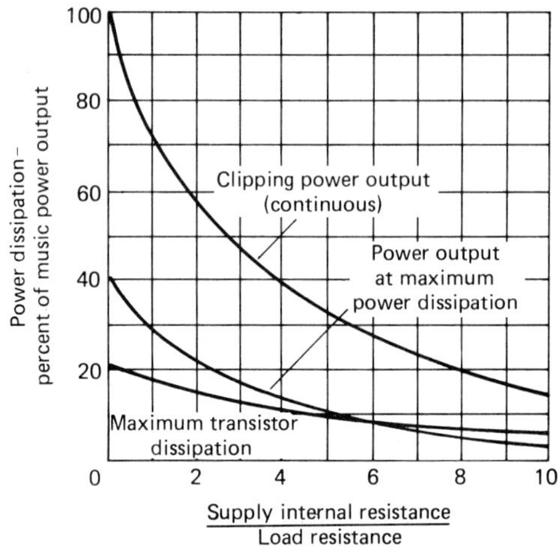

FIG. 9-5 Power output and dissipation as a function of the ratio of internal supply resistance to load resistance.

The above equations do not take into account the ripple effects on the power supply.

9.1.3 Amplifier Fidelity[1]

In designing an audio amplifier, especially in the high-fidelity field, the ratio R_S/R_L is important. To understand this we must remember that the aim of a hi-fi amplifier is to reproduce a peak sound pressure level of nearly 100 dB, which corresponds to about 0.4 W acoustic in a room with a capacity of around 3000 ft^3. If the efficiency of the loudspeaker is about 1 percent, a stereo amplifier must be able to supply about 20 W/channel.

In most cases the ratio between the output peak level and the average level is around 20 dB. Therefore an amplifier capable of providing a continuous level of 77 dB with peaks of 100 dB satisfies the hi-fi requirements.

To obtain the required output level, the supply voltage must not fall below the value needed for 100 dB of acoustic power while supplying the average current required for 77 dB. Since the passages may exceed the average level by 10 dB, the supply voltage cannot fall below the

[1]R. S. Hartz and F. S. Kamp, Power Output and Dissipation in Class B Transistor Amplifiers, *RCA Appl. Note 3576*, 1968. Reproduced with permission.

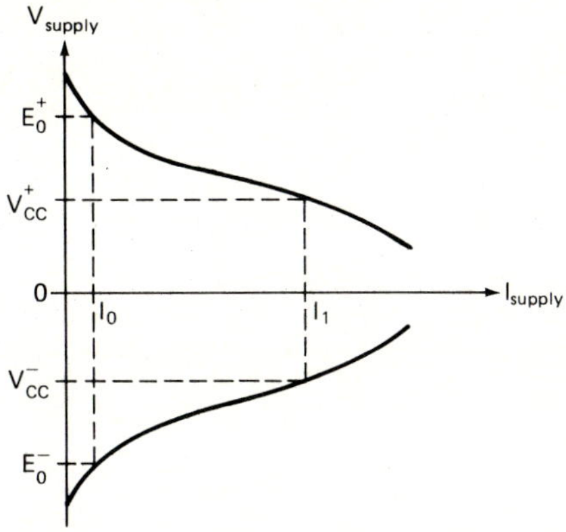

FIG. 9-6 Typical regulation curve for a transformer-rectifier power supply with a capacitive-input filter.

value required for an acoustic power of 100 dB while supplying 87 dB, which corresponds to about 1 W/channel.

9.1.4 Maximum Effective Series Resistance[1]

To determine the maximum R_S on the supply, on the power peaks the supply must not fall below the value corresponding to a $P_{0,\max} = 20$ W for short periods while the average power is 1 W/channel.

Figure 9-6 shows the supply regulation curve of the splitting power supply of the set of Fig. 9-4. E_0 is the voltage without signal. V_{CC} is the voltage at which the amplifier is able to supply 1 W/channel, supposing that the $V_{CE,\text{sat}}$ of the final transistors is negligible. Thus

$$E_0 = V_{CC} + (I_1 - I_0)R_S \tag{9-39}$$

where I_0 = zero-signal current
I_1 = steady-state current drain at a total power output of 1 W/channel, or 2 W total

$$I_1 - I_0 = \frac{I_{\text{rms}}\sqrt{2}}{\pi} \tag{9-40}$$

where

$$I_{\text{rms}} = \sqrt{\frac{P_{\text{out,rms}}}{R_L}} \tag{9-41}$$

[1]R. S. Hartz and F. S. Kamp, Power Output and Dissipation in Class B Transistor Amplifiers, *RCA Appl. Note 3576*, 1968. Reproduced with permission.

It is known that the music power is

$$P_{0,\text{music}} = \frac{E_0^2}{2R_L} = \frac{[V_{CC} + (I_1 - I_0)R_S]^2}{2R_L} \tag{9-42}$$

hence

$$R_S = \frac{\sqrt{2P_{0,\text{music}}R_L} - V_{CC}}{I_1 - I_0} \tag{9-43}$$

In the above example we have

$$I_1 - I_0 = 0.225 \text{ A} \qquad \text{with } R_L = 8\ \Omega$$

The V_{CC} required to get $P_{0,\text{music}} = 20$ W is $V_{CC} = \sqrt{20 \times 2R_L} = 17.9$ V; therefore

$$R_S = 17.9\sqrt{P_{0,\text{music}}} - 79.5\ \Omega \tag{9-44}$$

9.1.5 Amplifier Economics[1]

We have seen that it is economically advantageous to dimension an amplification system using fairly high values of the R_S/R_L ratio and a high ratio between music power and dissipated power. This means reducing the dimensions of the heat sinks, reducing the cost of the transformer and relative filter, and reducing the cost of the loudspeaker. Naturally this is possible if the amplifier is able to withstand high voltages at zero-signal condition.

At the present state of the art in the design of integrated monolithic audio amplifier, however, high voltage is still a considerable limitation.

9.1.6 Effects of $V_{CE,\text{sat}} > 0$ of Output Transistors

Up to now we have analyzed the behavior of an amplifier, ignoring the effect of the saturation voltage of the output stage.

The $V_{CE,\text{sat}}$ nonzero modifies the results for the ideal case both as regards the $\overline{P}_{\text{out}}$ and the $\overline{P}_{\text{diss}}$ and consequently the efficiency. To evaluate these effects we shall consider a clipped output waveform (Fig. 9-7) to get a certain distortion due to the clipping. The output power $\overline{P}_0$ in this case is

$$\overline{P}_0 = \frac{2}{\pi}\int_0^{\alpha} \frac{V_{CC}^2}{R_L}\sin^2 \omega t\, d(\omega t) + \frac{1}{\pi}\frac{V_{0,\text{max}}^2}{R_L}(\pi - 2\alpha) \tag{9-45}$$

[1]R. S. Hartz and F. S. Kamp, Power Output and Dissipation in Class B Transistor Amplifiers, *RCA Appl. Note 3576*, 1968. Reproduced with permission.

where $$\alpha = \arcsin \frac{V_{0,\max}}{V_{CC}} \tag{9-46}$$

$$V_{0,\max} = V_{CC} - V_{CE,\text{sat}} \tag{9-47}$$

By substituting (9-46) and (9-47) in (9-45), we get

$$\overline{P}_0 = \frac{2}{\pi} \int_0^\alpha \frac{V_{CC}}{R_L} \sin^2 \omega t \, d(\omega t) + \frac{1}{\pi R_L} (V_{CC} - V_{CE,\text{sat}})^2 \left(\pi - 2 \arcsin \frac{V_{CC} - V_{CE,\text{sat}}}{V_{CC}} \right) \tag{9-48}$$

The loss of output power is transduced into an increase in dissipated power and loss of efficiency. For example, with an output sinusoid clipped at 10 percent of distortion, and therefore with a $V_{0,\max} = 0.8 V_{CC}$, the output power is

$$\overline{P}_0 = 0.809 \overline{P}_{0,\max} \tag{9-49}$$

The dissipated power in the case of a clipped sinusoidal signal at the output is given by the equation

$$\overline{P}_d = \frac{2}{\pi} \int_0^\alpha (V_{CC} - V_0 \sin \omega t) \frac{V_0}{R_L} \sin \omega t \, d(\omega t) + \frac{1}{\pi} \frac{V_{CE,\text{sat}}}{R_L} (\pi - 2\alpha) \tag{9-50}$$

where the first term on the right represents the average power dissipated during sinusoidal operation, and the second term is the average power dissipated during the clipping time.

Example. Equation 9-50 makes it possible to calculate the $\overline{P}_d$ and the efficiency in the case of an amplifier with a supply of ± 15 V, a $V_{CE,\text{sat}} = 3$ V, distortion = 10 percent, and $R_L = 4\ \Omega$.

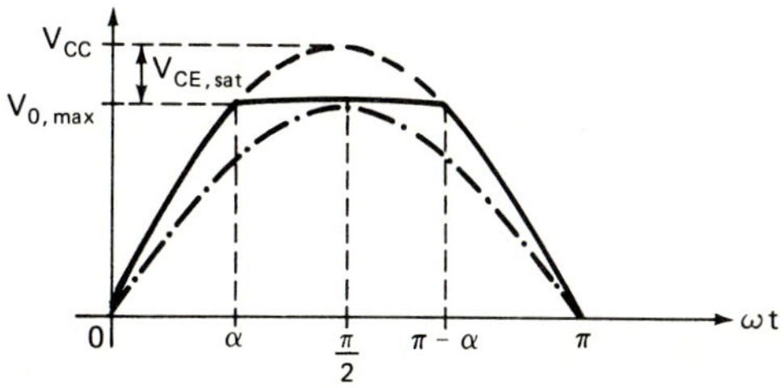

FIG. 9-7 Output clipped sinusoidal half-waveform due to $V_{CE,\text{sat}} > 0$.

Solution.

$$\overline{P}_d = \frac{2 \times 15 \times 12}{4\pi} \int_0^{0.927} \sin \omega t \, d(\omega t) + \frac{3^2}{4\pi} (\pi - 2 \times 0.927)$$

$$- \frac{2 \times 12^2}{4\pi} \int_0^{0.927} \sin^2 \omega t \, d(\omega t)$$

$$= 9.826 \text{ W}$$

$$\overline{P}_0 = \frac{2 \times 15^2}{4\pi} \int_0^{0.927} \sin^2 \omega t \, d(\omega t) + \frac{12^2}{4\pi} (\pi - 2 \times 0.927)$$

$$= 22.741 \text{ W}$$

$$\overline{P}_{\text{supply}} = \overline{P}_d + \overline{P}_0 = 32.567 \text{ W}$$

$$\eta = \frac{\overline{P}_0}{\overline{P}_{\text{supply}}} = 0.698$$

9.1.7 Thermal Problem in Audio ICs in the Presence of Real Load

The thermal problem in integrated audio amplifiers becomes particularly important in real load conditions, i.e., in driving a loudspeaker. It forms an ohmic-inductive load and obviously introduces a phase shift between the output voltage and the current, causing considerable dissipation in the IC. This condition deteriorates if the supply is single because in this case the loudspeaker is ac-coupled.

The presence of the output capacitor, especially at low frequency, causes an earlier phase shift of the current with respect to the voltage, as shown in Figs. 9-8 and 9-9, and this case is worse than the simple ohmic-inductive load because dissipation peaks are maintained for long periods.

In analyzing the phenomena connected with an ohmic-inductive load ac-coupled to the amplifier, we shall consider the most stressing case for the IC, that is, with single supply. In sinusoidal operation, using Z_L to represent the total impedance of the loudspeaker in series with the output capacitor, we get

$$v = V_0 \sin (\omega t + \varphi) \tag{9-51}$$

$$i = \frac{V_0}{Z_L} \sin \omega t \tag{9-52}$$

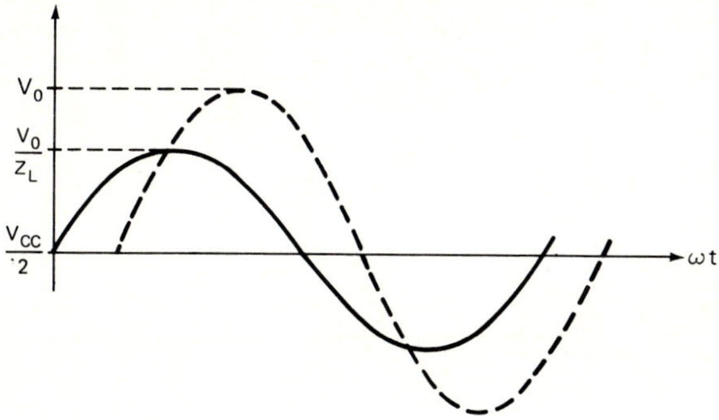

FIG. 9-8 Instantaneous output voltage and current.

where v = instantaneous output voltage
i = instantaneous output current
V_0 = output peak voltage
$\frac{V_0}{Z_L}$ = output peak current

and

$$\varphi = \arctan \frac{1}{R}\left(\omega L - \frac{1}{\omega C}\right) \tag{9-53}$$

The average output power $\overline{P}_0$ is obtained from

$$\overline{P}_0 = \frac{1}{\pi}\int_0^{\pi} vi\, d(\omega t) = \frac{V_0^2}{2Z_L}\cos\varphi \tag{9-54}$$

We can consider the effects of the output capacitor and of the loudspeaker inductance separately, considering two frequency ranges: (1)

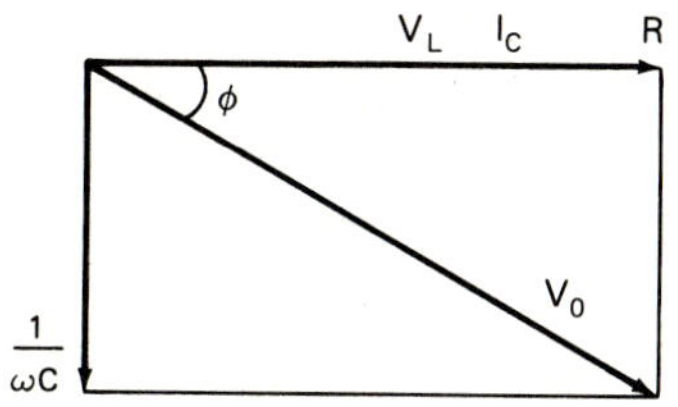

FIG. 9-9 Earlier phase shift of the output current with respect to the output voltage for a resistive-capacitive load.

with frequency low enough to consider the inductive effect negligible and (2) with frequency high enough to consider the capacitor a short circuit.

9.1.8 Effect of the Output Capacitor

In general in audio amplifiers with single supply and load ac-coupled the effects of the output capacitor C are particularly noticeable below 100 Hz. The output power $\overline{P}_0$ is that dissipated on resistor R_L. The starting equations are

$$v = V_0 \sin \omega t \tag{9-55}$$

$$i = I_0 \sin (\omega t + \varphi)$$

where

$$V_0 = V_L \sqrt{\frac{1}{(\omega RC)^2} + 1} = \frac{V_L}{\cos \varphi} \tag{9-57}$$

$$I_0 = I_L = I_C = \frac{V_L}{R_L} \tag{9-58}$$

$$\begin{aligned}\overline{P}_0 &= \frac{V_0^2}{|Z_L|} \frac{\omega}{\pi} \int_0^{\pi/\omega} \sin \omega t \sin (\omega t + \varphi)\, dt \\ &= \frac{V_0^2}{|Z_L|} \frac{\omega}{\pi} \int_0^{\pi/\omega} \sin^2 \omega t \cos \varphi\, dt + \frac{V_0^2}{|Z_L|} \frac{\omega}{\pi} \int_0^{\pi/\omega} \sin \omega t \cos \omega t \sin \varphi\, dt\end{aligned} \tag{9-59}$$

The first term in the second line represents the power dissipated on the load resistor R_L, while the second term is an average nul and represents the reactive power. Therefore

$$\overline{P}_0 = \frac{V_0^2}{2|Z_L|} \cos \varphi \tag{9-60}$$

Using V_L to indicate the peak voltage on the load,

$$V_L = V_0 \times \cos \varphi \qquad R_L = Z_L \times \cos \varphi \tag{9-61}$$

we get

$$P_0 = \frac{V_L^2}{2R_L} \tag{9-62}$$

The power taken from the supply is

$$\overline{P}_s = \frac{\omega}{\pi} \int_{-\varphi/\omega}^{(\pi-\varphi)/\omega} \frac{V_{CC}V_0}{|Z_L|} \sin(\omega t + \varphi)\, dt$$
$$= \frac{\omega}{\pi} \frac{V_{CC}V_0}{|Z_L|} \left[-\frac{1}{\omega} \cos(\omega t + \varphi) \right]_{-\varphi/\omega}^{(\pi-\varphi)/\omega} = \frac{2V_{CC}V_0}{\pi|Z_L|} \tag{9-63}$$

The dissipated power is therefore

$$\overline{P}_d = \frac{2V_{CC}V_0}{\pi|Z_L|} - \frac{V_0^2}{2|Z_L|} \cos\varphi \tag{9-64}$$

The maximum dissipated power is obtained from

$$\frac{\partial \overline{P}_d}{\partial V_0} = 0 \tag{9-65}$$

Hence

$$\overline{P}_{d,\max} = \frac{2V_{CC}^2}{\pi^2|Z_L| \cos\varphi} \tag{9-66}$$

where

$$V_0 = \frac{2V_{CC}}{\pi \cos\varphi} \tag{9-67}$$

The efficiency η is

$$\eta = \frac{\overline{P}_0}{\overline{P}_s} = \frac{\pi}{4} \frac{V_0}{V_{CC}} \cos\varphi \tag{9-68}$$

In maximum dissipation conditions, from Eqs. 9-60, 9-63, and 9-67 we get

$$\eta_{\text{max diss}} = \frac{1}{2} \tag{9-69}$$

while, placing $V_0 = V_{CC}$,

$$\eta_{\max} = \frac{\pi}{4} \cos\varphi \tag{9-70}$$

It is interesting to note the curve of the $\overline{P}_d$ and of η as a function of V_0 for a set frequency.

From Eqs. 9-64 and 9-68 we obtain the results in Fig. 9-10. As can

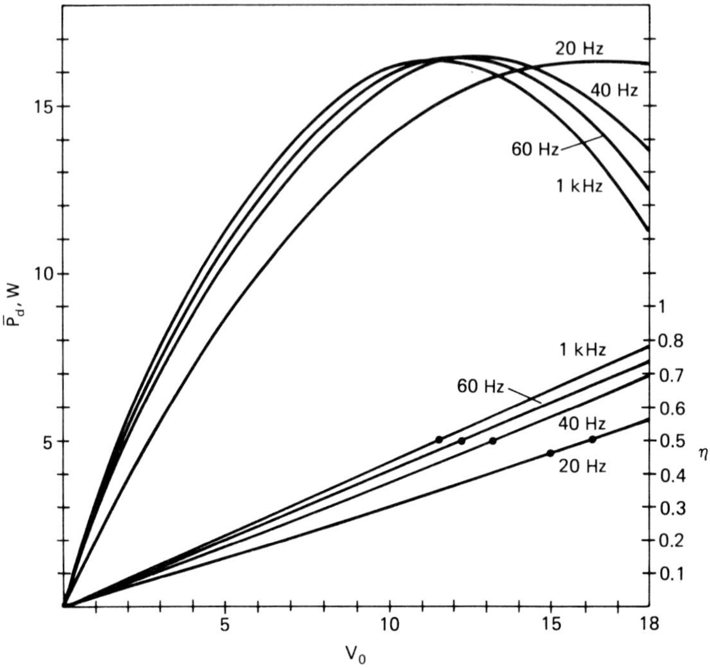

FIG. 9-10 Dissipated power and efficiency vs. output voltage swing as a function of frequency.

be seen $\overline{P}_{d,\max}$ does not depend on the frequency, as is obvious from Eq. 9-66, since it is $|Z_L| \cos \varphi = R_L$.

The thermal problem is important at low frequency due to the heat transmission time constants in the silicon. Since this time is around 1 ms, the instantaneous dissipated power acquires considerable importance.

The voltage-current phase shift on the output transistors can cause them to operate in conditions of simultaneous high current and high voltage for long enough times to cause failures due to secondary breakdown. In these cases it is therefore useful to evaluate the curve of the instantaneous dissipated power P_{di} when choosing the geometries of the power transistors of the integrated circuits.

$$v_{CE} = V_{CC} - V_0 \sin \omega t \tag{9-71}$$

$$i_C = I_0 \sin (\omega t + \varphi) \tag{9-72}$$

where

$$V_0 = \sqrt{\frac{2R_L\overline{P}_0}{\cos^2 \varphi}} \tag{9-73}$$

$$I_0 = \sqrt{\frac{2\overline{P}_0}{R_L}} \tag{9-74}$$

We obtain

$$v_{CE} = V_{CC} - \sqrt{2R_L\overline{P}_0\left[1 + \left(\frac{1}{\omega RC}\right)^2\right]}\ \sin \omega t \tag{9-75}$$

$$i_C = \sqrt{\frac{2\overline{P}_0}{R_L}}\ \sin(\omega t + \varphi) \tag{9-76}$$

where
$$\varphi = \arctan \frac{1}{\omega RC} \tag{9-77}$$

Therefore, from Eqs. 9-75 and 9-76,

$$P_{di} = v_{CE} i_C \tag{9-78}$$

Note that in a diagram v_{CE} and i_C Eqs. 9-75 and 9-76 give rise to an ellipse which represents the amplifier load curve, as seen in Fig. 9-11.

Only the upper section of the amplifier output stage is considered, ignoring the $V_{CE,\text{sat}}$ of the transistor and assuming that the supply is stabilized.

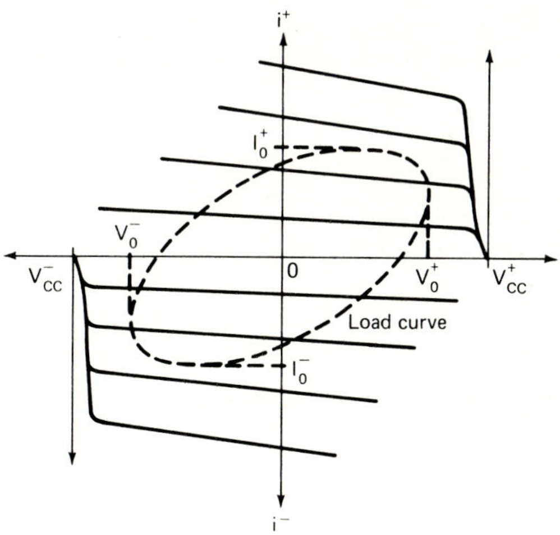

FIG. 9-11 Graphical representation of the amplifier load curve superimposed upon the output transistor characteristics.

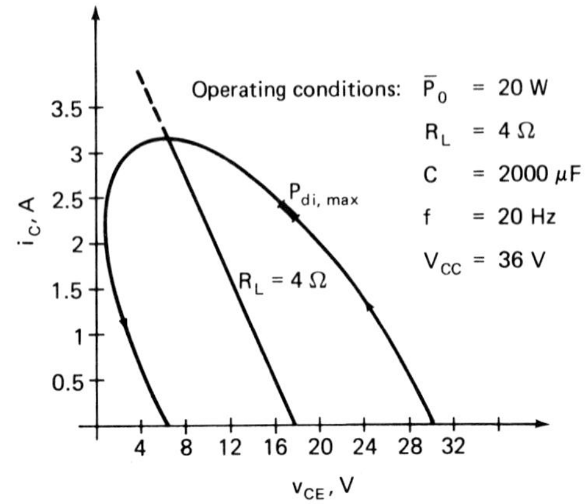

FIG. 9-12 Amplifier load curve (half-ellipse) for a resistive-capacitive load.

Figures 9-12 and 9-13 give the graphical solution of the load curve and of the instantaneous dissipated power for an audio amplifier, supplied at 36 V with a load $R_L = 4\ \Omega$, ac-coupled by means of a capacitor $C = 2000\ \mu F$, supplying a $\bar{P}_0 = 20$ W. The problem is examined at a frequency of 20 Hz, which puts the device under considerable stress.

With respect to dc coupling of the load, where $V_0 = \sqrt{2R_L\bar{P}_0} = 12.65$ V, we get $P_{di,\max}$ at $V_0' = 8.05$ V, from which $V_{CE} = 9.95$ V and $I_0 = 2.01$ A. We obtain $P_{di,\max} = 20$ W.

Note that with respect to the case of direct load coupling, when an

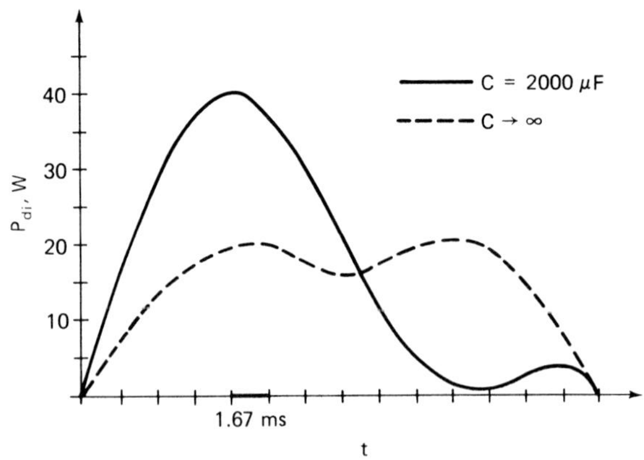

FIG. 9-13 Instantaneous dissipated power in a sinusoidal 20-Hz half-cycle for a resistive-capacitive load.

ohmic-capacitive load is present, the instantaneous dissipated power can be much greater, even double, depending on the frequency. In addition to this we can see that the time during which the output transistor dissipates a power near the maximum is a few milliseconds, long enough to cause the output transistor to operate in secondary breakdown condition.

It is therefore necessary to take into account these phenomena of strong dissipation when designing the layout of an integrated audio amplifier as regards the structure and geometry of the output stage transistors.

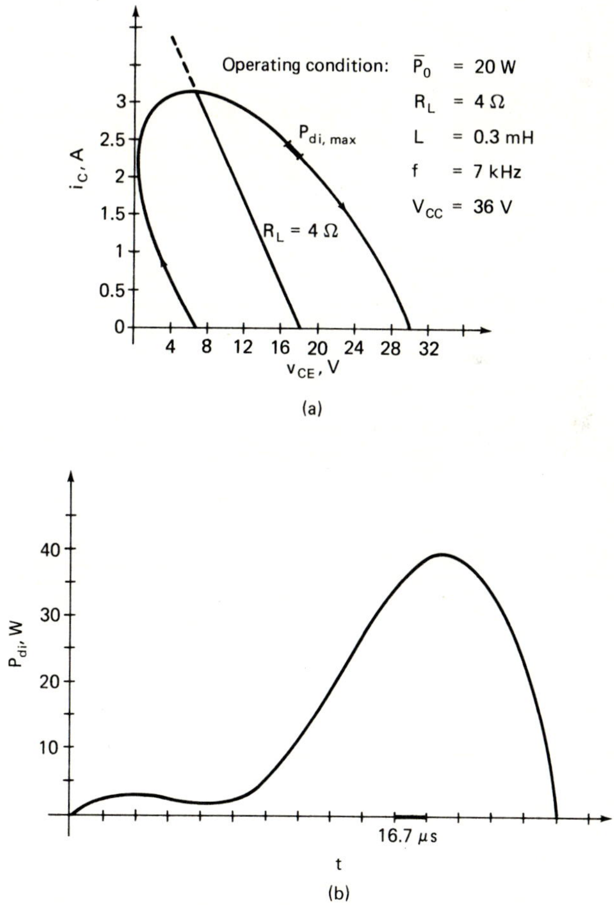

FIG. 9-14 (*a*) Amplifier load curve (half-ellipse) for an ohmic-inductive load. (*b*) Instantaneous dissipated power in a sinusoidal half-cycle for an ohmic-inductive load.

9.1.9 Effects of an Ohmic-Inductive Load

A similar reasoning can be made taking into account that the load of an audio amplifier is ohmic-inductive.

While at low frequency the effect of the inductance is negligible and the effect of the coupling capacitor prevails, at frequencies of more than 1 kHz the effect of the loudspeaker inductance prevails. In this case Eqs. 9-71 and 9-72 become

$$v_{CE} = V_{CC} - \sqrt{2R_L\overline{P}_0\left[1 + \left(\frac{\omega L}{R_L}\right)^2\right]}\ \sin \omega t \qquad (9\text{-}79)$$

$$i_C = \sqrt{\frac{2\overline{P}_0}{R_L}}\ \sin(\omega t + \varphi) \qquad (9\text{-}80)$$

The graphic solution for Eqs. 9-79 and 9-80 and their product P_{di} is shown in Fig. 9-14*a* and *b*. The instantaneous dissipated power is still very high, around 40 W, but for very short times, around 20 μs, so the secondary breakdown zone of the output transistors is not effected.

9.2 PERFORMANCE AND LIMITS OF A LOW-DISTORTION INTEGRATED OUTPUT STAGE

Bias Current Control and Thermal Stability

An output stage in class B (push-pull) is not suitable for driving a loudspeaker due to the crossover distortion. In fact the output stage of Fig. 9-15 has a transfer characteristic as shown in Fig. 9-16. It is therefore necessary to use output stages in class AB in which a bias current is

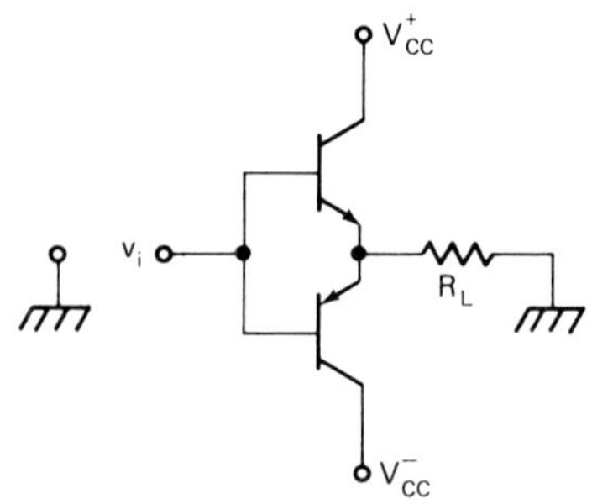

FIG. 9-15 Basic output stage.

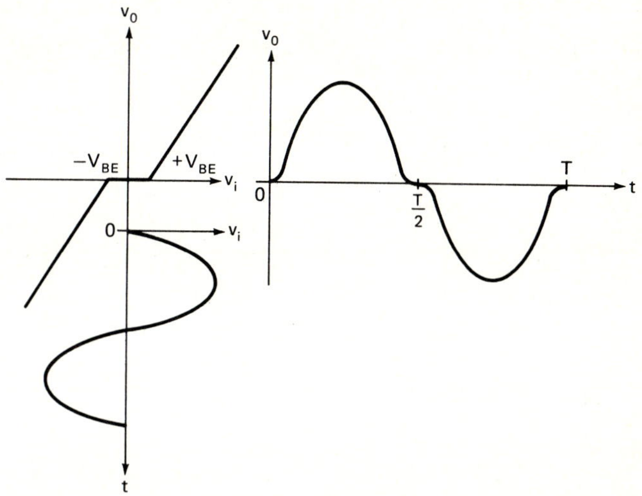

FIG. 9-16 Transfer characteristic of the output stage of Fig. 9-15.

maintained to prevent switching off the output transistors while the output signal is crossing zero.

At the present state of IC technologies it is impossible to have high-current capabilities by using PNP transistors, unless we have chips with very large areas and therefore high costs, due to the fact that the dop-

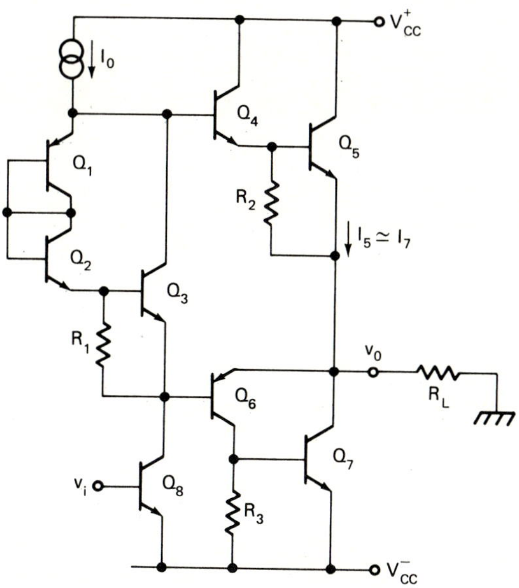

FIG. 9-17 Biased quasi-complementary class AB output stage.

ing levels of these transistors are not optimized for PNP structures. Therefore in audio ICs only NPN transistors are used as output transistors.

The configuration normally used, where powers of more than a few hundred milliwatts are required, is that with quasi-complementary symmetry, as in Fig. 9-17. The bias current of this stage is around a few milliamperes and is calculated by making a computation of the voltages in the bias network formed by transistors Q_1 to Q_6. This current can be made independent of the h_{FE} of the transistors, accepting a minimum value of the h_{FE}, which allows us to ignore the base currents of the transistors in the overall computation of the currents circulating in the transistors.

We can say that

$$V_{BE1} + V_{BE2} + V_{BE3} = V_{BE4} + V_{BE5} + V_{BE6} \tag{9-81}$$

since

$$V_{BE} = \frac{kT}{q} \ln \frac{I_C}{AJ_S} \tag{9-82}$$

where A = emitter area of the transistor.

By substituting in Eq. 9-81, we get

$$I_5 = I_{\text{fin}} = \frac{I_1 I_2 I_3}{I_4 I_6} \frac{A_4 A_5 A_6}{A_1 A_2 A_3} \tag{9-83}$$

This current is completely fixed, since

$$I_1 = I_2 = \frac{V_{BE}}{R_1} \tag{9-84}$$

$$I_3 = I_0 - I_1 = \frac{R_1 I_0 - V_{BE}}{R_1} \tag{9-85}$$

$$I_4 = \frac{V_{BE}}{R_2} \tag{9-86}$$

$$I_6 = \frac{V_{BE}}{R_3} \tag{9-87}$$

Therefore

$$I_{\text{fin}} = \frac{R_2 R_3}{R_1^3} (R_1 I_0 - V_{BE}) \frac{A_4 A_5 A_6}{A_1 A_2 A_3} \tag{9-88}$$

Without resistors R_1, R_2, and R_3, Eq. 9-88 is independent of the h_{FE} of the transistors and is

$$I_{\text{fin}} = I_0 \sqrt[3]{\frac{A_4 A_5 A_6}{A_1 A_2 A_3}} \tag{9-89}$$

The presence of resistors, R_1, R_2, and R_3 stabilizes currents independent of the h_{FE} in transistors Q_1, Q_2, and Q_6. Furthermore, currents which are too low in the above transistors limit the frequency response of the output stage.

We can see from Eq. 9-88 that I_{fin} does not depend on the absolute value of the resistors but only on the ratios. Since in ICs good matching of the resistors is possible with a spread not greater than 1 percent, I_{fin} does not depend on the resistors. It does, however, depend on the temperature, with a fixed I_0 = constant.

In fact Eq. 9-88 can be written

$$I_{\text{fin}} = BI_0 - B\frac{V_{BE}}{R_1} \tag{9-90}$$

where

$$B = \frac{R_2 R_3}{R_1^2} \frac{A_4 A_5 A_6}{A_1 A_2 A_3} \tag{9-91}$$

Since the variation of V_{BE} with the temperature is equal to -2 mV/°C and the variation of an integrated resistor is about 2.7 percent/°C, the second term of Eq. 9-90 varies by around -44 percent for ΔT = 100°C. Therefore I_{fin} will increase in relation to I_0 and B. The variation due to this parameter is usually not greater than 10–15 percent.

Effects greater than this are due to the thermal gradient inside the silicon chip when it drives a load during operation of the amplifier. The output transistors reach junction temperatures much greater than the rest of the circuit components, and since Q_5 is part of the bias chain, a decrease in its V_{BE} causes a corresponding increase in I_{fin}.

Furthermore, for reasons of layout commodity, Q_4 is situated near Q_5 and the decrease in its V_{BE}, and the effect of a temperature gradient with respect to the other components in the bias chain, leads to a further increase in I_{fin} according to the relationship 9-90. If this is not taken into account when preparing the layout, we get thermal runaway, and the device may fail due to the dissipation or secondary breakdown.

To reach stability, transistors Q_1, Q_2, and Q_3 must be as near as pos-

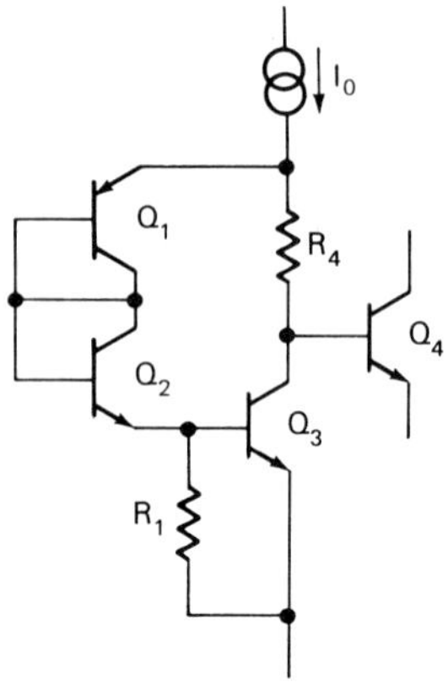

FIG. 9-18 Resistor inserted in the bias chain to increase the thermal stability.

sible to the area of the power transistors, while Q_6 must be at a suitable distance from it.

The stability of I_{fin} is also affected by the fact that the dissipator on which the chip is soldered has good thermal conductivity (generally it is copper), and this tends to reduce the thermal gradient along the chip.

One way to reduce the areas of transistors Q_1, Q_2, and Q_3 and increase the thermal stability of I_{fin} is to divide the V_{BE} of Q_3 by inserting a resistor, as in Fig. 9-18. In this case Eq. 9-90 becomes

$$I_{fin} = B\left(I_0 - \frac{V_{BE}}{R_1}\right) e^{-R_4 I_0/V_T} \tag{9-92}$$

Naturally with the introduction of R_4, I_{fin} decreases in temperature as R_4 increases. Nevertheless the drop on R_4 cannot exceed a certain value (for example, 100 mV) because the drop-out increases of the same amount and the variations of I_{fin} with the temperature become too noticeable.

Low-Drop Output Stage. Another output stage commonly used in the design of power amplifiers, due to the low drop, is shown in Fig. 9-19. In this case the transistors forming part of the bias network are Q_1 to Q_6. The voltage balance is given by

$$R_1 I_0 + V_{BE4} + V_{R2} + V_{BE5} + V_{BE6} = V_{BE1} + V_{BE2} + V_{BE3} \tag{9-93}$$

where

$$V_{R2} = R_2 I_4\left(1 + \frac{A_7}{A_4} + \frac{A_{11}}{A_{10}}\right) \tag{9-94}$$

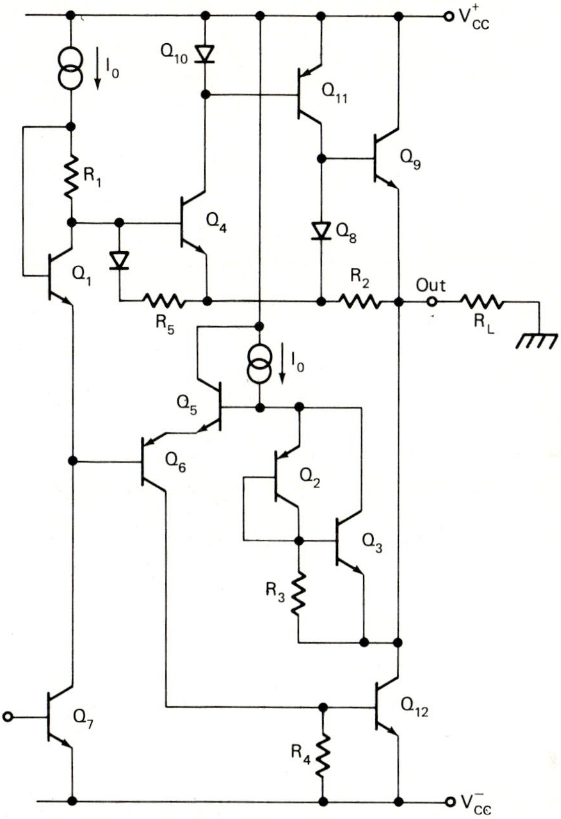

FIG. 9-19 Low-drop output stage.

Ignoring the transistor base currents, Eq. 9-93 becomes

$$I_4 = \frac{I_1 I_2 I_3}{I_5 I_6} \frac{A_4 A_5 A_6}{A_1 A_2 A_3} e^{-(R_1 I_0 + V_{R2})/V_T} \tag{9-95}$$

This equation is approximate since the gain of groups Q_4, Q_7 and Q_9, Q_8 is put equal to the ratio of the emitter areas, ignoring the drop on resistors R_5 and R_2.

This approximation of the current gain is good since the resistors are low in value (typically, $R_5 = 100\ \Omega$, and $R_2 = 50\ \Omega$).

The currents have the following values:

$$I_1 = I_0$$

$$I_2 = \frac{V_{BE}}{R_3} \tag{9-96}$$

$$I_3 = I_0$$

$$I_5 = I_6 = \frac{V_{BE}}{R_4}$$

By substituting in Eq. 9-95 and putting

$$D = \frac{R_4}{R_3} \frac{A_4 A_5 A_6}{A_1 A_2 A_3} \tag{9-97}$$

then Eq. 9-95 becomes

$$I_4 = D I_0^2 \frac{R_4}{V_{BE}} e^{-(R_1 I_0 + V_{R2})/V_T} \tag{9-98}$$

Therefore

$$I_{\text{fin}} = I_9 = \frac{A_9 A_1}{A_8 A_{10}} I_4 = D I_0^2 \frac{R_4}{V_{BE}} \frac{A_9 A_{11}}{A_8 A_{10}} e^{-(R_1 I_0 + V_{R2})/V_T} \tag{9-99}$$

Equation 9-99 expresses the current in the output transistors with both the generator I_0 currents and the areas of the bias network components fixed.

The thermal stability criteria described in the case of Fig. 9-17 are also valid in this case.

9.3 OPTIMIZING VOLTAGE DROPS IN OUTPUT POWER STAGES

As we said in Sec. 9.1 regarding the output power and dissipated power on an output transistor, when designing an audio amplifier particular attention must be paid to the voltage drop in the output swing because the output power depends on the voltage swing square. Therefore, even small voltage drops, around the value of the V_{BE}, have a considerable effect on the P_{out}, especially at low voltages.

9.3.1 Historical Evolution in Designing Output Stages

It is useful to examine the developments that have taken place in the design of the output stages of integrated audio amplifiers.

The TDA 2030, which is a 14-W audio power amplifier, has an out-

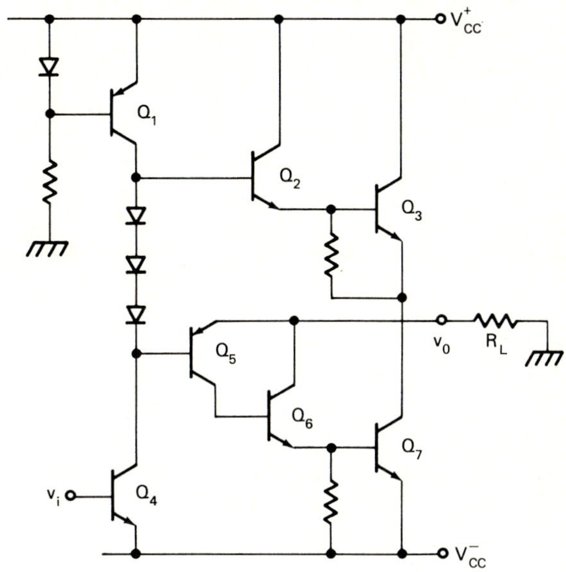

FIG. 9-20 The output stage of the TDA 2030.

put stage as shown in Fig. 9-20. The output voltage swing, i.e., the peak-to-peak voltage $V_{0,p-p}$, is

$$V_{0,p\text{-}p} = 2V_{CC} - V_{BE}(Q_2) - V_{BE}(Q_3) - V_{\text{sat}}(Q_1) - V_{\text{sat}}(Q_5) - V_{BE}(Q_6) - V_{BE}(Q_7) \quad (9\text{-}100)$$

Therefore the overall drop V_D is

$$V_D = 4V_{BE} - 2V_{CE,\text{sat}} \quad (9\text{-}101)$$

With an $I_{\text{out}} = 3$ A and a load of 4 Ω the V_{BE} of the power transistors is around 2.2 V and that of the driver transistors is around 1.3 V, while the $V_{CE,\text{sat}}$ of Eq. 9-101 is 0.5 V. The overall drop is therefore around 8 V. When the device is supplied at 32 V, the $V_{0,p-p}$ is 24 V. Therefore the maximum undistorted output power is

$$P_{\text{out}} = \frac{V_{0,p\text{-}p}^2}{8R_L} = 18 \text{ W} \quad (9\text{-}102)$$

An important improvement, which increased the current capability of the lateral PNPs by a factor of about 5, was achieved by doping the emitter and collector zones about 8 times more, making it possible to

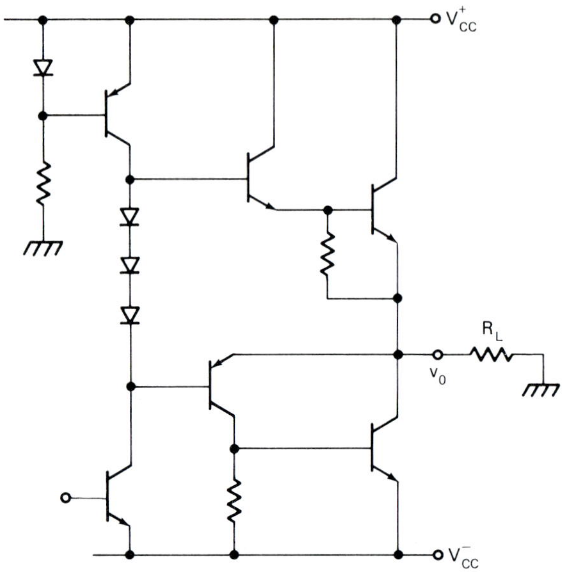

FIG. 9-21 Integrated output stage with a high-current-capability lateral PNP.

have a circuit such as the one shown in Fig. 9-21. This circuit makes it possible to gain one V_{BE} in the lower section. In the conditions of the previous example, with this configuration the P_{out} increases 2 W. The overall voltage drop is further reduced by using the diagram in Fig. 9-22. In this case the voltage swing of the upper section $V_{0,sup}$ and that of the lower section $V_{0,inf}$ is

$$V_{0,sup} = V_{CC} - \left(2V_{BE} + V_{CE,sat}\right) \tag{9-103}$$

$$V_{0,inf} = V_{CC} - V_{CE,sat} \tag{9-104}$$

Naturally this stage, like that of Fig. 9-21, is not symmetrical with the supply and requires adequate biasing of the output voltage at zero signal condition.

In the case of Fig. 9-21, the standby output voltage V_0 referred to ground will be

$$V_0 = -\frac{V_{BE}}{2} \tag{9-105}$$

and in the case of Fig. 9-22,

$$V_0 = -V_{BE} \tag{9-106}$$

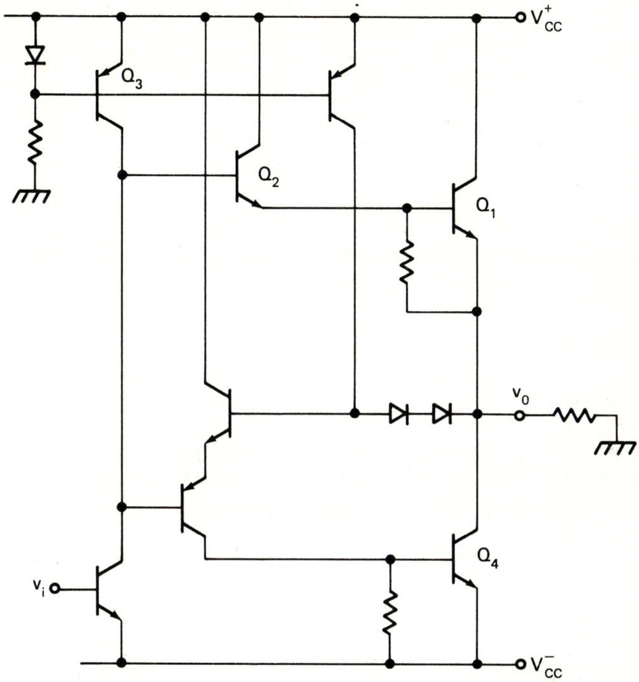

FIG. 9-22 Improvement with respect to the output stage of Fig. 9-21 in the lower-section voltage drop.

At present the solution with the lowest voltage drop without using the bootstrap seems to be the one considered in Fig. 9-19. The overall voltage swing is

$$V_{0,p\text{-}p} = 2V_{CC} - \left[V_{BE}(Q_9) + V_{\text{sat}}(Q_{11}) \right] - V_{\text{sat}}(Q_{12}) \quad (9\text{-}107)$$

With respect to the original solution of Fig. 9-20 there is an increase in the output voltage swing of $3V_{BE} + V_{CE,\text{sat}}$. By comparing the two P_{out} there is an increase of $\Delta P_{\text{out}} = 5$ W.

A drawback of the circuit in Fig. 9-19 is that it has overall storage times longer than the circuits in Figs. 9-20 to 9-22 and therefore requires higher standby currents to avoid crossover distortion.

9.3.2 The Bootstrap Solution

The best solution from the point of view of the minimum output voltage drop, even though it involves an extra external component, is the bootstrap solution.

The circuit in Fig. 9-22 has a total drop of

$$V_D = V_{BE1} + V_{BE2} + V_{CE,\text{sat }3} + V_{CE,\text{sat }4} \tag{9-108}$$

We shall now analyze the same circuit by applying the bootstrap solution (Fig. 9-23). To make the upper section of the output stage, which is the one with maximum drop, the voltage at point D must be increased during the output peak by an amount sufficient to bring the output transistor Q_1 into saturation. This is obtained by inserting capacitor C between pins A and B of the IC.

We shall see how the voltages vary in the various points of the output stage for the calculation of R and C (Fig. 9-24). During the positive peak, capacitor C supplies the maximum current to resistor R, transistor Q_2, and current generator Q_3, while during the negative peak it receives the maximum current from the supply through resistor R. Since the capacitor must remain charged at the minimum frequency, the medium current on the capacitor during an output waveform period must be zero.

Depending on the load resistor R_L, the V_b on the bootstrap pin varies its dc voltage as the signal varies, because during part of the positive

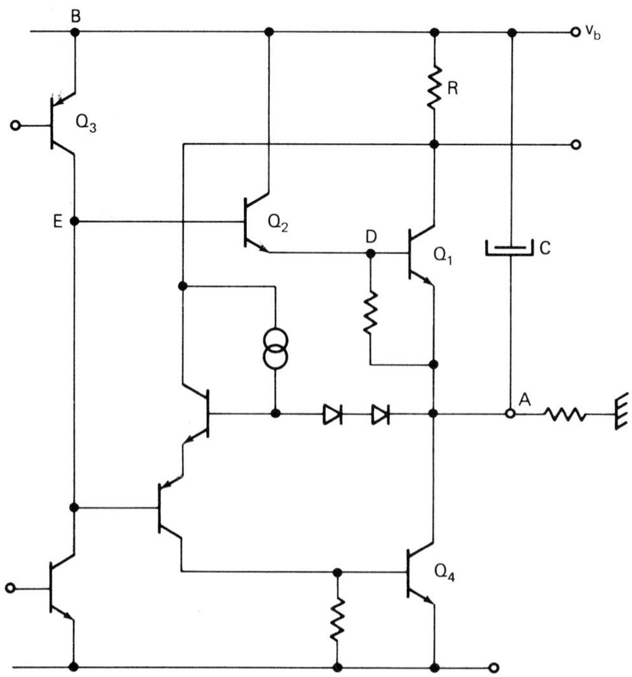

FIG. 9-23 Output stage including the bootstrap solution.

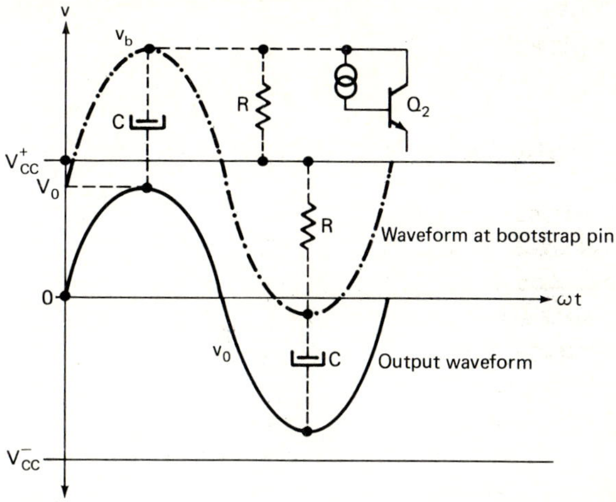

FIG. 9-24 Voltage waveforms in the various points of the output stage.

half-waves the bootstrap capacitor must not only supply current to R but also collector current to transistor Q_2. If it were to drive only R, the peak voltage on the bootstrap pin would be equal to V_{CC}. To calculate R, the total balance of currents in capacitor C must be zero. V_1 and V_2 are the peak voltages of Fig. 9-25, with I_1 the current peak supplied by

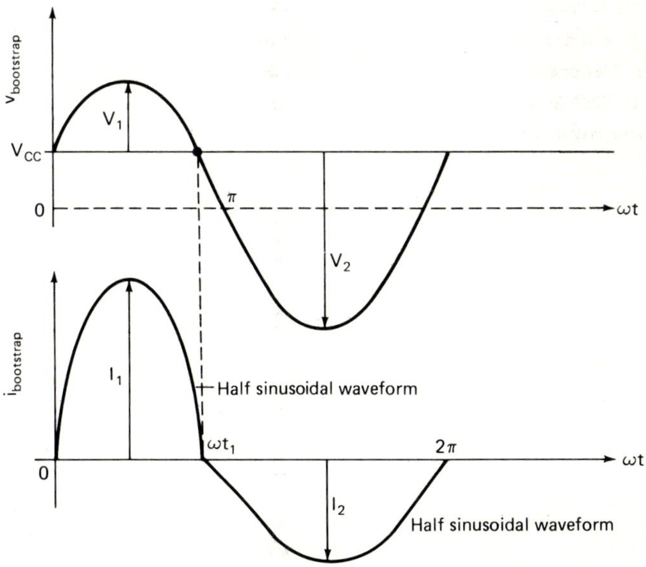

FIG. 9-25 Instantaneous voltage and current in the bootstrap pin in sinusoidal operation.

C during the positive output voltage peak and with I_2 the current peak supplied to capacitor C during the negative peak of the output voltage.

$$V_1 = V_{CC} - V_{\text{sat 1}} + V_{BE1} + V_{BE2} + V_{\text{sat 3}} \tag{9-109}$$

$$V_2 = V_{CC} - (V_{\text{sat 4}} + V_{BE1} + V_{BE2} + V_{\text{sat 3}}) \tag{9-110}$$

During the interval 0 to ωt_1, capacitor C supplies current to the bootstrap pin. The current I_1 is

$$I_1 = \frac{V_1}{R} + I_{E2} \tag{9-111}$$

where V_1 is given by Eq. 9-109, and I_{E2}, which is the base current of the power transistor and at the same time of the driver emitter, is

$$I_{E2} = \frac{V_{CC} - V_{\text{sat 1}}}{(1 + h_{FE1})R_L} \tag{9-112}$$

Therefore Eq. 9-111 becomes

$$I_1 = \frac{V_1}{R} + \frac{V_{CC} - V_{\text{sat 1}}}{(1 + h_{FE1})R_L} \tag{9-113}$$

During the interval ωt_1 to 2π, capacitor C receives current from the supply through resistor R. The maximum current I_2 is

$$I_2 = \frac{V_2}{R} \tag{9-114}$$

where V_2 is given by Eq. 9-110.

If C remains charged, and I_1 and I_2 have a sinusoidal trend, resistor R is found from the balance of charges on C given by

$$\int_0^{\omega t1} I_1 \sin \omega t \, d(\omega t) = -\int_{\omega t1}^{2\pi} I_2 \sin \omega t \, d(\omega t) \tag{9-115}$$

where I_1 and I_2 are given by Eqs. 9-113 and 9-114. The interval 0 to ωt_1 can be found by using a graph when V_1 and V_2 are known. All the other values of Eq. 9-115 are known with the exception of R, which is unknown.

R varies greatly as the electrical parameters of the active components in the stage vary. If R is not exact, the following drawbacks will occur:

1. *R much higher than calculated.* The problem occurs at clipping, since the voltage drop increases because the capacitor does not supply sufficient current to Q_2.

2. *R much lower than calculated.* The lower cutoff frequency increases because C discharges on R. The amplifier efficiency also decreases because R is always in parallel with R_L.

To dimension C we need to know the maximum current that must be supplied by C during the positive peak, taking into account that the peak of this current flows through R.

The total current is given by Eq. 9-113, while the voltage V_C at the terminals of C is

$$V_C = V_{BE1} + V_{BE2} + V_{\text{sat,gen}} - V_{\text{sat }1} \tag{9-116}$$

where $V_{\text{sat,gen}}$ is the saturation voltage of current generator Q_3. From the equation,

$$CV_C = I_1 T \tag{9-117}$$

and knowing the lower cutoff frequency of the amplifier, we can find C.

9.4 DISTORTION IN THE OUTPUT STAGES OF AUDIO CIRCUITS

9.4.1 h_{FE}-Difference Distortion[1]

The introduction of output stages with quasi-complementary symmetry into integrated circuits determines a nonlinear transfer characteristic due to the fact that the upper section has an h_{FE} different from that of the lower section. The circuit in Fig. 9-26 is considered typical.

As already said, in integrated circuits the h_{FE} of a lateral PNP like Q_3 is much lower than the corresponding driver Q_2 (5 to 10 times lower with high currents).

Since we shall be considering large signals here, we shall neglect the junction law distortion and assume the junction resistance to be zero for forward bias and infinite for reverse bias.

For v_1 positive, the transmission is given by

$$K_1 = \frac{v_0}{v_1} = \frac{R_L}{(1 - \alpha_1)R_i + R_L} = \frac{h_{FE1}R_L}{\alpha_1 R_i + h_{FE1}R_L} \tag{9-118}$$

[1]B. M. Oliver, Distortion in Complementary-Pair Class B Amplifiers, © copyright 1971, Hewlett-Packard Company. Reproduced with permission.

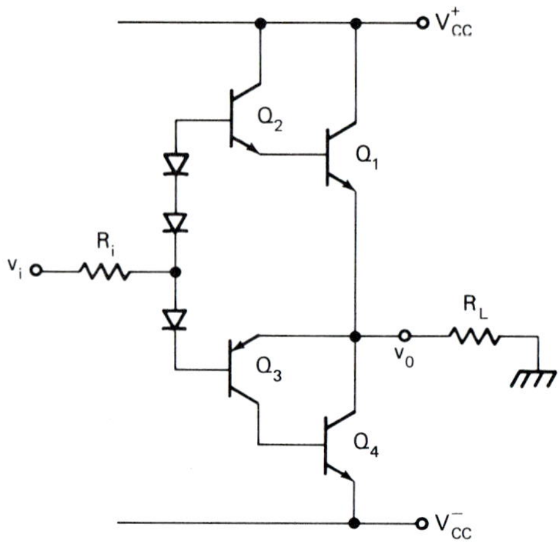

FIG. 9-26 Typical output stage with upper and lower sections having different h_{FE}.

where $\alpha_1 = \alpha$ of upper stage
$h_{FE1} = h_{FE}(Q_1) \cdot h_{FE}(Q_2)$
R_i = internal resistance of source
R_L = load resistor

Similarly, for negative v_0,

$$K_2 = \frac{v_0}{v_1} = \frac{R_L}{(1 - \alpha_2)R_i + R_L} = \frac{h_{FE2}R_L}{\alpha_2 R_i + h_{FE2}R_L} \tag{9-119}$$

where $h_{FE2} = h_{FE}(Q_3) \cdot h_{FE}(Q_4)$.

The ratio of these transmissions is

$$\frac{K_1}{K_2} = \frac{h_{FE1}}{h_{FE2}} \frac{\alpha_2 R_i + h_{FE2}R_L}{\alpha_1 R_i + h_{FE_1}R_L} \tag{9-120}$$

From this equation we can see that if the output stage is voltage-driven, R_i can be taken to be negligible and the ratio $K_1/K_2 = 1$. If it is current-driven, $R_i = \infty$ and Eq. 9-120 becomes

$$\frac{K_1}{K_2} = \frac{h_{FE1}}{h_{FE2}} \tag{9-121}$$

The ratio of R_L/R_i determines the amount of local feedback or degeneration and affects the ratio of the transmission for positive signals to that for negative signals.

Furthermore, we can see that if the output stage is voltage-driven the distortion due to the different h_{FE} is negligible, while it may become considerable if the stage is current-driven.

The point is that for any configuration one can always find the appropriate K_1 and K_2 and from these compute two normalized slopes

$$m_1 = \frac{K_1}{(K_1 + K_2)/2} \tag{9-122}$$

and

$$m_2 = \frac{K_2}{(K_1 + K_2)/2} \tag{9-123}$$

which have the properties

$$\frac{m_1 + m_2}{2} = 1 \tag{9-124}$$

$$\frac{m_1}{m_2} = \frac{K_1}{K_2} \tag{9-125}$$

We are now in a position to represent the actual amplifier, whatever its configuration, by a linear amplifier followed by a nonlinear device having the transfer characteristic shown in Fig. 9-27, where w is the input amplitude and v the output amplitude.

We neglect all curvature at the origin, present in actual devices, electing instead to simplify the analysis and obtain a slightly pessimistic result.

Figure 9-28 shows a feedback amplifier, which incorporates our nonlinear device. We have, for positive signals,

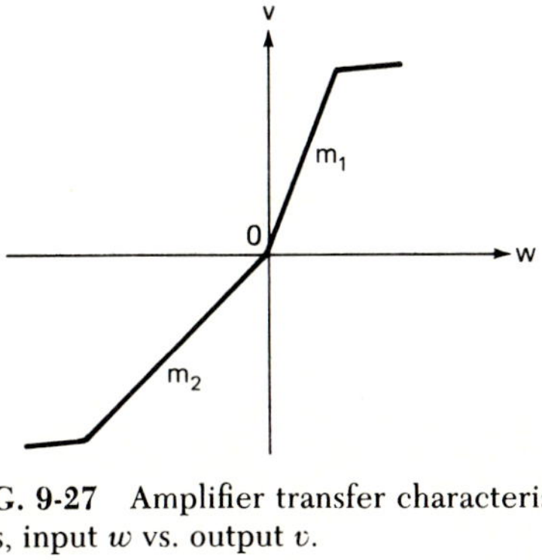

FIG. 9-27 Amplifier transfer characteristics, input w vs. output v.

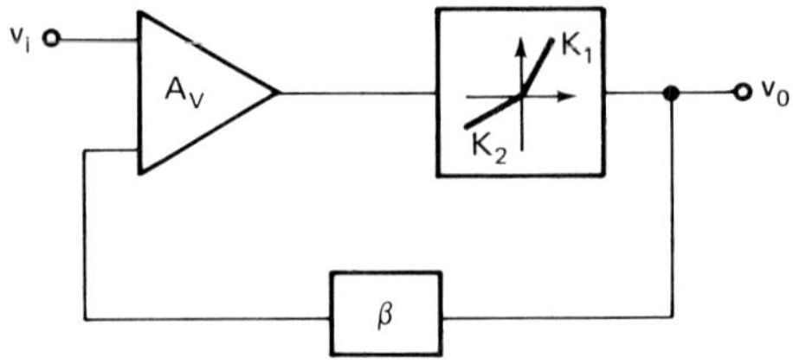

FIG. 9-28 Feedback amplifier incorporating nonlinear element.

$$v^+ = \frac{m_1\mu}{1 - m_1\mu\beta} s^+ \tag{9-126}$$

and similarly, for negative signals,

$$v^- = \frac{m_2\mu}{1 - m_2\mu\beta} s^- \tag{9-127}$$

where $\mu\beta$ is the loop gain if $v = u$ (unity slope). This β is the traditional feedback ratio.

If m were unity and w wished an output

$$y = a \sin \omega t \tag{9-128}$$

we would need an input

$$w = \frac{1 - \mu\beta}{\mu} a \sin \omega t \tag{9-129}$$

With this input and the actual transfer characteristic, the output will be

$$y = \begin{cases} as_1 \sin \omega t & y \geq 0 \\ as_2 \sin \beta t & y \leq 0 \end{cases} \tag{9-130}$$

where

$$s_i = \frac{1 - \mu\beta}{1 - m_i\mu\beta} m_i \qquad i = 1, 2 \tag{9-131}$$

s_1 and s_2 are the transfer characteristic slopes as modified by the feedback. Both approach unity as $\mu\beta \to \infty$ unless m_1 or m_2 is zero.

The amplitude of the fundamental component in the output is

$$a_1 = \frac{\int_{-\pi}^{\pi} y \sin \omega t \, d(\omega t)}{\int_{-\pi}^{\pi} \sin^2 \omega t \, d(\omega t)} = a \frac{s_1 + s_2}{2} \tag{9-132}$$

The mean square departure from the fundamental is

$$\overline{\delta a}^2 = \frac{\int_{-\pi}^{\pi} (y - a_1 \sin \omega t)^2 \, d(\omega t)}{\int_{-\pi}^{\pi} d(\omega t)} = \frac{a^2}{8} (s_1 - s_2) \tag{9-133}$$

The distortion d is given by

$$\begin{aligned} d &= \frac{\text{rms departure } \sqrt{\overline{\delta a}^2}}{\text{rms fundamental } a_1/\sqrt{2}} \\ &= \frac{|s_1 - s_2|}{|s_1 + s_2|} = \frac{|m_1 - m_2|}{2} \frac{1}{1 - m_1 m_2 \mu\beta} \end{aligned} \tag{9-134}$$

The distortion is reduced by feedback, but the effective loop gain is not $\mu\beta$ but $m_1 m_2 \mu\beta$. Thus, if either m_1 or m_2 is zero, the feedback is useless. On the other hand, if $m_1 = 1 + \Delta$ and $m_2 = 1 - \Delta$, where $\Delta \ll 1$, then $m_1 m_2 = 1 - \Delta^2 \approx 1$, and we can consider the normal feedback as being effective. We note that, for the sharp-cornered transfer characteristic we have assumed, the β-difference distortion is independent of amplitude. As we can see from Eq. 9-134 the higher the gain of the loop $m_1 m_2 \mu\beta$ the lower will be the distortion.

9.4.2 Crossover Distortion[1]

The distortion occurs from the law of nonlinearity of the emitter-base junction. Referring to Fig. 9-29, let us examine the mechanism of crossover distortion generation. The generator current I_g flowing through diodes D_1 and D_2 behaves as a bias voltage across these diodes so that a small current flows through Q_1 and Q_2, commonly called *quiescent current*.

In the presence of an output signal, the transistors' V_{BE} increase.

[1]S. Tanaka, New Biasing Circuit for Class-B Operation, an Audio Engineering Society Reprint, no. 1615(X2).

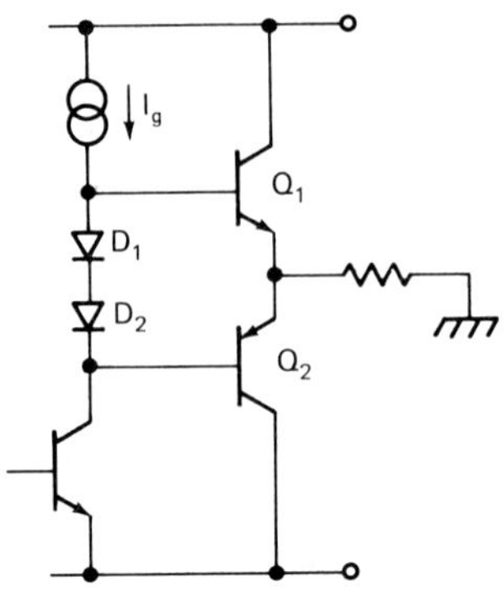

FIG. 9-29 Biased class AB output stage.

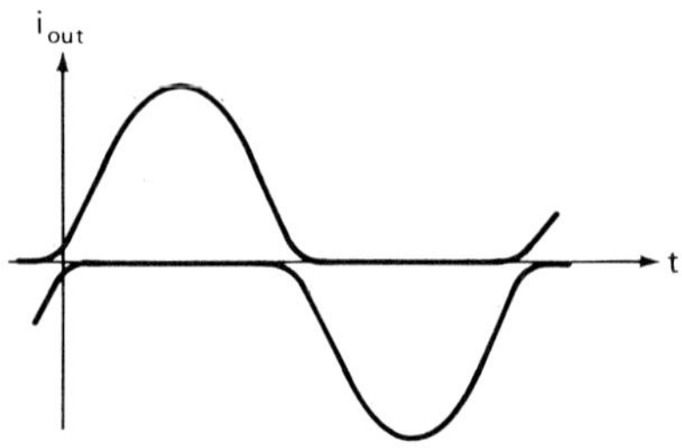

FIG. 9-30 Output transistors' current in class AB mode.

Thus, when either of the output transistors receives current, the other does not, since it is cut off.

The operation of the two output transistors at low level is shown in Fig. 9-30.

Increasing the power output waveforms of the currents of output transistors in class B operation becomes hard, leading to crossover distortion as seen in Fig. 9-31. It depends not only on the increase of V_{BE}, but when the output transistor is in cutoff condition, its V_{BE} is reverse-biased. This type of distortion increases by increasing the frequency. The phenomenon arises where charge and discharge occurs in the transistor's base carrier, especially in the presence of large output voltages.

By increasing the frequency, when large output voltages occur, the transfer characteristics of such nonlinear output stages as the one illus-

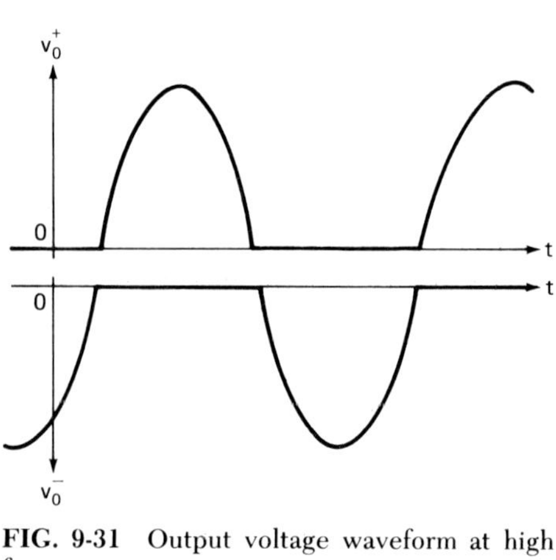

FIG. 9-31 Output voltage waveform at high frequency.

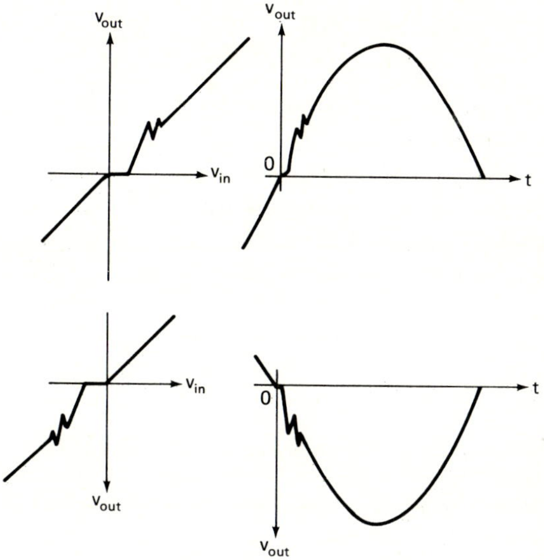

FIG. 9-32 Amplifier transfer characteristics in the presence of crossover distortion at high frequency.

trated in Fig. 9-29 are shown in Fig. 9-32. The corresponding positive and negative output half-waves are shown in the same figure.

The entire amplifying network can be represented as in Fig. 9-33 by a linear amplifier followed by a nonlinear transfer characteristic with slopes variable in frequency. The crossover distortion is reduced by the feedback, using the same criteria shown in the case of h_{FE}-difference distortion as seen in Eq. 9-134.

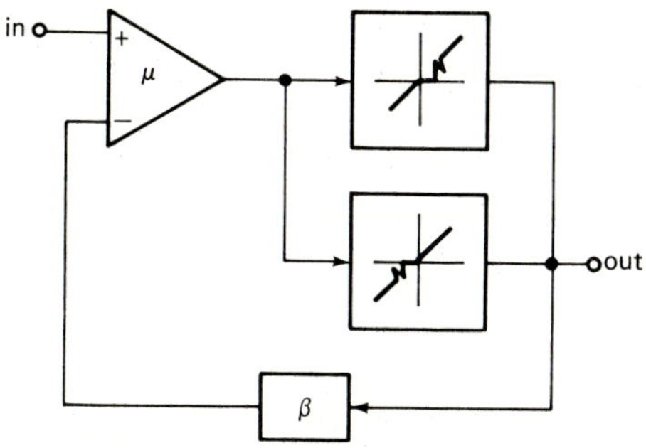

FIG. 9-33 Feedback amplifier including nonlinear elements due to the presence of separate output stages for positive and negative signals.

Presently several techniques are used to reduce crossover distortion output stages.

Good results are obtained by detecting the increased voltage V_{BE} of the conducting transistor and consequently increasing the voltage of the bias circuit. In this positive feedback, the gain should never exceed 1, because otherwise the output transistors could break down.

Another way to reduce crossover distortion is to apply a negative feedback by detecting the decrease of the nonconducting output transistor V_{BE} and by increasing the voltage of the bias circuit.

A fairly new technique to prevent crossover distortion uses both positive and negative feedback with bias circuits.

9.5 GENERAL AMPLIFIER ORGANIZATION

Since an audio amplifier is a special-purpose amplifier, the input stage must be designed to satisfy special requirements, such as low noise, the need to avoid switch-on pop, the required input dynamics which are not always high, and the compatibility of inputs with ground.

In certain cases, however, it may be useful to design an audio amplifier which can be used as a power op amp and therefore with differential input.

9.5.1 The Input Stage: Noise and Dynamic Performance[1]

It is known that a transistor, such as that in Fig. 9-34, has a diagram equivalent, from the point of view of noise, to that of Fig. 9-35. We have

$$e_{nc} = i_{nc} \frac{R_F + R_S + r_b + h_{FE} r_e}{h_{FE}} \tag{9-135}$$

$$e_{nb} = i_{nb}(R_S + R_F + r_b) \tag{9-136}$$

$$e_{nr} = \sqrt{4KTr_b \Delta f} \tag{9-137}$$

where Δf is the bandwidth. By transporting these generators to the base of the transistor, we get

$$e_{nr}^2 = 4KT\Delta f\,(R_S + r_b + R_F) \tag{9-138}$$

$$e_{nb}^2 = 2qI_b \Delta f\,(R_S + R_F + r_b)^2 \tag{9-139}$$

[1]G. Ghiringhelli, SGS-ATES Internal Technical Report CIL/10, 1975, unpublished.

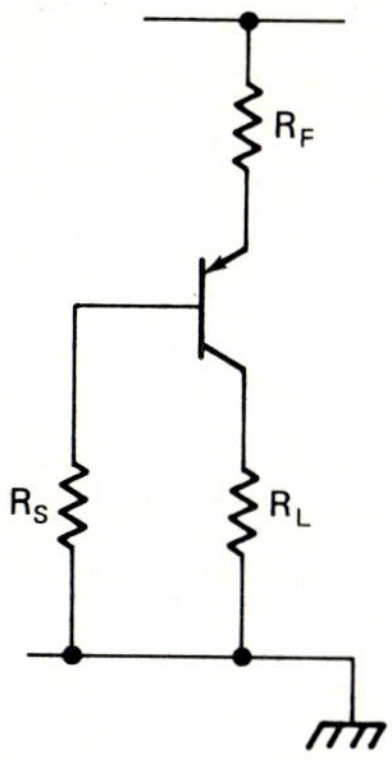

FIG. 9-34 Input transistor in common-emitter configuration.

$$e_{nc}^2 = 2qI_C\Delta f\left(\frac{R_F + R_S + r_b + h_{FE}r_e}{h_{FE}}\right)^2 \tag{9-140}$$

The total voltage noise at the input is

$$e_{n,\text{tot}} = \sqrt{e_{nr}^2 + e_{nb}^2 + e_{nc}^2} \tag{9-141}$$

In the case of a transistor with a common collector as shown in Fig. 9-36, the equivalent diagram is that of Fig. 9-37. In this case we get

$$e_{nb} = i_{nb}(R_S + r_b + r_e) \tag{9-142}$$

$$e_{nc} = i_{nc}\left(\frac{R_S + r_b}{h_{FE}} + r_e\right) \tag{9-143}$$

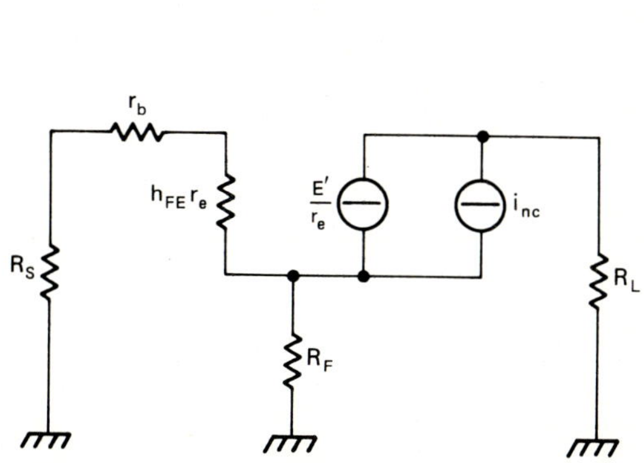

FIG. 9-35 Equivalent diagram of Fig. 9-34 from the point of view of noise.

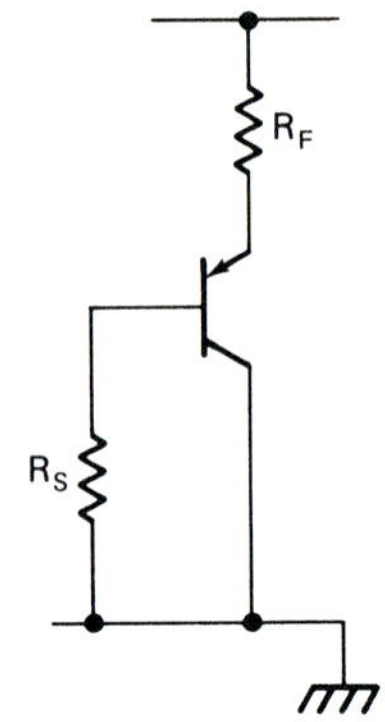

FIG. 9-36 Input transistor in common-collector configuration.

Having said this, let us examine the behavior of the input stages of two typical integrated audio amplifiers produced at present. The first, used in the TBA 800 and TBA 810, has the electrical diagram shown in Fig. 9-38 and the corresponding noise diagram of Fig. 9-39. On the basis of the previous consideration we get

$$e_{nr}^2 = KT\Delta f(R_S + 2r_b + R_F) \tag{9-144}$$

$$e_{nb1}^2 = 2q\frac{I_C}{h_{FE1}h_{FE2}}\Delta f\,(R_S + r_b + r_{e1})^2 \tag{9-145}$$

$$e_{nc1}^2 = 2q\frac{I_C}{h_{FE2}}\Delta f\left(\frac{R_S + r_b}{h_{FE1}} + r_{e1}\right)^2 \tag{9-146}$$

$$e_{nb2} = 2q\frac{I_C}{h_{FE2}}\Delta f\left(\frac{R_S}{h_{FE1}} + R_F + r_b\right)^2 \tag{9-147}$$

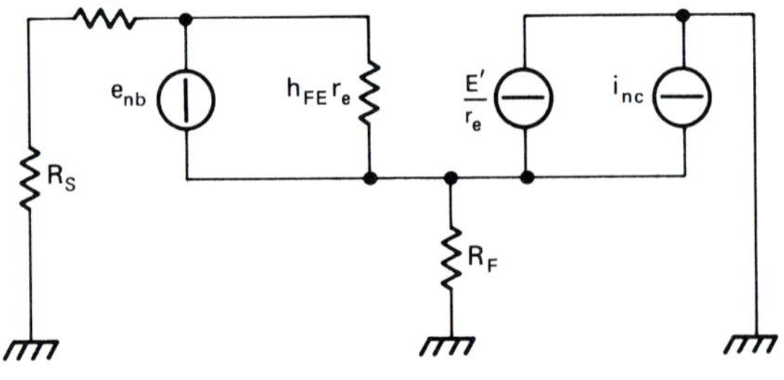

FIG. 9-37 Equivalent noise diagram of Fig. 9-36.

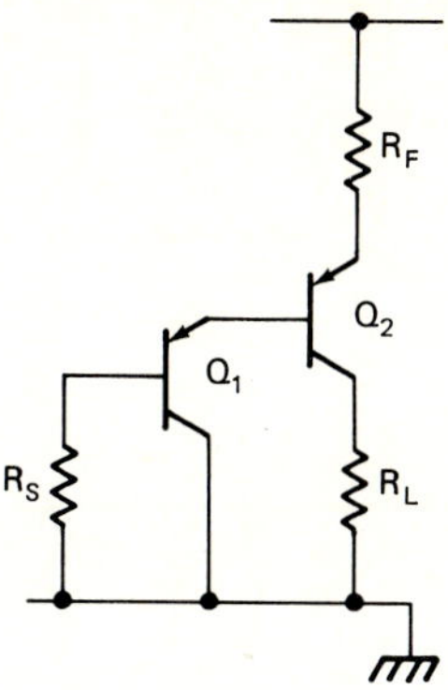

FIG. 9-38 Ground-compatible two-transistor input stage.

$$e_{ne2} = 2qI_C\Delta f\left(\frac{R_F + R_S/h_{FE1} + r_b + h_{FE2}r_e}{h_{FE2}}\right)^2 \tag{9-148}$$

$$e_{n,\text{tot}} = e_{nr}^2 + e_{nb1}^2 + e_{nc1}^2 + e_{nb2}^2 + e_{ne2}^2 \tag{9-149}$$

As a practical example, by using the following typical values of

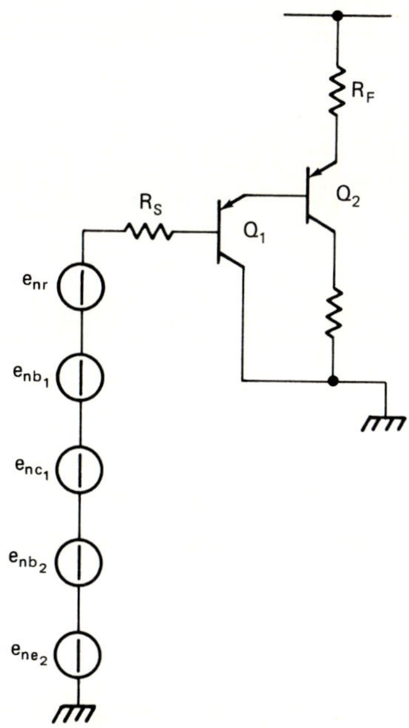

FIG. 9-39 Equivalent noise diagram of Fig. 9-38.

$$I_C = 1.5 \text{ mA}$$

$$h_{FE1} = 100$$

$$h_{FE2} = 30$$

$$R_S = 10 \text{ k}\Omega$$

$$R_F = 56 \ \Omega$$

$$r_{b1} = r_{b2} = 200 \ \Omega$$

$$\Delta f = 25 \text{ kHz}$$

from the previous equations we get

$$e_{n,\text{tot}} = 2.8 \ \mu\text{V}$$

The input stage of Fig. 9-38 is inserted in the amplifier chain as shown in Fig. 9-40. As can be seen it is a ground-compatible stage with input dynamics limited by the fact that if the input is brought to one V_{BE} under ground, generator R_L has zero voltage at its terminals, and there is the risk of inverting the signal at the input of amplifier *A*.

This type of input stage is suitable for feedback amplifiers with fairly high gain ($\geq$100) or operating at low voltage (low output swing). Furthermore since it is an asymmetrical stage, it is affected by the low-frequency thermal feedback.

Another problem arises due to the fact that since Q_2 operates at high current, a very high compensation capacitor ($\geq$1000 pF) is needed which cannot be integrated but must be connected externally.

An input stage which partly solves this problem is shown in Fig. 9-

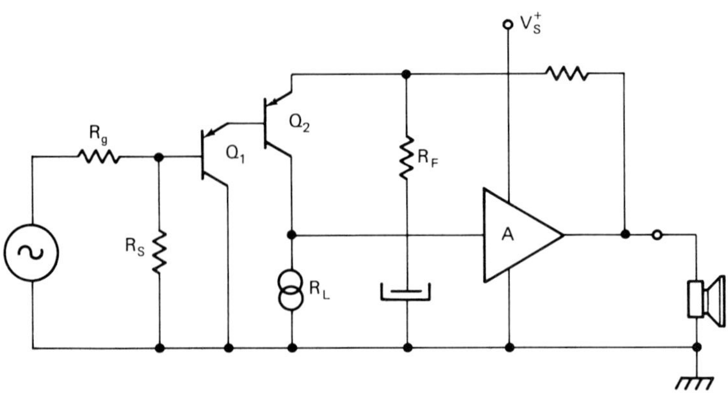

FIG. 9-40 The input stage of Fig. 9-38 inserted in the amplifier chain.

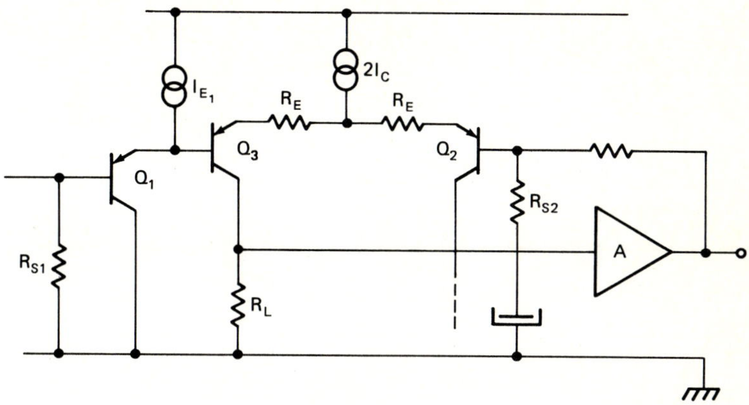

FIG. 9-41 The input stage of the TBA 930.

41. In this case, too, the voltage generators can be replaced at the input:

$$e_{nr}^2 = 4KT\Delta f\,(R_{S1} + R_{S2} + 2R_E + 3r_b) \tag{9-150}$$

$$e_{nb1}^2 = 2q\,\frac{I_{E1}}{h_{FE1}}\,\Delta f\,(R_S + r_b + r_{e1})^2 \tag{9-151}$$

$$e_{nc1}^2 = 2qI_{E1}\,\Delta f\left(\frac{R_{S1} + r_b}{h_{FE1}} + r_{e1}\right)^2 \tag{9-152}$$

$$e_{nb2}^2 = 2q\,\frac{I_C}{h_{FE2}}\,\Delta f\,(R_{F2} + R_{S2} + r_b)^2 \tag{9-153}$$

$$e_{nc2}^2 = 2qI_C\,\Delta f\left(\frac{R_{F2} + R_{S2} + r_b + h_{FE2}r_{e2}}{h_{FE2}}\right)^2 \tag{9-154}$$

$$e_{nb3}^2 = 2q\,\frac{I_C}{h_{FE2}}\,\Delta f\,(R_{F1} + R'_{S1} + r_b)^2 \tag{9-155}$$

$$e_{nc3}^2 = 2qI_C\,\Delta f\left(\frac{R_{F1} + R'_{S1} + r_b + h_{FE3}r_{e3}}{h_{FE3}}\right)^2 \tag{9-156}$$

where

$$R_{F1} = 2R_E + r_{e2} + \frac{R_{S2}}{h_{FE2}} \tag{9-157}$$

$$R_{F2} = 2R_E + r_{e3} + \frac{r_{e1}}{h_{FE3}} \tag{9-158}$$

$$R'_{S1} = \frac{R_{S1}}{h_{FE1}} + r_{e1} \tag{9-159}$$

In this case internal compensation is possible since $2I_C$ is around 100 μA.

The problems linked to poor input dynamics and partly to thermal feedback remain.

The noise is slightly higher than in the previous case, around 3 μV, having put

$$R_F = 200\ \Omega$$

$$I_{E1} = 20\ \mu\text{A}$$

$$I_C = 100\ \mu\text{A}$$

$$R_{S1} = 10\ \text{k}\Omega$$

$$R_{S2} = 56\ \Omega$$

and leaving the other parameters unchanged with respect to the previous case.

One of the input stages most commonly used for this type of low-input swing amplifier and which is better than the previous ones from the point of view of noise because it consists of only one transistor is that shown in Fig. 9-42 and used in the TDA 2003, TDA 2004, TDA 2005, TDA 2008, and TDA 2009.

From Eq. 9-141 and putting

$$R_S = R_{\text{in}} \| R_g = 10\ \text{k}\Omega$$

$$r_b = 200\ \Omega$$

$$R_F = 1.5\ \text{k}\Omega$$

$$I_C = 300\ \mu\text{A}$$

$$h_{FE} = 100$$

$$\Delta f = 25\ \text{kHz}$$

we find that $e_{n,\text{tot}} = 2.3\ \mu$V.

These are typical values for a certain type of application, but naturally the stage can be optimized as regards noise by changing some of the electrical parameters.

All the asymmetrical stages are affected by thermal feedback more than the differential stages. Furthermore, in the particular case with no ground compatibility, an input coupling capacitor is required.

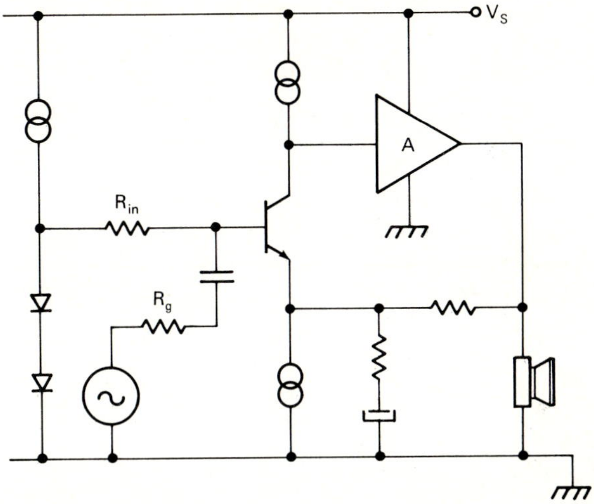

FIG. 9-42 Single-transistor input stage for low-input signals.

As regards differential input stages we shall not deal with the procedures for calculating noise and other features as these are already well known. We can, however, find such procedures from the considerations made regarding Fig. 9-41, removing Q_1 and generator I_{E1}.

9.5.2 Thermal Feedback Effects[1]

One of the most important dynamic features, especially in integrated audio amplifiers for the hi-fi market, is the thermal feedback problem. The phenomenon which causes it can be shown as follows. We shall consider an unclipped sinusoidal output signal. The instantaneous dissipated power P_{di} can be expressed as

$$P_{di} = \frac{V_0 V_{CC} - V_0^2}{R_L} \tag{9-160}$$

Suppose we have an integrated power amplifier on a silicon chip as in Fig. 9-43. For a dissipated power peak with a certain duration the curve of the temperature along the chip is shown in Fig. 9-44.

With low frequencies, 20 to 50 Hz, the temperature of the input transistors is therefore subjected to a temperature gradient ΔT given by

[1] J. E. Solomon, The Monolithic Op Amp: A Tutorial Study, *IEEE J. Solid-State Circuits*, vol. SC-9, no. 6, 1974. Reproduced by permission.

$$\Delta T = \pm K_T P_{di} \tag{9-161}$$

where K_T = constant, °C/W. The plus or minus sign is needed because the direction of the thermal gradient is unknown. In fact, the sign may reverse polarity during the output swing as the dominant source of heat shifts from one transistor to another.

Since the variation in the V_{BE} of the input transistor is around -2 mV/°C, it is equivalent to having an input signal V_{iT}, positive in the case of a single-transistor input.

If the dominant input components consist of the differential transistor pair, the thermally induced input voltage V_{iT} can be calculated as

$$V_{iT} = \pm K_T P_{di}(2 \times 10^{-3}) = \gamma_T P_{di} \tag{9-162}$$

where $\gamma_T = K_T(2 \times 10^{-3})$ V/W.

We shall now see how this thermal feedback affects the transfer characteristics of the amplifier and therefore the distortion.

To calculate the thermal effect we can consider the circuit in Fig. 9-45, inserting a supplementary feedback network in parallel with the existing one, which depends on the dissipated power P_{di}. The electrical form of the thermal feedback factor β_T is

$$\beta_T = \frac{\delta V_{iT}}{\delta V_0} = \pm \frac{\gamma_T}{R_L}(V_{CC} - 2V_0) \tag{9-163}$$

Therefore, taking into account the electrical and thermal feedback, the closed loop gain is

$$A_V(0) = \frac{\mu}{1 + \mu(\beta_e \pm \beta_T)} \tag{9-164}$$

where μ = open-loop gain in the absence of thermal feedback
β_e = applied electrical feedback as in Fig. 9-45

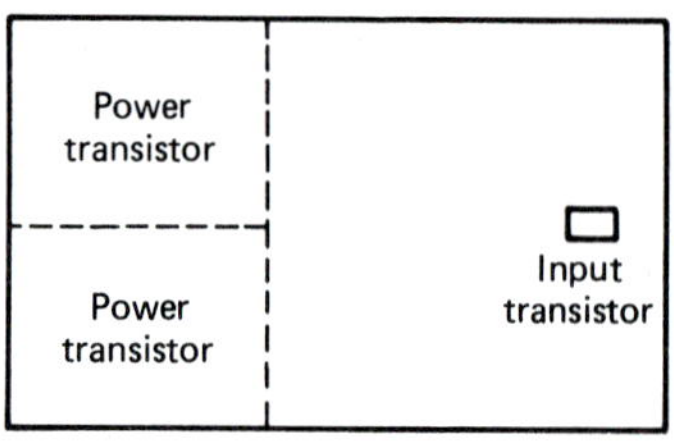

FIG. 9-43 Integrated power amplifier on a silicon chip.

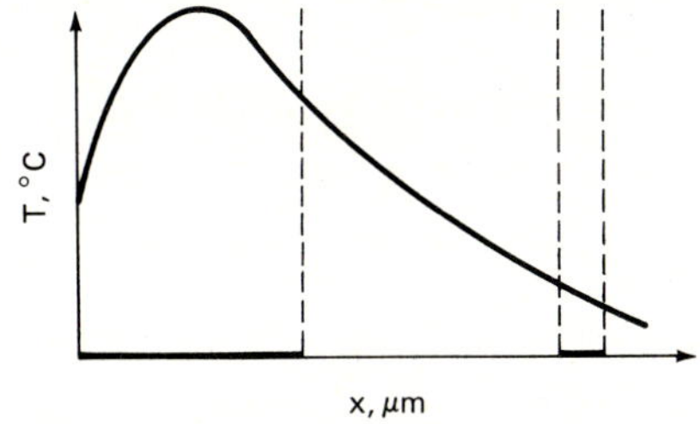

FIG. 9-44 Temperature distribution along the chip.

The reason for the double sign in front of β_T is that during both the positive and negative half-waves the V_{iT} signal is always positive.

If the electrical feedback β_e is high compared with the thermal feedback β_T, that is, $\beta_e \gg \beta_T$, the transfer characteristic depends only on β_e, and therefore

$$A_V(0) = \frac{1}{\beta_e} \tag{9-165}$$

If β_e is negligible, the thermal feedback is dominant, and we get a transfer characteristic determined by β_T. In this case by combining Eqs. 9-163 and 9-164, $A_V(0)$ becomes

$$A_V(0) = \frac{\mu}{1 \pm (\mu\gamma_T/R_L)(V_S - 2V_0)} \tag{9-166}$$

Recalling from Eq. 9-160 that V_0 ranges between zero and V_{CC}, we note that the incremental thermal feedback is greatest when $V_0 = 0$ or V_0

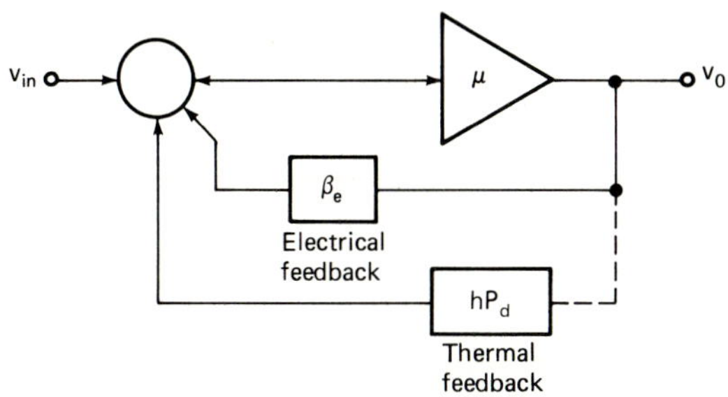

FIG. 9-45 Diagram used to calculate closed-loop gain with thermal feedback.

$= V_{CC}$, and it is at these points that the thermally limited gain is smallest.

To use the amplifier in a predictable manner, one must always apply enough electrical feedback to reduce the gain below this minimum thermal gain.

Thus, maximum usable gain can be defined as that approximately equal to the value of Eq. 9-166 with $V_0 = 0$ or $V_0 = V_{CC}$, which is

$$A_V(0)\Big|_{\max} \approx \frac{R_L}{\gamma_T V_{CC}} \tag{9-167}$$

or

$$A_V(0)\Big|_{\max} \approx \frac{1}{\gamma_T I_{\max}} \tag{9-168}$$

It was assumed in Eqs. 9-167 and 9-168 that thermal feedback dominates over the open-loop electrical gain μ. Finally, in Eq. 9-168, a maximum current $I_{\max} = V_{CC}/R_L$ was defined as the maximum current which would flow if the amplifier output could swing all the way to the supplies.

Equation 9-168 is a strikingly simple and quite general result which can be used to predict the expected maximum usable gain for an amplifier if we know only the maximum output current and the thermal feedback constant γ_T.

Equations 9-165 and 9-166 can be expressed graphically (Fig. 9-46) to see the effects superimposed in the transfer characteristics of the amplifier. A transfer characteristic of this type obviously gives rise to distortion. One way to reduce it is to make β_e prevail over β_T, that is, to

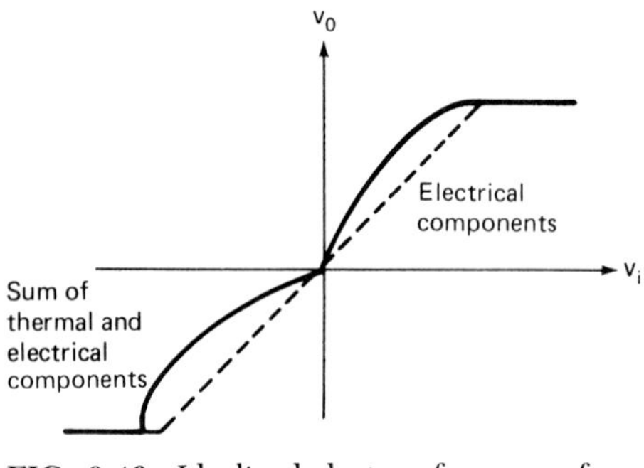

FIG. 9-46 Idealized dc transfer curve for a power amplifier showing its electrical and thermal components.

have high feedback ratios on the closed-loop amplifier. The situation naturally improves if instead of using a single transistor at the input a differential pair is used. With this differential pair in the chip at the maximum distance from the power transistors and on an isotherm, the phenomenon is greatly reduced.

9.5.3 Turn-On

A characteristic which depends on the circuit configuration of the input stage is the turn-on behavior. Depending on the input stage or on the choice of external components, a phenomenon known as *turn-on pop*, which is an output power peak on the loudspeaker at turn-on, may occur.

To explain this with an example let us look at Fig. 9-47. At switch-on, C_1 is discharged and voltage V_A is zero. The generator G keeps point D at $V_{sat,G}$, and Q_3 is inhibited. Therefore the output voltage V_B is at its maximum positive swing.

As C_1 discharges across R_1 and R_F, V_A increases until it turns on input stage Q_1 and Q_2 and V_B returns to the operating condition. The curves of voltages V_A, V_B, and V_D are shown in Fig. 9-48.

Differential stages not compatible with ground and having gain stages as in Fig. 9-47 have the same problem.

Turn-on pop is eliminated by using an input stage compatible with ground where the input stage goes to the operating condition before the output stage, or by using the input stage of the TDA 2003 (Fig. 9-49).

At switch-on, since C_1 and C_2 are discharged, $V_A = 0$ and $V_B = 0$; therefore Q_1 is inhibited, while the Darlington Q_2Q_3 is driven by generator G and keeps point D at its negative swing.

Then C_1 discharges across R_S, and Q_1 comes on and brings the ampli-

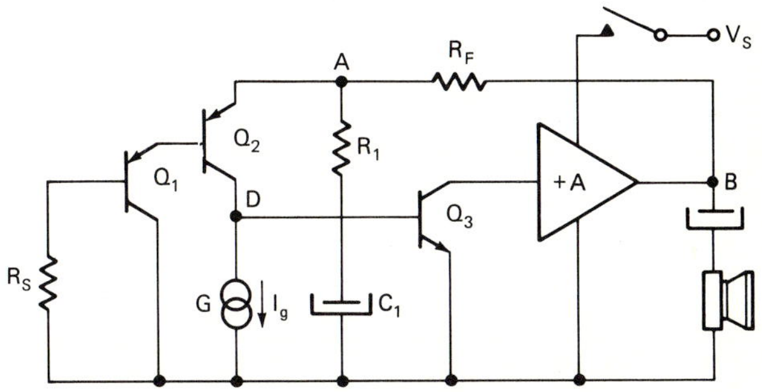

FIG. 9-47 Audio amplifier with PNP input stage.

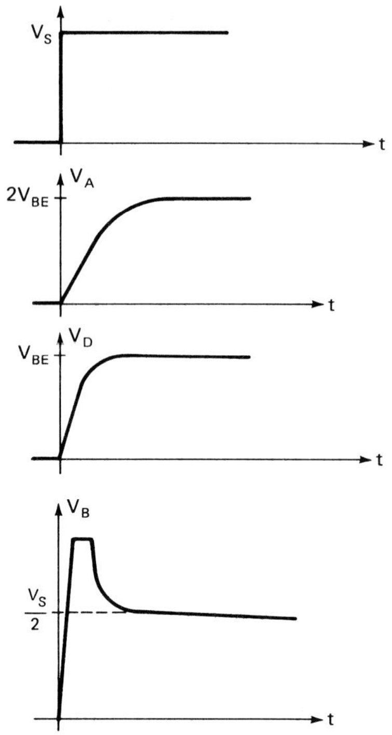

FIG. 9-48 Voltage variations at switch-on.

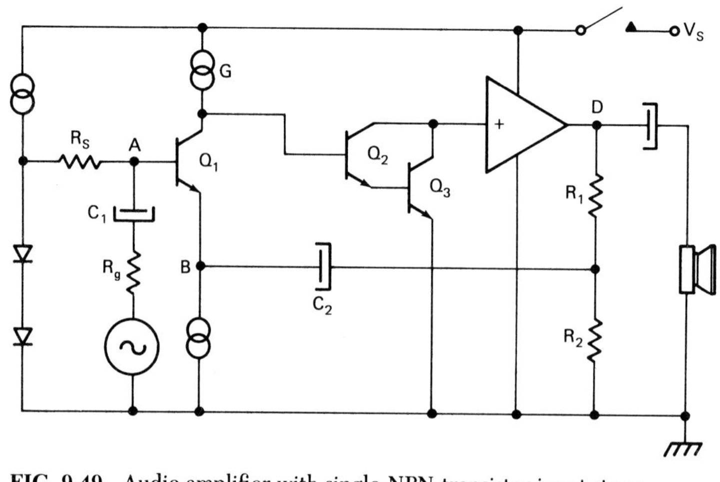

FIG. 9-49 Audio amplifier with single-NPN-transistor input stage.

fier to the operating condition. C_1 should be dimensioned as a function of C_2, because if C_1 is charged too fast Q_1 quickly goes into saturation due to the fact that in the meantime C_2 has held point B to ground, causing a positive-voltage transient at the output.

9.5.4 DC Biasing and Supply Voltage Rejection

Audio amplifiers not used as power op-amps generally do not have differential inputs and operate with single supply. It is therefore necessary to bias the output internally so that it goes to a voltage which will make the output swing symmetrical.

Single-Transistor Input Stage. As an example we shall analyze the case of Fig. 9-50. The dc output voltage V_0 is

$$V_0 = V_{BE} + \frac{R_2}{R_1}(V_{CC} - 2V_{BE}) \tag{9-169}$$

If $R_1 = 2R_2$, we have

$$V_0 = \frac{V_{CC}}{2} \tag{9-170}$$

The circuitry can be applied to the input stages of Figs. 9-49, 9-41, and 9-47, where the inverting input is at $2V_{BE}$, and therefore to bring V_0 to $V_{CC}/2$ the number of diodes in series with R_1 must be doubled.

The advantage of these polarization types is that the V_0 depends essentially on the matching of R_1 and R_2, which may be of the order of

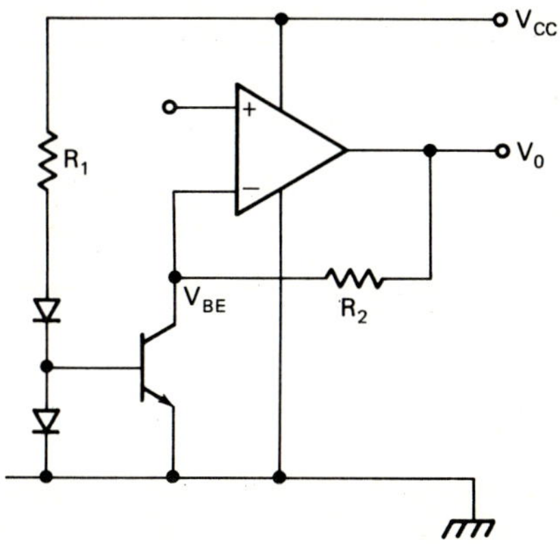

FIG. 9-50 DC biasing of an audio amplifier with a single-transistor input stage.

1 percent in integrated circuits. The main disadvantage is supply voltage rejection.

In fact, from an examination of Fig. 9-50, if $R_1 = 2R_2 = 2R$ and if V_1 is the interference on the supply and V_0 the interference at the output, we get

$$\frac{V_1}{2R} = \frac{V_0}{R} \qquad \text{and} \qquad \text{SVR} = \frac{V_0}{V_1} = \frac{1}{2} = -6 \text{ dB} \tag{9-171}$$

If the number of pins on the package allows, we can use the circuit shown in Fig. 9-51. In this case, supposing that a sinusoidal signal with frequency f is applied at the supply,

$$\text{SVR} = \frac{V_0}{V_1} = \frac{1}{1 + 2\pi fRC} \tag{9-172}$$

The influence of the supply ripple on the stabilized voltage which biases the noninverting input of the amplifier is negligible.

If there is no external pin to which the ripple rejection capacitor can be connected, an external circuit which can use the gain capacitor can be used, as in the case of the TDA 2003. In this case the circuit becomes more or less that of Fig. 9-52.

At the ripple frequency (normally 100 Hz in Europe and 120 Hz in the United States), capacitor C, which fixes the closed-loop gain, can be considered a short circuit. Furthermore $R_1 \ll R$; therefore, from Fig. 9-52,

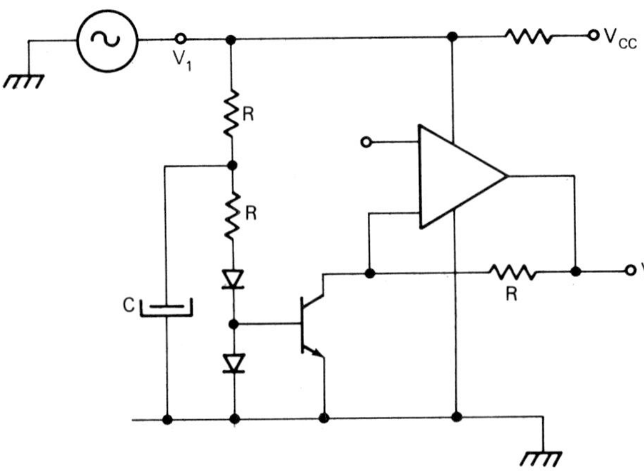

FIG. 9-51 SVR made by an added pin.

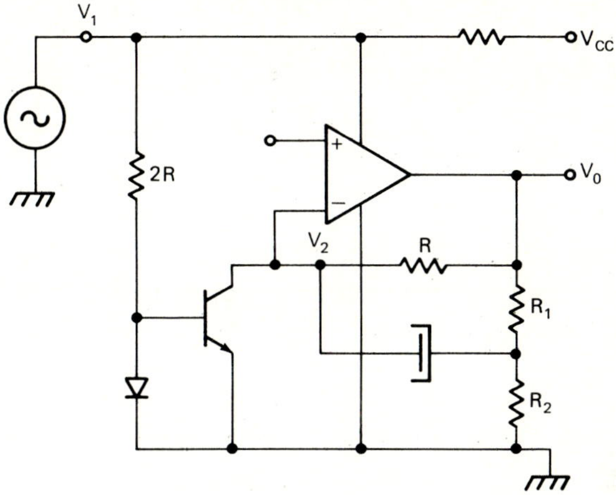

FIG. 9-52 SVR made by using a gain capacitor.

$$\frac{V_0}{V_1} = \mathrm{SVR} = \frac{R_1 + R_2}{2R} \tag{9-173}$$

Differential Input Stage. Another method of internal biasing, which makes it easy to obtain a good SVR with a small capacitor, is shown in Fig. 9-53. This is the method used in the TDA 1905. As regards bias, supposing that the input stage is differential, the resistors can be dimen-

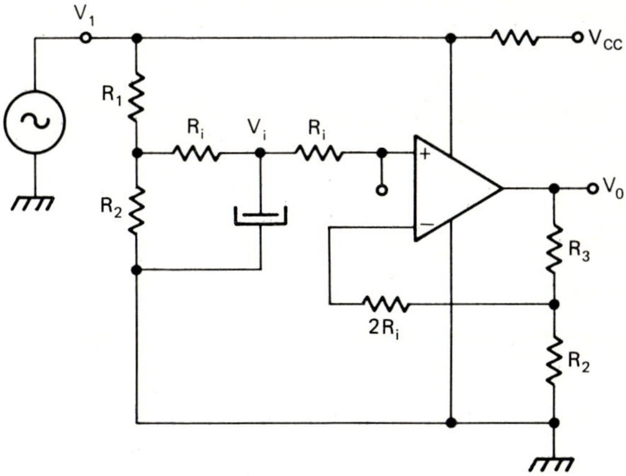

FIG. 9-53 Biasing and SVR in an amplifier with differential input stage.

sioned by taking into account the fact that a high dc gain cannot be given because of the offset of the input stage.

From Fig. 9-53 we get

$$\frac{R_2}{R_1 + R_2} V_{CC} = \frac{R_2}{R_2 + R_3} V_0 \qquad (9\text{-}174)$$

Putting $V_0 = V_{CC}/2$, we get

$$R_1 = R_2 + 2R_3 \qquad (9\text{-}175)$$

In the TDA 1905, $R_2 = 6$ kΩ and $R_3 = 10$ kΩ, thus $R_1 = 26$ kΩ. The dc gain is therefore $G_V = 1 + {}^{10}/_{6} = 8.5$ dB, low enough to avoid excessive variations of V_0 as a function of the input voltage offset.

To calculate the SVR, using V_1 and V_0 to indicate the ripple signals in Fig. 9-53, we get

$$\text{SVR} = \frac{V_0}{V_1} = \frac{R_2 + R_3}{R_2} \frac{R_2}{R_1 + R_2} \frac{1}{(1 + j\omega R_i C)}$$

$$= \frac{R_2 + R_3}{(R_1 + R_2)(1 + j\omega R_i C)} \qquad (9\text{-}176)$$

In the case of the TDA 1095, $R_i = 100$ kΩ, so for a frequency of 100 Hz and putting $C = 10\ \mu$F we get SVR = 62 dB, not taking into account the ripple introduced in the other parts of the circuit through the start-up current generators.

9.6 DIFFERENT SYSTEM CONFIGURATIONS

Audio amplifiers have three common configurations: mono, stereo, and bridge.

When designing an amplifier we have to take into account the maximum output current, the ambient temperature at which the circuit must operate, the output power required at a certain distortion, the type of loudspeaker to be driven, the system in which it will operate, etc. This allows us to select the most suitable technology, to dimension the power transistors, and to include overvoltage, thermal and short-circuit protection, as well as protection for the loudspeaker, etc. This provides considerable flexibility of application in all sectors requiring the driving of a loudspeaker.

9.6.1 Mono Amplifier

The mono amplifier has a variety of uses in many different systems such as in portable radios, cassette recorder and players, car radios, TVs, compacts, etc. The tendency today is to make special products for specific applications and to reduce the number of external components, making the circuitry inside the chip more complicated.

Two typical applications of mono circuits with a single supply are the TDA 1905 and the TDA 2008. Both are medium-voltage amplifiers (30 V). The first, mounted in an 8 + 8 leads plastic dip, can dissipate a $P_0 = 5.5$ W with $d = 10$ percent on a load of 4 Ω, while the second, mounted in a Pentawatt plastic package, can dissipate a P_0 of 12 W with $d = 10$ percent on 4 Ω.

In the first case the bootstrap and ripple rejection capacitor are present, while in the second the ripple rejection is carried out by using the gain capacitor. The trend today, however, is toward stereo and bridge configurations, which are required by most of the systems using audio amplifiers.

9.6.2 Monolithic Stereo Amplifier

When designing an integrated stereo audio amplifier, several important parameters must be taken into account, e.g., the cross talk and the thermal feedback of one amplifier in relation to another. It is a well-known fact that, in producing an amplifier with a silicon chip, if the various active and passive components forming the circuit are adjacent to one another, they are connected by parasitic elements such as capacitors.

Furthermore there is also a transmission of heat from certain areas of the chip, where there is greater dissipation, to others, causing temperature gradients which generate distortion, as in the case of thermal feedback. Since the input stages are naturally the most sensitive points as regards cross talk both for electrical transmission across parasitic elements and for thermal transmission, special attention must be given to the layout of a monolithic stereo amplifier. We can take the TDA 2004, a stereo amplifier for a car radio with a P_0 of 10 + 10 W, as an example. The components are distributed in the layout shown in Fig. 9-54. The thermal and electrical influence of one channel on another is thus minimized.

The 1-kHz cross talk is typically 60 dB. An application diagram is shown in Fig. 9-55.

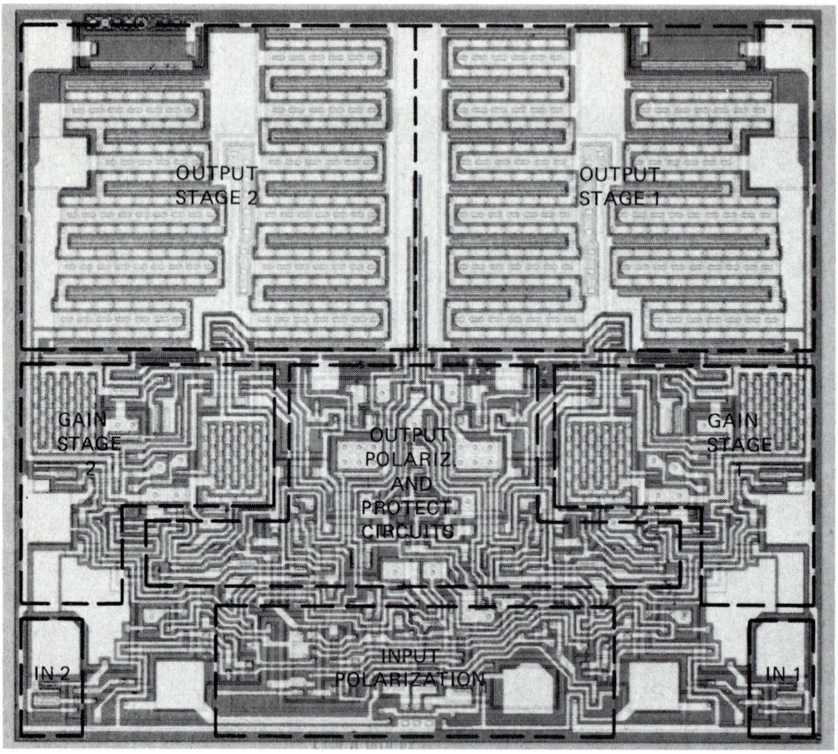

FIG. 9-54 Photo of the TDA 2005 monolithic dual audio amplifier.

9.6.3 Bridge Configuration

The bridge configuration is used to increase the power with the same load resistance and supply voltage. It requires two amplifiers having the same characteristics driven at the input with signals of the same amplitude but with opposite phase. By inserting the load between the two outputs, it will be subjected to double the voltage of the mono configuration. The $P_{out,max}$ is therefore increased by a factor of 4. This would naturally be true only if the V_{sat} of the output stages is independent of the I_{out} currents. In reality the P_{out} (bridge) is lower than $4P_{out}$ (mono) with the same R_L.

The external network to obtain the during required in the bridge can be realized simply as shown in Fig. 9-56 (refer to the TDA 2004 in Fig. 9-55). The gains of the two amplifiers are

$$A_{V1} = \frac{R_1 + R_2 \| R_3}{R_2 \| R_3} = \frac{R_1R_2 + R_1R_3 + R_2R_3}{R_2R_3} \tag{9-177}$$

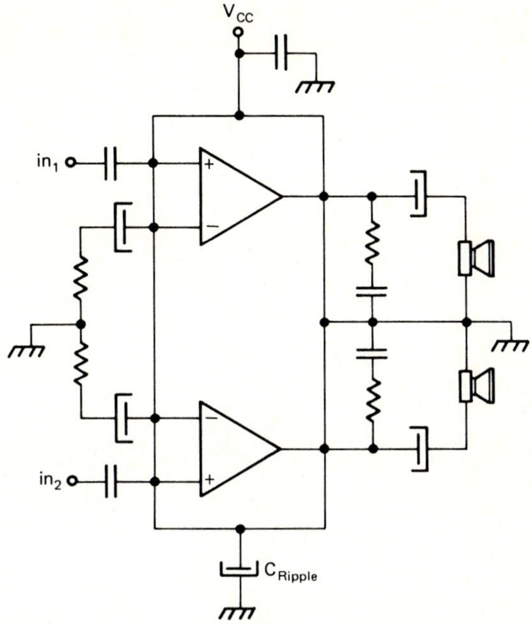

FIG. 9-55 Stereo configuration application circuit.

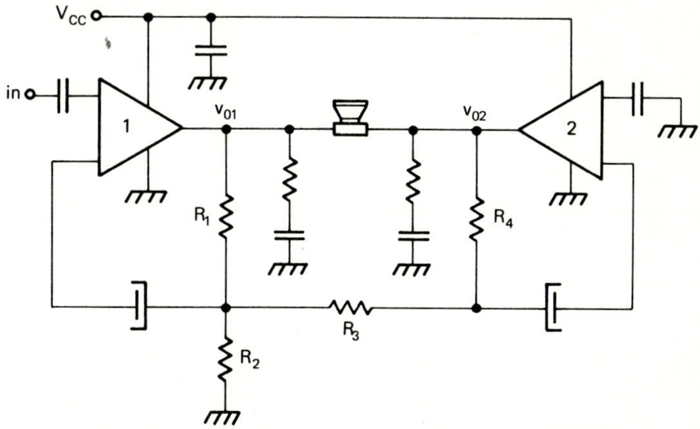

FIG. 9-56 Bridge configuration application circuit.

and if $R_2 = R_3 = R$, Eq. 9-177 becomes

$$A_{V1} = 1 + \frac{2R_1}{R} \tag{9-178}$$

while

$$A_{V2} = -\frac{R_4}{R_3} = -\frac{2R_1}{R} \tag{9-179}$$

having put $R_4 = 2R_1$ and $R_3 = R$, hence

$$A_{V1} = -A_{V2} + 1 \tag{9-180}$$

The total gain of the bridge, as a ratio between the voltage at the terminals of the load and the input voltage, is obviously

$$A_{V1} - A_{V2} = 2A_{V1} - 1 \tag{9-181}$$

9.7 REDUCTION OF EXTERNAL PARTS

One of the present problems in audio integrated amplifiers is that there are too many external components and therefore too high a cost. A very simple applications diagram like that of the TDA 2008, for example, includes three resistors and five capacitors. It is used in systems where compactness is vital.

The present trend in the design of integrated audio amplifiers is to satisfy this requirement of compactness even though the possibility of greater circuit complexity is thus introduced.

At present the components which it seems impossible to eliminate are the filter capacitor on the supply and the gain capacitor. As regards the other external components, the input capacitor can be eliminated by using an input stage compatible with ground, and the two gain resistors can be eliminated by fixing the gain internally.

The output capacitor can be removed by using a bridge configuration with a load connected directly between the two outputs of a dual monolithic amplifier. This, however, would require two *RC* stabilization cells at the output instead of one.

The *RC* cell at the output can be eliminated by using output stages with positive local gain feedback and preventing total switch-off of the lower stage when the upper one is operating, and vice versa.

In other words, an audio amplifier with a minimum number of components, as in Fig. 9-57, can be obtained. The external components are reduced to two: gain capacitor and supply filter capacitor.

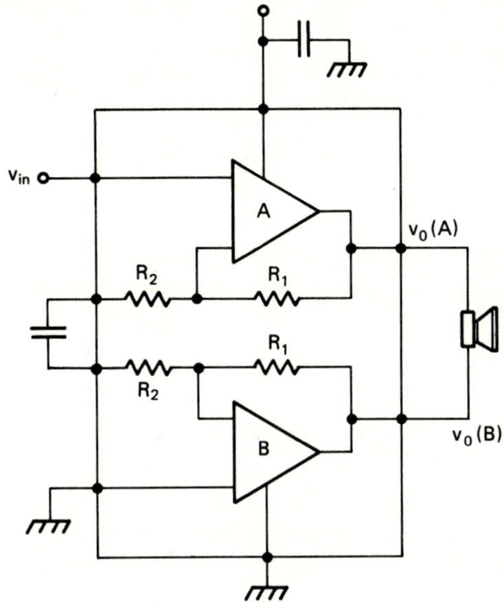

FIG. 9-57 Audio amplifier with a minimum number of external components.

The gain of amplifier *A*, considering that it sees a ground at the inverting input of amplifier *B*, is

$$A_V(A) = 1 + \frac{R_1}{2R_2} \tag{9-182}$$

The gain of amplifier *B*, bearing in mind that it is driven by the inverting input, is

$$A_V(B) = -\frac{R_1}{2R_2} \tag{9-183}$$

The total gain is given by

$$\frac{V_0(A) - V_0(B)}{V_i} = 1 + \frac{R_1}{R_2} \tag{9-184}$$

The difference in gain between channels *A* and *B* is negligible as regards the output swing and distortion.

The applications circuit in Fig. 9-57 has a drawback: between the two amplifiers there is also a common-mode gain which is obviously a unit gain. The system is therefore unstable if the amplifier is unstable with unit gain. To avoid this drawback a low-value capacitor (for example, 1 nF) must be inserted between one terminal of the gain capacitor and ground to open the frequency coil.

9.8 PROTECTION TECHNIQUES

Audio integrated circuits undergo different types of stress, both thermal and electrical, depending on the type of system in which they are used. Although this stress is often accidental the integrated circuit must be designed so that it does not fail under stress, which although undesired can be predicted fairly accurately.

The stress may be *overvoltage* of a certain duration and definite value, an abnormal *increase* in the ambient *temperature,* an increase in the chip temperature due to a *poor connection* between the package and dissipator, or *overloads* or *short circuits* at the output.

We shall now analyze in detail the short-circuit protection of an integrated audio amplifier by applying it to the different system configurations.

9.8.1 Short-Circuit Protection in Power Transistors

Integrated power transistors have a clearly defined SOA which depends on the fabrication process and on the geometry used. The SOA is also a function of the time during which the transistor is stressed and of the junction temperature T_j. Since the SOA is defined on a graph with axes V_{CE} and I_C, to protect the transistor both these values must be measured. To make the most of the transistor's characteristics it must be protected according to a curve which is as close as possible to the SOA shown in Fig. 9-58.

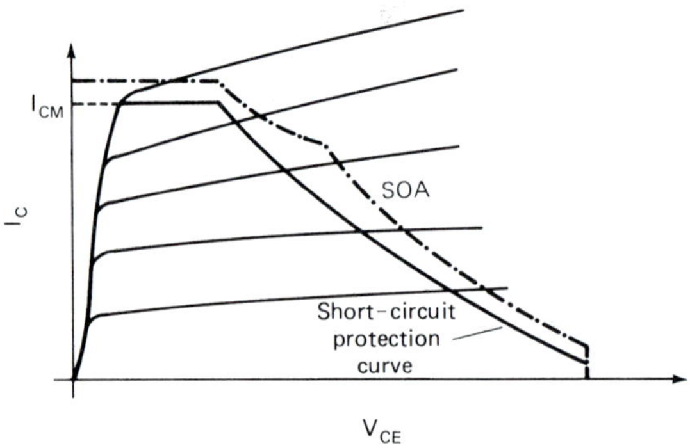

FIG. 9-58 Safe operating area and short-circuit protection curve of an integrated output transistor.

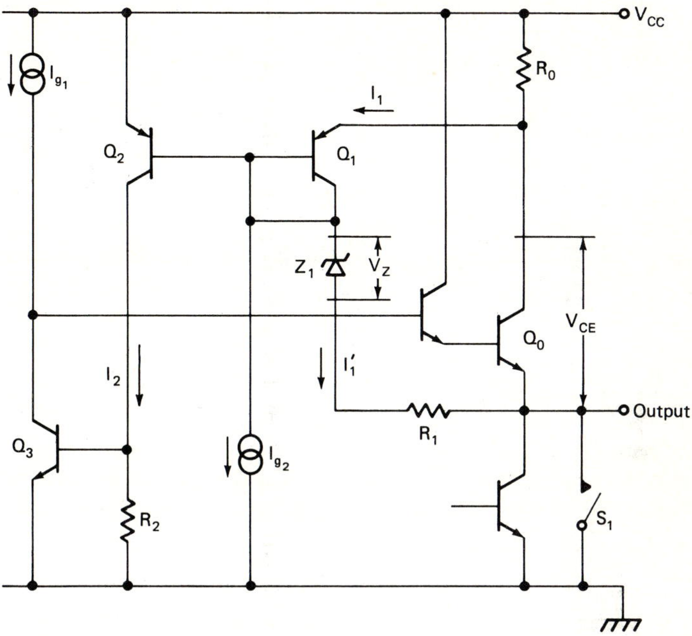

FIG. 9-59 Circuit diagram of the short-circuit protection for the upper section of an integrated output stage.

One of the protection circuits which best meets these requirements is shown in Fig. 9-59. Resistor R_0 is formed on the metal connecting the power transistor collector to the relative pad. The circuit operates as follows. Transistors Q_1 and Q_2 have different emitter areas (for example, $A_1/A_2 = 5$). Therefore $I_1 = 5I_2$ in the case of $I_0 = 0$. As I_0 increases (short-circuit currents) the voltage R_0I_0 increases V_{BE2} and therefore I_2, causing Q_3 to switch on. This removes part of the base current I_{g1} of the Darlington Q_1Q_0 from the base of Q_1, thus limiting I_0.

With reference to Fig. 9-59 the following equations are valid:

$$V_{BE1} + R_0I_0 = V_{BE2} \tag{9-185}$$

hence

$$I_0 = \frac{kT}{q} \ln \left(\frac{A_1}{A_2} \frac{I_2}{I_1} \right) \tag{9-186}$$

$$I_1 = I_{g2} \tag{9-187}$$

$$I_2 = \frac{V_{BE3}}{R_2} \tag{9-188}$$

By substituting Eqs. 9-186 and 9-187 in Eq. 9-185, we get

$$I_0 = \frac{kT}{qR_0} \ln \left(\frac{A_1}{A_2} \frac{V_{BE3}}{R_2 I_{g2}} \right) \tag{9-189}$$

This is the expression of the maximum short-circuit current (V_{BE3} is the value of V_{BE} which transistor Q_3 assumes to limit the current at the base of the Darlington Q_1Q_0).

For $V_{CE} > V_Z + V_{BE2}$, Eq. 9-187 becomes

$$I_1 = I_{g2} + \frac{V_{CE} - V_Z - V_{BE2}}{R_1} \tag{9-190}$$

and therefore Eq. 9-185 becomes

$$I_0 = \frac{KT}{qR_0} \ln \left[\frac{A_1}{A_2} \frac{V_{BE3}/R_2}{I_{g2} + (V_{CE} - V_Z - V_{BE2})/R_1} \right] \tag{9-191}$$

By plotting Eqs. 9-189 and 9-191 on a graph with axes I_0 and V_{CE}, we get the protection curves of Fig. 9-58.

By suitable dimensioning the electrical parameters of the circuit in Fig. 9-59 can make the protection curve follow the SOA of power transistor Q_0 fairly closely. The SOA of the transistor is obviously the dc SOA. This puts a fairly tight limit on the use of the transistor's characteristics since it is known that the SOA increases as the overload or short-circuit time decreases, while the protection curve of Fig. 9-58 remains constant.

The main drawback of this type of protection occurs when the amplifier drives a highly reactive load.

9.8.2 Protection Techniques in Presence of Reactive Loads

A loudspeaker is an ohmic-inductive load. It has a single supply and is coupled to the amplifier by means of a capacitor of 1000 to 2000 μF. Therefore at frequencies of 20 to 30 Hz the load is capacitive, while it becomes inductive at frequencies of around 1 kHz.

In normal operating conditions, for wide output signals and at certain frequencies, the situation of Fig. 9-60 may occur. Consequently the output voltage waveforms are as shown in Figs. 9-61 and 9-62 for ohmic-capacitive and ohmic-inductive loads, respectively.

For many applications this distortion cannot be tolerated, but the protection curve cannot be raised above the SOA in direct current because a prolonged short circuit would cause failure of the power

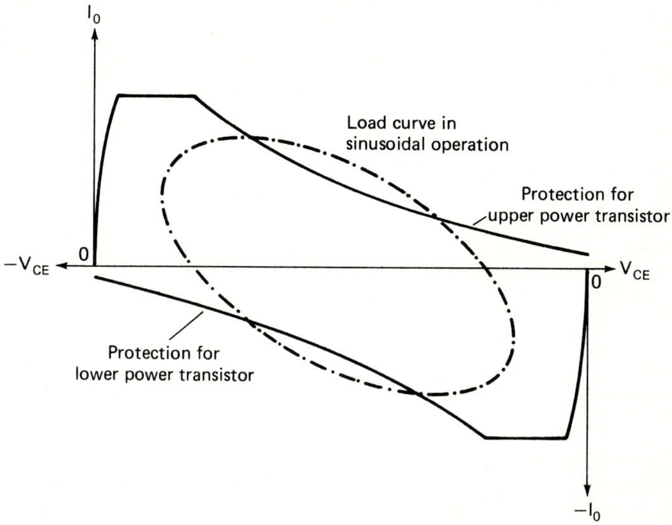

FIG. 9-60 Protection curves for the full final stage plotted on the load curve for reactive load operating.

transistor. As can be seen in Fig. 9-60 the load curve leaves the protection curve for short periods. If it were possible to modulate the protection curve as a function of the intervention duration so as to expand it for a short period, the distortion problems of Figs. 9-61 and 9-62 would be solved, and the protection circuit would continue to operate because the transistor's SOA varies as already described. This is possible if the components of the circuit in Fig. 9-59 are located suitably in the layout.

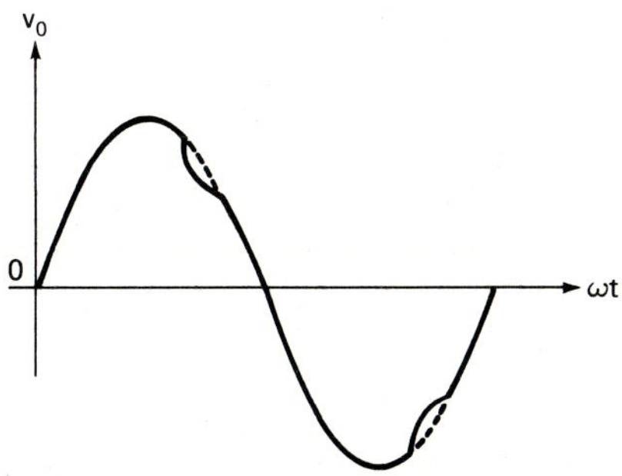

FIG. 9-61 Output waveform for ohmic-capacitive load with intervention of protection circuit.

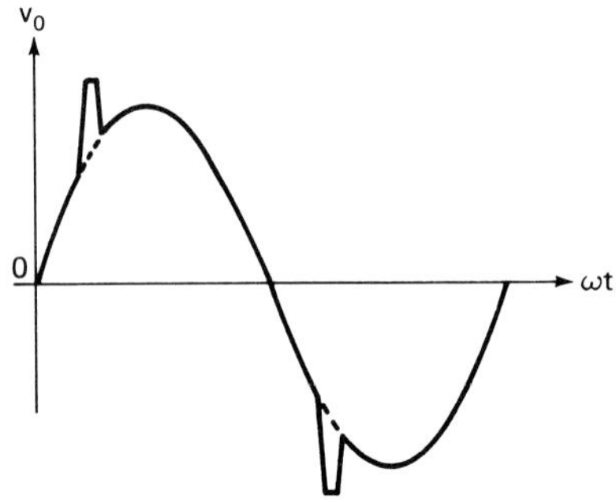

FIG. 9-62 Output waveform for ohmic-inductive load with intervention of protection circuit, showing severe distortion.

9.8.3 Delayed Short-Circuit Protection[1]

Equation 9-186 of the protection curves can also be written

$$I_0 = \frac{K}{qR_0}\left[T_2 \ln \frac{I_2}{A_2J_{S2}} - T_1 \ln \frac{I_1}{A_1J_{S1}}\right] \tag{9-192}$$

where T_1 and T_2 are the absolute temperatures of Q_1 and Q_2 and J_{S1} and J_{S2} are their reverse saturation currents.

Equation 9-192 can be written

$$I_0 = f(V_{CE}, T) \tag{9-193}$$

Furthermore from Eq. 9-192 since the influence of J_S prevails on T as the temperature increases, I_0 decreases as T increases if $T_2 > T_1$.

It is therefore necessary to place Q_2 nearer to the power device than Q_1 and to have a suitable distance between Q_1 and Q_2 so that a clearly defined gradient $\Delta T = T_2 - T_1$ corresponds to a certain power peak dissipated by Q_0. These values can be calculated fairly exactly by using the laws for propagation of heat in the silicon during a dissipated power transient.

Let us consider a monolithic amplifier mounted on a copper dissipator as shown in Fig. 9-63. K_1 is the thermal conductivity of the silicon and K_2 that of the copper. The temperature distribution in the chip due to a dissipated power step is found from the equation

[1]P. Antognetti, G. R. Bisio, F. Curatelli, and S. Palara, Three-Dimensional Transient Thermal Simulation: Application to Delayed Short Circuit Protection in Power ICs, *IEEE J. Solid-State Circuits*, vol. SC-15, no. 3, 1980. Reproduced by permission.

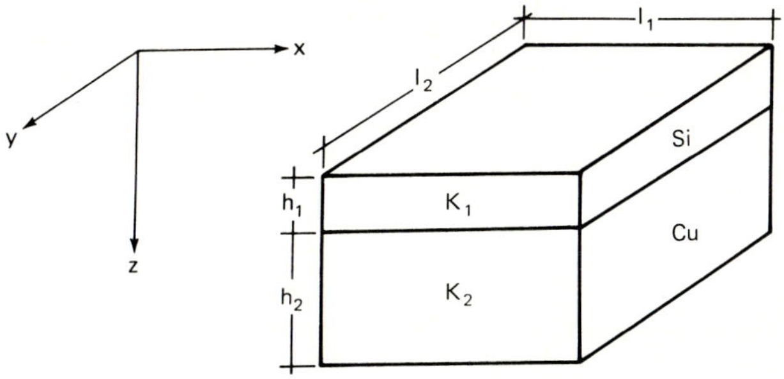

FIG. 9-63 Schematic representation of a power IC (simplified structure).

$$\nabla^2 T(x, y, z, t) = \frac{1}{D}\frac{\partial T}{\partial t} \tag{9-194}$$

considering the diffusivity D to be constant with the temperature.

We shall impose the following boundary conditions on Eq. 9-194 with reference to Fig. 9-63:

1. The power dissipation on the top surface of the chip has an arbitrary but known distribution $P(x, y, 0)1(t)$.
2. The lateral sides of the two-layer structure are adiabatic.
3. At the silicon-copper interface the temperature and the Z components of the heat flux are continuous.
4. The bottom surface of the substrate has a uniform temperature T_0.

The solution of Eq. 9-194 is given by

$$T(x, y, z, t) = \sum_{0}^{\infty} m \sum_{0}^{\infty} n \left\{ \sum_{1}^{\infty} i R_{mn}(z)\Big|_{s_i} \right.$$

$$\left. (1 - e^{s_i t}) \cos\left(\frac{m\pi}{L_1} x\right) \cos\left(\frac{n\pi}{L_2} y\right) \right\} \tag{9-195}$$

where s_i are the poles of the mn Fourier component $T_{mn}(z, s)$ of the thermal transfer function and are real negative quantities and $R_{mn}(z)|_{s_i}$ are the residues of $T_{mn}(z, s)e^{st}$ computed at s_i.

A computer program based on this solution of Eq. 9-195 gives the three-dimensional temperature distribution on the IC chip surface. Figure 9-64 shows the temperature profile along the broken line, which goes from Q_1 to Q_2 deriving from a constant dissipated power peak from

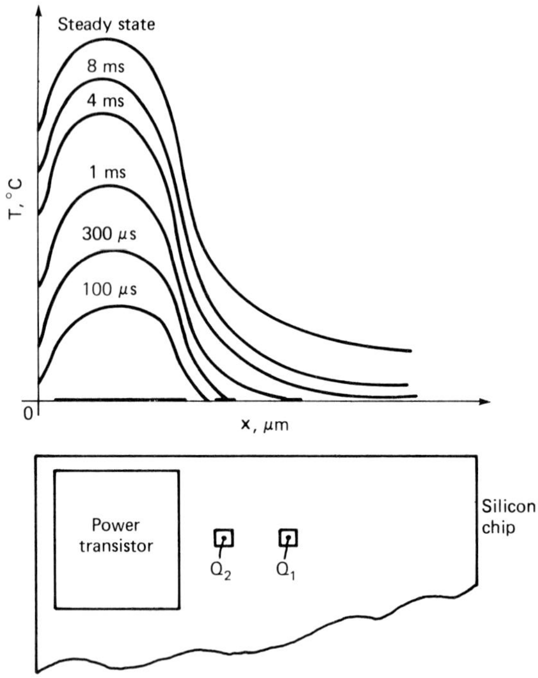

FIG. 9-64 Simulated transient temperature profile; on the x axis, the black regions indicate the positions of the power transistor and the sensors.

the power transistor, as the duration of this peak varies. An essential fact is that the thermal gradient ΔT between Q_2 and Q_1 increases as the duration of the dissipated power peak increases following the curve shown in Fig. 9-65.

We shall now return to the short-circuit protection equation, Eq. 9-

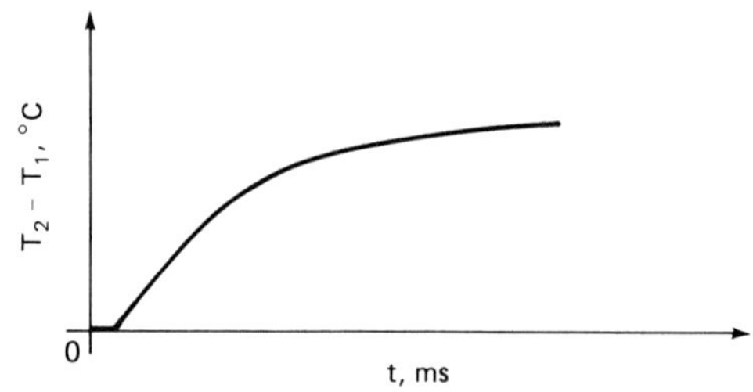

FIG. 9-65 Simulated thermal gradient existing between sensors Q_1 and Q_2 as a function of time.

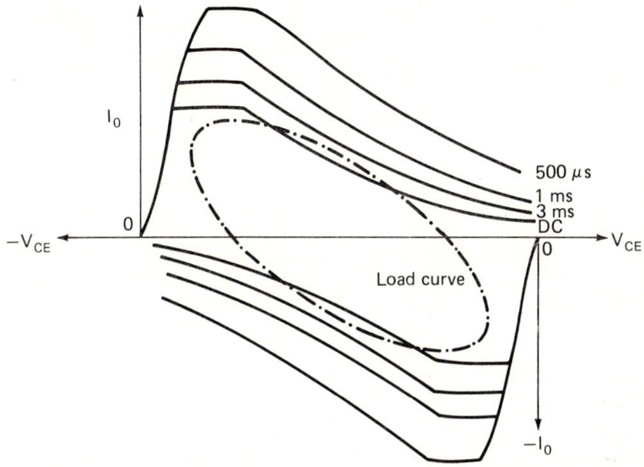

FIG. 9-66 Protection characteristics $I_0 = I_0(V_{CE})$ for different temperature gradients show a delayed short-circuit protection.

192. As we have already seen, I_0, which is the short-circuit current, decreases as the temperature gradient $\Delta T = T_2 - T_1$ increases, and this is exactly what happens in Fig. 9-65.

The protection curve in this case is modulated as the short-circuit duration t varies, and this is shown in the graph of Fig. 9-66. In the particular case of Fig. 9-66 the protection would intervene if it were not modulated at the reactive load represented by the broken line. In reality the protection curve corresponding to the instantaneous power dissipated by the power transistor is wider and does not have the distortion phenomena indicated in Figs. 9-61 and 9-62.

The most important result is that the protection curve automatically

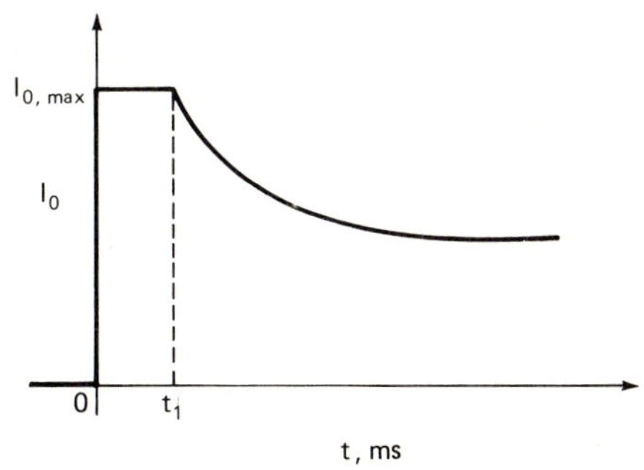

FIG. 9-67 Variation of short-circuit current I_0 as a function of time for a delayed short-circuit protection.

adapts to the extended SOA limits as the duration of the stress undergone by the power transistor varies.

From Figs. 9-65 and 9-66 we can see the variation of I_0 as the short-circuit duration varies (Fig. 9-67). This current transient can only be compared as regards quality with the temperature gradient variations shown in Fig. 9-65 since Fig. 9-65 refers to a constant dissipated power, while in Fig. 9-67 the dissipated power is a function of time.

With reference to Fig. 9-67, time t_1 is determined by the distance of Q_2 from the power transistor and is the time required for the dissipated power to cause a temperature gradient between Q_2 and Q_1. During this time the power transistor conducts the maximum current in accordance with its SOA determined by t_1.

The distance between Q_2 and Q_1 determines the reduction rate of the short-circuit current due to the thermal gradient between these two transistors.

9.8.4 Loudspeaker Protection Technique

The type of short-circuit protection described can be used in all mono and stereo amplifiers but is not sufficient for bridge configurations.

In fact we shall examine the short-circuit condition of Fig. 9-68, closing switch *s*, and supposing that both amplifiers *A* and *B* have the type of protection described above.

Amplifier *A* will have a short-circuit current I_{0A}, while amplifier *B* does not have a short circuit but an overload caused by the loudspeaker. Therefore the direct current I_{OB}, which may be a few ampères, flows through the loudspeaker and may damage or even break it. Therefore, today's bridge circuit users require loudspeaker protection against short circuits of the type shown in Fig. 9-68.

This means that it is necessary to switch off the final stages of amplifiers *A* and *B* simultaneously when S_1 is closed. The TDA 2005, a dual

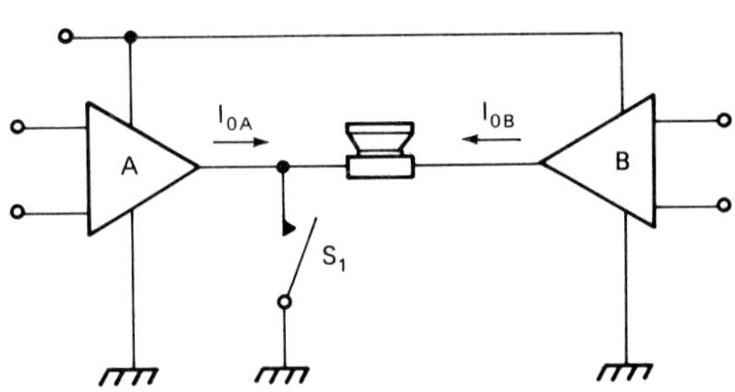

FIG. 9-68 Short-circuit condition for bridge configuration application circuit.

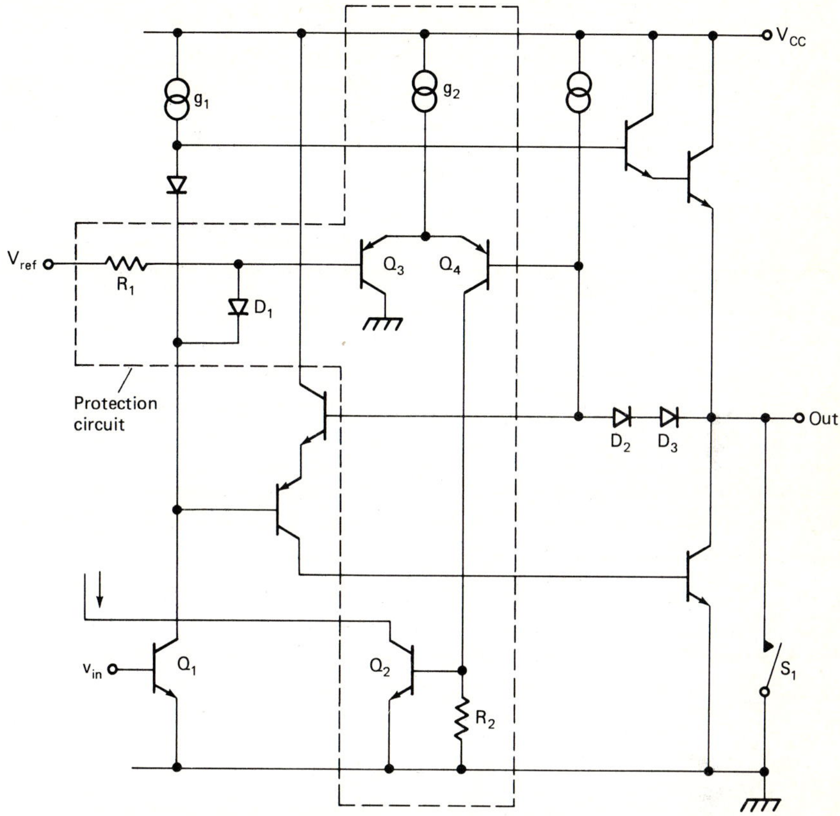

FIG. 9-69 Short-circuit protection diagram with switch-off of the amplifier.

monolithic power amplifier, has a protection circuit which meets this requirement. Figure 9-69 shows a final stage with a protection circuit inside the broken line.

During normal operation, the output voltage waveform is more or less the same in phase and amplitude on the collector of Q_1. Diode D_1 modulates the base of Q_3 with a voltage equal to the output voltage but shifted up by one V_{BE}. The same voltage curve is detected on the base of Q_4 but shifted up by one V_{BE} with respect to the base of Q_3. Consequently the differential stage formed by Q_3 and Q_4 is always unbalanced by one V_{BE} and only Q_3 conducts. Therefore Q_2 is always inhibited. As soon as switch S_1 closes, the amplifier is unbalanced, Q_1 is inhibited, and the current of g_1 supplies the upper final section. Inhibition of Q_1 no longer causes modulation on the base of Q_3. Diode D_1 is inhibited so the voltage at the base of Q_3 is equal to a suitable $V_{ref} > 2V_{BE}$.

In the specific case of the TDA 2005, $V_{ref} = 4$ V. At the same time the base of Q_4 is at $+2V_{BE}$ with respect to ground.

The differential stage Q_3Q_4 immediately changes state with respect to the previous condition of normal amplifier operation, and T_4 injects all the current of generator g_2 onto resistor R_2 which direct-biases transistor T_2. The collector of Q_2 is connected to a point in the circuit by means of which the final stages of the two amplifiers are switched off. Therefore during the short circuit neither of the two outputs supplies current, and the loudspeaker is not damaged.

Furthermore in this condition the chip is not subjected to the strong dissipation which occurs when it is simply limited by the SOA-type protection curve, so its reliability is improved.

9.8.5 Protection Circuit with Inhibit

With an SOA-type protection circuit only the short-circuit current is limited. This, although it does not cause failure of the power transistors, determines a condition of strong dissipation which limits device reliability, especially if the short circuit lasts for a considerable length of time.

At present the problem of reliability in integrated circuits is becoming extremely important, more so even than the electrical performance. Therefore the traditional concept of short-circuit protection, taken simply to be limitation of maximum current, has been revolutionized. New protection circuit concepts have been introduced. These completely switch off the final stages for any type of short circuit, whether directly at the output or after the capacitor for coupling the loudspeaker in the case of mono and stereo or between the two outputs in the case of the bridge configuration. All these short circuits are shown in Fig. 9-70.

The amplifiers must switch off if one or more of the short-circuit conditions shown in Fig. 9-70 occur.

SGS recently produced a dual monolithic amplifier which can be used in stereo and bridge applications and which uses the protection described above. Let us consider the upper section of a final stage (the same considerations apply to the lower section). The protection circuit is realized as shown in Fig. 9-71.

The first block, the short-circuit sensor, is basically the circuit already examined in Fig. 9-59, which with the components suitably positioned in the layout produces a protection curve of the type shown in Fig. 9-66 modulated by the duration of the short circuit so as to explore the dynamic SOA limits of the power transistor. The output of this block is the collector of Q_3 in Fig. 9-59, and it is an on-off type of signal.

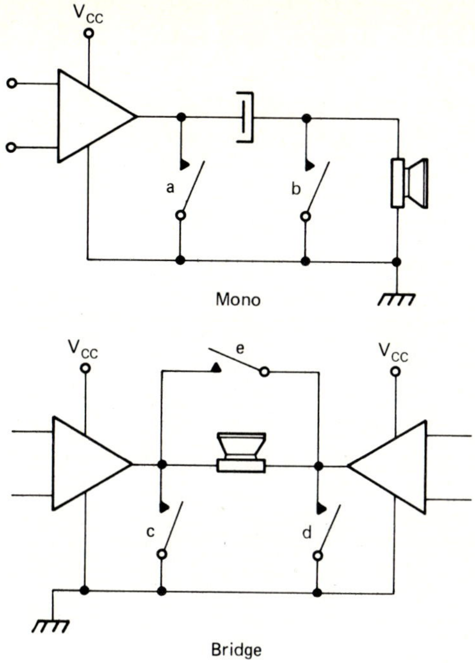

FIG. 9-70 Short-circuit conditions for single-supplied mono amplifier and for bridge configuration.

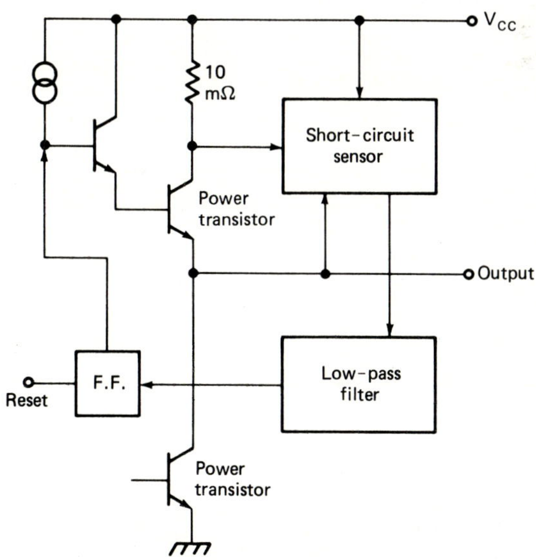

FIG. 9-71 Schematic diagram of short-circuit protection with inhibit.

When the final transistor operates within certain limits fixed by the protection curve of Fig. 9-66, transistor Q_3 is off. It goes into conduction when the load curve intersects the protection curve. If Q_3 were connected directly to the flip-flop, the intervention of the protection would become very critical since any interference, even for only a microsecond, would cause switching of the flip-flop, resulting in a switch-off of the amplifier during its normal operation. It is therefore essential to make switching of the flip-flop insensitive to interference or accidental short voltage or current spikes which do not damage the power transistors.

Furthermore the load curve may go beyond the protection curve for short periods. This may happen at switch-on or with highly reactive loads. In these conditions the amplifier must not go off. This is achieved by preventing short signals at the output of the short-circuit sensor from reaching the flip-flop. These signals can be blocked by a low-pass filter which allows pulses of duration greater than a few tens of microseconds to pass.

In an integrated circuit this can be realized simply as shown in Fig. 9-72. If there is no short circuit, capacitor C is maintained charged at a voltage of around V_{ref} by means of Q_1; therefore Q_2, Q_3, and Q_4 are inhibited. Current I_g is = 5 to 6 μA, while V_{ref} is = 5 V and V_1 = 2 V.

When there is a short circuit, input A is brought to ground by the collector of the output transistor of the short-circuit sensor. Capacitor C discharges onto current generator I_g with the law

$$t = \frac{C\,[V_{ref} - (V_1 - V_{BE2} - V_{BE3})]}{I_g} \tag{9-196}$$

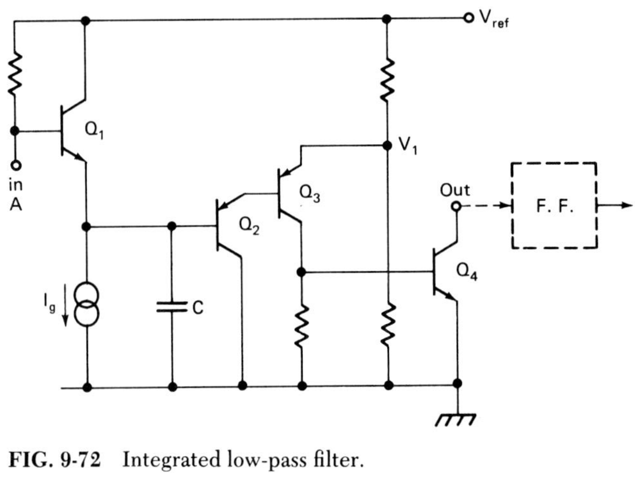

FIG. 9-72 Integrated low-pass filter.

Therefore Q_2 and Q_3 activate Q_4 which, acting on the flip-flop, causes switching of the state that leads to switching off of the final stages of the amplifier. The time selected for t is about 30 to 50 μs, too short for the power transistors to be damaged by an overload but long enough to make the protection intervention insensitive to interference.

REFERENCES

1. R. S. Hartz and F. S. Kamp, Power Output and Dissipation in Class B Transistor Amplifiers, *RCA Appl. Note 3576*, 1968.

CHAPTER 10

Power Integrated Circuits for Industrial Applications

P. G. Erratico
SGS Microelettronica
Milan, Italy

10.1 INTRODUCTION 10.2
10.2 POWER SWITCHING COMPONENTS IN MONOLITHIC FORM 10.3
10.2.1 NPN Transistor Voltage and Current Ratings 10.3
10.2.2 Commutation Speed 10.6
10.2.3 Monolithic Diodes 10.7
10.2.4 Power PNP Configurations 10.8
10.3 ENERGY TRANSFER EFFICIENCY AND IC LIMITS 10.10
10.3.1 Saturation Losses 10.10
10.3.2 Turn-On Transition 10.11
10.3.3 Turn-Off Transition 10.12
10.3.4 Class B Stages 10.14
10.4 FINAL-STAGE CONFIGURATIONS 10.15
10.4.1 Half-Bridge Structure 10.15
10.4.2 Push-Pull Amplifiers 10.17
10.4.3 Full-Bridge Structures 10.20
10.5 APPLICATIONS 10.22
10.5.1 Driving DC Motors 10.22
10.5.2 Solenoid-Current Controllers 10.27
10.5.3 Stepping Motor Applications 10.30
10.5.4 Analog Current-Control Techniques 10.35

10.1 INTRODUCTION

In many applications, it is necessary to provide some sort of motion under electronic control; this is the case in dc or stepping motors; relays or solenoids for driving various types of valves, for which the actuation may be the simple on-off scheme; or more sophisticated current-mode control.

The majority of these applications concern moderate power levels, where the current is on the order of a few amperes and the supply voltage lower than 50 V, mainly for safety reasons. In all these cases, the load is highly repetitive, and so, as would be expected, a power module is required which is able to drive the actuator and which in turn is controllable from a powerful microprocessor.

Power integrated circuit technology can satisfy all these conditions: the power can be delivered at the most efficient rate, often by using switching energy-transfer techniques; and modularity and high volume of application are bread and butter for semiconductor design.

Moreover, new technologies allow highly sophisticated control circuitry to be used with power driving elements, leading to two fundamental results:

1. The interface to the central controller (e.g., a microprocessor-based system) is simplified and the CPU time required for managing the load actuation is reduced.
2. High accuracy is achieved with analog circuits in load parameter control (mainly current, voltage, and timing); such accuracy is not so easy to reach with digital control techniques.

These are the reasons for the success of power control chips, which have thus moved from very simple structures (e.g., monolithic Darlington arrays) to true LSI devices. However, the key element is a high-speed, high-power switching device, which will be the starting point of our analysis. We will also consider the basic power-stage arrangements suitable for monolithic implementation and we will review the key factors affecting energy-conversion efficiency. The chapter will be concluded with an overview of the more popular power drivers available in IC form, in order to give an application outline to power system designers.

In this way the reader may have an idea of present-day realities and the potentialities for integrated power circuits.

10.2 POWER SWITCHING COMPONENTS IN MONOLITHIC FORM

As stated in the introduction, the most popular supply voltage for industrial applications is less than 50 V. We will therefore consider a bipolar ion-implanted process, able to guarantee a sustaining voltage of more than 50 V both for NPN and PNP.

As far as current capability is concerned, NPN transistors are by far superior. Therefore we will assume for this discussion that all the high-current stages use an NPN transistor.

In this section we will carry on our investigation in order to find the best arrangement for both current-sinking and current-sourcing power stages.

10.2.1 NPN Transistor Voltage and Current Ratings

It is well known that the current capability of a bipolar transistor is a quadratic function of the sustaining voltage (see Sec. 4.2). In the 50-V process, it is worth remembering that quasi-saturation phenomena play an important role in power transistor design (see Sec. 2.2.4).

As an example, Fig. 10-1 shows the dc characteristics for a power element with an emitter area of 900 mil^2, equivalent to an overall size of 3000 mil^2 (isolation to isolation).

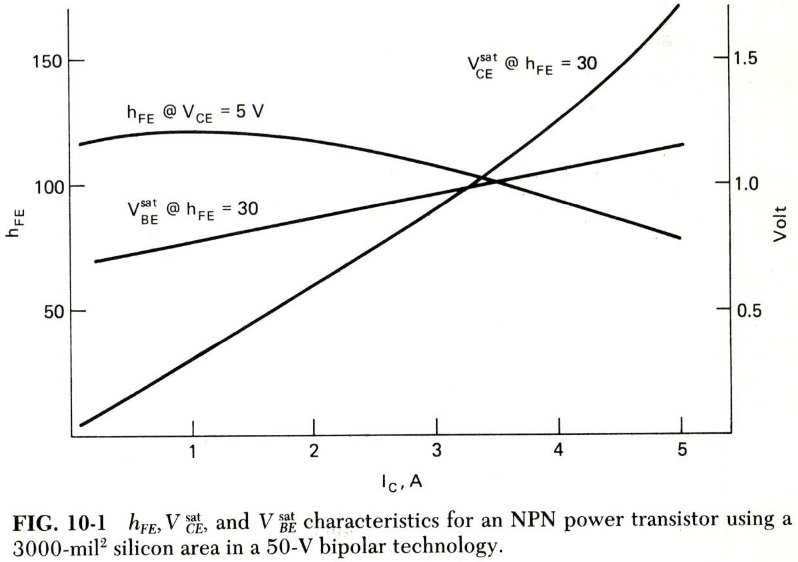

FIG. 10-1 h_{FE}, V_{CE}^{sat}, and V_{BE}^{sat} characteristics for an NPN power transistor using a 3000-mil^2 silicon area in a 50-V bipolar technology.

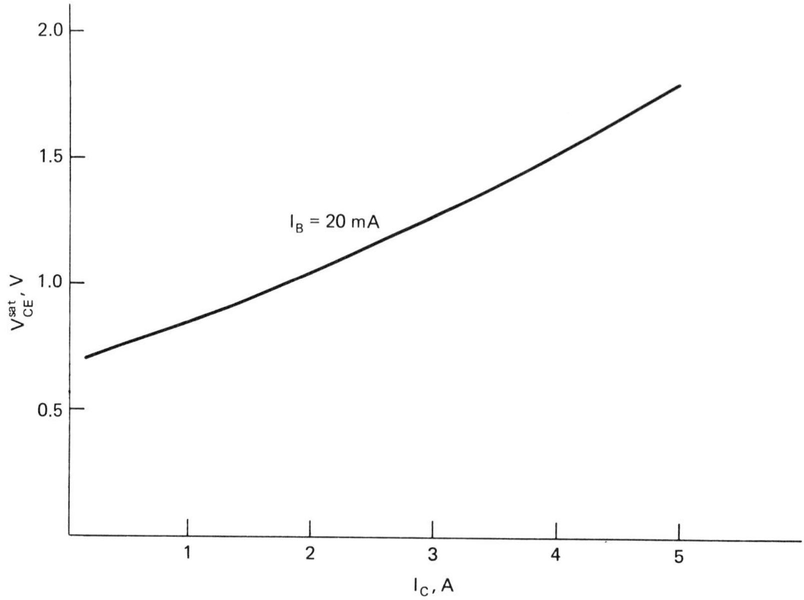

FIG. 10-2 V_{CE}^{sat} characteristics for an NPN Darlington structure using 3000 mil^2 of silicon area in a 50-V bipolar technology.

We can see that, while with $V_{CE} = 5$ V, $h_{FE}/2$ is reached for a current of 6 A, the saturation voltage is higher than V_{BE}^{sat} (at $h_{FE} = 30$) just starting from 3 A. In this condition, a Darlington configuration can be a satisfactory solution, mainly because of the great power savings in the base biasing circuitry. The V_{CE}^{sat} characteristic for a Darlington stage with an overall silicon area of 3000 mil^2 is shown in Fig. 10-2.

We can conclude that, with a monolithic 50-V process, it is possible to achieve good switching devices with $V_{CE}^{sat} \leq 2$ V at a collector current of 4 A using a reasonable silicon area.

As far as voltage ratings are concerned, the main problem is encountered when switching off a saturated transistor loaded with an inductance. In this case the conventional sustaining voltage ratings have to be taken into consideration. Moreover, in order to speed up the commutation time, a high extraction current is needed. Integrated circuits with no negative supply voltage are almost mandatory, so the realization shown in Fig. 10-3 is commonly used.

Depending on the value of resistor R_1, different trade-offs for commutation speed and reverse second breakdown E_{sB} can be obtained. Figure 10-4 shows different safe operating area (SOA) boundaries for

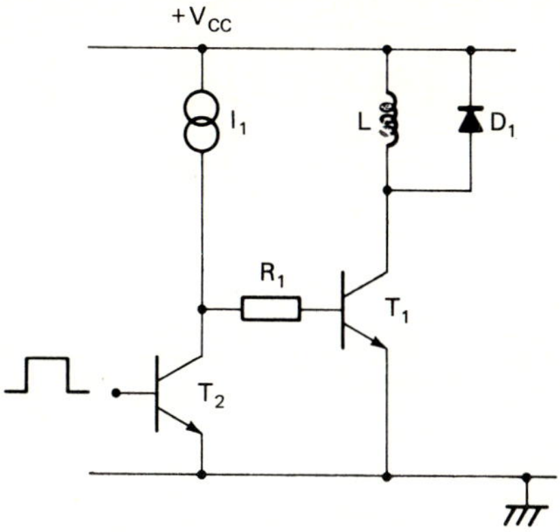

FIG. 10-3 Active extraction circuit.

two extraction impedances. The transistor geometry is the same as that considered in Fig. 10-2.

It is a common practice to regard the point where the SOA curve of $R_1 = 50\ \Omega$ crosses that for BV_{CEO} as an indication of a power transistor's switching capabilities. In a 50-V planar process this condition is often more demanding in terms of silicon surface than the residual voltage at saturation.

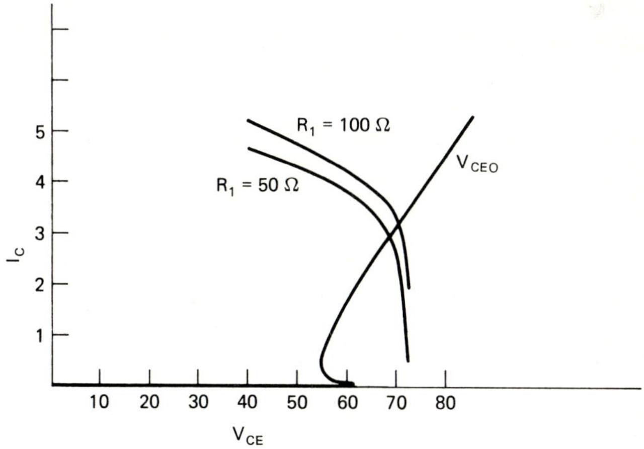

FIG. 10-4 Reverse second breakdown safe operating area for the same transistors as in Fig. 10-3, for different extraction impedances between base and ground.

10.2.2 Commutation Speed

Once the current and voltage problems have been solved, we must concentrate on reducing the storage time and the rise and fall time. These are responsible for commutation delay and power losses, which in turn limit the maximum commutation frequency at which the circuit can be operated.

From classical charge control theory [1], we know that τ_S, the parameter that controls the storage time, is given by

$$\tau_S = \frac{\alpha_F(\tau_F + \alpha_R\tau_R)}{1 - \alpha_F\alpha_R} \tag{10-1}$$

where α_F, α_R = common-base current gain in direct and reverse biasing
τ_F, τ_R = base transit time in direct and reverse biasing

The storage time t_S, as defined in Chap. 2, is given by

$$t_S = \tau_S \ln \frac{I_{B2} + I_{B1}}{I_{B2} + I_B^{\text{sat}}} \tag{10-2}$$

where I_{B1} = base drive current
I_{B2} = base extraction current
$I_B^{\text{sat}} = I_C/h_{FE}$ = minimum base current for driving the transistor in saturation at collector current I_C

For the rise time t_r and fall time t_f we have

$$t_r = \frac{I_C\tau_F}{I_{B1}} \qquad t_f = \frac{I_C\tau_F}{I_{B2}} \tag{10-3}$$

The analysis of these formulas gives useful information on the behavior of switching speed in integrated transistors, as follows:

1. As for discrete components, in order to have short storage times we must avoid overdrive ($I_{B1} \gg I_B^{\text{sat}}$) and use $I_{B2} > I_{B1}$. Obviously this last condition can affect reverse second breakdown phenomena.
2. For monolithic planar NPN transistors, we have, as an order of magnitude: $\tau_F = 10^{-9}$ and $\tau_R = 10^{-7}$ s; we can then simplify Eq. 10-1, assuming $\alpha_F = 1$ and $\alpha_R = 0.90$:

$$\tau_S = \frac{\alpha_R\tau_R}{1 - \alpha_R} \simeq 10^{-6}\ \text{s} \tag{10-4}$$

The current rise and fall times are mainly controlled by τ_F and the base drive conditions; therefore they can be very small compared with τ_S.

3. For PNP transistors $\tau_F = \tau_R = 10^{-7}$ s, and so, still assuming as typical figures $\alpha_R = 0.9$ and $\alpha_F = 1$, we obtain $\tau_S = 2 \times 10^{-6}$, not very different from the former case. But the rise and fall times are quite different because the actual value of τ_F is by far bigger. As far as PNP power devices are concerned, storage time is a problem since it puts an upper limit on the maximum commutation frequency. But the rise and fall times are much more critical in that they directly affect the efficiency of the energy conversion. This fact shows that, while the design of an NPN power stage is mainly a matter of calculating the correct drive and extraction currents, for PNP stages a direct base drive will not be enough in most applications. This will be discussed in more detail in Sec. 10.2.4.

10.2.3 Monolithic Diodes

It is well known that in power applications diode performance (mainly turn-off time) usually places severe limitations on the circuit efficiency. When speaking of integrated circuits, this is even more important, mainly because of the intrinsic limitations of these components, discussed in Chap. 4; there it was described how in any integrated diode with a breakdown voltage higher than BV_{EBO} the phenomenon of parasitic current losses to ground must always be considered.

In practice, this current is on the order of a few percent of the diode current. This fact has several important consequences:

1. Reduction of overall system efficiency.
2. Increase in the power dissipated on the chip itself; this affects system reliability via an increase in the junction temperature.
3. The system ruggedness is also affected because of the stress resulting from the simultaneous presence of voltage and current in the parasitic element.

Moreover, in terms of silicon area, at the same current level, a high-voltage diode is equivalent to an NPN power transistor. In terms of speed, there appears to be an upper limit around 500 ns (diode turn-off time); this adversely affects efficiency (through commutation losses at turn-on). It would therefore appear that in monolithic devices up to now, discrete recirculation diodes are the most effective solution for currents higher than 0.5 A, both from a technical and an economic point of view.

New processing techniques, based on an extensive use of ion implantation (see Chap. 4), enable most of these problems to be overcome, even if special care must be devoted to the overall ruggedness of these components.

10.2.4 Power PNP Configurations

In the conventional 50-V ion-implanted process, it is generally considered that the only power element available at reasonable cost is an NPN transistor. This is convenient for current-sinking elements, but it means that a more elaborate circuit structure is needed for current sourcing.

There are conventional solutions, like the one shown in Fig. 10-5, commonly called a composite PNP-NPN Darlington transistor. In this configuration, the NPN will not affect the commutation time of the stage, as it never saturates, and its transistor turn-on time is fast, as allowed by PNP current rise time; in any case it is the best low-saturation solution that uses only transistors (no resistances) in the final stages. Unfortunately, at turn-off one must accept the full storage and fall times of the PNP driver transistor, which are on the order of several microseconds.

Better solutions must be found that will avoid the obviously expensive choice of a complementary PNP-NPN process; circuit complexity can be accepted in a trade-off for improved process economy and good overall performance.

A good solution to the problem is presented in Fig. 10-6. The current path T_1, T_6, T_4—that is, the power element—implements the main

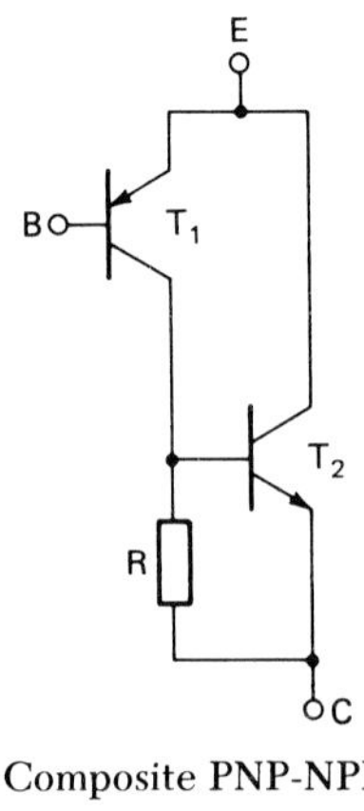

FIG. 10-5 Composite PNP-NPN structure.

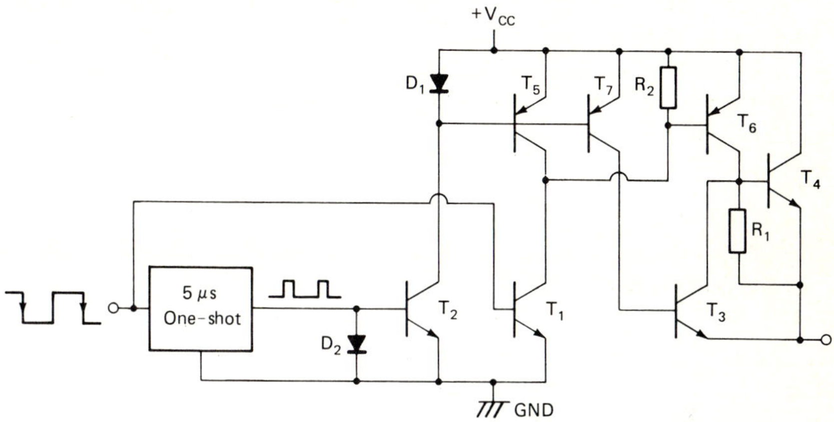

FIG. 10-6 Monolithic structure for PNP high-speed output stage.

On-Off actuations and is reminiscent of the classical configuration shown in Fig. 10-5. It is also the pattern in which the stage works at turn-on.

At turn-off it is necessary to find a high-speed method of extracting current from the base of T_4. At first glance we can turn on T_3 (via T_2 and T_7), which short-circuits the emitter and base terminal of T_4. T_3 must obviously have more current capability than T_6.

In order to minimize overall power dissipation, T_6 must in any case be turned off as soon as possible. Since T_6 is a saturated PNP, this path will be slower than the former one; in any case the best solution is to extract current from the base of T_6 by means of an active element, here, transistor T_5.

After the current extraction is completed, we have a stage with T_4 in the Off condition, but with two elements (T_3 and T_5) in saturation and T_2 on. This will not affect power dissipation in this state to a great extent, but it is responsible for a noticeable delay at the subsequent turn-on transition, which can affect the maximum frequency of operation. In any case, after the storage capability of T_6 has been exhausted, T_2 (and its associated transistor) need no longer conduct: T_4 and T_6 can be maintained off by R_1 and R_2. Therefore the T_2 command can be a one-shot pulse, triggered by the negative transition of T_1, with typically a 5-μs width.

This solution, at first glance a bit complex, is quite convenient in monolithic form, since it uses only a small area in addition to that for the power element, at least when the current is in excess of 1 A; and more important, it allows commutation times (both storage and current rise and fall) quite similar to those of NPN current-sinking stages.

10.3 ENERGY TRANSFER EFFICIENCY AND IC LIMITS

As we have seen in Chap. 9, output power is in some ways limited in ICs by dissipation and heat sink efficiency. Therefore in this section we will analyze the elements that will limit the power in the load for switch-mode devices, where the problem is faced in a different way. Let us consider the simplest schematic configuration, shown in Fig. 10-7; here the power load is shown as a resistor, representing, for example, a dc motor, or electronics to be supplied. The base of transistor T_1 is switched with a duty cycle controlled by a regulating circuit, in order to obtain a current I_L on the load.

10.3.1 Saturation Losses

If it is working at low frequency, the only significant power losses in the circuit are the saturation losses on T_1 and D_1. With good approximation, we have

$$P_L = V_{CC} \times \text{D.C.} \times I_L \tag{10-5}$$

$$P_{T1} = V_{CE}^{\text{sat}} \times \text{D.C.} \times I_L \tag{10-6}$$

$$P_{D1} = V_F(1 - \text{D.C.})I_L \tag{10-7}$$

where V_{CC} = supply voltage
D.C. = conduction duty cycle
I_L, P_L = continuous current and power on the load
V_F = forward voltage drop on the diode

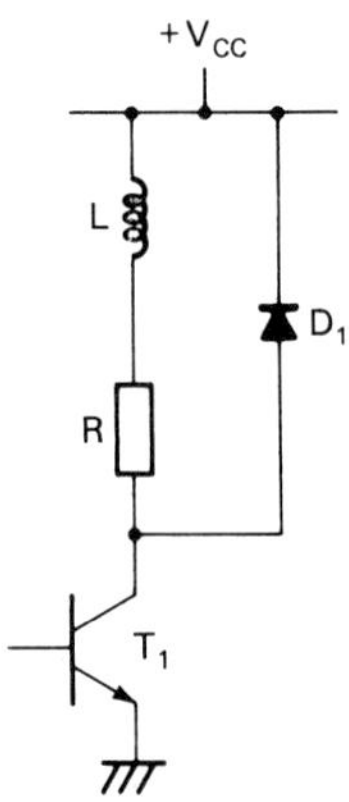

FIG. 10-7 Elementary switching circuit.

As discussed in Sec. 10.2.3, diode D_1 is outside the IC mainly for reasons of speed and parasitic losses. It is possible to define an efficiency η_{sat} as seen by the IC, not considering power losses in the diode; thus we have for any load current level, assuming D.C. = 1,

$$\eta_{sat} = 1 - \frac{P_{T1}}{P_L} \tag{10-8}$$

$$\eta_{sat} = 1 - \frac{V_{CE}^{sat}}{V_{CC}} \tag{10-9}$$

where P_1 is the dissipated power and P_L is the power delivered to the load. A practical example could be: $V_{CE}^{sat} = 2.0$ V, $V_{CC} = 50$ V; then

$$\eta_{sat} = 1 - \frac{2.0}{50} = 96\% \tag{10-10}$$

If 15 W of power dissipation is available, then a lot of power is available on the load (more than 350 W).

Obviously this efficiency will be limited by the dc biasing and driver current (which can be reduced when using a Darlington configuration), and by commutation losses when the frequency is more than a few kilohertz. We can divide these losses into two types: turn-on and turn-off losses, which are described next.

10.3.2 Turn-On Transition

Still considering Fig. 10-7, the current is now flowing in diode D_1, and the current I_L in the coil is assumed to be constant at its static value. At T_1 turn-on, the transistor current rises with its natural rise time, decreasing the current in diode D_1 down to zero (see Fig. 10-8). At this point the current in transistor T_1 further increases to provide extraction current for the diode, which is turned off. From this time the collector voltage can decrease very quickly at a rate limited only by parasitic capacitances.

It is clear that this turn-on phenomenon is fairly complex, since it is influenced by many parameters, including those external to the IC (e.g., diode turn-off time). However, an estimation of the resulting efficiency can be made from experience.

Up to time t_r, the rise time is mainly controlled by the transistor: a typical figure for an NPN transistor is 0.2 μs; considering one dissipation time any cycle that is repeated with a frequency f and assuming triangular waveforms, we have a dissipated power P_1:

$$P_1 = \tfrac{1}{2} t_r f V_{CC} I_L = (0.1 \times 10^{-6}) f V_{CC} I_L \tag{10-11}$$

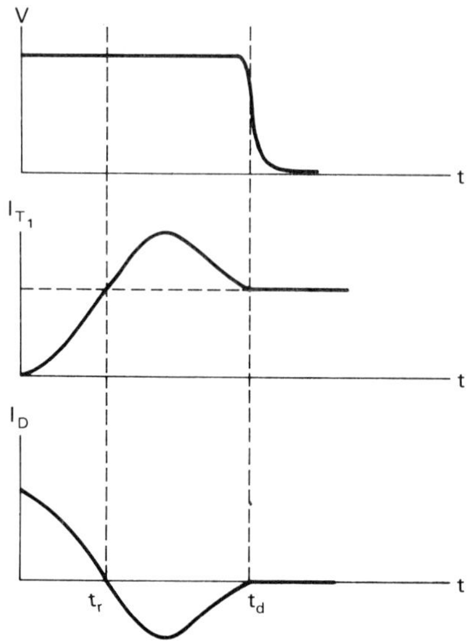

FIG. 10-8 Turn-on waveforms.

From t_r to t_d, a current twice I_L is common. Assuming also here a time of 0.2 μs, we have a dissipated power P_2:

$$P_2 = (0.2 \times 10^{-6}) f V_{CC} I_L \tag{10-12}$$

with

$$P_L = V_{CC} I_L \times \text{D.C.}$$

we get a dissipation factor related to the rise time t_r

$$(1 - \eta)_{t_r} = \frac{(0.3 \times 10^{-6}) f}{\text{D.C.}} \tag{10-13}$$

As an example, for $f = 50$ kHz, D.C. $= 0.5$, we have

$$(1 - \eta)_{t_r} = \frac{0.3 \times 5 \times 10^4 \times 10^{-6}}{0.5} = 3\% \tag{10-14}$$

The order of magnitude is similar to that of the saturation losses.

10.3.3 Turn-Off Transition

At turn-off, the situation is simpler, as shown in Fig. 10-9. Storage time gives rise to no dissipation problem. The diode has no function in this phase. Thus we have, for the wasted power P_D,

$$P_D = \tfrac{1}{2} t_f f I_L V_{CC} \tag{10-15}$$

$$(1 - \eta)_{tf} = \frac{(10^{-7})f}{\text{D.C.}} \tag{10-16}$$

For the same case as before, we have

$$(1 - \eta)_{tf} = \frac{10^{-7} \times 5 \times 10^{4}}{0.5} = 1\% \tag{10-17}$$

Which is quite a bit less than for turn-on.

The total efficiency in the example considered here is

$$(1 - \eta)_{\text{tot}} = (1 - \eta)_{\text{sat}} + (1 - \eta)_{t_r} + (1 - \eta)_{tf} = 8\% \tag{10-18}$$

$$\eta_{\text{tot}} = 92\% \tag{10-19}$$

From this example, offered as a rule-of-thumb explanation, we can see that 90 percent efficiency can be obtained with switching techniques; thus power conversion on the order of 150 W appears possible, if we are able to dissipate 15 W of thermal power. As final considerations we can now add:

1. We have always discussed power dissipation under static conditions; usually in industrial applications, power is needed for short periods (the start-up time of a motor, for example), so these limits

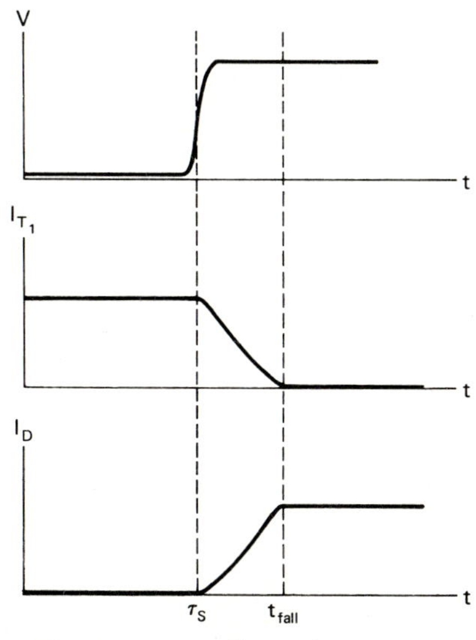

FIG. 10-9 Turn-off waveforms.

can be surpassed, mainly by the inverse ratio of transient to steady-state thermal resistance of the system.

2. In many situations (such as with a solenoid driver or a needle printer), it is not possible to define an efficiency, since one cannot obtain a regulated current on a purely reactive load.

The method for calculating dissipated power set forth here still applies and makes it easy to determine the size of the heat sink for extracting the heat generated on the silicon surface.

10.3.4 Class B Stages

When dealing with switching techniques, class B output stages are quite different from the ones used for analog applications, where the key point is power efficiency and crossover distortion.

For a switch-mode class B stage, there are no problems arising from crossover distortion; rather, power efficiency is always the biggest problem. Usually class B output stages are used in two cases:

1. When there is a double supply system that feeds a load connected to ground
2. When there is a bridge output stage that supplies bipolar current to a balanced load in a single-supply system

We can apply the following considerations:

1. In dual-supply systems, saturation and commutation losses may be the same as for unipolar stages, while in bridge configurations these losses are doubled. So, for the standard case considered (15 W of power dissipated on the silicon), the overall output power is limited to 75 W.
2. For saturation reasons, common-collector stages are not the best solution when using switch-mode techniques, as they might be in the case of audio amplifiers. A typical solution for these stages combines the two complementary NPN and PNP output stages shown in Secs. 10.2.2 and 10.2.4. But at this point a key problem arises: simultaneous conduction of the two power devices at commutation, i.e., the possibility that a current will be shunted from V_{CC} to ground, directly via the two power elements, when the one in conduction turns off some time after the turn-on of the complementary element. In these conditions the peak current is not limited and it reaches a very high value, even when low currents are flowing in the load.

Let us suppose there is a 1-μs spike of 4 A in a system switched at a commutation frequency of 20 kHz and supplied at 50 V. The average power dissipated (P_D) due to the spike at this frequency is

$$P_D = \frac{4 \text{ A} \times 50 \text{ V} \times 1 \ \mu\text{s}}{50 \ \mu\text{s}} = 4 \text{ W}$$

Considering that in a bridge structure there are four transitions per cycle, the total wasted power is

$$P_D = 4 \times 4 \text{ W} = 16 \text{ W} \tag{10-20}$$

This number is obviously not compatible with a situation in which total power losses, for different and nonimprovable limiting phenomena, are 15 W, as previously considered. Any power loss for simultaneous conduction is in any case very detrimental, so circuit techniques must be studied to avoid the problem.

Different solutions will be discussed in Sec. 10.4.3. As a conclusion, we have seen that a lot of useful power can be available from an IC device, while the most difficult problem to solve is that of simultaneous conduction in class B stages.

10.4 FINAL-STAGE CONFIGURATIONS

Until now, we have mainly discussed how the conventional discrete solutions for current sinking and sourcing have to be modified when working with monolithic bipolar technology. IC structures also require the availability of more control circuitry to improve power-stage performance, which will be analyzed in their classical forms in this section, which will cover complex structures using more than one power element.

On the other hand, IC configurations are extremely sensitive to parasitic elements; thus we will also discuss the particular care which must be given in order to avoid their effects, where they have not already been suppressed.

10.4.1 Half-Bridge Structure

A typical half-bridge configuration is shown in Fig. 10-10 [3]. This circuit operates as follows:

1. In standby, transistors T_1 and T_2 are off.

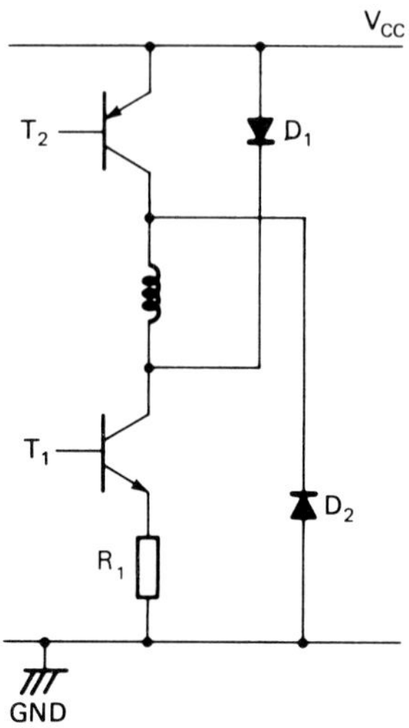

FIG. 10-10 Half-bridge structure.

2. In conduction, T_1 is on while T_2 switches with a duty cycle sufficient to give the desired current in the load. Monitoring of this current is always available on R_1, regardless of the phase of T_2.
3. Recirculation of current in the switching phase occurs on D_2 and therefore with low reverse voltage on the coil L (= 3 V). This guarantees a low duty cycle for transistor T_2, that is, low power losses in the IC, if D_2 is external.
4. At turn-off, both T_1 and T_2 are switched off. The energy accumulated in L is thus returned to the battery via diodes D_1 and D_2.
5. If the above structure is realized using an NPN Darlington for T_1 and a compound Darlington as shown in Fig. 10-6, there are no fundamental problems linked to parasitic transistors. The only part of the circuit that has to withstand negative voltages is the collector of T_2, which in effect consists of an NPN emitter and is therefore not able to inject current into the substrate.

The obvious complexity of this stage, which needs two power transistors to drive a load with a unipolar current, is counterbalanced by significant advantages particularly important for its realization in integrated form. In fact with the same supply voltage it enables particularly high current-switching speed and efficiency to be obtained. This means that in integrated form with 50-V technology, performance can be obtained that is equivalent to discrete transistors having V_{CE} sustaining of 100 V.

10.4.2 Push-Pull Amplifiers

Using switching techniques, a push-pull amplifier can in theory be obtained by connecting two complementary stages driven in opposition. Once the most appropriate configurations for the NPN and PNP stages in integrated form have been determined, it may appear simple to realize a stage of this type.

Nevertheless, as discussed in Sec. 10.3.4, it is necessary to avoid the phenomenon of simultaneous conduction of the two complementary transistors, if the efficiency of the system is to be kept high and stressful situations for the transistors avoided. The design of good class B amplifiers is therefore based on research in solutions for driving the final-stage transistors at nonoverlapping times, in order to avoid any possibility of simultaneous conduction.

A first solution [2] to the problem is presented in Fig. 10-11, where in particular it is applied to a bridge structure. Let us suppose that at point V_0 we have a triangular wave; this signal is compared with two different thresholds obtained from the same voltage level V_{TH}, from which it is subtracted. Control pulses are thus obtained for transistor pairs T_1, T_4 and T_2, T_3 with no overlap, as shown in Fig. 10-12.

There is a time τ, commonly defined as dead time, in which all four transistors are off. This time is a function of the voltage ramp slope and of the offset between the two thresholds. This time must certainly be greater than the switch-off time of the saturated transistor in order to avoid simultaneous conduction phenomena.

On the other hand, it cannot be too high, as it would cause problems of linearity and control dynamics in the dc load-current conversion. A typical value for integrated circuits is $\tau \simeq 3.0\ \mu s$.

This solution, based on circuit tricks, has proved to be very effective in systems using a local oscillator: it has the advantage of being programmable and can therefore, by adjusting τ, be adapted to all transistors. On the other hand, it involves the comparison between two completely uncorrelated times, one linked to an oscillation frequency (dead time τ) and the other represented by the storage time of a power transistor. Furthermore it requires the presence of an oscillator (with a specific waveform) at the power stages. Therefore from the point of view of integration it would seem more promising to adopt a solution based on one of the basic advantages of integrated technologies, i.e., coupling between components of the same type. Moreover, this solution is very interesting in that it does not take into account the driver waveform, which can therefore be considered as a simple digital signal.

A typical example of a solution based on this latter concept is shown in Fig. 10-13 [4]. The input stage consists of an analog comparator with a threshold of 1.75 V. The upper transistor is controlled in conduction

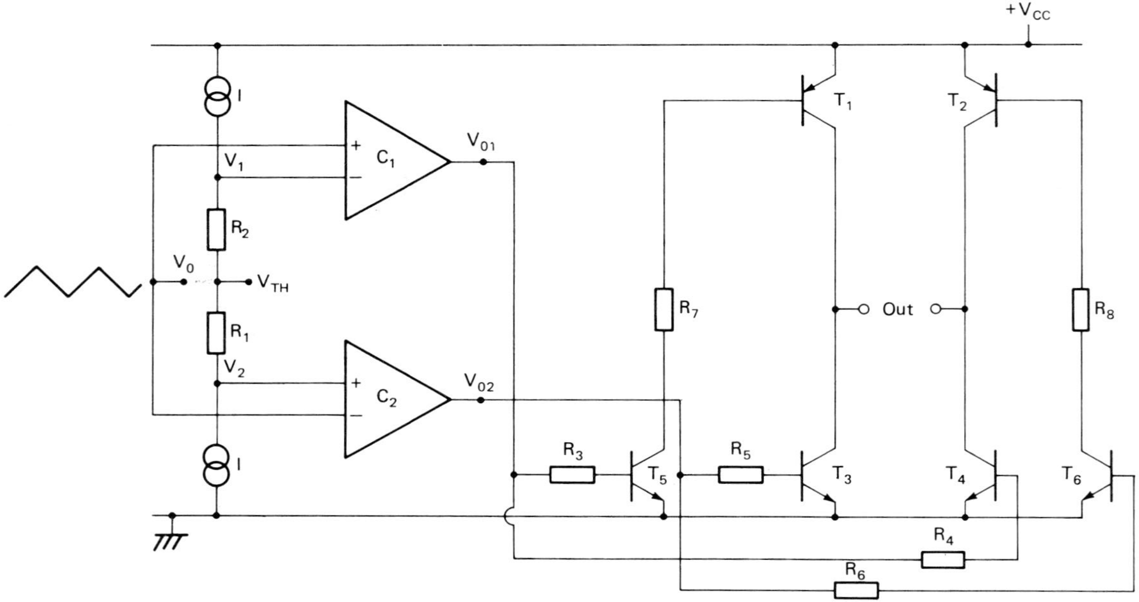

FIG. 10-11 Bridge-structure delayed drive.

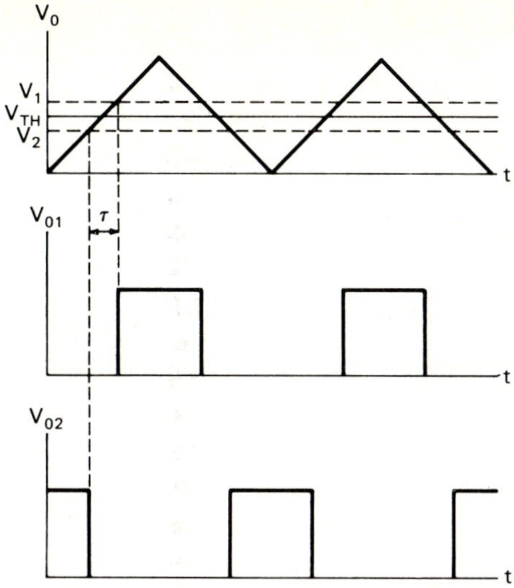

FIG. 10-12 Waveforms of the circuit shown in Fig. 10-11.

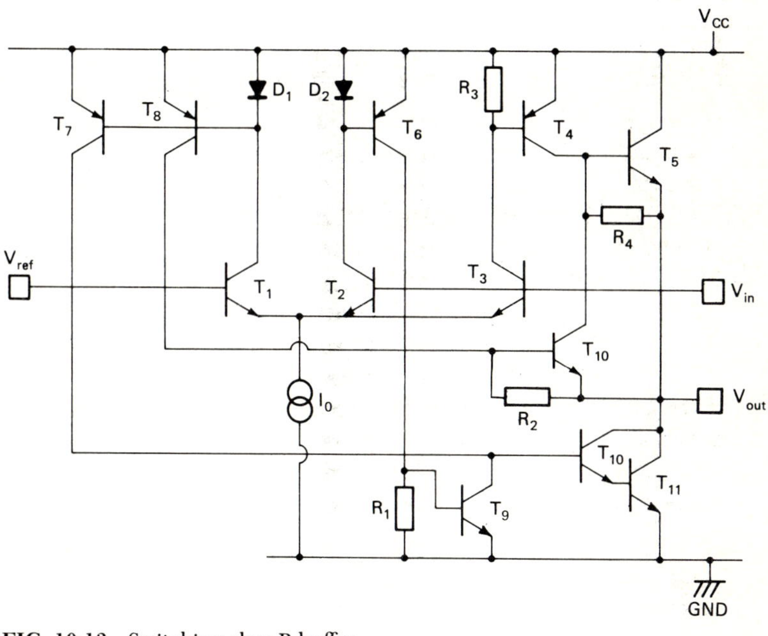

FIG. 10-13 Switching class B buffer.

via T_3 and T_4 and in switch-off by T_8 and T_{10}. The lower transistor is controlled at switch-on by T_7 and at switch-off by T_6 and T_9. The current capability of the switch-off transistors (T_9, T_{10}) is higher than that of the transistors which control switch-on (T_4, T_7).

Let us consider the high-low transition. With V_{in} high, T_4 and T_9 are saturated; the high-low input transition causes T_8 and T_{10} to switch on.

As we said, T_{10} has a current capability greater than T_4, so T_5 is immediately switched off, regardless of the storage time of T_4. On the contrary, T_7 can force Darlington pair T_{10} and T_{11} into saturation only after the storage charge on T_9 has been discharged. In this way T_5 is switched off before T_{11} can conduct.

During the low-high transition, transistors T_{10} and T_9 exchange roles: T_9 immediately switches off the lower stage, while T_5 can come on only after the stored charge on T_{10} has been discharged.

Obviously the dimensioning of resistors R_1 and R_2 plays an important role in defining the storage times of transistors T_9 and T_{10}, which form the safety margins on which the circuit operates.

The solution of Fig. 10-13 allows delay times on resistive loads of less than 1.0 μs with load current rise and fall times of less than 200 ns, without the use of any components outside the integrated circuit. This solution is excellent in terms of both performance and cost: in fact, with this type of configuration it is possible to switch a load of 2 A, 50 V—i.e., equal to 100 W of output power—using a silicon area of 4000 mil^2 and with a conversion efficiency of nearly 80 percent for chopper frequencies greater than 50 kHz. In any case efficiency is not limited by losses due to simultaneous conduction of the final transistors.

10.4.3 Full-Bridge Structures

For driving inductive loads one of the most attractive solutions is the bridge structure shown in Fig. 10-14, where resistor R_1 forms the sensing element for the current flowing into the load. This structure offers considerable advantages:

1. It enables the use of a single supply to obtain bipolar current in the load.
2. It avoids any problems of energy transfer from one power source to another as may happen when a system with dual supply voltage is used.
3. It requires transistors with maximum operating voltage equal to V_{CC}, as opposed to structures with dual supply, which for the same currents and voltages applied to the load require operating voltage of $2V_{CC}$. This is particularly important for integration purposes.

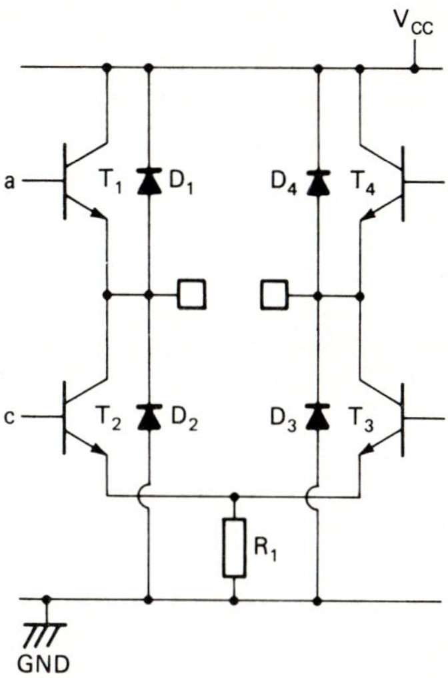

FIG. 10-14 Bridge configuration.

For all these reasons, bridge structures are widely used in monolithic solutions. In general, different methods of operation can be identified. In the most general case, transistors T_1 and T_3 of Fig. 10-14 are considered to be controlled in conduction in one phase with T_2 and T_4 off. In the complementary phase the situation is inverted: T_1 and T_3 are off and T_2 and T_4 are on. The direction of the current in the load (usually inductive) is determined by the predominance in time of one phase over the other. Good linear performance is also obtained for low current values, corresponding to a duty cycle of around 50 percent.

Due to simultaneous conduction of the stages in class B and to current-voltage crossing on the power transistors at turn-on and turn-off, dissipation problems are at their highest level. If the current is unipolar for a long period of time, we can decide to keep T_3 on and T_4 off permanently and to switch pair T_1, T_2 at high frequency, thus reducing the risk of simultaneous conduction by half. It is also possible to maintain the load with zero voltage at its terminals but with a low drive impedance in standby condition.

Generally in this condition transistor T_2 does not conduct, so the most efficient solution seems to be the following: T_2 and T_4 always off, T_3 always conducting, and T_1 switched at high frequency to obtain the desired value of current in the load. In this way there is no risk of simultaneous conduction. Current continuity is guaranteed by the recirculation diodes.

Obviously the situation described can be changed into a completely symmetrical one with the current flowing in the opposite direction in the inductance.

This configuration has only one potential problem: with low current values in the load, at least one load terminal may be terminated with a high impedance, making it impossible to define a precise forcing voltage. Normally this does not cause any practical problems, in particular in the case of stepping motors, for which there is always a current in the coils even when there is no motor movement.

Current inversion does not give rise to any particular problem, since in the transient phase conduction of the recirculation diodes is involved.

In the case of bridge structures, as previously, the battery voltage can be used to force current switch-off, thus increasing the possible operating speed.

10.5 APPLICATIONS

10.5.1 Driving DC Motors

DC motors represent the most widely used solution for speed and/or position control of moving parts, especially because of their energy conversion efficiency and overall cost (even if more control electronics are needed). Figure 10-15 shows the equivalent diagram of a dc motor characterized by an electrical time constant (or armature) equal to L_M/R_M and a mechanical time constant (J/F) linked to the inertia and friction of the motor and the load.

One of the most common configurations uses current driving of the motor to make the electrical time constant negligible, to accelerate system actuation, and finally to simplify design of a speed (or position) con-

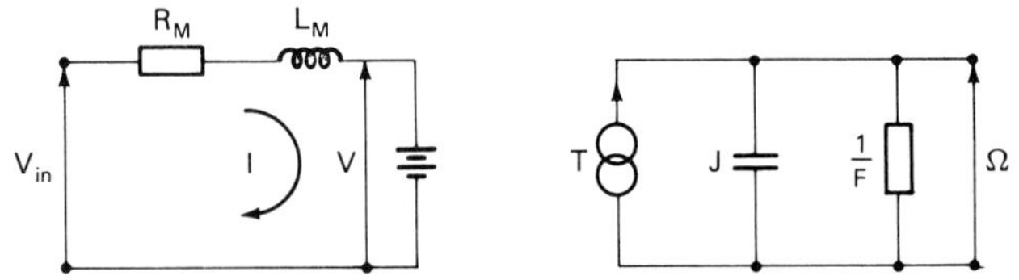

$T = k_M I$: Motor torque $V = k_V \Omega$: Induction voltage

J : System inertia   F : System friction

Ω : Velocity $\tau_{EL} = L_M/R_M$ $\tau_M = J/F$

FIG. 10-15 DC motor equivalent circuit.

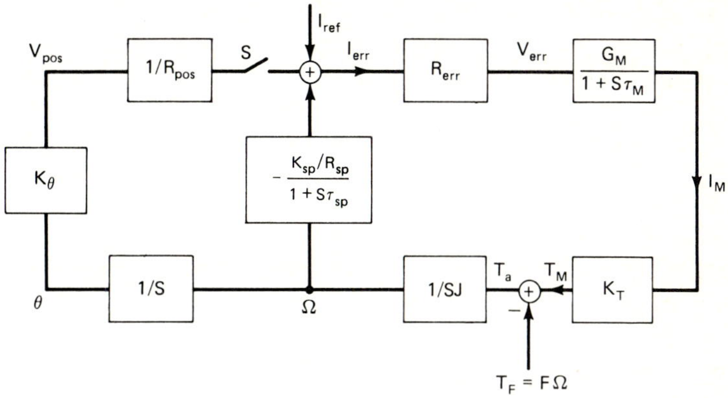

FIG. 10-16 Speed and position loop-block diagram.

trol loop. Figure 10-16 [2] shows a typical example of speed (and position) control, based on current driving of a dc motor. With switch S open, speed control is implemented; the reference current I_{ref} is compared with a tachometrical signal equal to $\Omega K_{sp}/R_{sp}$: the error signal thus obtained is transformed into voltage on the resistor R_{err} and then drives a transconductance amplifier, which forces a current I_M on the motor. Using the torque constant K_T, an accelerating torque T is obtained on the motor, which causes an acceleration

$$\frac{d\Omega}{dt} = \frac{T}{J} - F\Omega$$

where J is the moment of inertia of the load, while $F\Omega$ represents the torque due to friction. The speed analog variable Ω is then translated into a tachometrical voltage V_{tach}, which is fed back to close the loop. With switch S closed, position control is added in parallel to the speed loop.

The integrated analog variable Ω gives the position information θ. This is translated into voltage with a suitable constant K_θ. The feedback information is generated via resistor R_{pos} and flows into the principal node of the system.

From this analysis it emerges that the only mechanical parameters affecting the design of this system are the motor torque constant K_T, the moment of total inertia J (rotor + load), and the friction coefficient F.

The dc motor armature circuit parameters do not enter into consideration. A completely integrated realization of this system, based on

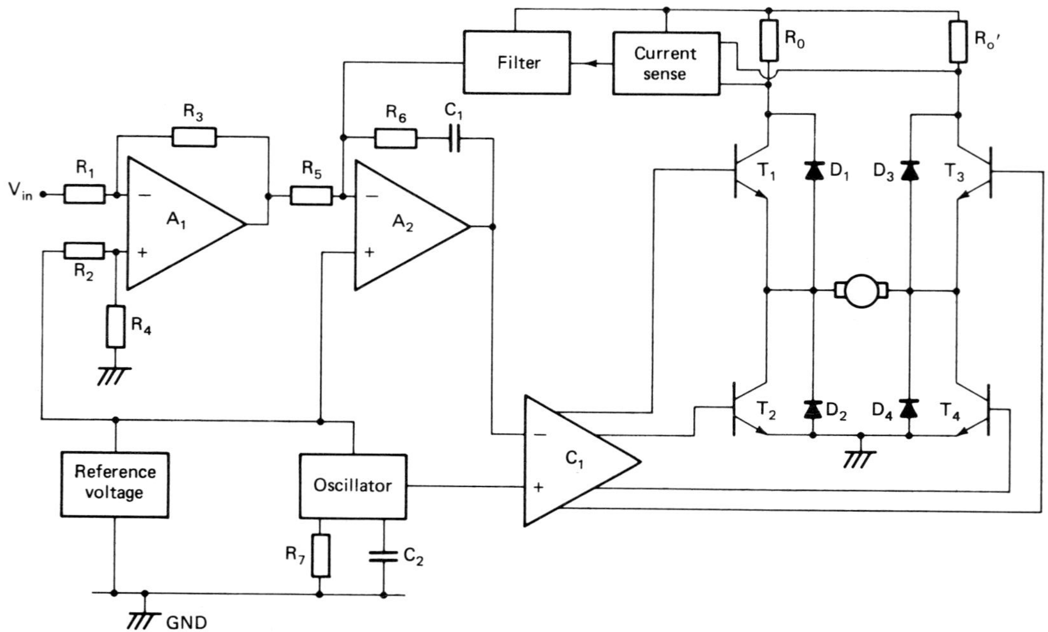

FIG. 10-17 L 292 block diagram.

three bipolar chips and an optical encoder, is presented in the references [2].

For power actuation, the key element in the system is a transconductance amplifier able to drive an essentially inductive load on wideband. It is possible to realize an integrated driver capable of providing a bipolar current of 2 A, from a supply voltage of 44 V, with a high power-conversion efficiency and active braking capability. Figure 10-17 shows the block diagram of such a device. A current-sensing stage reconstructs the analog information of the current circulating in the motor. This signal is compared with the input signal V_{in} to generate an error signal. This error signal is amplified and converted into a pulse-width-modulated signal with $f \approx 20$ kHz, which drives a bridge amplifier (H structure).

From a circuit point of view, the bridge structure is the most delicate part of the device. It is realized using a diagonal switching structure; in other words the transistor pairs T_1, T_4 and T_2, T_3 are controlled simultaneously. This is particularly useful in obtaining good linearity of the current-voltage conversion even around zero (no crossover distortion), in order to improve system stability.

Special care must be taken to avoid simultaneous conduction of transistors on the same branch of the bridge; in this case the problem was solved by setting a dead time during which all four transistors are off, as described in Sec. 10.4.2.

As regards fabrication technology and integration, special care must be given to the optimization of final-transistor performance in reverse second breakdown. The layout of the device must also be optimized in order to avoid dangerous injection of parasitic current when one of the lower branches of the bridge goes into recirculation, forcing the collector of T_2 or T_4 to a negative potential with respect to the substrate. In a situation of this type, a parasitic lateral NPN transistor is generated. This transistor affects the low-signal circuits on the chip (see Sec. 4.7).

The structure used in the integrated circuit shown in Fig. 10-18, which implements in silicon the block diagram of Fig. 10-17, solves this problem by placing transistors T_1 and T_3 in the layout between transistors T_2 and T_4, on the one hand, and the low-signal circuit, on the other. In fact, the collectors of transistors T_1 and T_3, basically connected to $+V_{CC}$, are able to collect most of the current that is injected toward the islands of the integrated circuit.

Another critical point of the circuit is the way in which feedback information on the current circulating in the motor is acquired. Since the current always runs diagonally along the bridge, the direct current flowing into the motor can be reconstructed by means of two equal resistors, in series with transistors T_1 and T_3 or with T_2 and T_4, con-

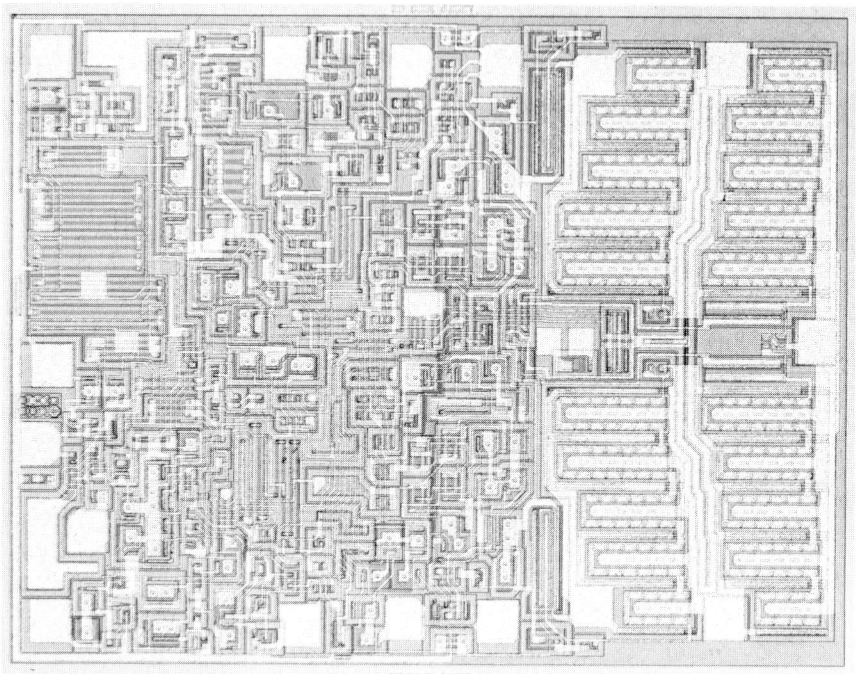

FIG. 10-18 L 292 chip microphotograph. (*Courtesy of SGS Microelettronica*)

nected to a differential amplifier. During the recirculation phases, however, either voltages higher than V_{CC} will be obtained if the sensors are placed in series with T_1 and T_3, or voltages negative with respect to ground will be obtained if the sensors are placed in series with the emitters of T_2 and T_4. The second solution is not acceptable, as it would force a voltage equal to $V_{BE} + IR = 1.2$ V on the collectors of T_2 and T_4, which would not be compatible with the presence of substrate collector diodes between the amplifier outputs and ground. It is therefore necessary to put the current sensors toward V_{CC}, as shown in Fig. 10-17. At this point a differential amplifier with input voltage range greater than the supply becomes necessary.

The solution is shown in Fig. 10-19. The input stage is obtained with the two emitters of the PNP transistors, T_1 and T_2, which can withstand voltages exceeding V_{CC}. An active feedback circuit brings the signal present on the sensing resistors back to points *A* and *B*, where it is converted into current and sent to an NPN current mirror which converts the differential signal into a single-ended signal.

The circuit configuration shown here makes it possible to realize a complete driver stage for current driving of dc motors up to 80 W,

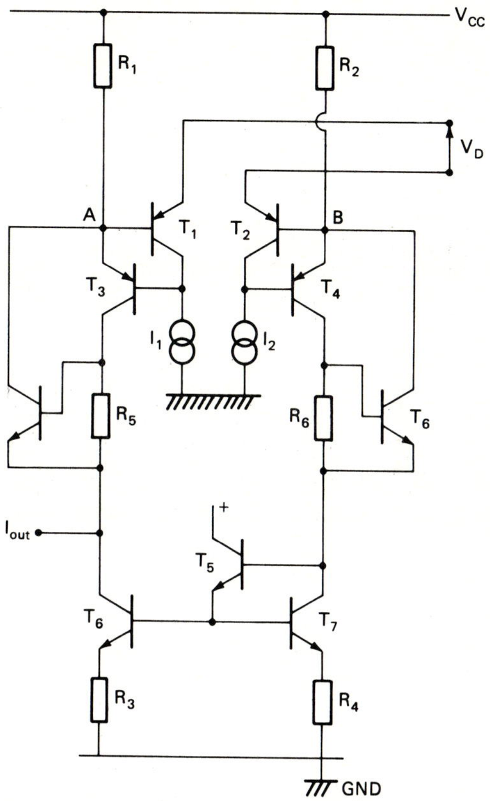

FIG. 10-19 L 292 current sense amplifier.

completely bipolar, in the sense that it allows inversion of motion and active braking. The complexity of the control part of the final stage makes it a perfect example of how much monolithic bipolar technologies are able to provide in terms of high-efficiency output power and accuracy of analog performance.

10.5.2 Solenoid-Current Controllers

Solenoid driving is one of the most common industrial electronics applications: the solenoid may take the form of a simple relay, an element of a needle or parallel printer, an electronic valve, or an electric injector for automotive applications, among other things. In each of these cases the analog variable to be controlled is basically the current in the load.

While it is difficult to define a useful power, it is important to con-

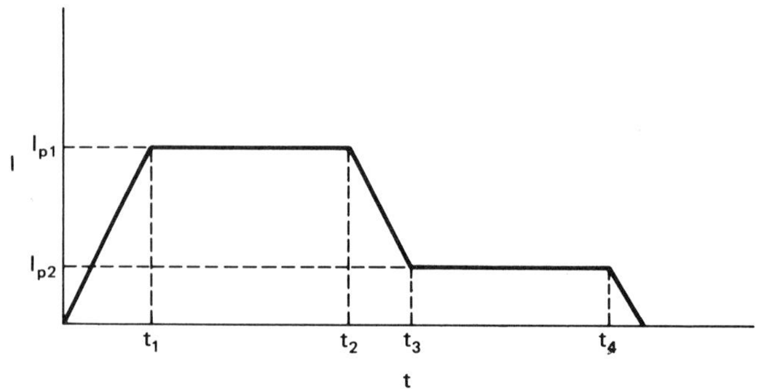

FIG. 10-20 Two-level current drive.

sider the overall efficiency of the system by reducing to a minimum the dissipation required to obtain control of the output variables. It is also essential to get high precision for the current in the load.

A switching-driver configuration is therefore recommended. This enables high supply voltages to be used to reduce the rise time of the current in the load. Considerable advantages can often be obtained by

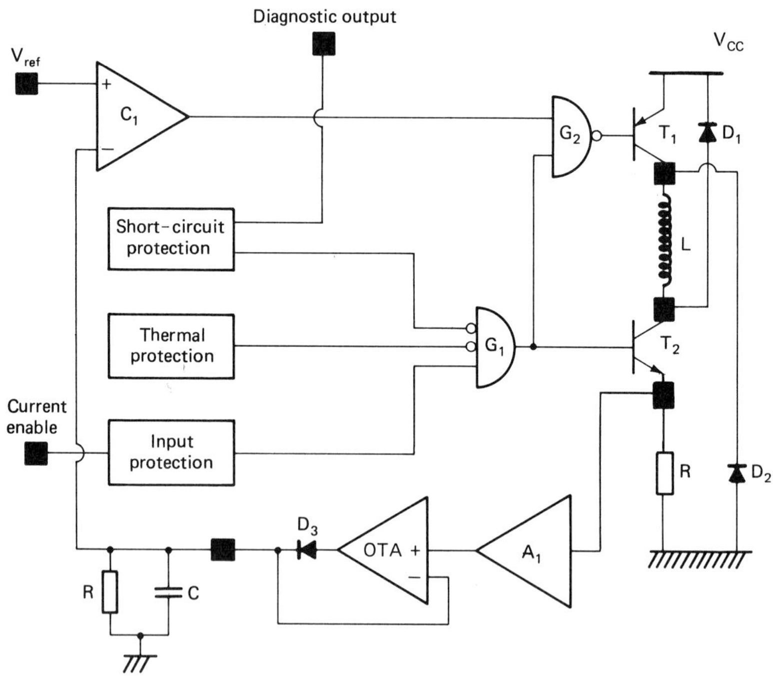

FIG. 10-21 L 294 block diagram.

selecting a suitable waveform for the current in the load. The most typical solution is shown in Fig. 10-20, which shows the conventional two-level control diagram, the first level for switch-on and the second for hold.

There are various configurations for use, so the drive systems must be as flexible as possible. As regards realization as an integrated circuit, this leads toward the choice of a programmable circuit in the sense that the current in the load can be controlled with an input voltage.

Figure 10-21 shows a block diagram especially suitable for this type of control. The design centers around a half-bridge output stage. As described in Sec. 10.4.1, it consists of two power elements, the first a sink which always conducts in the On phase and on whose emitter the current is sensed. The PNP stage, on the other hand, is switched to obtain regulation via the following mechanism: when the input signal goes high, both stages are in conduction, and the current in the load increases, as does the sensing voltage.

The sensing voltage, amplified and rectified, is compared with a threshold set from the outside; when this is exceeded the PNP transistor is switched off. The load current therefore recirculates through diode D_1. The voltage at pin 4, which followed the load current during the current increase phase, decreases with a time constant RC until it reaches the lower threshold of hysteresis comparator C_1.

At this point the cycle is repeated, giving rise to a new conduction phase of source transistor T_1.

If the current value set in the load is reduced, there is a long phase in which current flows across R, T_2, L, and D_2, then with low dissipation, as seen in the previous analysis of the half-bridge final stage.

At switch-off both power transistors stop conducting: the load current recirculates across diodes D_1 and D_2, so that the energy stored in inductance L is restored to the supply, contributing considerably to the energy balance of a system with high duty cycles, as may be the case with a printer or an electric injector in automotive applications. Furthermore this recirculation mechanism makes it possible to shut off the current in the load quickly, since the same voltage that forces the turn-on is used with inverted sign to force the current turn-off; this can improve the maximum frequency at which the circuit can be used.

Figure 10-22 shows a photo of SGS chip L 294, which is based on this operating principle; the chip includes, in addition to the basic function, special circuits for protection and diagnosis in the case of load short circuits (toward either ground or V_{CC}), as well as overtemperature [3]. This chip, assembled in a multiwatt package, is able to drive inductive loads up to a maximum current of 4 A using a supply voltage

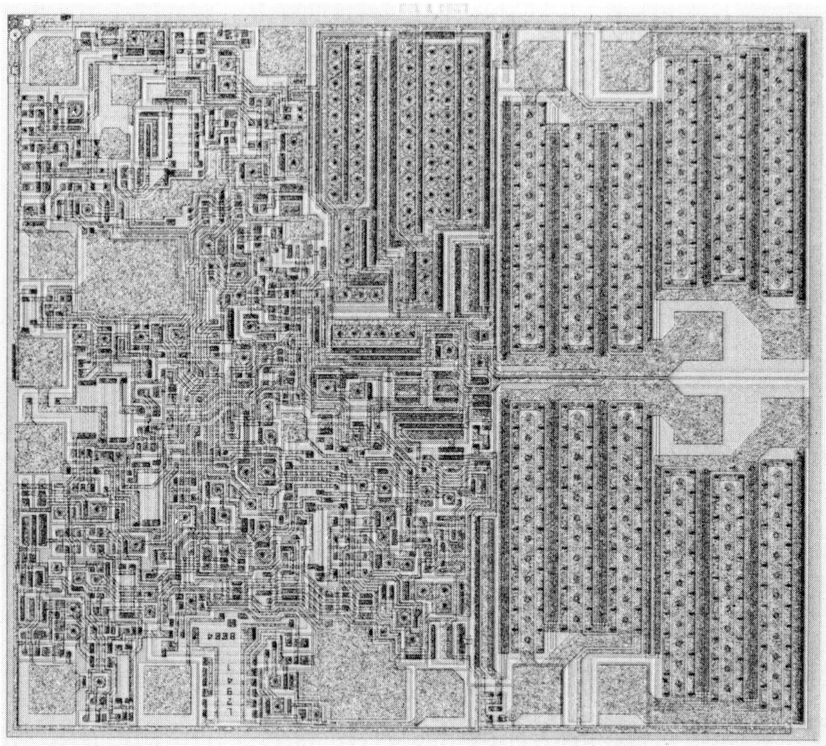

FIG. 10-22 L 294 chip microphotograph. (*Courtesy of SGS Microelettronica*)

between 12 and 50 V. It is important to note that an integrated circuit of around 10 mm^2 is able to drive a peak power of 200 W while providing accurate control of the load current to within a few percent of set value.

10.5.3 Stepping Motor Applications

The use of a stepping motor as a design basis for a position control is a natural choice, since it allows matching one-to-one the input control pulses and the angular output displacement. In most cases stepping motors are low-power applications, so a solution using integrated circuits is not only possible but offers considerable size and cost advantages.

The various types of command (full-step, half-step, progressive-wave, etc.) do not affect the drive circuit configuration, which, in fact, depends on the windings, whether they are unipolar or bipolar.

Unipolar Stepping Motor. In this case four coils have to be driven. These may have zero current or phase current present in one coil only or two simultaneously. Often the current in the On phase is limited by the resistance of the coil, so a switching driver is not necessary. On the other hand it is useful to have a high reverse voltage to obtain rapid shutoff of the current at switch-off.

Let us look at the drive circuit in Fig. 10-23. Assuming that V_{CC} = 40 V and the breakdown voltage of Z_1, or V_{Z1}, is 40 V, in order to have almost symmetrical switching sources at turn-on and turn-off, Darlington transistors with a sustaining voltage of more than 80 V are required. Normally the current is lower than 1 A (peak).

These structures can be well-integrated, as shown in Fig. 10-24, which shows a quadruple Darlington using a technology with isolation and sinker implants, as discussed in Chap. 4.

In Fig. 10-23 diodes D_5 and D_8 clamp the negative voltages induced by the magnetic coupling between coils on the same magnetic circuit in turn-off conditions.

The general solenoid driver presented in Sec. 10.5.2 is particularly suitable for driving unipolar stepping motors, providing all the advantages previously described. It does, however, require the availability of both ends of the four coils of a unipolar motor, which is not so common, especially in low-power motors. When possible, this solution presents obvious advantages because it allows dual–current level driving with-

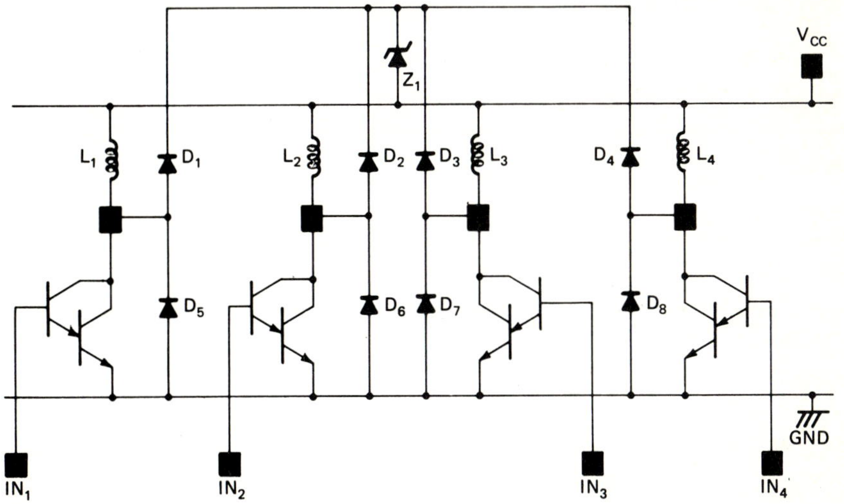

FIG. 10-23 Circuit for driving unipolar stepping motors.

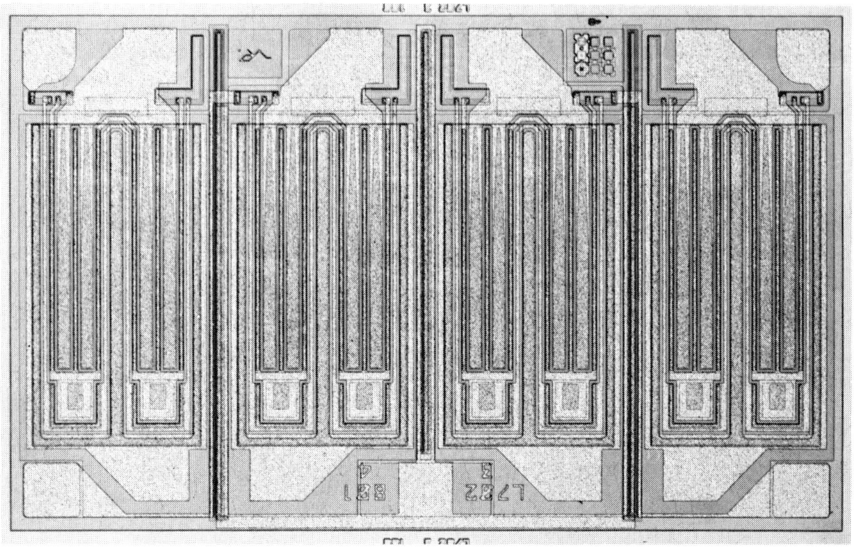

FIG. 10-24 L 702 chip microphotograph. (*Courtesy of SGS Microelettronica*)

out switching the supply voltage, which is the most popular method of reducing consumption with the motor in position.

Bipolar Stepping Motor. From an electromagnetic point of view, bipolar stepping motors provide the advantages of reducing the number of contacts required and in particular of reducing the power losses in the copper of the motor coils. They do, however, require a more complex drive circuit; this has often caused an increase in the cost of the drive electronics in solutions using discrete devices, so that in the end the application is rejected in favor of unipolar solutions.

To overcome this problem in recent years an integrated circuit of fundamental importance for these applications has been realized. This circuit is a quad TTL-compatible 50-V, 2-A push-pull buffer. It consists of four class B amplifiers based on the principles described in Sec. 10.4.2 to avoid simultaneous conduction. It is a very flexible device, suitable for single- and double-supply systems, relay drivers, and dc motors with clockwise and counterclockwise movement. It accepts loads connected to ground, $+V_{CC}$, or bridges. The application configuration which is of most interest here is shown in Fig. 10-25.

With one package it is possible to realize a double bridge which can control the two coils of a bipolar stepping motor. In this application it is necessary to realize three states of the motor current: positive, negative, and zero. The two inhibit signals, which can be applied to two

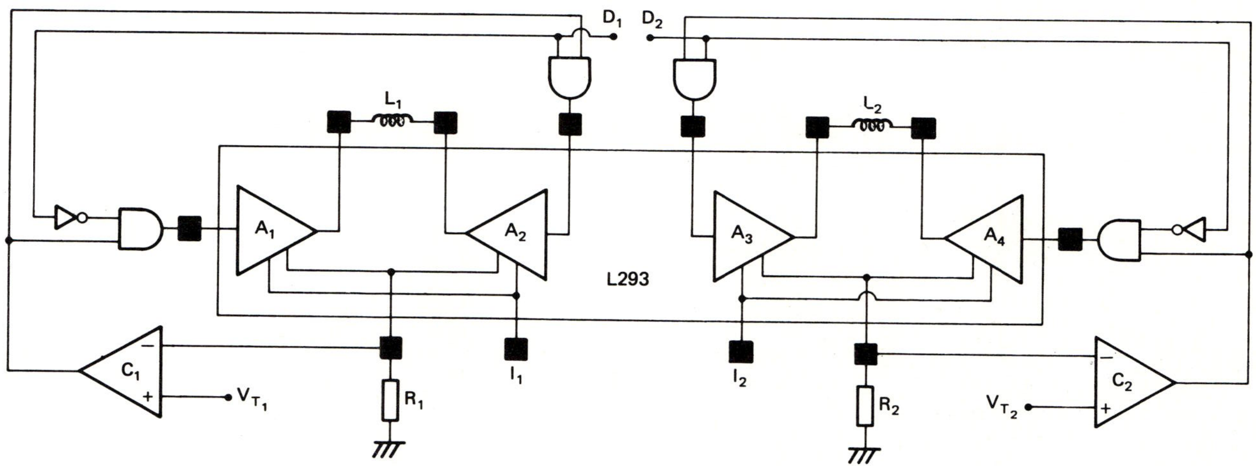

C_1, C_2 : Hysteresis comparators

D_1, D_2 : Current direction

I_1, I_2 : Channel 1 or 2 inhibit

V_{T_1}, V_{T_2} : Current threshold

R_1, R_2 : Current sense

FIG. 10-25 Quadruple push-pull driver application.

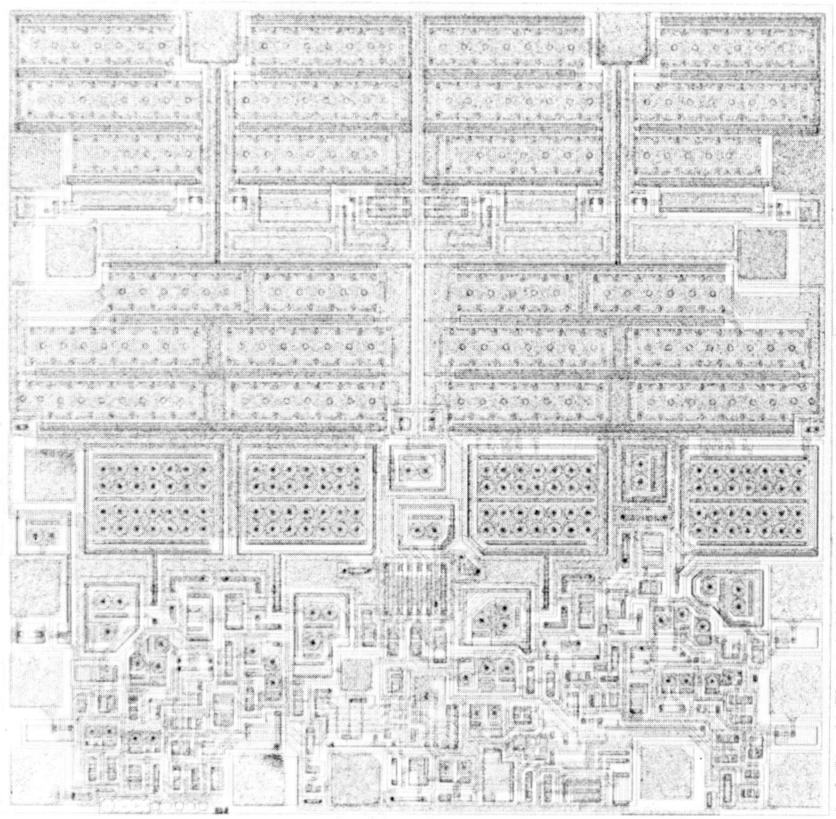

FIG. 10-26 L 293 chip microphotograph. (*Courtesy of SGS Microelettronica*)

pairs of buffers, are able to switch off the stages regardless of any other external command, thus realizing the zero current condition in the motor. If a current different from zero is present, the inhibit signal instantly causes recirculation of this current across the battery voltage. The simple conditioning logic in Fig. 10-25 provides switching control of the motor current level.

The signal diodes D_1 and D_2 set the current direction, for example, sending A_1 high and A_2 low; the level of current present in the load is sensed by resistor R_1 (or R_2). If the set current level is exceeded, a low level will be generated at the chopper input, causing low voltage at outputs A_1 and A_2. The desired current regulation is obtained through suitable hysteresis on comparators C_1 and C_2.

Figure 10-26 shows a photo of the chip whose schematic was shown in Fig. 10-25. Note that in an area of about 18,000 mil^2, the chip contains eight 2-A, 50-V transistors, yielding a current-voltage product in the chip of 800 VA; therefore it is by far the best-performing power integrated circuit ever produced until 1983.

10.5.4 Analog Current-Control Techniques

In some applications, especially when the supply voltage is low (e.g., the 14 V of a car battery), analog current-control circuits can be realized for reactive loads up to values of around 1 A, since the peak dissipated power, around 10 W, can easily be extracted from silicon using a good heat sink with a thermal resistance near 5°C/W, as described in Chap. 9.

Figure 10-27 shows a commonly used circuit diagram. If a voltage signal is sent to the V_{ref} input, current waveforms on two levels can be obtained as we saw in Sec. 10.5.2 (Fig. 10-19). An equivalent structure with load referred to ground can be used employing a compound PNP transistor as the power element.

As regards ruggedness, transistor T_1 of Fig. 10-27 must have good characteristics, especially in direct second breakdown; in general in this case the same design criteria as for the pass transistors of monolithic voltage regulators are used.

One of the most interesting applications of this kind is the control circuit for the vertical deflection current in cathode ray tubes. Starting from a relatively low breakdown voltage ($LV_{CEO} \geq 36$ V), it is possible to obtain a high discharge voltage on the inductance using the breakdown characteristics of the final transistors in BV_{CES} conditions and implementing a supply voltage, doubled with respect to V_{CC}, during the retrace phase.

The circuit diagram used is shown in Fig. 10-28: a voltage ramp at

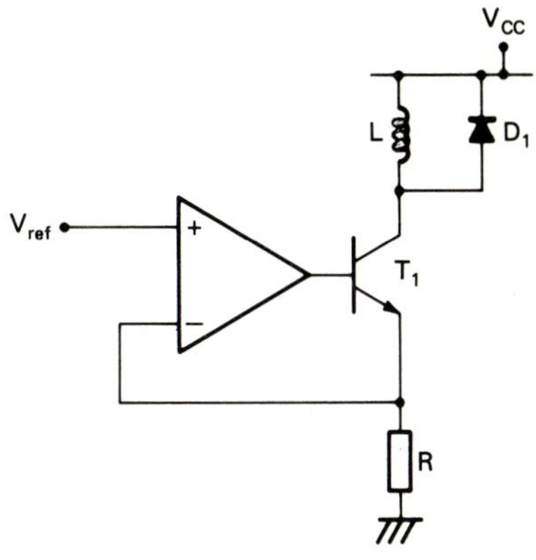

FIG. 10-27 Analog current control.

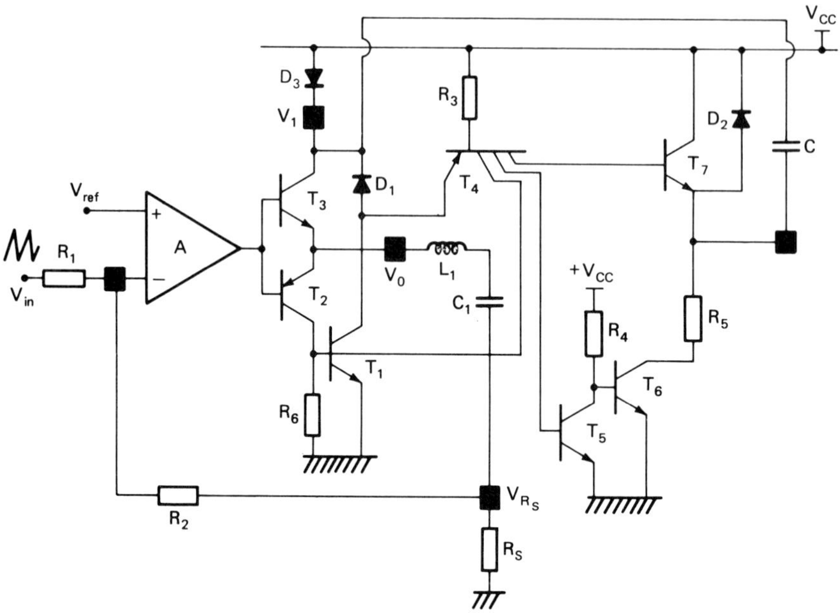

FIG. 10-28 Vertical deflection circuit diagram.

the input forces a similarly shaped current in the load, which is ac-coupled and inductive in nature (see Fig. 10-29).

At time t_1, the current in the load is at a maximum and flows into the amplifier, i.e., transistor T_2 of Fig. 10-28 is sinking the current. An instant after the input commutation, exactly the reverse current is required. In this condition the output voltage increases, triggering diode D_1, which conducts all the current in the load. At this point transistor T_4 comes on, thus taking the hold off the lower terminal of capacitor C_1, which is thus connected to $+V_{CC}$.

Transistor T_1 is now put into the BV_{CES} condition by means of T_4, and a voltage of around $2V_{CC}$ is available at point V_1, thus generating a good bias for the coil during the flyback phase. This voltage discharges the inductance in the first phase (D_1, C, and D_2 conducting) and rapidly inverts the current direction (T_3, C, and T_2 conducting). The moment the sensing current balances the feedback loop, the amplifier returns to the dynamic condition with an output voltage lower than V_{CC}. In this case T_6 turns off and capacitor C is charged to V_{CC} by D_3 and T_6. From now on, the circuit works as the one shown in Fig. 10-27, but with operation in alternating current and class B.

As we said, this principle has been applied successfully to vertical deflection circuits such as the TDA1170 and the TDA1670 [5], but it

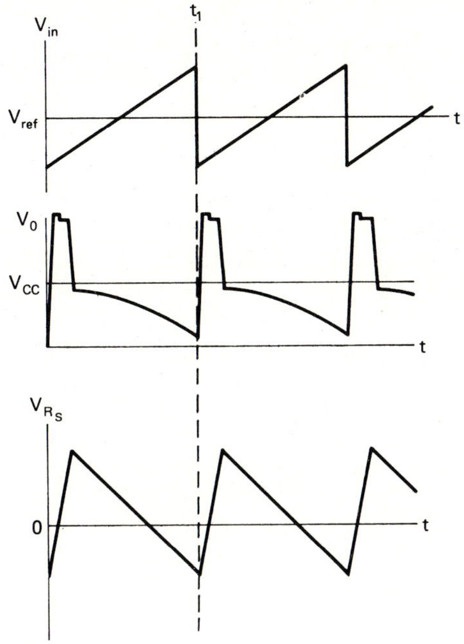

FIG. 10-29 Waveforms for the circuit shown in Fig. 10-28.

can be used whenever it is necessary that a high overvoltage invert a current rapidly. The fact that the load can be supplied in the normal conduction phase starting from a relatively low voltage makes it possible to keep the dissipation within acceptable limits, so that analog monolithic techniques can be used, as shown in the following example.

Example. Determine the supply voltage and the dissipation of a drive circuit of a vertical deflection yoke, such as that shown in Fig. 10-27, with an $L = 10$ mH, $R = 10\ \Omega$, and operating current of 2 A peak to peak ($f = 50$ Hz). The flyback time must be less than 0.1 ms.

The voltage excursion on the yoke during conduction is:

$$V_{pp} = RI_{pp} + L\frac{\Delta I}{\Delta t} \tag{10-21}$$

$$= 10 \times 2 + (10 \times 10^{-3})\frac{2}{20 \times 10^{-3}} = 21.0\ \text{V}$$

Considering the losses due to saturation, to the sensing resistor R_S, and to the inaccuracy of the bias circuit on the operating point, we can choose $V_{CC} = 26$ V.

The voltage applied during the retrace phase is around $\frac{3}{2}V_{CC}$, since one terminal of the yoke is at $V_{CC}/2$ while the other goes to $2V_{CC}$. Therefore

$$t_{\text{flyback}} = \frac{L\,\Delta I}{V} = \frac{10 \times 10^{-3} \times 2}{39} = 0.5 \text{ ms} \qquad (10\text{-}22)$$

which is well within the limits of the project.

Leaving aside the retrace phase, in which the current on the supply is on average zero, the circuit absorbs current from $+V_{CC}$ only during a semiperiod, when the current has a triangular waveform with peak value 1 A. Therefore the average value of the current during the period is 0.25 A.

The power absorbed by the supply, P_S, is therefore

$$P_S = 0.250 \times 26 = 6.5 \text{ W}$$

This is in addition to the power required by the circuit itself, estimated at about 1 W.

The power transferred to the load is in this case

$$P_L = \frac{2}{T}\int_0^{T/2} I^2(t) \times R\,dt = \frac{I_p^2 R}{3} = 3.3 \text{ W} \qquad (10\text{-}23)$$

The dissipated power is therefore

$$P_D = 7.5 - 3.3 = 4.2 \text{ W} \qquad (10\text{-}24)$$

Assuming that the maximum ambient temperature is 70°C, a heat sink that sets a total thermal resistance of 15°C/W between junction and ambient brings the maximum junction temperature to 133°C, well below the value of 150°C, which is considered the maximum for an integrated circuit in a plastic package.

REFERENCES

1. J. N. Harris, P. E. Gray, and C. L. Searle, *Digital Transistor Circuits,* Wiley, New York, 1966.
2. P. Erratico, A Complete Integrated Kit for Microprocessor Interfacing in Speed and Position Control of Motors, *Int. Microelectronic Cong.*, Munich, 1980, p. 230.
3. C. Cini, P. Erratico, and F. Stefani, A Monolithic, High Efficiency Solution for a 50V, 4A Switch-Mode Driver, *ESSCIRC '82 Dig. Tech. Papers.*

4. A. Alzati, F. Stefani, G. Copperi, and V. Pomo, A Power Interface Solution for Driving Stepping Motors, *PCI/Motorcon '82*, Geneva.

5. A. Romano and L. Venutti, A Monolithic Integrated Circuit for Vertical Deflection in Television Receivers, *Chicago Consumer Fall Conf.*, 1973; also *IEEE Trans. Consumer Electronics*, February 1975, Vol. CE-21, p. 85.

CHAPTER 11

Three-Terminal Voltage Regulators and Their Applications

A. A. Adamian
Fairchild Semiconductor
Mountain View, California

11.1 INTRODUCTION 11.2
11.2 INTERNAL CIRCUIT FEATURES 11.2
11.2.1 Voltage Reference 11.3
11.2.2 Error Amplifier 11.6
11.2.3 Feedback Elements 11.7
11.2.4 Series-Pass Element 11.7
11.2.5 Protection Circuits 11.8
11.2.6 Start-Up Circuit 11.11
11.3 CIRCUIT DESCRIPTION 11.11
11.3.1 Fixed-Positive Regulators 11.12
11.3.2 Fixed-Negative Regulators 11.16
11.3.3 Adjustable Three-Terminal Regulators 11.24
11.4 THREE-TERMINAL REGULATOR APPLICATIONS 11.30
11.4.1 Selecting the Correct Type of Regulator 11.30
11.4.2 Design Precautions 11.31
11.4.3 Positive and Negative Regulator Applications 11.35

11.1 INTRODUCTION

Considerable effort has gone into the development of three-terminal voltage regulators since the first fixed-positive three-terminal regulator was introduced in the early 1970s [1]. Three-terminal regulators have reached a mature stage in that today users have at their disposal positive as well as negative regulators covering output currents from the milliampere range up to several amperes. Fixed-output devices are available in popular output voltage configurations from 2.6 to 24 V, whereas adjustable devices can be configured for virtually any output level. On-card voltage regulation has gained wide acceptance because IC voltage regulators are economical and easy to use, because they are available in the popular transistor power packages such as TO-3, TO-5, TO-92, TO-202, and TO-220, and hence because they are easy to handle and can compete with designs using discrete components, and have added features and increased reliability besides. The protection features built into these regulators have simplified the design of the power supply in a system.

In conventional monolithic ICs, NPN transistors are far more efficient than PNPs and hence most regulators use NPN output stages. The use of an NPN output stage and the polarity of the regulators then dictate the output configuration. In general, positive regulators utilize emitter-follower configurations and are "current-sourcing" types, whereas negative regulators use common-emitter configurations and are "current-sinking" types. The output configuration must be kept in mind in certain regulator applications such as circuits requiring more than one supply or regulators operating with a battery or batteries.

11.2 INTERNAL CIRCUIT FEATURES

Figure 11-1 shows the block diagram of a typical series-type fixed three-terminal regulator. It consists of a voltage reference V_{ref}, an error amplifier A_α, feedback elements R_A and R_B, and a series-pass transistor Q_T. The error amplifier compares the reference with a fraction of the output voltage, derived from the resistive divider formed by the feedback resistors, and provides the necessary base drive to the series-pass output transistor to maintain an output voltage proportional to the reference regardless of changes in input voltage, output current, or ambient temperature. In addition to these basic regulator elements, three-terminal IC regulators have on-chip protection circuits—such as thermal shutdown, current-limit, and safe operating area (SOA) protection—which make the regulators immune to certain types of overloads.

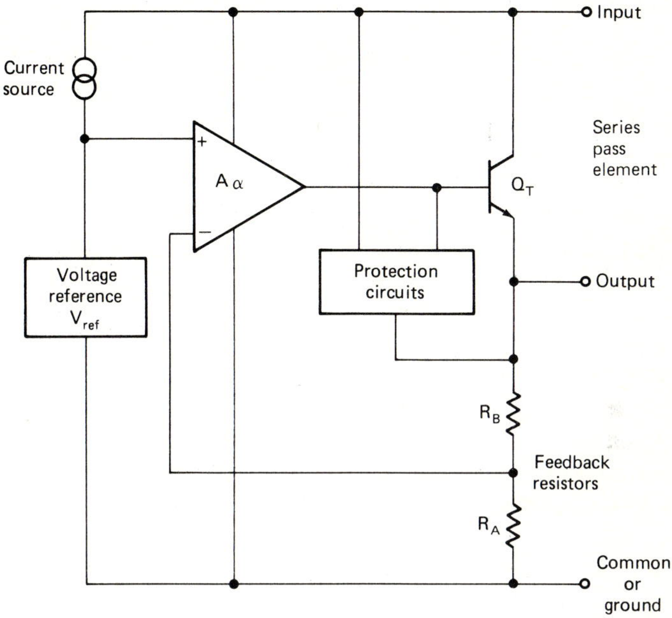

FIG. 11-1 Block diagram of a series-type fixed three-terminal regulator.

The inclusion of these protective features coupled with the ease of their use is one of the main reasons why three-terminal regulators have become so popular.

11.2.1 Voltage Reference

The reference is the heart of a regulator. Since any abnormality exhibited by the reference will be amplified by the error amplifier and seen at the output, the reference must be stable with time and temperature as well as with input-voltage and output-current variations. In addition, it must be free of noise. Band-gap reference has become the most widely used reference in IC regulators, although some popular devices do use zener-derived references.

There are various ways of obtaining a temperature-compensated zener reference, but the basic principles are the same (see Fig. 11-2). The technique is to compensate the positive temperature coefficient of a zener diode with the negative temperature coefficient of a base-emitter diode or diodes [2]. Zener diodes are normally base-emitter junctions of NPN transistors operating in the avalanche breakdown mode, typically 6 to 7 V. They are relatively easy to implement but suffer from

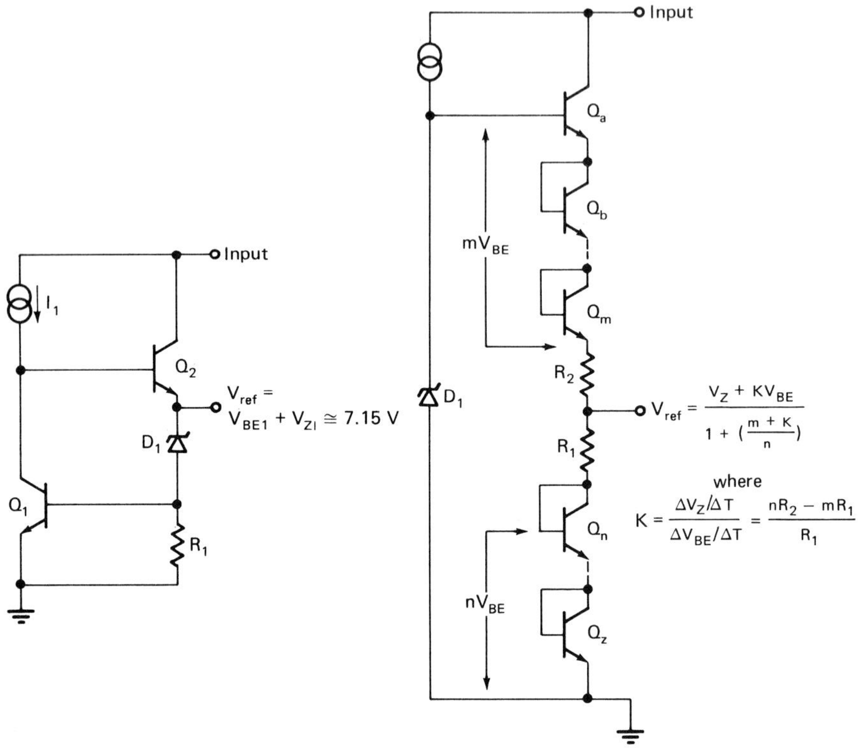

FIG. 11-2 Temperature-compensated zener references.

the following disadvantages: they require a minimum of 8 to 9 V for operation; they are noisy; and because the breakdown occurs at the surface, they are susceptible to surface charges and hence can drift. Noise and drift can be reduced, however, by using a subsurface breakdown zener, normally achieved by extra processing steps, such as ion implantation [3]. The band-gap reference, also referred to as the Widlar reference, overcomes these limitations and relies on the excellent matching and tracking characteristics achievable with integrated circuits [4, 5].

The band-gap reference, shown in Fig. 11-3, is derived from the predictable temperature, current, and voltage relationship in base-emitter junctions of integrated transistors. To obtain a temperature-compensated reference voltage, the positive temperature coefficient of the base-emitter voltage differential between two transistors, operating at different current densities, is added to the negative temperature coef-

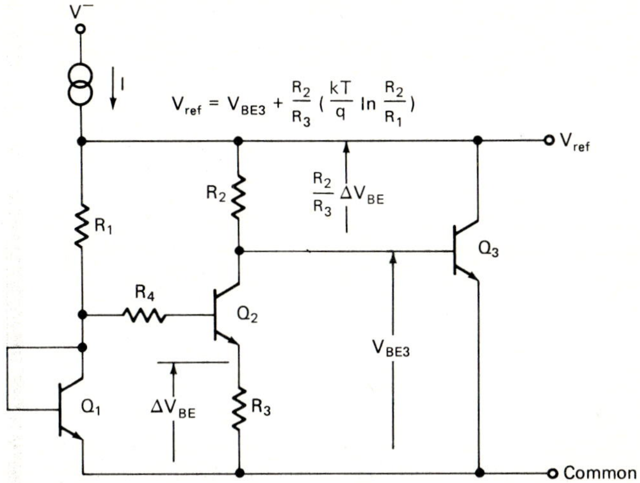

FIG. 11-3 Band-gap reference.

ficient of the base-emitter voltage. In Fig. 11-3, diode-connected transistor Q_1 operates at a higher current density than Q_2, typically 10 times. The base-emitter differential voltage ΔV_{BE} appears across R_3. Transistor Q_2 is a gain stage and thus develops a voltage drop across R_2 that is proportional to the drop across R_3. Transistor Q_3 provides additional gain and yields a reference voltage V_{ref} equal to the base-emitter voltage of Q_3 plus the voltage drop across R_2. The reference voltage can thus be expressed as:

$$V_{ref} = V_{BE3} + \frac{R_2}{R_3}\frac{kT}{q} \ln \frac{J_{C1}}{J_{C2}} = V_{BE3} + \frac{R_2}{R_3}\frac{kT}{q} \ln \frac{R_2}{R_1}$$

To obtain a temperature-stable reference,

$$\frac{\Delta V_{ref}}{\Delta T} = \frac{\Delta V_{BE3}}{\Delta T} + \frac{R_2}{R_3}\frac{k}{q} \ln \frac{R_2}{R_1} = 0$$

must be satisfied. $\Delta V_{BE3}/\Delta T$ is the negative temperature coefficient of the Q_3 base-emitter voltage, and $(k/q) \ln (R_2/R_1)$ is the positive temperature coefficient of the differential base-emitter voltage of Q_1 and Q_2. Therefore, by appropriately selecting the geometries of Q_1, Q_2, and Q_3 and the ratios R_2/R_1 and R_2/R_3, a temperature-compensated reference can be obtained [6]. Resistor R_4 is included to compensate for the transistor beta nonlinearity with temperature and errors due to base currents. First-order effects of the base currents can be minimized [7] by selecting

$$\frac{R_4}{R_3} = \frac{I_{B3}}{I_{B2}} = \frac{I_{C3}}{I_{C2}}$$

It should be obvious that in the design and selection of the components in the reference, matching and tracking of the transistors as well as the resistors is very critical. Placement of these components with respect to the heat source(s) on isotherms is equally important and normally is not as easily achievable because of complexities and interactions involved. As a result, IC regulator designers take great pains in the layout not only by placing the reference circuit as far as practical from the output stage, which is the major source of heat on the die, but also by arranging on isotherms the components that require matching. If these precautions are not taken the regulator will suffer from poor thermal regulation [8, 9].

11.2.2 Error Amplifier

The error amplifier in a regulator is a special operational amplifier used in a negative feedback mode where the reference is connected to the noninverting input while the feedback signal from the regulated output is divided through the pair of feedback resistors R_A and R_B (see Fig. 11-1). The output voltage in this feedback configuration is

$$V_{out} = V_{ref} \frac{\alpha}{\beta(1 + \alpha)}$$

where α is the amplifier gain and $\beta = R_A/(R_A + R_B)$ is the fraction of the output returned as feedback. Since the error amplifier has a relatively high gain, the output voltage gain equation can be rewritten as:

$$V_{out} = \frac{V_{ref}}{\beta} = V_{ref} \frac{R_A + R_B}{R_A} = V_{ref} \left(1 + \frac{R_B}{R_A}\right)$$

It is obvious that the lowest output voltage possible with this configuration is $R_B = 0$, which results with $V_{out} = V_{ref}$ (unity gain, or $\beta = 1$). To obtain a wide range of regulated output voltages, then, a low value of V_{ref} is an advantage. However, to obtain a high output voltage from a low reference, the loop gain beta must be low (i.e., the value of R_B must be high). As β decreases, i.e., as V_{out} increases, the regulator performance degrades, affecting such parameters as line regulation, load regulation, ripple rejection, and output impedance. The effects of line-voltage changes on the error amplifier can normally be minimized by biasing critical components from the regulated output.

The error amplifier can be a simple gain stage or can consist of several amplifier stages, depending on the desired regulator performance. Nevertheless, it requires compensation for stability. All three-terminal fixed regulators have on-chip compensation and, depending on the type or the application, may require additional external compensation for stability. Excessive compensation, however, causes the error amplifier gain to fall off at high frequencies, thus deteriorating regulator performance.

11.2.3 Feedback Elements

The simplest part of the regulator is the feedback network. It consists of resistors R_A and R_B (Fig. 11-1), whose sole purpose is to sample the output voltage and provide the feedback necessary for the error amplifier to maintain a constant output voltage. In the layout of the IC, the feedback resistors, just like the reference components, are kept on isotherms to avoid thermal problems. In the manufacture of three-terminal fixed regulators, resistor R_B normally consists of a matrix of resistors matched to R_A. The metallization pattern is normally "programmed" to pick the correct value of R_B to give the desired output voltage, thus avoiding the need for a different mask set for each output voltage configuration. Note that the feedback resistors can be external to the IC if the inverting input of the error amplifier is made available to the user to select the output voltage. Some three-terminal regulators are offered in this configuration as adjustable devices, but in four-lead packages.

11.2.4 Series-Pass Element

The output stage in a regulator is normally a Darlington-connected pair of transistors. The size of the series-pass transistor is determined by the output-current and input-voltage requirements of the regulator. Once the output-current requirement of a regulator approaches 1 A, the output transistor and its metallization pattern become the major contributors to the die size, as can be seen on the die photomicrographs in this chapter. One of the benefits of including the series-pass transistor on the chip is that the circuit thus needs only three terminals and, hence, an ordinary, inexpensive transistor package can be used. Another important advantage is that thermal as well as current overload protection can be incorporated on the chip in close proximity to the series-pass device.

Since the regulator must survive all sorts of fault conditions, considerable thought must go into the design and layout of the output geometry. Care must also be taken in the placement of the critical regulator components, such as the reference, error amplifier, and feedback resistors, with respect to the output stage to avoid performance degradation due to thermal gradients across the chip. For optimum design of the output geometry, the peak-current and safe operating area requirements of the regulator must be taken into consideration (see Chap. 2). Of the various geometries used on power devices, interdigitated base and emitter structures have become the most commonly used in IC regulators because of their large emitter periphery and reduced parasitic elements. Regardless of the structures used, all designs utilize emitter degeneration or ballast resistors in series with the emitters. These resistors, which are made up of an N^+ emitter diffusion, help equalize the currents through the individual emitter sites and thus prevent localized heating that may destroy the device. The voltage drop across one of these degeneration resistors is normally used as part of the sense circuit for short-circuit protection, as will be explained in the next section. Since these resistors are in series with the emitters, they increase the dropout voltage. The decision the IC designer is faced with is a compromise between improved safe operating area and a degradation in the dropout voltage, load regulation, and output impedance.

11.2.5 Protection Circuits

Protective circuits (Fig. 11-4) are added on-chip to improve reliability and to make regulators immune to certain types of overloads. These features, which have made monolithic regulators so popular, require relatively small die area and yet protect the regulator against short-circuit conditions (through current-limit protection), against excessive input-output voltage differential conditions (safe operating area protection), and against excessive junction temperatures (thermal-limit protection) [10].

a. Current-Limit and Short-Circuit Protection

The most commonly used protection scheme is current limiting for guarding the output series-pass transistor against excessive output currents or short-circuit conditions (see Fig. 11-4). With low input-output conditions when zener diode D_1 is not conducting, there is no current flowing through R_1. The current-limiting transistor Q_2, then, senses the voltage drop across the current-limiting resistor R_3. As the output current increases, the drop across R_3 increases, and as a result the Q_2 base-emitter voltage increases until Q_2 begins to conduct, thereby removing

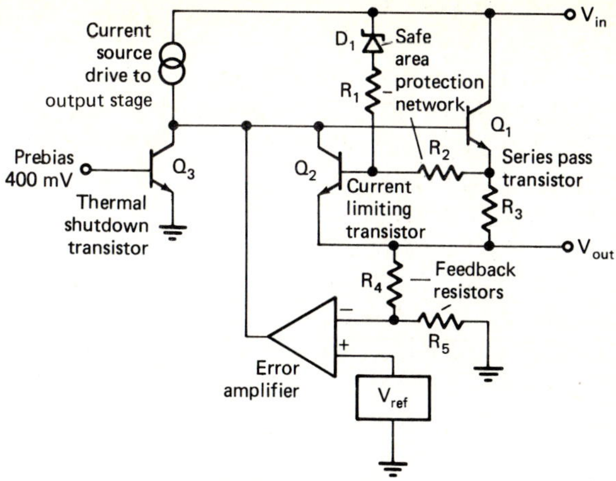

FIG. 11-4 Basic regulator protection circuits.

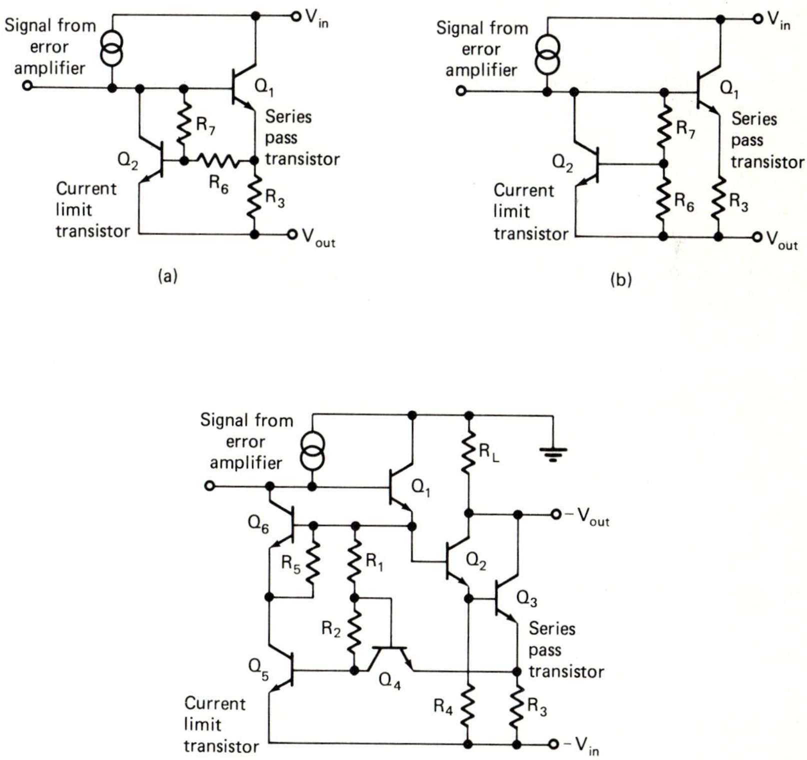

FIG. 11-5 Current limit circuits.

the base drive of the series-pass transistor Q_1. No additional output current can be pulled out, since any increase in the output current will cause Q_2 to conduct "harder."

This current-limit circuit has a slight disadvantage: the voltage developed across R_3 adds to the regulator dropout voltage and degrades the load regulation and output impedance. A simple way of minimizing the effects of this problem is to prebias the base of the current-limit transistor Q_2 with a fraction of the base-emitter voltage of the series-pass transistor Q_1 via R_6 and R_7, as shown in Fig. 11-5*a* and *b*.

In those current-limit schemes discussed so far, the output current decreases with increasing temperature since the base-emitter voltage of the current-limit transistor Q_2 is the threshold for preventing the flow of additional regulator output current. Temperature-independent short-circuit current can be achieved with a slight increase in circuit complexity. Figure 11-5*c* shows the scheme used in fixed-negative and adjustable regulators. At low to medium output current levels, Q_4 is on and Q_5 and Q_6 are off. When the voltage drop across R_3 reaches a predetermined level, Q_4 begins to turn off and current is diverted to Q_5, which in turn causes Q_6 to turn on. The Q_1 and consequently Q_2 base currents are thus diverted and the output current is prevented from increasing any further. By proper choice of R_1 and R_2 and of the emitter area ratio of Q_4 and Q_5, the peak current is set to be independent of temperature.

b. Safe Operating Area Protection

Safe area protection is included in IC regulators to protect the series-pass transistor against excessive power dissipation by reducing the collector current as the collector-to-emitter voltage is increased (Fig. 11-4). When the input-output voltage differential exceeds the breakdown voltage of zener diode D_1, current flows from the input to the output through D_1, R_1, R_2, and R_3. The voltage drop across R_3 and, therefore, the Q_2 base voltage become a function of not only the output current but also the input-output voltage differential. Hence, maximum output current is available when the input-output voltage differential is less than the D_1 breakdown voltage. The safe area protection network reduces the available output current I_{out} as the input-output differential $V_{in} - V_{out}$ increases. This rate of reduction (in amperes per volt) is determined by

$$\frac{\Delta I_{out}}{\Delta(V_{in} - V_{out})} = -\frac{R_2}{R_1 R_3}$$

which is the slope of the safe area protection network. The safe area protection network thus reduces the available output current as the input-output differential increases and limits the regulator operation to within the safe operating area of the series-pass transistor. Note that the safe area protection network is lumped in with the short-circuit protection network and consequently both have the same temperature characteristics.

c. *Thermal Overload Protection*

A regulator consisting of discrete components usually relies on current limiting for overload protection since there is no practical way to sense junction temperature in a separate series-pass transistor. The dominant failure mechanism of this type of regulator, then, is excessive heating of the series-pass transistor. In monolithic regulators, the base-emitter junction of a transistor placed as close as practical to the series-pass transistor is used to sense chip temperature. The thermal shutdown transistor Q_3, in Fig. 11-4, is normally biased below its activation threshold so that it does not affect normal operation of the circuit. However, if the chip temperature rises above its maximum limit as a result of an overload, inadequate heat sinking, or another condition, the thermal shutdown transistor turns on, removing the base drive to the output transistor Q_1 and shutting down the regulator to prevent any further chip heating.

11.2.6 Start-Up Circuit

Whenever practical, current sources for the reference and/or error amplifier are derived from the regulated output to improve regulator performance, and as a result these elements may not be self-starting when power is first applied to the regulator or when the output is shorted to ground. Simple start-up circuits are therefore included to initiate and if necessary to maintain the flow of current independent of output condition.

11.3 CIRCUIT DESCRIPTION

Three-terminal regulators have become so popular that most major linear-IC manufacturers now offer a variety of devices, both positive and negative and in both fixed and adjustable configurations, and with output currents of 100 mA, 500 mA, 1 A, 3 A, and 5 A. The fixed regulators are available in six to eight popular output voltage options from 2.6 to 24 V. The adjustable three-terminal devices have references of 1.25 V.

It would be impossible to cover in this chapter circuit details of every device available. Just as in most other ICs, of the few designs that have made it to the marketplace, normally one design has become the most popular and has been second-sourced by the other linear-IC houses. The internal circuitry of three-terminal regulators is relatively simple and it should be obvious that similarities as well as commonalities do exist not only between devices offered by the various manufacturers but also between the different families offered by the same manufacturer. For example, the positive 0.5-A μA78M00 series is a scaled-down version of its predecessor, the 1-A μA7800 device; similarly for the negative μA79M00 series. A close examination of the three-terminal adjustable 317, 317L, 317M, 350, and 338 will reveal an almost identical control circuit and a corresponding output-geometry scaling for the rated output current. In the descriptions that follow, some of the most popular devices will be covered in detail, with die photomicrographs showing some of the similarities.

11.3.1 Fixed-Positive Regulators

The most popular three-terminal positive regulator is the μA7800 series of 1-A regulators, schematically shown in Fig. 11-6. Transistors Q_1 through Q_7 and their associated resistors constitute a temperature-compensated band-gap reference of 5 V. The Darlington configuration of Q_3 and Q_4 is the error amplifier lumped into the voltage reference. The error amplifier gain is increased by the vertical PNP transistor Q_{11}, which acts as a buffer to drive the active collector load formed by the PNP current-source transistor pair Q_8 and Q_9. Capacitor C_1 is included on the chip for stability. Note that the voltage across C_1 will be equal to the output voltage less a V_{BE} drop and therefore it must be a thin or thick oxide capacitor but not a junction capacitor.

The current through Q_8 and Q_9 is set up by the current through resistor R_1. During regulator turn-on, the current in R_1 flows first through transistor Q_{13}, which is part of the start-up circuit containing zener diode D_1, transistors Q_{12} and Q_{13}, and resistors R_5, R_6, and R_7. After the device is in regulation, Q_{13} is biased off and the regulator takes over setting the current in R_1, which then flows through Q_5 and Q_{10}.

Thermal protection is provided by transistor Q_{14}, the base of which is clamped by the resistive divider string R_5, R_6, and R_7 in the start-up circuit. At a junction temperature of about 175°C, Q_{14} turns on and removes the base current to the output stage, thus turning the series-

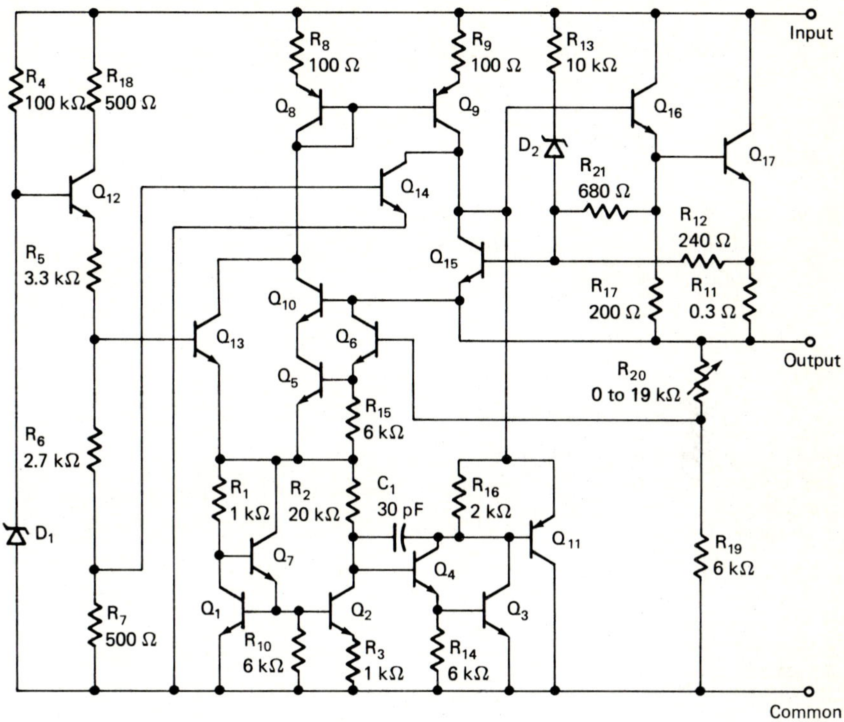

FIG. 11-6 μA7800 schematic diagram. *(Courtesy Fairchild Semiconductor)*

pass transistor Q_{17} off and preventing further increase in chip temperature.

The short-circuit protection transistor Q_{15} is prebiased from the base-emitter voltage drop of the series-pass transistor Q_{17} via resistors R_{12} and R_{21}. As the output current increases, the drop across R_{11} increases, causing the base voltage of Q_{15} to increase also. When the output current reaches the preset level, Q_{15} begins to turn on, shunting the base current of the output stage, and prevents any further increase in output current.

Safe area protection, accomplished via resistors R_{11}, R_{12}, and R_{13}, zener diode D_2, and Q_{15}, limits the instantaneous power through the series-pass transistor to a safe value by decreasing the maximum regulator current as the input-output voltage differential is increased.

The μA7800, available in TO-3 and TO-220 packages, is rated to deliver a minimum output current of 1 A at an input-output differential voltage of 15 V—that is, 15 W—at a junction temperature of 125°C. At lower junction temperatures, the device can deliver higher output currents as long as the power dissipation is kept at or below 15 W.

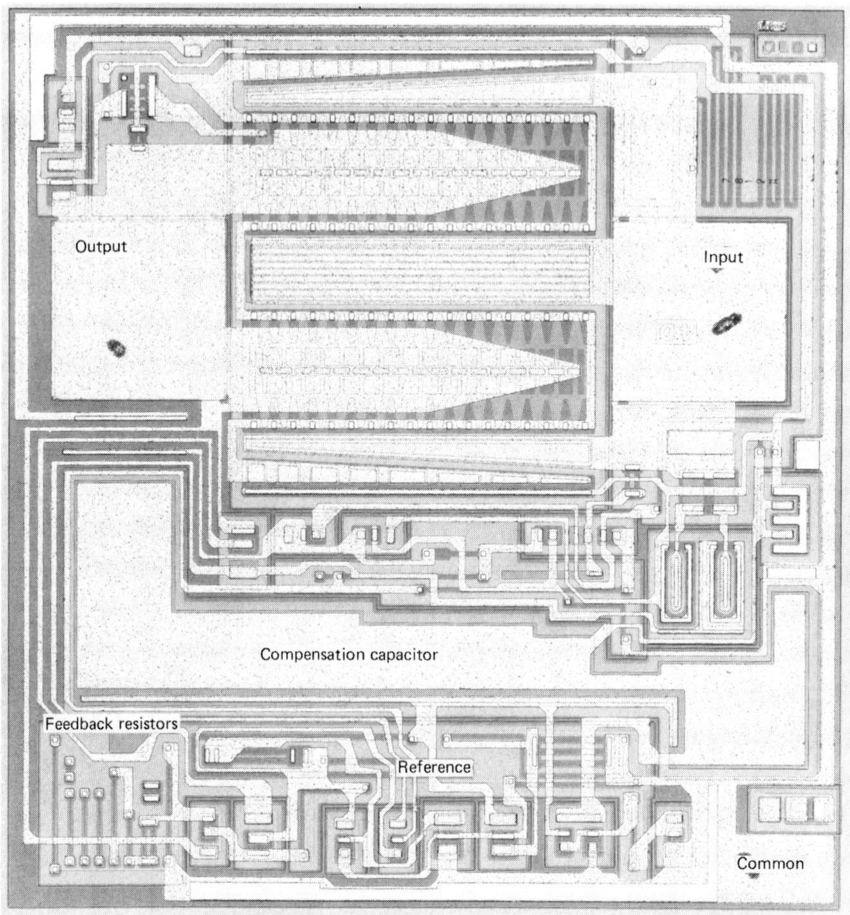

FIG. 11-7 Photomicrograph of the μA7800 1-A regulator, die size 89 mil × 93 mil. *(Courtesy Fairchild Semiconductor)*

Figure 11-7 is a photomicrograph of the μA7800 regulator. Note that the output geometry and its metallization occupy more than half of the die area. Figure 11-8 is a photomicrograph of the μA78M00 0.5-A regulator. Note the similarities in the layout. Except for minor circuit modifications, mainly in the output structure and safe operating area protection circuit, they are identical. This device is offered in the TO-5 and TO-220 packages.

A schematic diagram of the LM123 3-A, 5-V regulator is shown in Fig. 11-9. Increased power capability is achieved with the rather large output geometry, which occupies the major portion of the die (see Fig. 11-10). Circuit operation is similar to the μA7800, with notable excep-

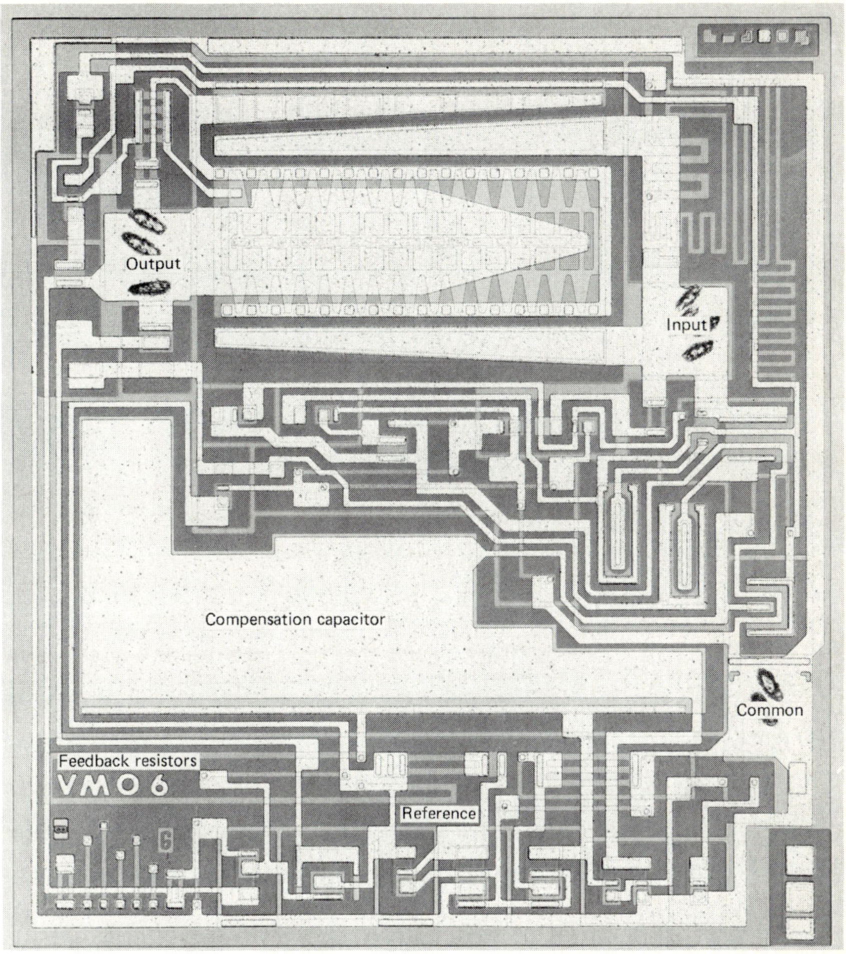

FIG. 11-8 Photomicrograph of the μA78M00 0.5-A regulator, die size 70 mil × 77 mil. *(Courtesy Fairchild Semiconductor)*

tions in the current-limit and short-circuit protection network, which consists of D_2 and Q_{18} and Q_{19} and their associated components. The resistance ratio of R_{15} and R_{16} along with the emitter area ratio of Q_{18} and Q_{19} is selected to give the peak current (4 A at 25°C) a slight negative temperature coefficient, 0.05%/°C. The use of this temperature-compensated current-limit scheme becomes almost mandatory in high-current devices. A noncompensated scheme with a temperature coefficient of 0.25%/°C would have required a peak current in excess of 5 A to guarantee 3 A at 125°C.

The lowest-current three-terminal regulators are those of the

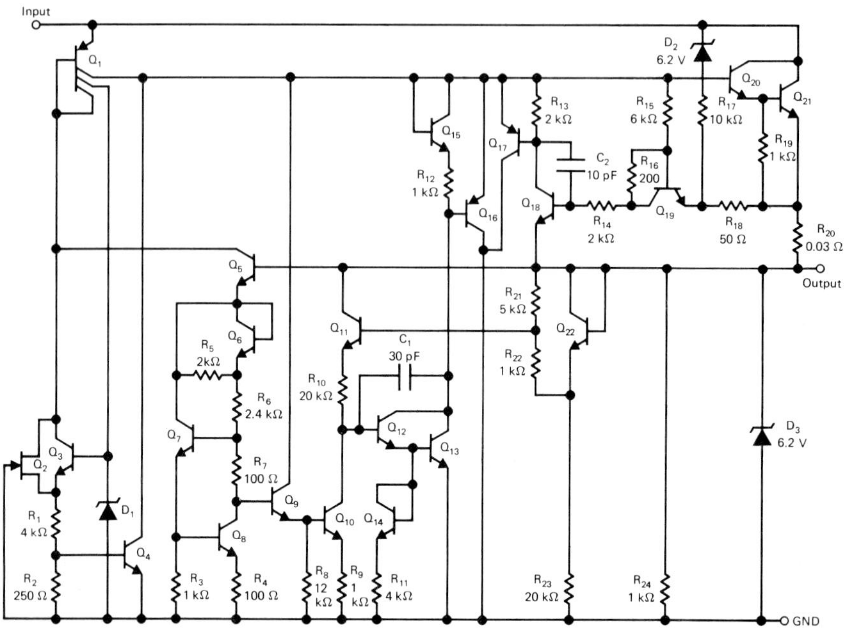

FIG. 11-9 Schematic diagram of the LM123 3-A, 5-V regulator. *(Courtesy National Semiconductor)*

μA78L00 family. Available in the conventional TO-5 and TO-92 transistor packages, this device is capable of delivering output current in excess of 100 mA. Several different μA78L00 or equivalent designs exist on the market, some with band-gap reference and others with a zener-derived reference. Figure 11-11 shows a photomicrograph of one of the designs with a band-gap reference. Note that the output geometry to handle the 100-mA current occupies a small portion of the die area.

11.3.2 Fixed-Negative Regulators

The fixed-negative three-terminal regulators are complements of the positive regulators discussed above; they are available in the same output voltage options, are offered in the same packages, have the same protective features, and can be used by themselves or in combination with positive regulators. A circuit diagram of the popular μA7900 regulator is shown in Fig. 11-12. The start-up circuit consists of zener diode D_1, transistor Q_2, and FET Q_1. As D_2 begins to conduct, Q_2 turns off and the reference zener diode D_2 becomes biased from the constant–current source transistor Q_7. The current through Q_7 in turn is

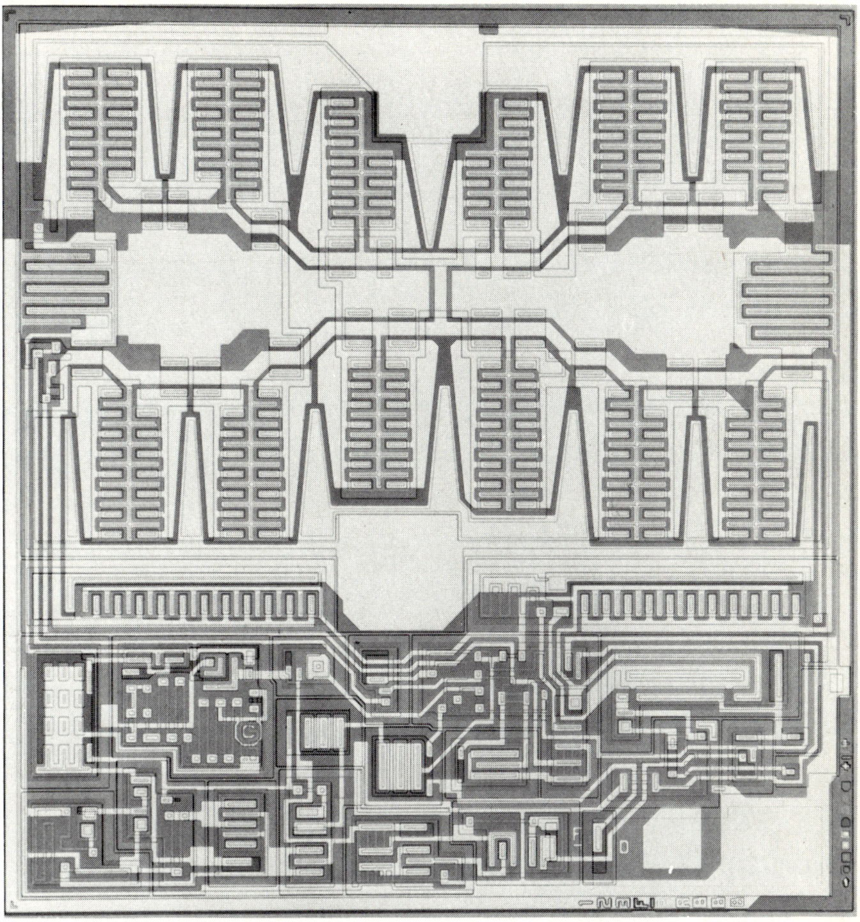

FIG. 11-10 Photomicrograph of the LM123 3-A, 5-V regulator, die size 95 mil × 98 mil. *(Courtesy National Semiconductor)*

set by D_2, Q_3, Q_4, and the reference resistors R_1, R_2, and R_{20}. The reference circuit has two temperature-compensated levels: a −2.23-V reference derived from D_2 and compensated by transistors Q_3 and Q_4, and a −6.2-V reference also derived from D_2 but compensated by Q_6. The −2.23-V reference is used in the lower–output voltage options, whereas the −6.2-V option is used in the −12-V and higher options to improve the performance of the device as explained earlier in this chapter (Sec. 11.2.2).

The voltage reference circuit also provides the bias for the error amplifier, which consists of Q_{10} through Q_{13} via Q_9 and R_8. The error amplifier is a differential pair with an active load, Q_{14} and Q_{15}. The ref-

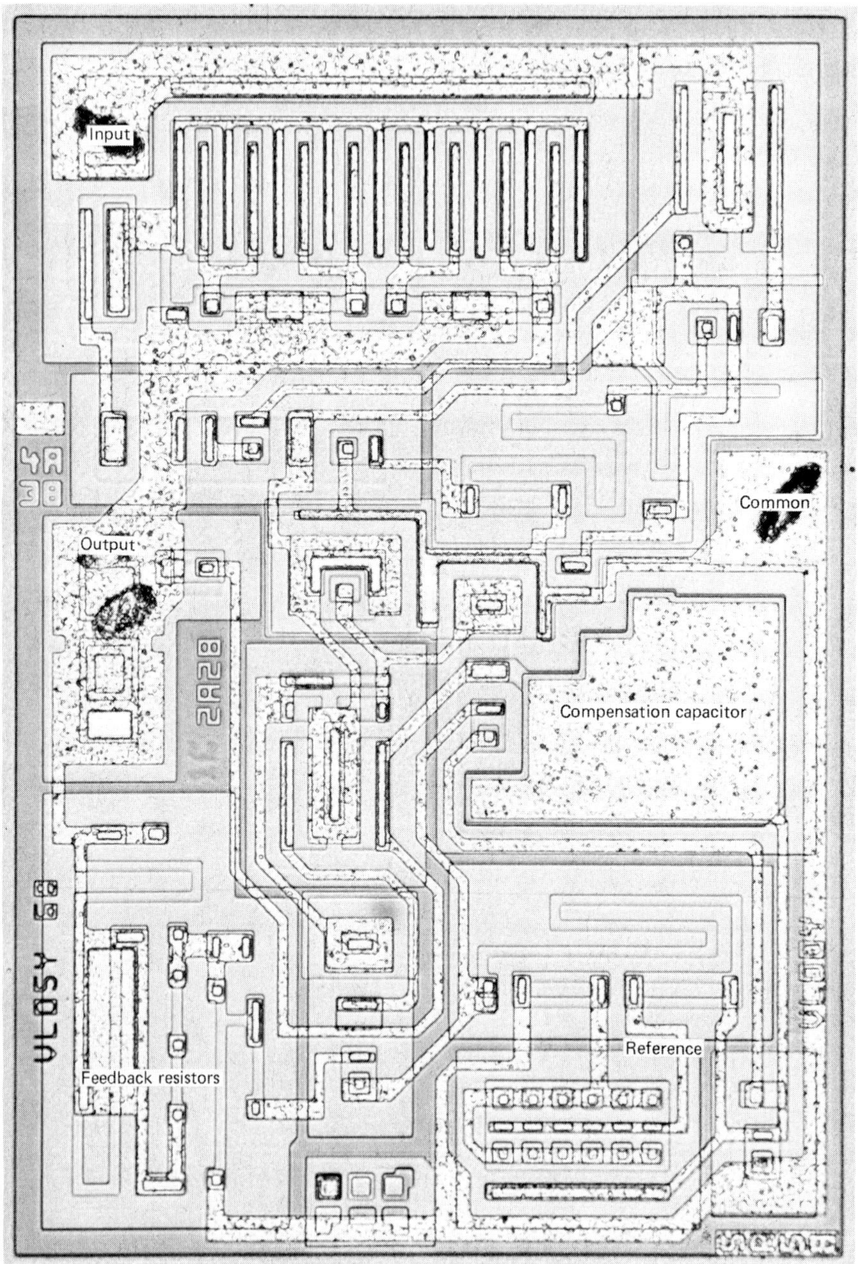

FIG. 11-11 Photomicrograph of a μA78L00 100-mA regulator, die size 36 mil × 50 mil. *(Courtesy Fairchild Semiconductor)*

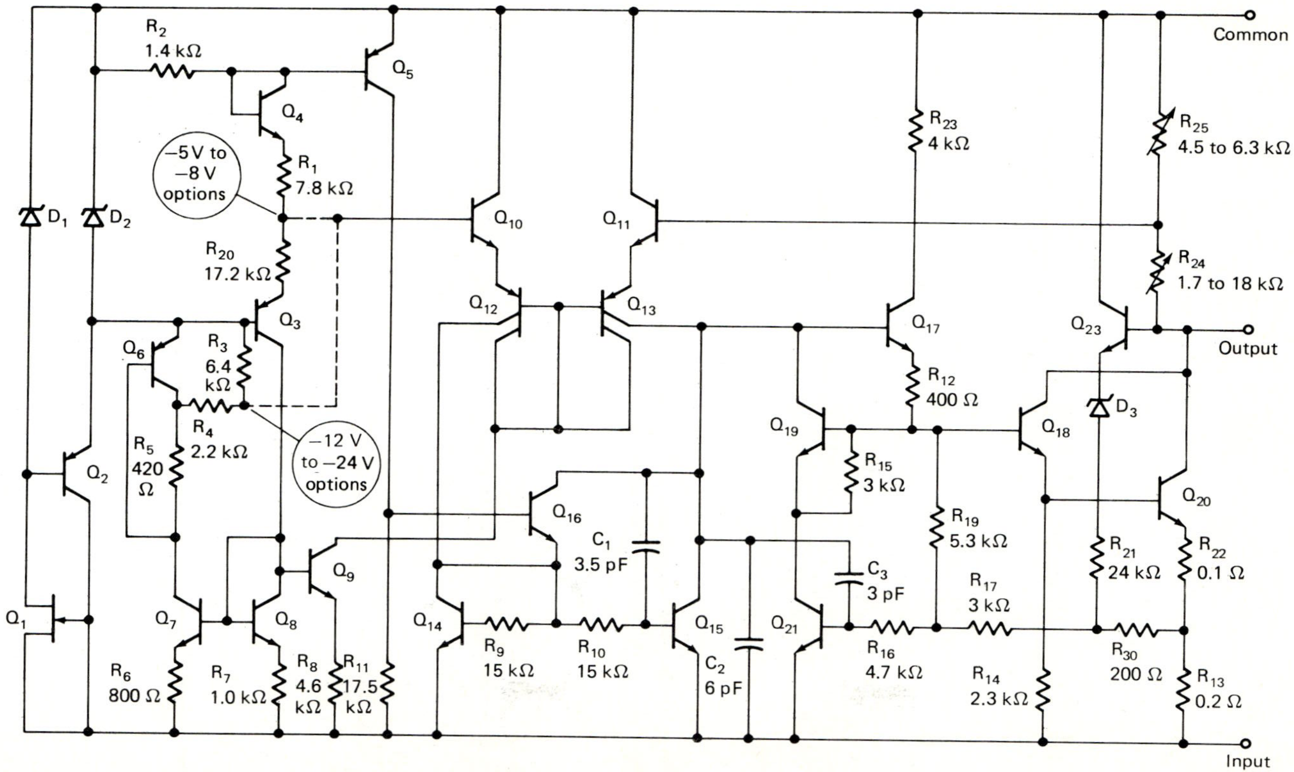

FIG. 11-12 μA7900 schematic diagram. *(Courtesy Fairchild Semiconductor)*

erence input to the error amplifier is applied to the base of Q_{10} (as a noninverting input), while the error voltage is applied from the feedback network of R_{24} and R_{25} to the base of Q_{11} (inverting input). The output of the error amplifier is fed to the series-pass transistor Q_{20} via Q_{17}, R_{12}, and Q_{18}. Frequency compensation of the error amplifier is accomplished with resistors R_9 and R_{10} and capacitors C_1 and C_2. In addition, an external capacitor (from output to ground) is necessary since the series-pass transistor is operating in the common-emitter mode.

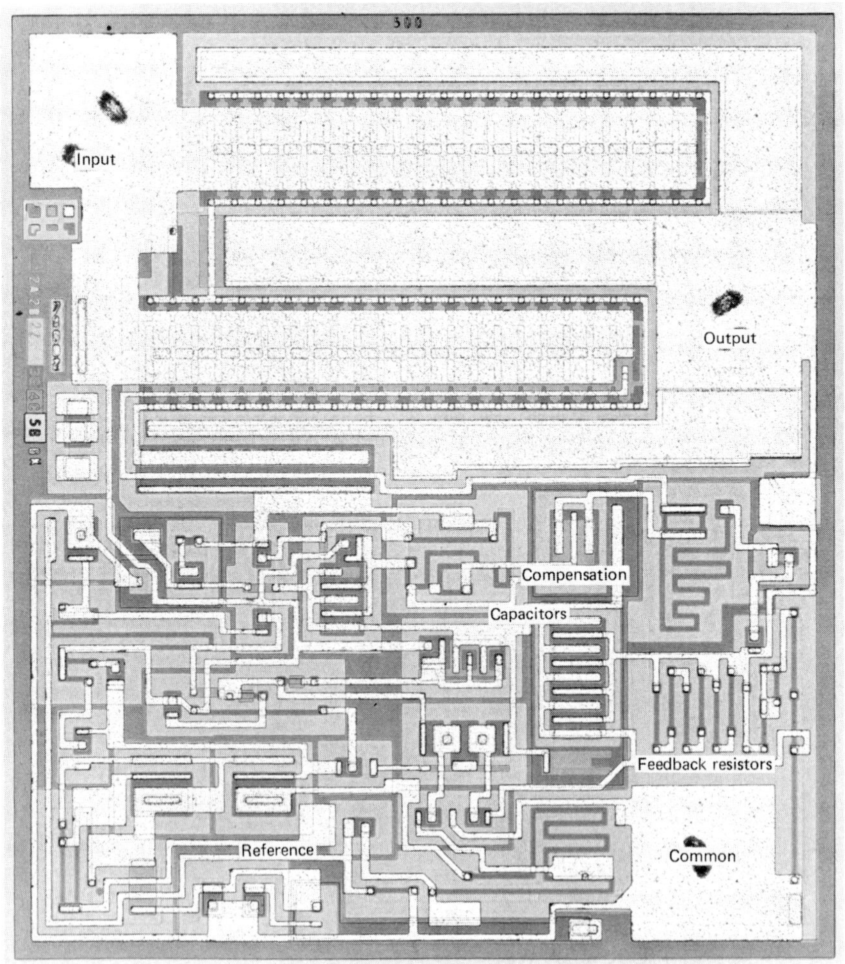

FIG. 11-13 Photomicrograph of the μA7900 1-A negative regulator, die size 92 mil × 73 mil. *(Courtesy Fairchild Semiconductor)*

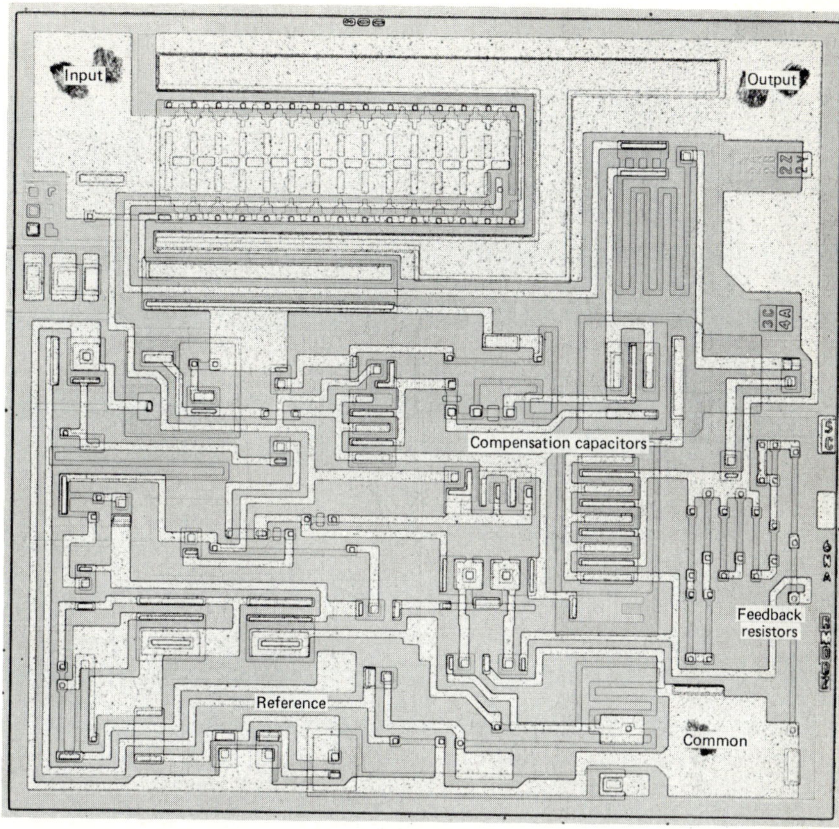

FIG. 11-14 Photomicrograph of the μA79M00 0.5-A negative regulator, die size 70 mil × 66 mil. *(Courtesy Fairchild Semiconductor)*

Thermal overload protection is provided by Q_5, Q_{16}, R_2, and R_{11}. Q_5 is placed in close proximity to the series-pass transistor Q_{20}, and when the V_{BE} of Q_5 decreases sufficiently with rising temperature, Q_{16} turns on and shunts the base drive to Q_{17}.

Short-circuit protection is provided by sensing the voltage drop across R_{13}. As this voltage increases, current-limit transistor Q_{21} begins to turn on, shunting the base drive from Q_{17} via Q_{19}. R_{16} and C_3 provide compensation for the short-circuit protection network and prevent the device from oscillating during short-circuit conditions. Safe area protection is provided by Q_{23}, D_3, and resistors R_{21} and R_{30}. The base drive of the series-pass transistor is cut off gradually, via Q_{17}, Q_{18}, and Q_{19}, as the collector-to-emitter voltage of Q_{20} increases.

A photomicrograph of the μA7900 is shown in Fig. 11-13. Note that, just as in its positive version, the series-pass transistor occupies nearly

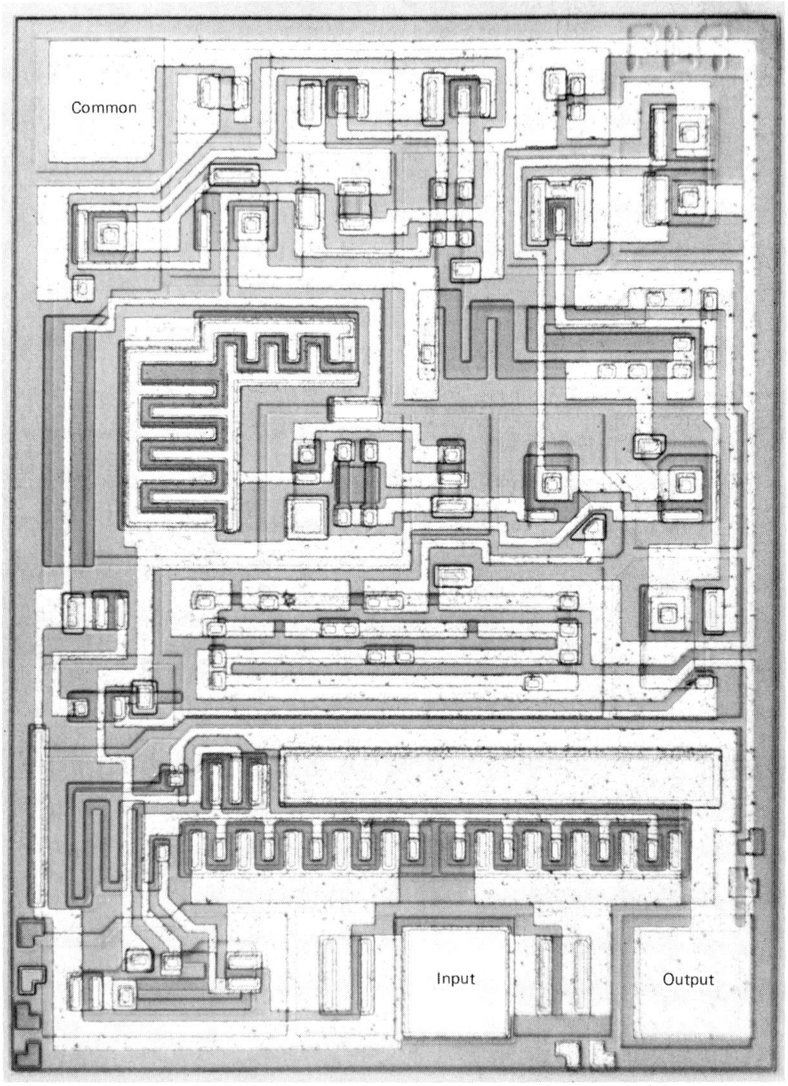

FIG. 11-15 Photomicrograph of the MC79L00 100-mA negative regulator, die size 38 mil × 50 mil. *(Courtesy Motorola Semiconductor)*

half of the die area. In the 0.5-A version (μA79M00), shown in Fig. 11-14, as well as in the 100-mA version (MC79L00), shown in Fig. 11-15, the output devices occupy a small portion of the die, the greater part of it being devoted to the control circuit. In the 3-A version (LM145), shown in Fig. 11-16, the output series-pass transistor and the bonding area occupy nearly two-thirds of the die area. Just as is the case with

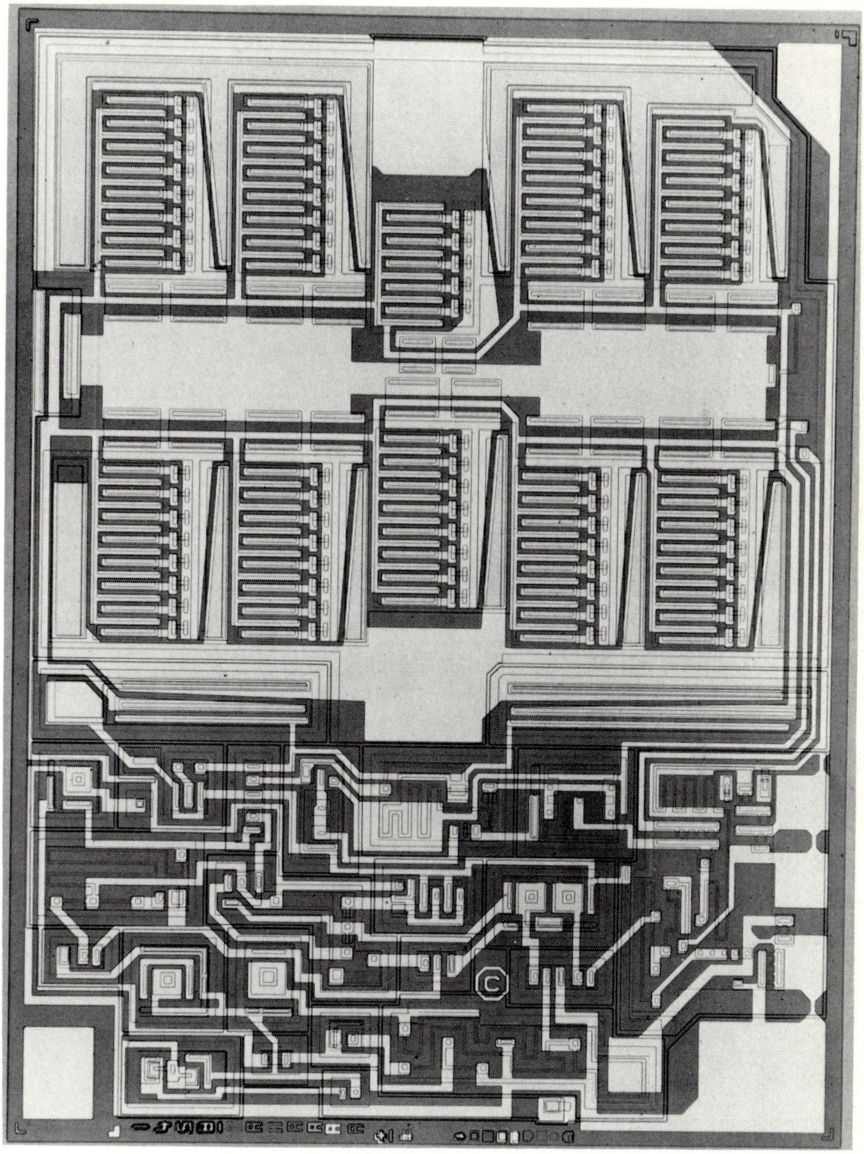

FIG. 11-16 Photomicrograph of the LM145 3-A negative regulator, die size 81 mil × 105 mil. *(Courtesy National Semiconductor)*

the positive regulators, the basic control circuitry of the negative regulators is fairly similar in most of the designs, with minor exceptions, mainly in the protection circuits. Some of the designs, including the 3-A LM145 device, utilize a compensated current-limit scheme as shown in Fig. 11-17. There are also some designs that have two 6- to 7-V

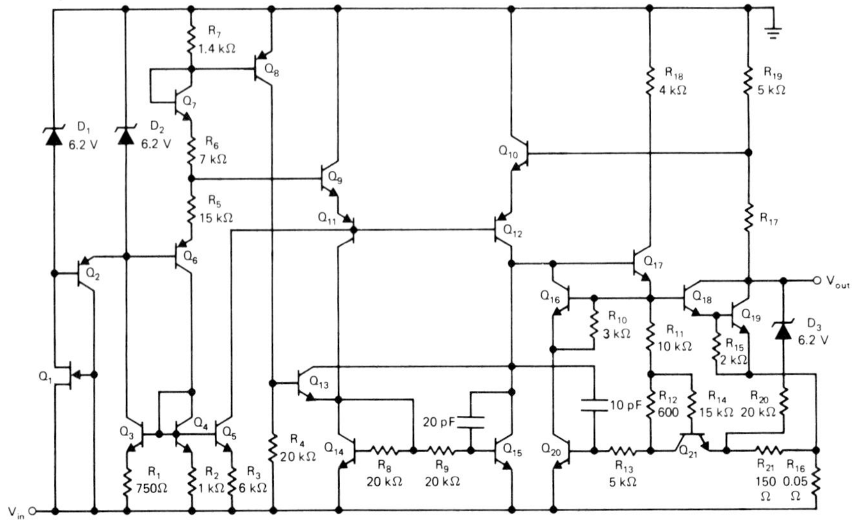

FIG. 11-17 LM145 schematic diagram. *(Courtesy National Semiconductor)*

zeners, instead of one, in the safe area protection network, thus extending the range of the peak current of the device.

11.3.3 Adjustable Three-Terminal Regulators

With the lead limitation imposed by the power transistor packages, one might think that an adjustable regulator with three leads would be

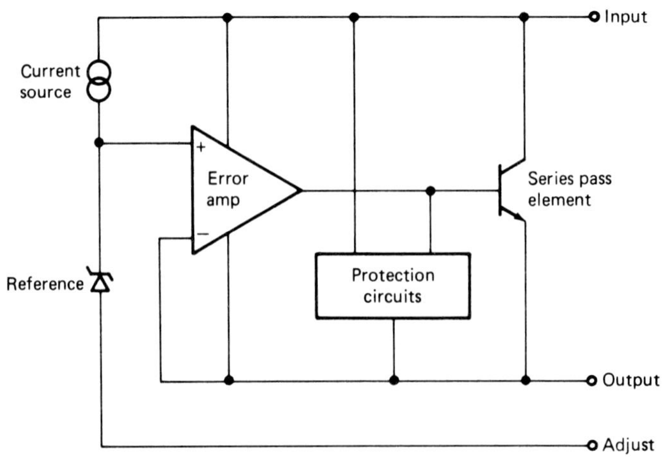

FIG. 11-18 Block diagram of basic adjustable three-terminal positive regulator.

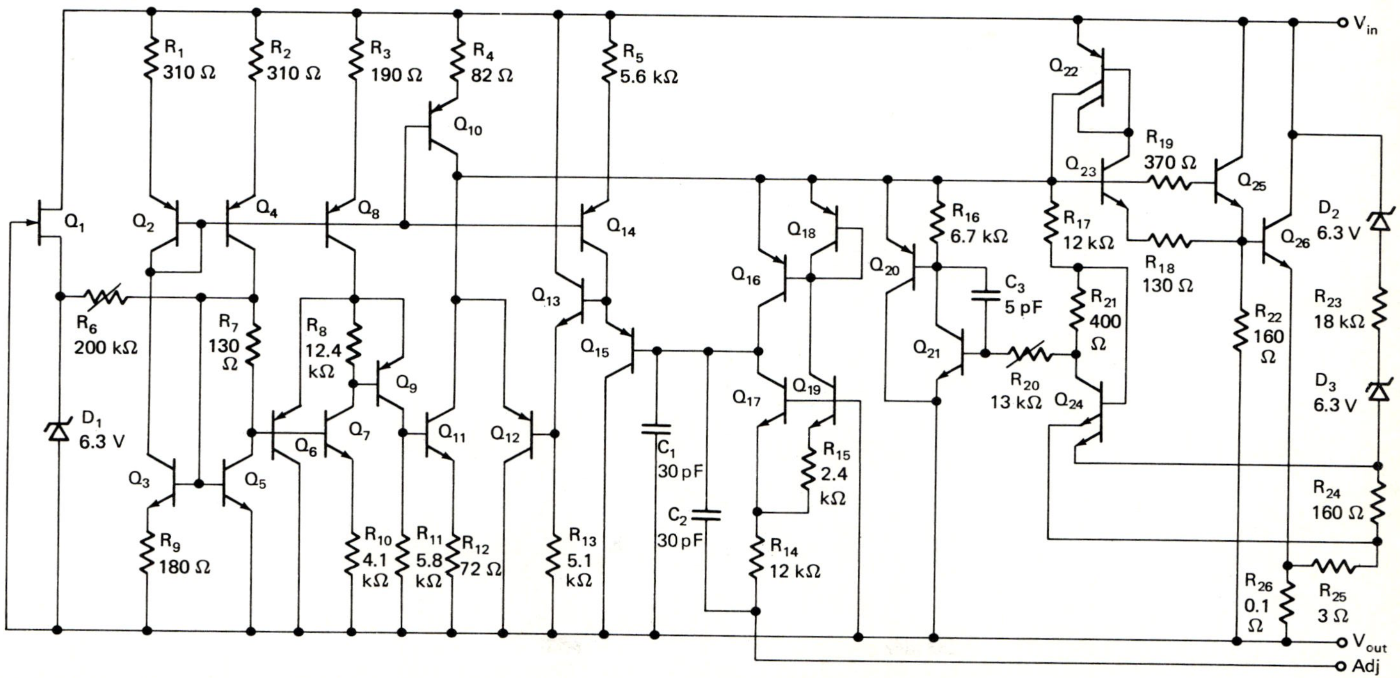

FIG. 11-19 LM117 adjustable three-terminal positive regulator: schematic diagram. *(Courtesy National Semiconductor)*

impossible to design since two of the leads would be dedicated to input and ground, with the third one reserved for the output. Three-terminal adjustable regulators get around this limitation by using the output as the common reference point or substrate (see Fig. 11-18). Note the similarity and differences between the block diagram of the three-terminal adjustable regulator, shown in Fig. 11-18, and that of the fixed regulator in Fig. 11-1. The block diagrams are essentially the same except for the swapping of the adjustment and common terminals. In the adjustable configuration, the low side of the reference is brought out as the control or adjustment terminal, and the error amplifier is referenced to the output, hence eliminating the common terminal. It should be obvious that in this configuration, the quiescent current flows through the output terminal and therefore imposes a minimum load current on the device operation—certainly not a bad trade-off for the added versatility of the device!

A schematic diagram of the LM117 adjustable three-terminal regulator is shown in Fig. 11-19. The start-up circuit consists of Q_1 and D_1. When power is first applied, Q_3 turns on via Q_1 and R_6, thus initiating the current flow through Q_2. Q_5 sets the current through the lateral PNP current sources Q_4, Q_8, Q_{10}, and Q_{14}.

The temperature-compensated band-gap reference, derived from the voltage drop across resistor R_{14} and the base-emitter voltage drop of Q_{17}, is referenced to the output terminal. The emitter-follower transistors Q_{15}, Q_{13}, and Q_{12} are part of the error amplifier. Voltage changes on the base of Q_{15} will cause a corresponding change in the drive current provided by Q_{10} to the output stage consisting of Q_{25} and Q_{26}.

Current-limit transistor Q_{24} is normally conducting. As the output current increases, the voltage drop across the current-limit sense resistor R_{26} increases and causes Q_{24} to turn off, thus turning on Q_{21} and Q_{20} and removing the drive to the output stage. With high input-to-output differential, zener diodes D_2 and D_3 turn on, causing Q_{24} to turn on earlier and thus removing the drive to the output stage at a lower peak current.

Thermal overload protection is accomplished by Q_9 and Q_{11}. Under normal operating conditions, these transistors are in the Off state. The base-emitter voltage of Q_9 is biased in such a manner that at elevated temperatures it turns on, causing Q_{11} to also conduct and thus shunting the drive current to the output stage of Q_{25} and Q_{26}.

The adjustable positive regulator family covers an output current range from 100 mA to 5 A. Figures 11-20 and 11-21 are photomicrographs of the μA317 1.5-A and μA338 5-A adjustable regulators, respectively. Note the similarities in the layout. In the 5-A version, just

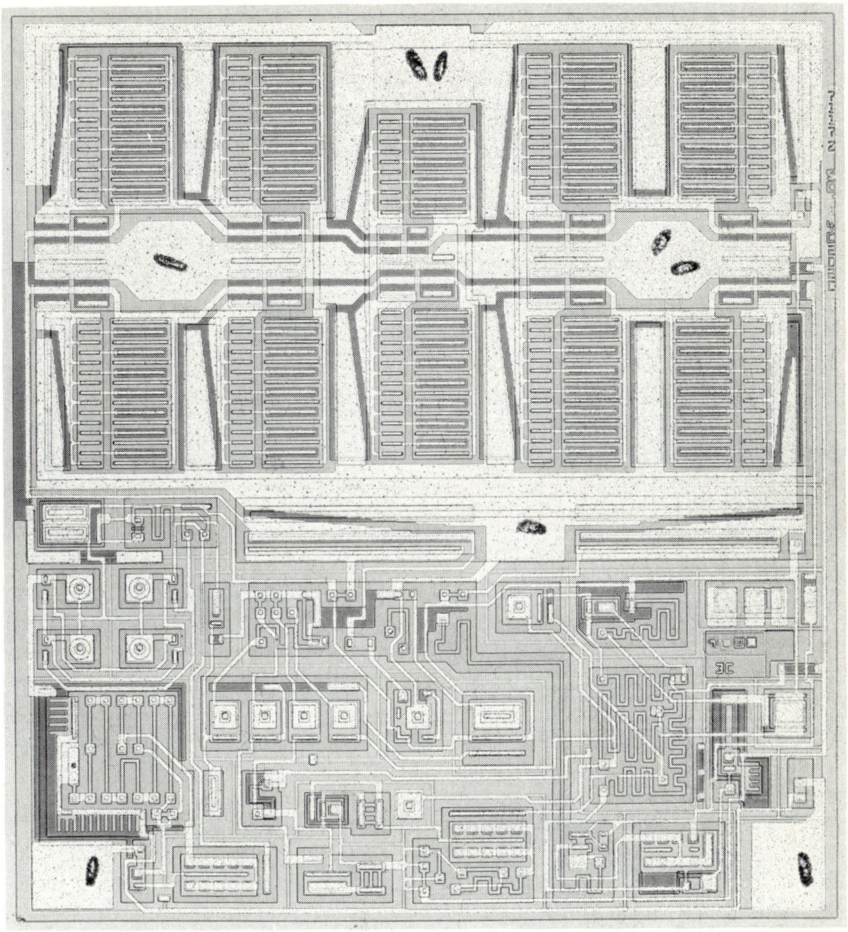

FIG. 11-20 Photomicrograph of the μA317 1.5-A adjustable regulator. *(Courtesy Fairchild Semiconductor)*

as observed in the fixed configurations, the series-pass device occupies the major part of the die.

The proliferation of products among negative adjustable devices has not gone as far. A schematic diagram of a three-terminal adjustable negative regulator, the LM137, is shown in Figure 11-22.

Just as in its positive counterpart, in this device the 1.25-V band-gap reference is generated between the adjust and output terminals via Q_1 and Q_2, which operate at 10:1 current ratios. Start-up is initiated as current flows through D_1, R_7, R_8, R_{22}, R_{34}, R_{35}, and R_{37}. Current sources

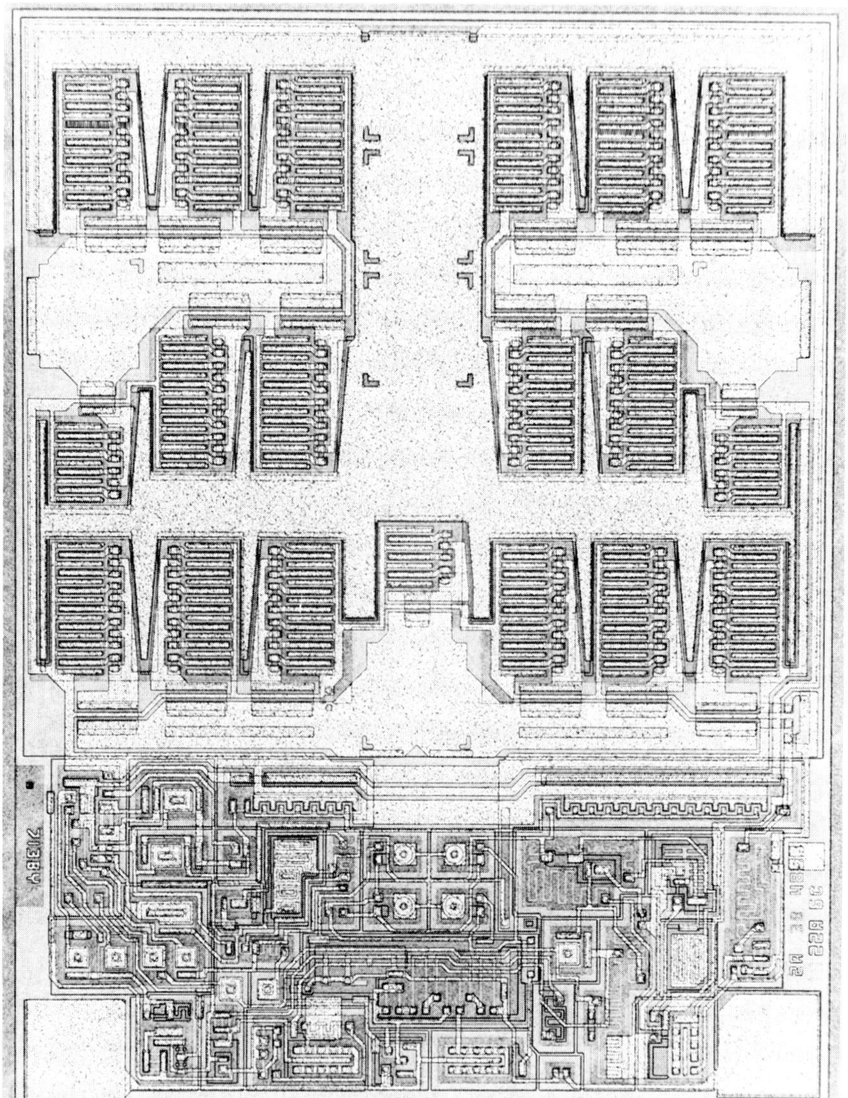

FIG. 11-21 Photomicrograph of the μA338 5-A adjustable regulator. *(Courtesy Fairchild Semiconductor)*

Q_6 and Q_7 set the bias currents for proper operation of the regulator. Under normal operating conditions current-limit transistor Q_{24} is on and Q_{22} and Q_{23} are off. With heavy load conditions, Q_{24} begins to turn off, causing Q_{22} and Q_{23} to turn on and thus removing the drive from the Darlington output stage of Q_{25} and Q_{26} and limiting the output cur-

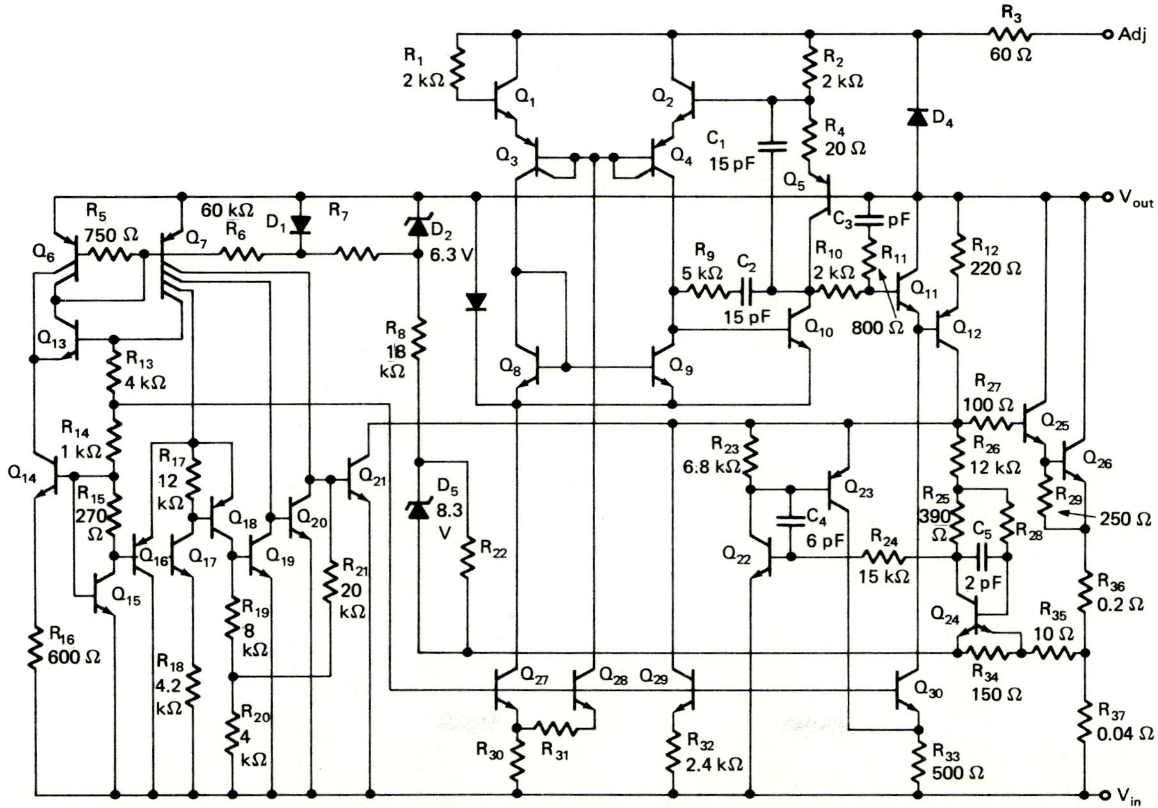

FIG. 11-22 LM137 adjustable three-terminal negative regulator schematic diagram. *(Courtesy National Semiconductor)*

rent. At input-output differentials greater than 15 V, safe area protection diodes D_2 and D_5 along with R_8 limit the output current further by turning Q_{24} off at lower peak currents. Thermal protection is achieved as Q_{18} turns on at elevated junction temperatures, causing Q_{19} to turn on and Q_{20} to turn off. With Q_{20} off, Q_{21} turns on and removes the drive from the output stage.

11.4 THREE-TERMINAL REGULATOR APPLICATIONS

Normally, the output current and output voltage requirements in a system dictate the type of regulator circuit to use. In addition, other considerations such as output polarity, and other regulator parameters such as line and load regulation, dropout voltage, and ripple rejection, must be taken into consideration before a decision is made about the particular circuit suited for the application.

11.4.1 Selecting the Correct Type of Regulator

The selection of the proper regulator is not a simple decision and very much depends on the system's requirements and its imposed limitations. There are no set rules; the selection process will vary from one application to another and even from one designer to the next. In general, however, input voltage, output voltage and its tolerance, and output current requirements will narrow the choice. If the output current requirement of the system is at or below 5 A, the chances are that the needs can be met by using an off-the-shelf three-terminal regulator. System requirements (output voltage and current) and cost trade-offs normally determine the optimum choice of whether a fixed or an adjustable regulator is best-suited for a particular application. Adjustable regulators eliminate the need to stock a variety of regulators and in general exhibit better performance than the fixed versions. However, they require two external resistors to set the output voltage (if it is other than 1.25 V) and a floating heat sink or isolation of the device from the heat sink with an insulator. Therefore, it is probably more economical to use a fixed-output device if the output voltage is a standard value covered by these devices. On the other hand, adjustable regulators are better-suited for nonstandard output configurations or for applications requiring finite trimming.

For output currents in excess of those covered by three-terminal regulators or for tighter parameters and improved performance, other possibilities have to be explored, such as external boost circuits, four-terminal adjustable regulators, or other regulator building blocks such as the popular LM104, LM105, and μA723, not covered in this text. There are undoubtedly several applications for which more than one "best solution" exists, but the end result (assuming the basic system requirements are met) generally comes down to economics.

After the output voltages, the input voltage requirement is the next crucial parameter that influences the choice of the regulator suited for the application. Most fixed and adjustable positive regulators require a minimum of 2 to 2.5 V of differential voltage between the input and the output because of their Darlington emitter-follower output stage. This requirement is slightly lower for negative regulators. For proper operation of the regulator, the filtered dc input supply should under no circumstance fall below the maximum output voltage of the regulator plus the dropout voltage at the low end, nor should it exceed the maximum input voltage of the regulator at the high end. It is important to keep in mind that the full output current capability of the regulator will not be realized if the regulator is operated at or near its maximum input voltage level, because of the internal safe operating area protection circuit of the device. The roll-off of the safe area curve, provided in the device data sheet, should be checked to ensure that the device is capable of delivering the output current needed under high–input voltage conditions.

Ambient temperature, maximum operating temperature, and heat sinking should not be overlooked when selecting a regulator. The thermal properties and limitations of voltage regulators are extremely important and they must be taken into consideration. The effects of temperature on the available output current must not be forgotten.

11.4.2 Design Precautions

There are several commonsense precautions that are often overlooked when a regulator circuit is designed and laid out. Ground leads should be as short as practical. Ground conductors must be sufficiently large to handle fault currents and to reduce unwanted voltage drops across the leads. Single-point grounding at the regulator common terminal should be used to prevent circulating currents or ground loops.

a. Bypassing

As mentioned earlier in this chapter, positive regulators use NPN emitter-follower output stages, whereas negative regulators use NPN common-emitter stages with the load connected to the collector. Because of their output stage configurations, positive regulators are more stable than negative regulators; therefore, bypassing of positive regulators can be omitted in certain applications. It is a good practice, however, to use bypass capacitors at all times. Manufacturers' data sheets should be consulted for the proper bypass capacitors. For positive regulators, input bypass capacitors (typically 0.33 μF) are required if the regulator is located an appreciable distance from the power supply filter. Output bypass capacitors (0.1 μF) are not normally needed, but they do improve the transient response of the regulator. On the other hand, bypass capacitors are a *must* for stable operation of negative regulators over the input voltage and output current ranges. This requirement is caused by their common-emitter output stage. The bypass capacitors (2 μF on the input, 1 μF on the output) should be mylar, ceramic, or tantalum and should have good high-frequency characteristics. If more than one bypass capacitor source or more than one type is used, stability should be checked on each source or type; the same applies for IC manufacturers. Stable operation with one capacitor or one regulator from one vendor may not necessarily result in stable operation with devices of the same type from different vendors, since the characteristics of the capacitors and regulators may vary.

Regulator output impedance is on the order of 100 mΩ or less and increases as a function of frequency above 10 kHz as a result of the gain roll-off of the error amplifier. A tantalum electrolytic bypass capacitor connected to the regulator output will maintain low impedance for frequencies up to 1 MHz. A ceramic capacitor should be placed in parallel with the tantalum capacitor for driving fast switching loads to compensate for the rising impedance of the electrolytic capacitor above 1 MHz. If switching loads are distributed over a large area, additional ceramic bypass capacitors should be located at the loads. Very-large-value output bypass capacitors should not be used unless adequate measures are taken to prevent the output from rising above the input, or to avoid discharging the bypass capacitor through the series-pass transistor of the regulator if the input is momentarily grounded by accident. A reverse-biased diode connected from input to output is normally sufficient to achieve this protection.

b. Protection Circuits

Internal protection circuits are provided in voltage regulators to improve reliability and to make these regulators immune to certain types of overloads. These on-chip components protect the regulators against short-circuit conditions (by current limiting), excessive input-output voltage differential conditions (by safe area limiting), and excessive junction temperatures (by thermal limiting). The protection circuits protect the device against abuse and fault conditions that may be encountered occasionally: continuous operation of the device under fault conditions such as a short circuit or in a thermal shutdown mode is not a recommended procedure.

Proper attention must be paid to the safe area protection network when regulators are operating with excessive input voltage or excessive input-output differential-voltage conditions. In addition to reducing the available output current with high input-output differential conditions, the safe area protection network may, under certain conditions, cause the device to latch up if the output is shorted to ground. This situation is aggravated as the input voltage, load current, or operating junction temperature is increased. Operation under latch-up conditions does not damage the device, but power (input voltage or load current) must be interrupted momentarily for the device to recover from the latched condition.

Care must also be taken when adjustable three-terminal regulators are operating under high–input voltage conditions. Since, under normal conditions, the regulator is floating and sees only the input-output differential, output voltages significantly greater than those achievable with fixed regulators are possible. High–input voltage applications require external circuitry to clamp the input voltage to the regulator to a safe value when the output voltage is shorted to ground. In the absence of a clamp circuit, the regulator is likely to be destroyed when a short circuit is encountered since the full input voltage will appear across the device. Precautions must also be taken to avoid operation of a regulator beyond its maximum ratings. Switching transients exceeding the maximum input voltage rating of a regulator, for instance, can destroy a regulator. These transients, which occur especially if the regulator input voltage is switched instantaneously rather than ramped by the natural smoothing provided by the ac line and the filter capacitors, are usually hard to track and are normally caused by lead inductance and fast switching currents. Good-quality bypass capacitors that have low series resistance cause the in-rush current to increase further, thereby causing a higher-magnitude transient at the input of the regulator. In such cases, a lower-quality, cheaper bypass capacitor may be the answer.

Because of their output stage configurations, positive regulators source current and negative regulators sink current. These restrictions should be kept in mind, and under no circumstance should a regulator output terminal be allowed to go more than a few volts higher than the required output of the regulator. Power should be turned off first before removing or inserting a regulator into a test socket. However, if it is necessary to insert a regulator into a "live" socket, care must be taken to ensure that the common terminal connection is made prior to, or simultaneously with, the input terminal connection. In the absence of the common terminal connection, the output voltage of the regulator will rise to 2 V below the input voltage. Not only will this type of fault condition cause an excessive output voltage that may adversely affect the circuits supplied by the regulator, but when the common terminal is connected, the regulator can be destroyed by the discharging of the bypass capacitor through the output and common terminals.

c. *Thermal Considerations*

The thermal properties and limitations of voltage regulators are extremely important in circuit design. Whether or not a heat sink is required should be determined before the circuit is laid out. Heat sinking is necessary to keep the operating junction temperature of the regulator below the maximum value specified by the manufacturer. Since semiconductor reliability improves as operating junction temperature is lowered, a reliability versus cost compromise must be made during the design. To minimize heating of the regulator by external means, heat-dissipating components such as power resistors should be kept as far away as practical from the regulator. When using low-dissipation packages such as TO-5, TO-39, and TO-92, lead lengths should be kept to a minimum and areas as large as possible should be used on the printed circuit board, especially for the ground lead, to provide additional heat dissipation paths for the package. When using larger packages, the heat sink surface must be flat and free from ridges or high spots. Regulator packages should be checked for burrs or peened-over corners. Thermal grease or lubricant should be used between the regulator and the heat sink material for good thermal conduction.

In some applications, such as three-terminal adjustable regulators as well as negative regulators, it is desirable and sometimes mandatory to electrically insulate the regulator case from the heat sink. Hardware kits for this purpose are commercially available for such packages as the TO-3 and the TO-220. They generally consist of a 0.003- to 0.005-in-thick piece of mica or bonded fiberglass to electrically isolate the two surfaces yet provide a thermal path between them. Naturally, the thermal resistance will increase, but, as in direct metal-to-metal joints, some improvement can be realized by using thermal lubricant on each side of the insulator.

If the regulator is mounted on a heat sink with fins, the most efficient heat transfer takes place when the fins are in a vertical plane, as this type of mounting forces the heat transfer from fin to air in a combination of radiation and convection.

If it is necessary to bend any of the regulator leads, they should be handled carefully to avoid straining the package. Lead bending should be limited, as repeated bending will fatigue and eventually break the leads.

11.4.3 Positive and Negative Regulator Applications

Apart from current and power considerations, most three-terminal regulators of the same polarity are interchangeable within the same family; that is, μA7805 is interchangeable with μA78M05 and μA78L05, and LM338 with LM350 and LM317, etc. Therefore, in the circuit applications that follow, no component values are given since they would vary depending on the output current requirements. Furthermore, since most applications derived for the fixed regulators can be reconfigured for their adjustable counterparts, for simplicity the regulators are shown as fixed devices. It should also be noted, again, that since negative regulators are complements of the positive types, almost all circuits for one type can be converted to those of the other by appropriate changes in the polarities, including those of external components (NPNs for PNPs; diodes and capacitor polarities reversed; etc.).

The basic configurations for three-terminal regulators with input and output bypass capacitors and single-point grounding are shown in Fig. 11-23*a* and *b*. While the configuration for the fixed devices (Fig. 11-23*a*) is straightforward, the adjustable devices merit some explanation (Fig. 11-23*b*). In operation, these devices develop a compensated reference voltage of 1.25 V between the output and the adjustment terminal. This reference voltage causes a constant current of V_{ref}/R_1 to flow through the output-setting resistor R_2 to ground, causing the device to produce an output voltage

$$V_{out} = V_{ref}\left(1 + \frac{R_2}{R_1}\right) + I_{adj}R_2$$

It is worth noting here that bypassing the adjustment pin to ground with a 5- to 10-μF capacitor will improve the ripple rejection, a feature not available with fixed regulators. This capacitor filters the reference voltage and prevents the ripple from being amplified as the output voltage is increased. If the ripple capacitor is used, it may be necessary to use protection diodes to prevent the capacitor from discharging through internal low-current paths and damaging the device [11].

Currents in excess of the output capability of the basic three-terminal regulators can be obtained with the current-boost circuits in Fig. 11-24. Resistor R_1 determines the point at which the external transistor

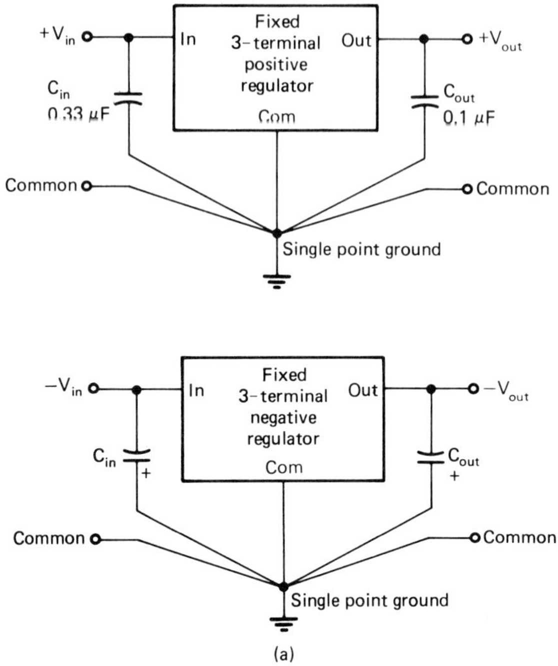

FIG. 11-23 Basic three-terminal regulator with bypass capacitors and single-point grounding. (*a*) Fixed regulators. (*b*) Adjustable regulators.

begins to conduct and is also used to determine the sharing of the output current between the regulator and its boost transistor. Although these circuits are not short-circuit-protected, protection can be achieved by adding another external transistor, Q_2, as shown in Fig. 11-25.

If a regulator is operating with a large input-output differential voltage, an external dropping resistor may be used in series with the input terminal to "share" the power dissipation, thus reducing the input voltage to the regulator, as shown in Fig. 11-26. The value of the dropping resistor R_1 must be low enough that under worst-case conditions of lowest supply voltage, highest output voltage, and heaviest load, the device remains in regulation. For a constant load, the regulator input voltage varies by the same amount as the supply voltage and consequently the line regulation of the device remains the same. For load regulation, assuming constant supply voltage, the combined effects of the change at the input due to the voltage change across R_1 must be taken into

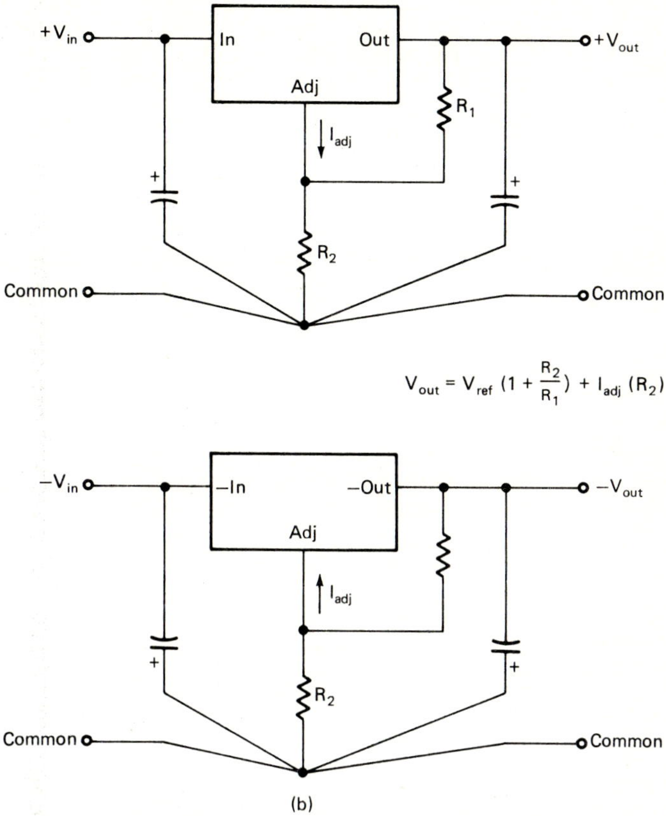

FIG. 11-23 *(Continued)*

consideration. In this configuration, as the load is increased, the regulator input voltage decreases and the net result, in most cases, is a slight degradation in the performance of the regulator since these two effects are additive.

Nonstandard regulated output voltages greater than the regulator voltage can be obtained from a fixed three-terminal regulator as shown in Fig. 11-27. There are also other methods of achieving either an adjustment or a nonstandard output from a fixed regulator. Before any of these are considered, it is recommended that three- or four-terminal adjustable regulator possibilities be explored first. Nevertheless, in the application shown in Fig. 11-27, the regulator output voltage V_{out} is riding on top of a pedestal voltage developed across R_2, just as was the case with the adjustable regulators in Fig. 11-23*b*. When choosing a

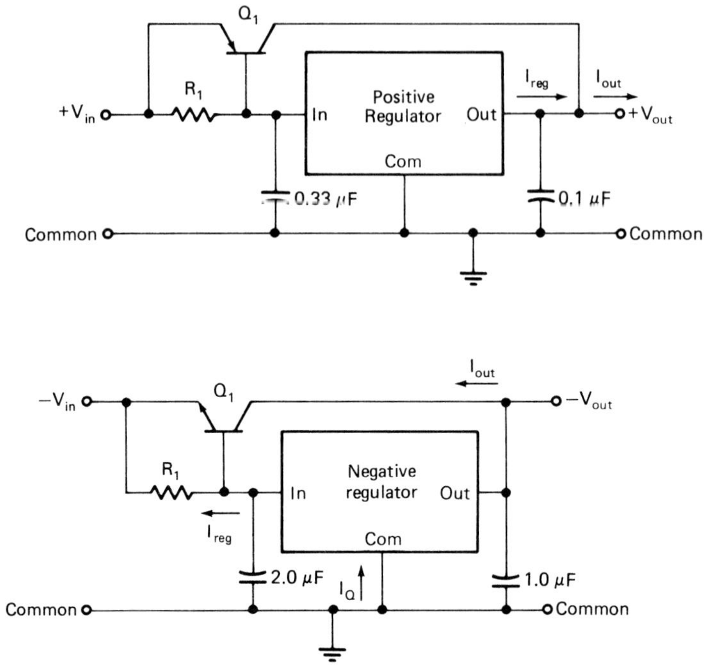

FIG. 11-24 Increased–output current circuit for three-terminal fixed regulators.

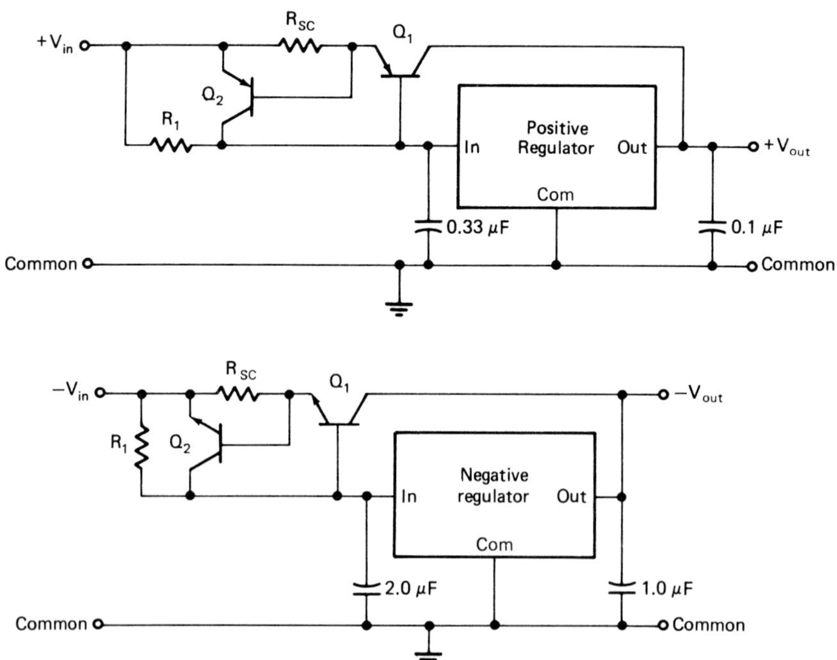

FIG. 11-25 High–output current circuits with short-circuit protection.

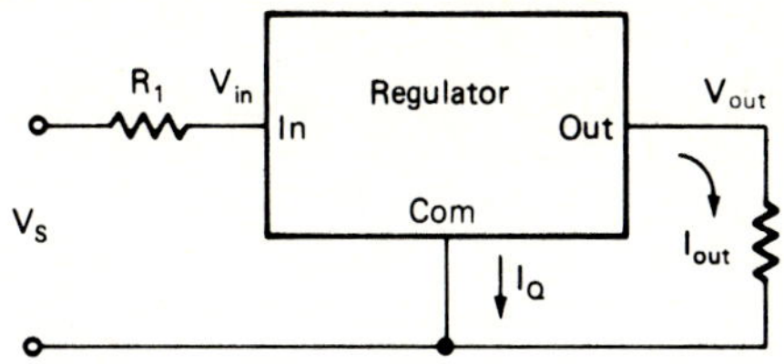

FIG. 11-26 Reducing power dissipation in a regulator with an external resistor.

regulator for this application, the changes of quiescent current with input voltage, load, and temperature must be taken into consideration since the regulator current through R_1 and the quiescent current set the pedestal level across R_2. The effects due to quiescent current can be minimized if the regulator current through R_1 is chosen to be 5 to 10 times larger than the quiescent current of the regulator.

Any three-terminal regulator can be used as a constant-current reg-

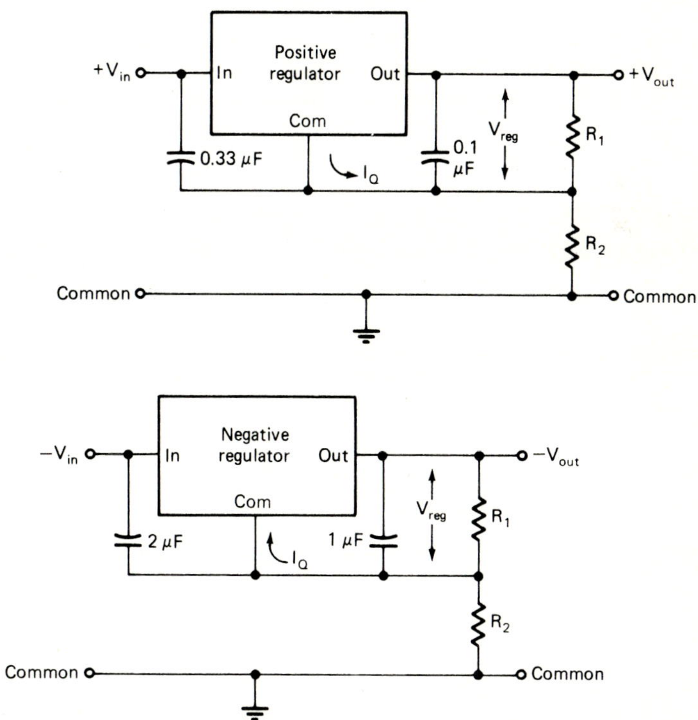

FIG. 11-27 Circuits for obtaining nonstandard fixed output voltages from fixed regulators.

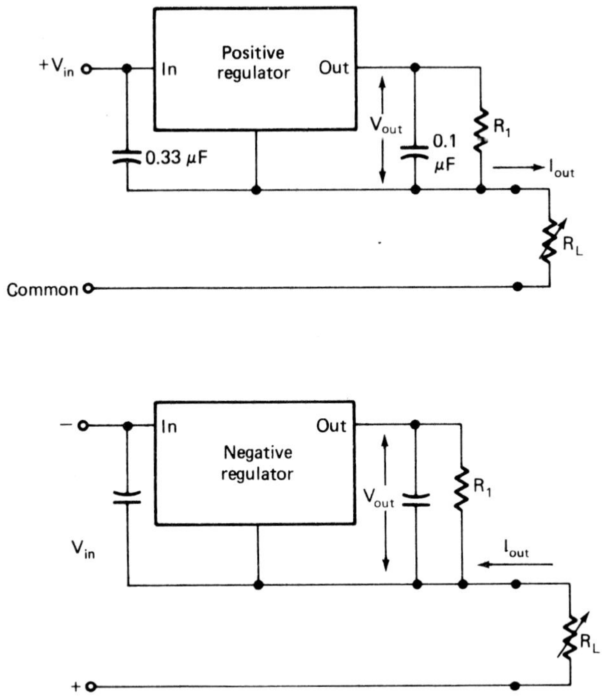

FIG. 11-28 Constant-current regulators.

ulator if connected as shown in Fig. 11-28. Note that the current through the load comprises the regulator output current and its quiescent current for a fixed regulator, and the adjustment pin current for an adjustable regulator.

Dual regulators, or dual power supplies, are normally used for applications requiring two output voltages of opposite polarities that do not

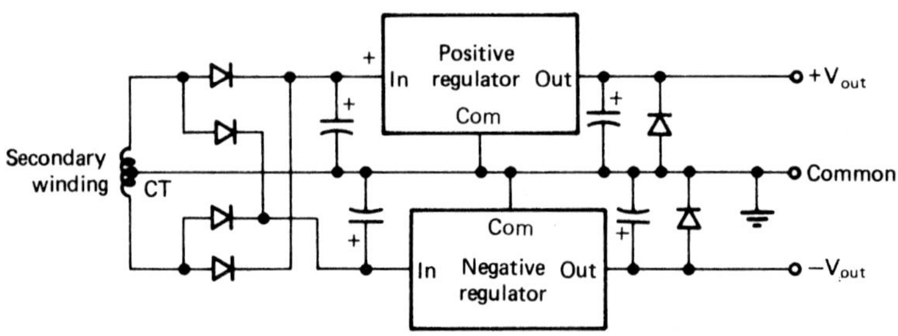

FIG. 11-29 Dual supply using a center-tapped transformer with a positive and a negative regulator.

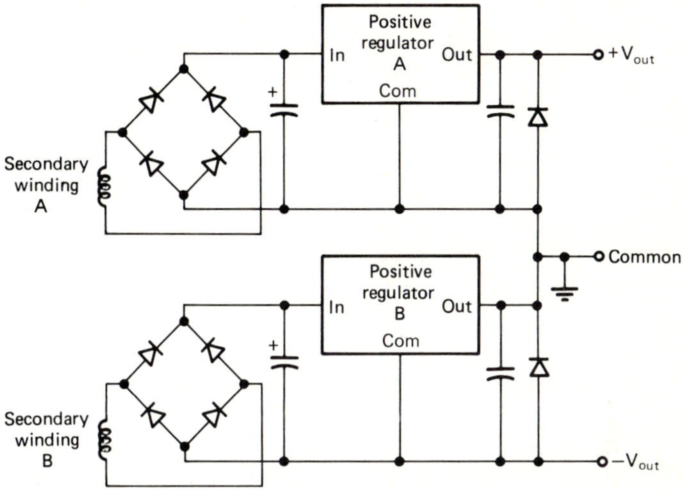

FIG. 11-30 Dual supply using a transformer with two isolated windings and two positive regulators.

necessarily have equal magnitudes, for example, +12 V, −5 V. However, the word *dual* can also imply two supplies of the same polarity but of different magnitudes, such as +5 V, +12 V. With dual tracking, not only are the output voltages of different polarities, but one output voltage always follows the other one; i.e., an increase in the positive output voltage results in a decrease in the negative output voltage, and vice versa. The simplest dual-polarity supply can be obtained by using a positive and a negative three-terminal regulator with a center-tapped transformer as shown in Fig. 11-29. The same type of dual supply can be achieved with two positive (or two negative) regulators if a transformer with two isolated windings is used as shown in Fig. 11-30. The reverse-biased diodes connected across the respective outputs of the dual regulator circuits ensure proper start-up of the regulators by preventing a parasitic action from taking place when power is turned on. This parasitic action can cause the regulators to latch up.

There are other methods available for achieving dual supplies by stacking the common of one regulator on top of the output of another. System ground must obviously be watched when dealing with dual supplies.

Figure 11-31 illustrates a power supply for applications that require tracking between dual supplies. This circuit has a positive and a negative regulator, an operational amplifier used as a comparator, and two matched resistors. Tracking is accomplished by connecting the two reg-

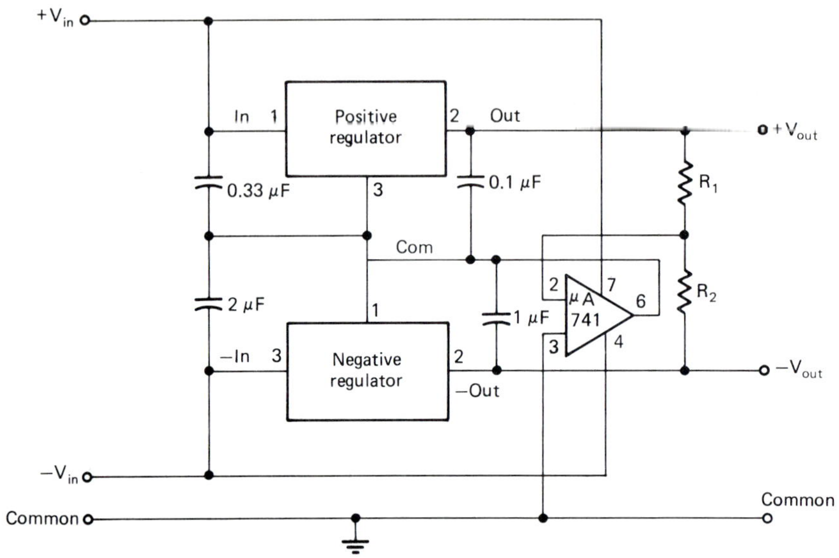

FIG. 11-31 Fixed dual tracking supply using fixed positive and negative regulators.

ulator common terminals to the output of the μA741, which provides a potential on which the common terminals of the regulator float. The summed regulator outputs, $+V_{out}$ and $-V_{out}$, are then compared to the power supply common. Another method of obtaining a dual tracking supply is to use a power operational amplifier and a regulator as shown in Fig. 11-32. The power operational amplifier μA759 is used in the inverting mode. The output of the μA759 tracks the regulator output because any change in the regulator output causes a proportional change in the inverting input of the μA759 and results in a corresponding change in the complementary output. Other methods of obtaining dual tracking supplies include the use of adjustable four-terminal regulators by themselves as well as with operational amplifiers and boost transistors.

The applications covered so far apply to fixed as well as adjustable regulators. In fact almost all circuit configurations applicable to fixed regulators can be configured for adjustable devices by appropriately adding the reference and voltage-setting resistors R_1 and R_2 of Fig. 11-23*b*. The reverse, however, is not true. The added versatility of the adjustable devices allows certain applications not readily realizable with the fixed devices. Most of these are made possible by the low reference of 1.25 V of the device. Figure 11-33, for instance, shows a 5-V regulator with electronic shutdown. Under normal conditions, with

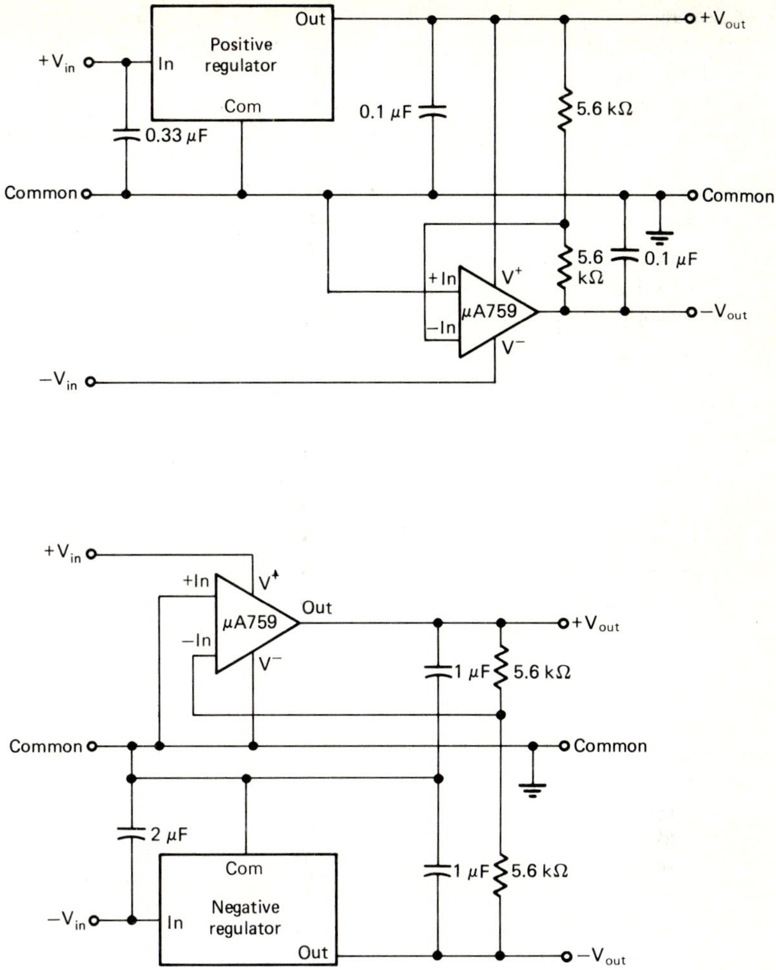

FIG. 11-32 Dual tracking supplies using a regulator and a power op amp.

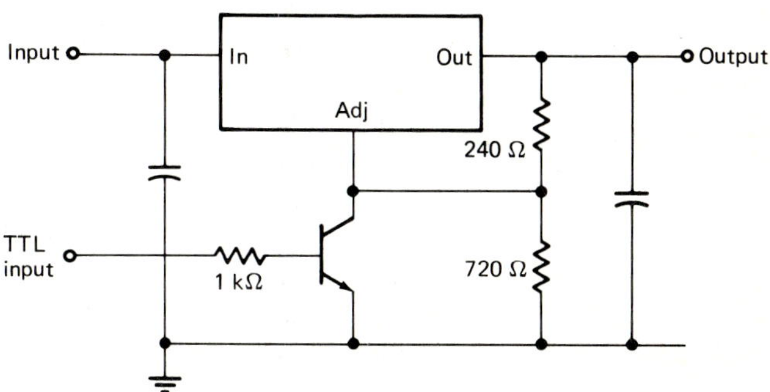

FIG. 11-33 Adjustable regulator with electronic shutdown.

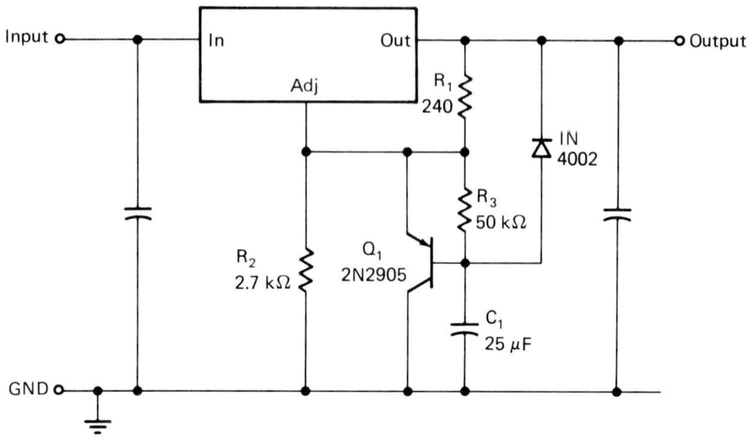

FIG. 11-34 Adjustable regulator circuit for slow turn-on.

no signal applied to the base of the external NPN transistor, the output voltage is set to a nominal value of 5 V. When the NPN transistor turns on, the output voltage will be switched down to 1.25 V plus the saturation voltage of the NPN transistor.

Figures 11-34 and 11-35 show two interesting applications utilizing the low reference of the adjustable regulators [11]. Figure 11-34 shows

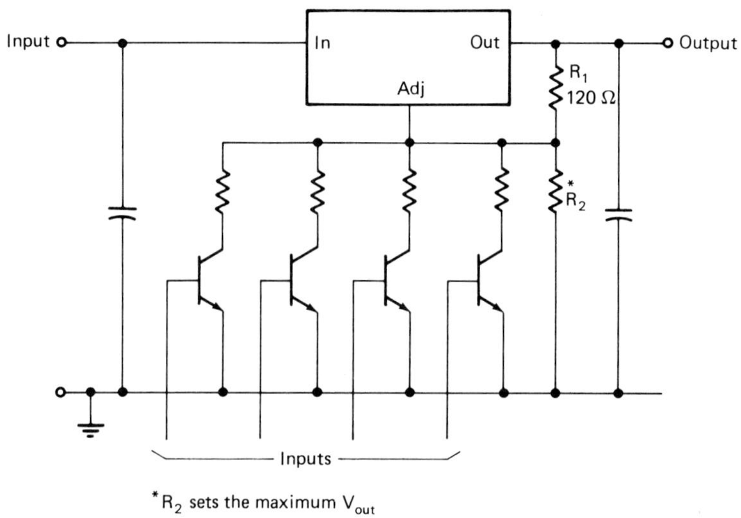

FIG. 11-35 Regulator allowing selectable output voltage.

a slow-turn-on regulator circuit. The PNP transistor and its associated capacitor cause the output voltage to rise slowly from 1.25 V toward its final value when power is first applied to the regulator. The diode discharges the capacitor through the load and resets the circuit when power is turned off. In Fig. 11-35, the four NPN transistors allow adjustment or selection of the output voltage in increments between the maximum value, determined by resistor R_2 when the NPN transistors are off, and the lowest value (when the NPN transistors are on), achieved with the parallel combination of R_2 and the four collector pull-up resistors.

In selecting the applications included in this chapter, an attempt has been made to include the most popular circuits that would cover areas of general interest. To include all or most applications possible with three-terminal regulators would have been an impossible task. For specific applications not covered here, the reader is referred to IC manufacturers' data sheets, their voltage regulator handbooks, and the references and bibliography that follow at the end of this chapter.

REFERENCES

1. R. J. Widlar, "New Developments in IC Voltage Regulators," *ISSCC Dig. Tech. Papers,* February 1970, pp. 158–159.
2. A. B. Grebene, *Analog Integrated Circuit Design,* Robert E. Kreger, New York, 1978, pp. 124, 125.
3. S. Lui, R. G. Meyer, and N. Kwan, "An Ion-Implanted Subsurface Monolithic Zener Diode," *IEEE J. Solid-State Circuits,* Vol. SC-14, August 1979, pp. 782–784.
4. R. J. Widlar, "New Developments in IC Voltage Regulators," *IEEE J. Solid State Circuits,* SC-6, February 1971, pp. 2–7.
5. A. P. Brokaw, "Simple Three Terminal IC Bandgap Reference," *IEEE J. Solid State Circuits,* SC-9, December 1974, pp. 388–393.
6. J. W. Chu and R. D. Ricks, "The μA7800 Series, Three Terminal Positive Voltage Regulators," *Fairchild Application Note 312,* 1971.
7. W. K. Owens, "Temperature Compensated Voltage Regulator Having Beta Compensating Means," U.S. Patent No. 3,731,648, Dec. 25, 1973.
8. K. Fukahori and P. R. Gray, "Computer Simulation of Monolithic Circuit Performance in the Presence of Electrothermal Interactions," *ISSCC Dig. Tech. Papers,* February 1976, pp. 112–113.
9. R. C. Dobkin, "5A Regulator with Thermal Gradient Controlled Current Limit," *ISSCC Dig. Tech. Papers,* February 1979, pp. 228–229.
10. A. A. Adamian, *Voltage Regulator Handbook,* Fairchild Semiconductor Publication, March 1978, pp. 1:8–1:10.

11. Voltage Regulator Handbook, National Semiconductor Corporation Publication, 1982, pp. 10:98–10:100.

BIBLIOGRAPHY

Adamian, A. A.: "A Simple Approach to Cooling Hot IC's with Heat Sinks," *Machine Design*, May 26, 1977, pp. 106–109.

Adamian, A. A.: "Thermal Considerations . . . When Designing with IC Voltage Regulators," *Progress* (Fairchild Semiconductor Publication), September–October 1976, pp. 16–19.

Adamian, A. A.: *Voltage Regulator Handbook*, Fairchild Semiconductor Publication, March 1978.

Brokaw, A. P.: "A Simple Three Terminal IC Bandgap Reference," *IEEE Solid-State Circuits*, Vol. SC-9, December 1974, pp. 388–393.

Chu, J. W., and R. D. Ricks: "The μA7800 Series, Three Terminal Positive Voltage Regulators," *Fairchild Application Note 312*, December 1971.

Davis, W. F.: "A Five Terminal, Complementary Tracking ±15V ±200mA Monolithic Voltage Regulator," *ISSCC Dig. Tech. Papers*, February 1971, pp. 188–189.

Davis, W. F.: "A Five-Terminal ±15V Monolithic Voltage Regulator," *IEEE J. Solid-State Circuits*, Vol. SC-6, December 1971, pp. 366–376.

DC Power Supply Handbook, Hewlett-Packard Co., *Application Note 90A*, 1970.

Dobkin, R. C.: "Break Loose from Fixed IC Regulators," *Electronic Design*, Apr. 12, 1977, pp. 118–122.

Dobkin, R. C.: "Build High-Stability Regulators," *Electronic Design*, Feb. 3, 1972, pp. 62–63.

Dobkin, R. C.: "Designer's Guide to: IC Voltage Regulators—Part 1," *EDN*, Aug. 20, 1979, pp. 93–98.

Dobkin, R. C.: "Designer's Guide to: IC Voltage Regulators—Part 2," *EDN*, Sept. 5, 1979, pp. 199–204.

Eatock, F. E.: "Band Gap Voltage Regulator Circuit Including a Merged Reference Voltage Source and Error Amplifier," U.S. Patent No. 4,064,448, Dec. 10, 1977.

Frederiksen, T. M.: "A Monolithic High-Power Series Voltage Regulator," *IEEE J. Solid-State Circuits*, December 1968, pp. 380–387.

Frederiksen, T.: "The Limits of Power IC's," *Electronic Products*, August 1970, pp. 22–28.

Fukahori, K., and P. R. Gray: "Computer Simulation of Monolithic Circuit Performance in Presence of Electrothermal Interactions," *ISSCC Dig. Tech. Papers*, February 1976, pp. 112–113.

Hoffart, F.: "Use Simple External Circuitry to Extend Regulator Capabilities," *EDN*, Nov. 24, 1983, pp. 131–140.

Kuijk, K. E.: "A Precision Reference Voltage Source," *IEEE J. Solid-State Circuits*, Vol. SC-8, June 1973, pp. 222–226.

Long, E. L., and T. M. Frederiksen: "High-Gain 15-W Monolithic Power Amplifier with Internal Fault Protection," *IEEE J. Solid-State Circuits*, SC-6, February 1971, pp. 35–44.

Lui, S., R. G. Meyer, and N. Kwan: "An Ion-Implanted Subsurface Monolithic Zener Diode," *IEEE J. Solid-State Circuits*, Vol. SC-14, August 1979, pp. 782–784.

Nelson, C. T.: "Power Distribution and Regulation Can Be Simple, Cheap and Rugged," *EDN*, Feb. 20, 1973, pp. 52–58.

Owens, W. K.: "Temperature Compensated Voltage Regulator Having Beta Compensating Means," U.S. Patent No. 3,731,648, Dec. 25, 1973.

Palmer, C. P., and R. C. Dobkin: "A Curvature Corrected Micropower Voltage Reference," *ISSCC Dig. Tech. Papers*, February 1981, pp. 58–59.

Pascoe, B.: "Four-Terminal Voltage Regulators—The Next Generation," *IEEE Intercon Conf. Record*, Session 24-2, April 1975, pp. 1–6.

Sass, F.: "Little, Low Cost Monos Make High Class Regulators," *Electronic Products*, December 1975, pp. 41–44.

Spencer, J. D., and D. E. Pippenger: *The Voltage Regulator Handbook*, Texas Instruments, Inc., Publication, 1977.

Voltage Regulator Handbook, National Semiconductor Publication, 1982.

Widlar, R. J.: "New Developments in IC Voltage Regulators," *ISSCC Dig. Tech. Papers*, February 1970, pp. 158–159.

Widlar, R. J.: "New Developments in IC Voltage Regulators," *IEEE J. Solid State Circuits*, Vol. SC-6, February 1971, pp. 2–7.

Wurzburg, H.: *Voltage Regulator Handbook, Theory and Practice*, Motorola Semiconductor Products, Inc., Publication, 1976.

CHAPTER 12

BiMOS Power Integrated Circuits

P. R. Emerald
Sprague Electric
Worcester, Massachusetts

12.1	INTRODUCTION	12.2
12.2	BiMOS TECHNOLOGY LIMITATIONS	12.3
12.3	COMPETING TECHNOLOGIES	12.5
12.4	TECHNOLOGY SELECTION	12.7
12.5	BiMOS INTEGRATED CIRCUIT FABRICATION	12.7
12.6	DESIGN OF BiMOS POWER ICs	12.10
12.6.1	High-Voltage PNP (Lateral)	12.11
12.6.2	NPN Darlington (50 mA)	12.11
12.6.3	Darlington (500 mA)	12.11
12.6.4	NPN High-Power Darlington (1.5 A)	12.13
12.6.5	High-Voltage, Low-Current NPN (5 mA)	12.14
12.6.6	CMOS Latch Cell	12.14
12.6.7	Shift Register Cell	12.15
12.6.8	Input Protection Circuitry	12.16
12.6.9	Circuit Resistors	12.16
12.7	BiMOS POWER INTEGRATED CIRCUITS	12.18
12.8	BiMOS APPLICATIONS	12.20
12.9	BiMOS TESTING	12.26
12.10	BiMOS TEMPERATURE LIMITATIONS	12.27
12.11	MILITARY AND AVIONICS USES	12.28
12.11.1	Circuit Evaluation	12.29
12.11.2	Screening	12.29
12.12	HERMETIC PACKAGING	12.29
12.13	CHIP SIZE (COST) REDUCTIONS	12.30
12.14	NEW-PRODUCT EVALUATION	12.32
12.15	ALTERNATE SOURCING	12.32

12.16 THE FUTURE OF BiMOS POWER INTEGRATED CIRCUITS 12.34
12.17 "THE INFLUENCE" 12.34

12.1 INTRODUCTION

Until very recently virtually all power integrated circuits (PICs) either were fabricated using linear, bipolar processes for monolithic ICs, or were constructed as hybrids. Other technologies are increasing in use and importance, and BiMOS power ICs are rapidly becoming an extremely viable technology for ICs for interface from logic (very often microprocessor) to a widening variety of peripheral loads.

BiMOS power ICs are fabricated by combining the bipolar (*Bi*) process with CMOS circuits (*MOS*); hence, the acronym BiMOS. The normal and separate fabrication of either bipolar or CMOS circuits is such that the two process technologies are fundamentally incompatible. The most basic distinction is the substrate material; bipolar ICs begin with P-doped substrates, while CMOS circuits begin with N-type wafers (substrates).

The BiMOS devices made at Sprague Electric are fabricated around bipolar starting material (P substrate); sufficient additional photomask and diffusion steps have been added to create the CMOS circuits (usually logic functions only). This combination of steps provides the basic characteristics and advantages of the linear, bipolar technology for the power portions of the device design while also taking advantage of the very low power of CMOS logic.

The use of BiMOS technology for PICs (also for high-voltage, low-current interface ICs) evolved from a need to satisfy industry needs. The recent past has brought on an increased demand for circuits combining logic with power outputs, and further evolution is bringing about the incorporation of combinations of logic, power, analog functions, and protective circuitry on the same chip.

A tremendous stimulant to the development of BiMOS interface circuits has been the ubiquitous microprocessor. The design revolution resulting from the microprocessor (μP) has brought about an increasing groundswell of interest in further reducing component count and cost, along with improved quality and reliability. Additionally, systems have become much smaller while, simultaneously, also increasing in complexity (space is of great concern, and improvements in throughput and assembly have also become critical). Designers have increasingly utilized the microprocessor and thus require new devices to further minimize the number of discrete components (often hand-inserted) and/or combinations of ICs. This pressure is ongoing, and will bring about many new, innovative "intelligent" power ICs.

BiMOS technology allows the configuration of potent and "intelligent" PICs very suitable for microprocessor-based systems. As a result of the use of CMOS for the logic functions there is minimal, almost minuscule, power consumption in the logic or control portion of the device. Since the device output capability may be considerably affected by power dissipation in the control functions, CMOS offers a great advantage. Power handling should be associated with device outputs, rather than being significantly reduced by the power dissipated in the logic—important both from an efficiency standpoint and from that of reliability (because of concerns for chip temperature).

Bipolar PICs have been available for some years and have proven reliability and other advantages, and the BiMOS PIC outputs take advantage of this. An added factor is the cost relationship of competing PIC technologies; no other technology presently in use affords the current ratings per unit area associated with bipolar power outputs. The cost of virtually all ICs is much more related to silicon area than to any other parameter.

However, such advantages do not come gratis. Each choice of BiMOS as a circuit technology is based upon a determination of possible alternatives and apparent trade-offs. BiMOS is not, nor is it ever likely to be, a technology to be used indiscriminately. The choice of BiMOS is, and should be, restricted to those IC designs where the apparent advantages considerably outweigh the disadvantages (of which more discussion will follow).

BiMOS technology affords solutions to many microprocessor/interface/load problems, and also is becoming a factor in "*forcing technological* (rather than incremental) change with *vastly superior* solutions to *customer problems and needs.*"

12.2 BiMOS TECHNOLOGY LIMITATIONS

Listing limitations pertaining to the fabrication of BiMOS is like shooting at a moving target. The semiconductor industry has a tradition that its only natural constant is continual change. However, the limitations and feasibility of any prospective product must be determined by matching its requirements against a set of electrical and dynamic limits—most of which are expandable with time and resource.

The set of present limits for BiMOS would provide the features listed in Table 12-1, while extrapolating toward the mid-1980s (or slightly

TABLE 12-1 Present BiMOS Limits

Output breakdown voltage V_{out}	to 100 V (max)
Output current I_{out}	to 1 A (max)
Logic supply range	standard CMOS (5–15 V)
Logic switching speed (shift register)	to 1 MHz at 5 V to 2MHz at 12 V
Channels per package	20 (28 LDIP)
Packaging	predominantly DIP
Temperature range	predominantly 0 to +70°C (some have extended temperature range)
Device chip size	25,000–30,000 mil^2

beyond) should produce expansion and extension reflected in Table 12-2.

BiMOS competes with other technologies, and in so doing, it should not be chosen without due consideration of its technical and cost limitations. For BiMOS to compete favorably the advantage versus disadvantage determination must be well in favor of BiMOS for high-power, high-current, or (even) high-voltage ICs.

The general heading of "power ICs" might not include some of the present uses of BiMOS (depending upon the individual subjective

TABLE 12-2 Projected BiMOS Capabilities (Beyond the Mid-1980s)

Output breakdown voltage V_{out}	to 150 V (perhaps 200 V)
Output current I_{out}	2–4 A (5 A peak)
Logic supply range	5 V to 15 V (unchanged)
Logic switching characteristics (shift register)	to 5 MHz at 5 V to 10–12 MHz at 12 V
Channels per package (or chip)	32 (40 LDIP) 64 channels (84 lead "quad pack")
Packaging	DIP SIP leadless carriers hermetic power (TO-66, etc.)
Temperature range	predominantly −20 to +85°C and −40 to +85°C; (automotive, telecommunications, and industrial) also −55 to +125°C (military)
Device chip size	30,000–40,000 mil^2

determination), but high-voltage BiMOS interface ICs dramatically extend the ability to interface from microprocessors to loads such as high-voltage displays. Some of these displays (such as plasma and electroluminescent) require peak currents of a much higher value than the average and do pose a problem for devices having very small output transistors. Hence, they will be included as part of the discussions of BiMOS PICs.

12.3 COMPETING TECHNOLOGIES

Under the general heading of competing technologies we may place the following semiconductor technologies:

Linear bipolar

I^2L

DMOS

CMOS/DMOS

Dielectric isolation processes (bipolar and MOS)

BiMOS

Within these competing process technologies there is a considerable variety of capabilities, characteristics, costs, etc., and addressing them is no easy task. However, some basic comparisons may be made (somewhat subjective and prone perhaps to rapid change). Table 12-3 provides and attempts to quantify important characteristics.

At present in the higher-power (greater than 1 A) circuits the BiMOS competition stems mainly from linear technologies (bipolar linear or I^2L), but also competes rather head-on with CMOS/DMOS (labeled BIDFET by Texas Instruments) in the high-voltage circuits with low to modest current outputs (usually less than 50 mA).

I^2L is a factor in the 40- to 60-V (perhaps to 80-V) levels, but this requires double-epitaxial I^2L (nearly as complex as BiMOS). Lower operating voltages (less than 30 V) are often much better left the province of standard I^2L technology, and competing in this region would usually offer an advantage to I^2L (a simpler, lower-cost approach to the processing of logic-plus-power combinations).

The higher-voltage applications may be more a determination of linear bipolar, dielectric isolation (DI), DMOS, or CMOS/DMOS as a competitor. In the high-power designs the BiMOS approach appears to have a definite edge, since the others either cannot readily accommodate high currents (greater than 1 A) or require much greater silicon

TABLE 12-3 Competing Power IC (PIC) Technologies

Technology	Device outputs		Logic				Relative process complexity
	BV°	I†	Complexity	Speed	Supply	Power	
Linear bipolar	200 V	2–4 A	MSI	1 MHz	5 V	high	easiest
I^2L	20–60 V‡	2–4 A	LSI	3–6 MHz	5 V	modest	medium
DMOS	200 V	500 mA	LSI	2–5 MHz	10 V	low	difficult
CMOS/DMOS	200 V	500 mA	LSI	2–5 MHz	10 V	low	difficult
DI	300 V	100 mA§	MSI	1 MHz	5 V	high	very difficult
BiMOS	200 V	2–4 A	MSI/LSI	5–10 MHz	5–15 V	low	difficult

°Breakdown voltage limits based upon available (or near-term) estimates.

†Current rating based upon device estimates and power ratings; MOS output ratings may be increased through larger chip size.

‡Double-epitaxial I^2L process for 30–60 V.

§DI process allows very little opportunity for high current.

area (hence a greater cost factor). Designs requiring combinations of logic and power in reasonable proportions seem to favor BiMOS. Circuits targeted for interface to motors (stepper, servo, brushless dc), hammer or solenoid drivers (printers), and capacitive display loads (ac plasma, ac electroluminescent, etc.) appear to be particular prospects for BiMOS devices.

12.4 TECHNOLOGY SELECTION

As yet there are no easy "cookbook" techniques to determine a suitable technology for a new PIC design. None of the semiconductor manufacturers have in their "bag" all, or even most, of the potential PIC technologies; thus the initial determination must be the approaches available within the organization. At Sprague the choice to use BiMOS stems from weighing a composite of the technology parameters (voltage, current, speed, power, cost, etc.) in order to determine if BiMOS should be used. As among most other suppliers, such a choice is generally made from no more than three or four possibilities.

The BiMOS interface ICs from Sprague Electric evolved from earlier power arrays and other components that were simple yet versatile. Adding latches, shift registers, and similar functions was a natural evolution to continue satisfying the market needs for power interface ICs which dictated adding suitable logic circuitry. Thus another set of industry standard building blocks began to evolve.

12.5 BiMOS INTEGRATED CIRCUIT FABRICATION

Sprague Electric's BiMOS technology has been developed around a linear-process bipolar foundation and adds necessary process steps to the sequence to include the metal-gate CMOS circuitry. Several of the process steps are common or shared, but BiMOS requires additional masking and process operations. While a rather typical bipolar design requires eight or nine photomasking operations, typical BiMOS devices require 12 to 14 masking steps.

A P-doped substrate (wafer) is used as the starting material and the diffusion and masking process follows the sequence below; for a cross section of the NPN, PNP, and N- and P-channel MOS refer to Fig. 12-1 and 12-2. Steps 1 to 7 of the following outline list the various elements of the IC formed in the corresponding process steps, and the

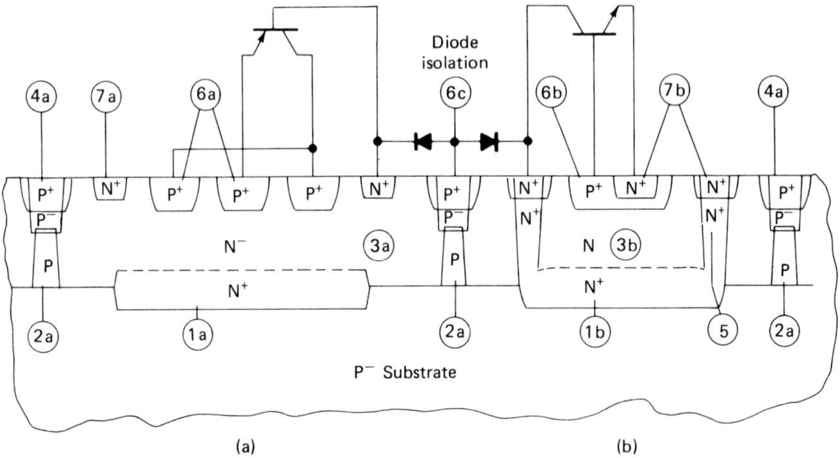

FIG. 12-1 Bipolar integrated transistor cross section. (*a*) PNP and (*b*) NPN.

purpose they serve in the NPN, PNP, NMOS, and/or PMOS technologies.

1. N^+ buried layer
   a. For lateral PNP, reduction in parasitic vertical PNP beta (where the substrate is the parasitic collector)
   b. For NPN, reduces collector resistivity and $V_{CE(sat)}$
   c. P-channel MOS: reduction of vertical parasitic PNP beta (same as for lateral PNP)

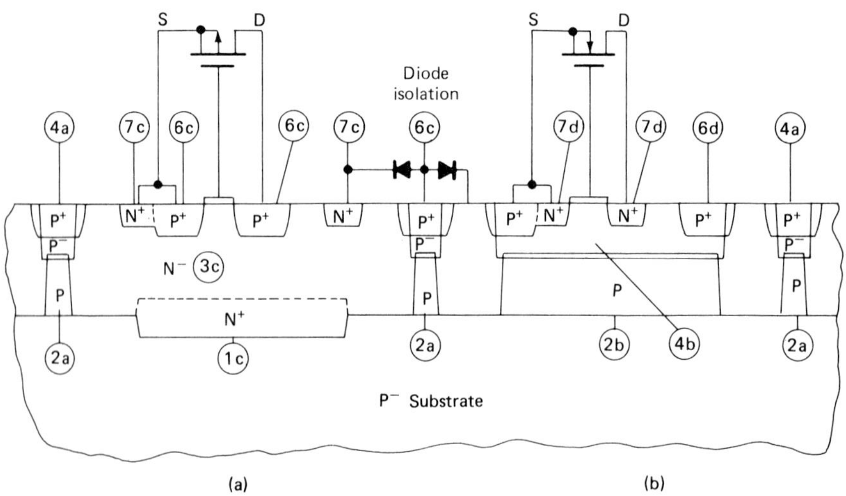

FIG. 12-2 MOS integrated transistor cross section. (*a*) PMOS and (*b*) NMOS.

2. P buried isolation or isolation well (updiffusion)
   a. PN diode isolation between NPN, PNP, PMOS, and NMOS transistors
   b. Isolation (P) well for formation of NMOS transistors
3. N^- epitaxial deposition
   a. PNP: N^- base region of lateral PNPs
   b. NPN: collector "tub" for all NPNs
   c. P-channel MOS: substrate region for PMOS transistors (normal CMOS would use N substrate)
4. P^- isolation or P well (implanted dopant/diffusion driven)
   a. PN diode isolation between transistors (combined with 2*a*, above)
   b. N channel: formation of P well (substrate) for NMOS
5. N^+ collector plug (sinker): provides low-resistivity collector for NPNs; also prevents inversion of lightly doped N^- epitaxial layer (used as guard ring)
6. P^+ base and resistor diffusion
   a. PNP: emitter and collector of lateral PNPs
   b. NPN: base region of NPNs
   c. PN isolation: redope of isolation region to prevent inversion of lightly doped isolation (2*a* and 4*a* and 6*c* together form diode isolation)
   d. P channel: source and drain diffusions for the PMOS
   e. N channel: guard ring for N channel
7. N^+ emitter diffusion
   a. PNP: base contact and N^+ guard ring (to prevent inversion)
   b. NPN: emitter of NPNs and low-resistivity contact for NPN collectors
   c. P channel: guard ring for PMOS transistors
   d. N channel: source and drain diffusions

The following process sequence is not shown in Figs. 12-1 and 12-2:

8. *Optional step,* dependent upon device (it may include 8*a* or 8*b* or be omitted entirely)
   a. Ion implant adjustment of P-channel depletion loads (if used)
   b. Incorporation of thin-film (polysilicon) resistors (if used)

9. Gate oxide etch: removal of circuit oxides (to suitable depth) over gate regions of P- and N-channel transistors
10. P-channel threshold adjust: ion implantation adjustment of PMOS transistors—largely for speed enhancement
11. Etch contact windows: removal of circuit oxides to allow contact and connection to all bipolar and MOS transistors, diodes, resistors, etc.
12. Metallization: deposition of aluminum for circuit interconnections; also etching of pattern and bonding pad areas
13. Passivation: deposition of passivation material and subsequent etching of vias (contact windows) allowing connection to circuit bonding pads

Such a fabrication sequence adds considerable manufacturing cycle time to wafer processing (compared to bipolar power ICs). The obvious implication of the added cycle time, number of steps, diminished throughput, and increased complexity is an increased cost associated with manufacturing BiMOS ICs.

12.6 DESIGN OF BiMOS POWER ICs

As mentioned earlier the foundation underlying BiMOS PICs is a linear bipolar technology. Given a history of early activity in bipolar power ICs such as peripheral/power drivers, power arrays, etc., Sprague has evolved proven elements in power (output) circuitry. Through utilization of well-established transistor designs and proven processes, the difficulty of producing BiMOS was reduced. It was not necessary to create both logic functions (CMOS) and the power structures (bipolar) concurrently; existing power transistor designs were stored in the computer-aided design (CAD) system and were used as needed (and newer versions were added as new designs proceeded).

Much of the logic and control circuitry has been created (then stored) as required by the design. Little early activity designing a cell "library" assortment of gates, latches, shift registers, etc., was provided. Designs evolved and components (primarily bipolar transistor designs) and cells (latches, shift registers, gates, etc.) were added (even improved upon) as required. Random logic portions of new circuits are created "real time," and then added around the "library" components. The most frequently used (to date) cells and components will be shown and their relevant characteristics presented.

12.6.1 High-Voltage PNP (Lateral)

The most commonly used PNP for any high-voltage and high-power circuit is shown in Fig. 12-3. In large part because of the extremely large base width (20 to 25 μm) these lateral PNPs are "PNP functions" *only* and in PICs virtually always require the addition of either an NPN or an NPN Darlington to provide sufficient current gain and current handling capability to be meaningful and useful (schematically represented in Fig. 12-4*a* and *b* are the often-used compound PNP/NPN combinations).

12.6.2 NPN Darlington (50 mA)

The lower-current NPN Darlington shown in Fig. 12-5 has most frequently been part of high-voltage display interface circuits, but such ICs have also been used on occasion for thermal and ink jet printers, etc. Most of the applications of this output structure have been in conjunction with the high-voltage PNP (see Fig. 12-4*b*) and are found in sourcing outputs.

12.6.3 Darlington (500 mA)

One of the most frequently utilized output structures for BiMOS PICs is a 500-mA Darlington, which has been used at various voltage ratings (largely via test selection). The initial BiMOS ICs from Sprague Electric

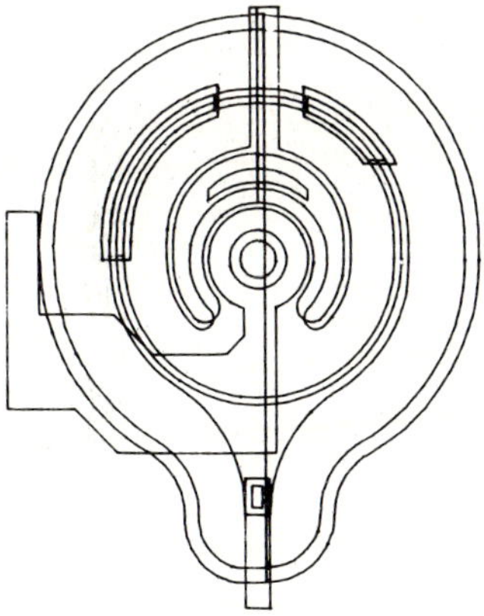

FIG. 12-3 High-voltage lateral PNP (<5 mA).

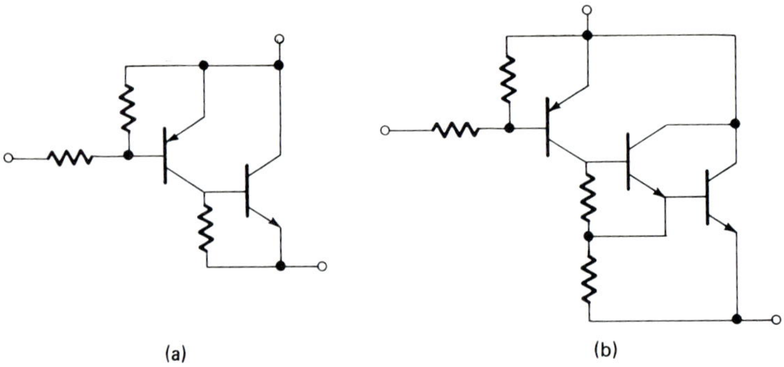

FIG. 12-4 (*a*) High-voltage composite PNP-NPN. (*b*) High-voltage composite PNP-NPN Darlington.

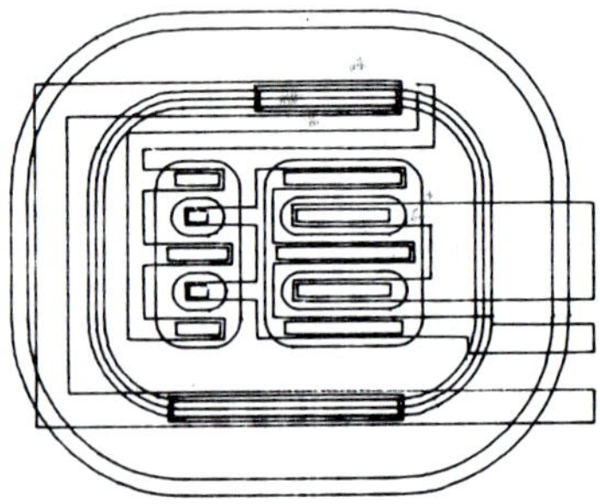

FIG. 12-5 High-voltage NPN Darlington (50 mA).

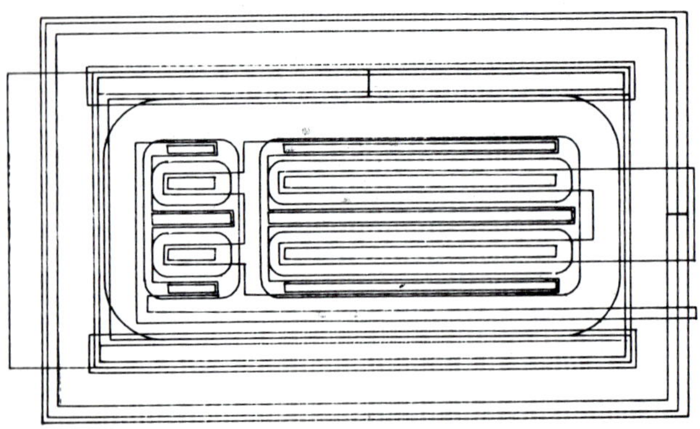

FIG. 12-6 High-voltage NPN Darlington (500 mA).

employed this output structure (originally used in the ubiquitous ULN-2003A) in conjunction with a level-triggered latch. It is a well-established structure and is presented in Fig. 12-6.

12.6.4 NPN High-Power Darlington (1.5 A)

Although not initially used in BiMOS PICs, the 1.5-A NPN structure is coming into use as devices oriented toward motor and hammer inter-

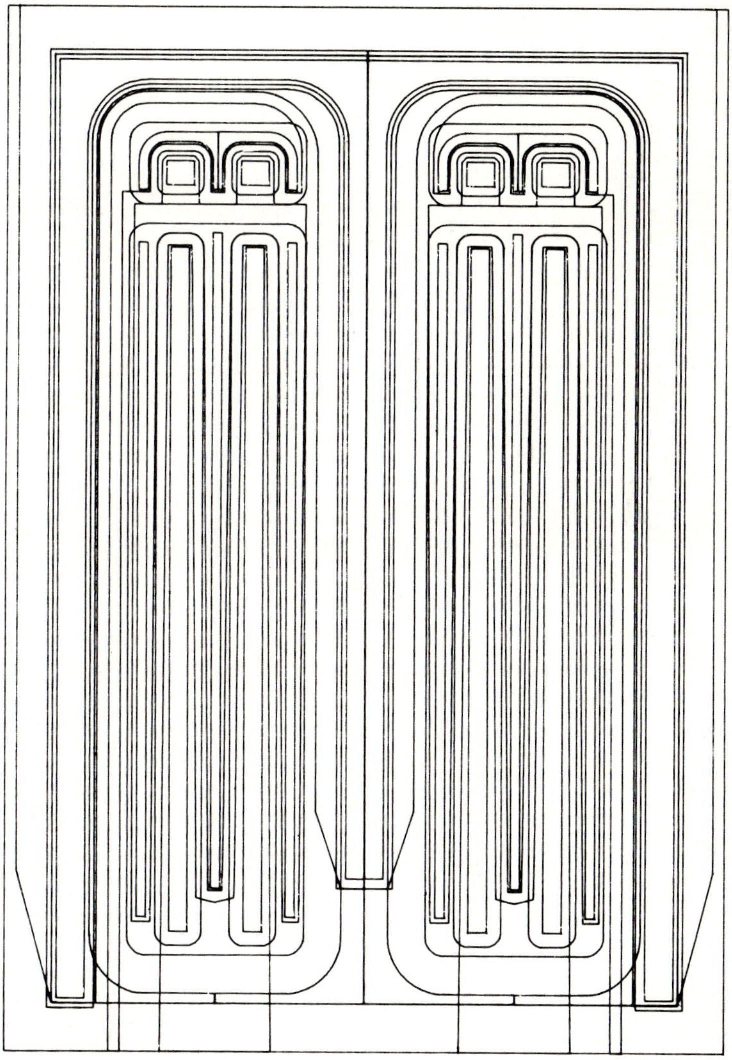

FIG. 12-7 High-voltage NPN Darlington (1.5 A).

face applications are designed. Here again is a well-proven Darlington that has been extensively used in various quadruple arrays (shown in Fig. 12-7).

12.6.5 High-Voltage, Low-Current NPN (5 mA)

Many designs require use of high-voltage components for level shifting purposes. Such transistors are usually very low-current (less than 100 μA), and minimum geometries are required for the basic high-voltage transistor. However, due to the high-voltage requirements it is necessary to use N^+ guard rings in the collector "tub" to prevent surface inversion. Such a transistor is shown in Fig. 12-8.

12.6.6 CMOS Latch Cell

The first Sprague BiMOS PICs were 4- and 8-bit latch and driver functions. These were a customer-stimulated extension of the octal Darlington arrays, and provided a single-package combination of what many users were previously accomplishing with a pair of DIPs. The latch is a level-triggered D type and is shown in Fig. 12-9. This cell is considered one of the standard building blocks for generating new BiMOS ICs. The relevant truth table and timing conditions for the latch cell are shown in Fig. 12-10*a* and *b*, respectively.

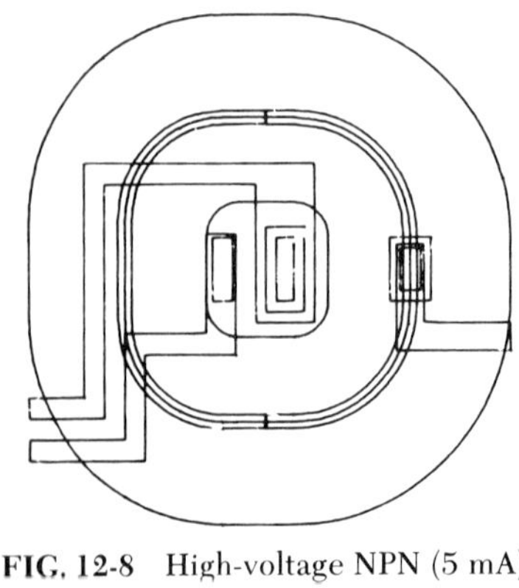

FIG. 12-8 High-voltage NPN (5 mA).

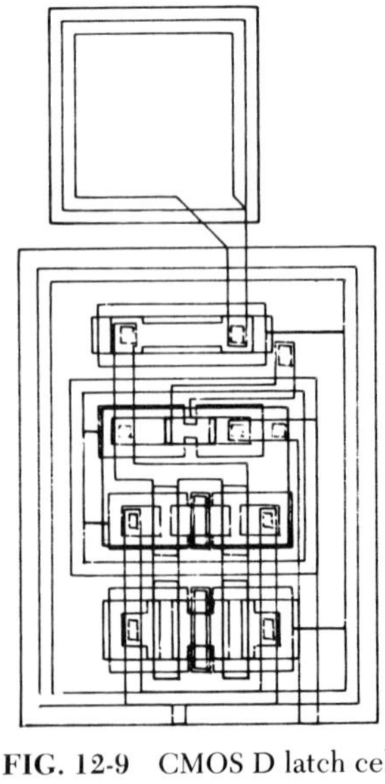

FIG. 12-9 CMOS D latch cell.

Timing conditions
(Logic levels are V_{DD} and V_{SS})

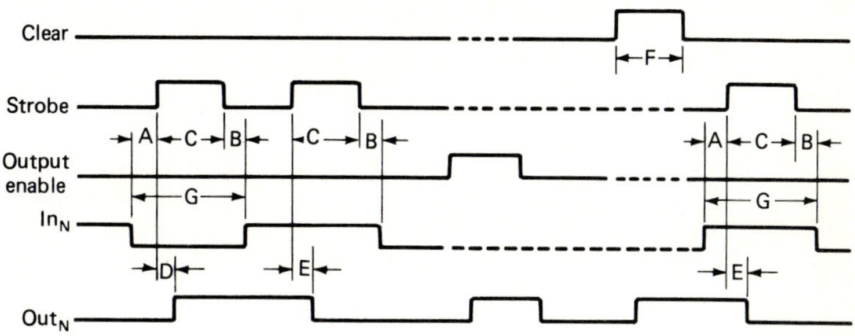

A. Minimum data active time before strobe enabled (data set-up time) .... 100 ns
B. Minimum data active time after strobe disabled (data hold time) .... 100 ns
C. Minimum strobe pulse width .... 300 ns
D. Typical time between strobe activation and output on to off transition .... 500 ns
E. Typical time between strobe activation and output off to on transition .... 500 ns
F. Minimum clear pulse width .... 300 ns
G. Minimum data pulse width .... 500 ns

(b)

Truth table

In_N	Strobe	Clear	Output enable	Out_N t-1	Out_N t
0	1	0	0	X	Off
1	1	0	0	X	On
X	X	1	X	X	Off
X	X	X	1	X	Off
X	0	0	0	On	On
X	0	0	0	Off	Off

X = irrelevant
t-1 = previous output state
t = present output state

(a)

FIG. 12-10 (*a*) D latch truth table. (*b*) D latch timing diagram.

12.6.7 Shift Register Cell

The CMOS shift register (S/R) is an edge-triggered design; the data present (single bit per clock) at the input are transferred to the shift register during a positive-going transition of the Clock input. Succeeding clock pulses advance the input data toward the data output of the shift register. Many BiMOS PICs using the S/R make considerable use

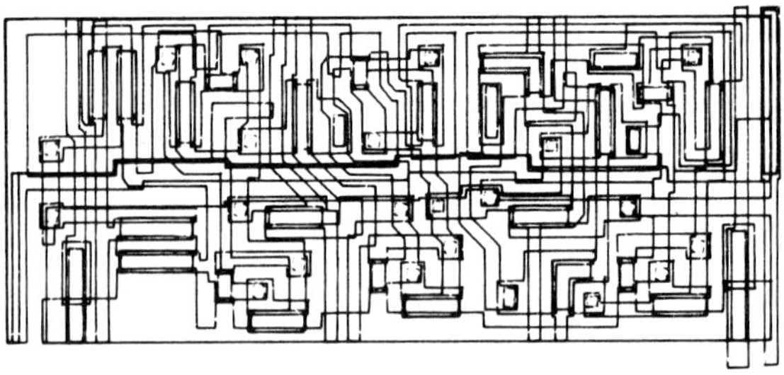

FIG. 12-11 CMOS shift register cell.

of the ability to cascade stages—thus reducing input-output (I/O) lines in the system and resulting in a reduction in interconnects, space, and complexity.

The shift register cell is shown in Fig. 12-11; improvements and size reductions in this function have been made. A typical truth table for a serial-to-parallel BiMOS PIC is shown in Fig. 12-12*a*, while Fig. 12-12*b* provides timing conditions. The basic improvements in this cell design have been evolutionary and provide improved speed and timing, along with size reduction.

12.6.8 Input Protection Circuitry

Since most BiMOS PICs have metal-gate CMOS input circuitry, they are, like any MOS logic, static-sensitive and require input protection. All CMOS input-output lines have static protection; the circuitry in use is shown in Fig. 12-13 and is very standard in the industry. As newer BiMOS circuit designs evolve they may be executed using bipolar inputs, and hence CMOS logic may be "sandwiched" between the bipolar inputs and the bipolar power outputs. Such an approach (already utilized in a 20-channel high-voltage driver) isolates the static-sensitive CMOS and could dramatically reduce static problems associated with handling, testing, shipment, assembly, etc. Note that there is no current effort underway to utilize the "sandwich" approach.

12.6.9 Circuit Resistors

Three basic techniques are available for implementing BiMOS IC resistors: diffused B&R; ion-implanted resistors; and polysilicon (thin-film) resistors. Of the three possibilities the B&R diffused resistors and the ion-implanted polysilicon types are both commonly used. Ion-implanted silicon resistors with their PN junctions and larger space requirements (larger than polysilicon) have not (as yet) been useful to

Series UCN-4820A truth table

Serial data input	Clock input	Shift register contents $I_1\ I_2\ I_3 \ldots I_8$	Serial data output	Strobe input	Latch contents $I_1\ I_2\ I_3 \ldots I_8$	Output enable	Output contents $I_1\ I_2\ I_3 \ldots I_8$
L	↑	$H\ R_1\ R_2 \ldots R_7$	R_7				
H	↑	$L\ R_1\ R_2 \ldots R_7$	R_7				
X	↓	$R_1\ R_2\ R_3 \ldots R_8$	R_8				
		$X\ X\ X \ldots X$	X	L	$R_1\ R_2\ R_3 \ldots R_8$		
		$P_1\ P_2\ P_3 \ldots P_8$	P_8	H	$P_1\ P_2\ P_3 \ldots P_8$	L	$P_1\ P_2\ P_3 \ldots P_8$
					$X\ X\ X \ldots X$	H	$H\ H\ H \ldots H$

L = Low logic level
H = High logic level
X = Irrelevant
P = Present state
R = Previous state

(a)

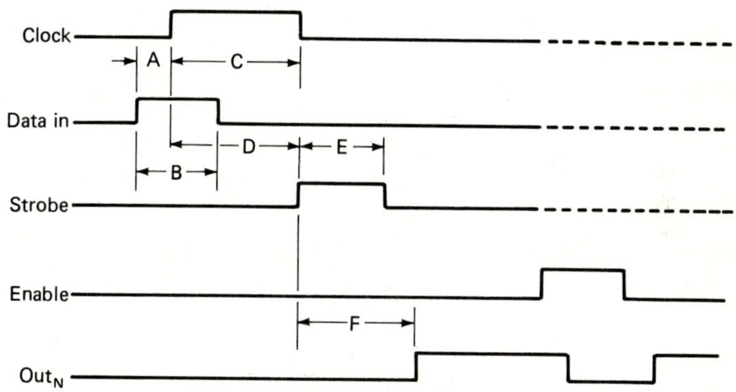

Timing conditions
(Logic levels are V_{DD} and V_{SS})

	V_{DD} = 5.0 V	V_{DD} = 15 V
A. Minimum data active time before clock pulse (data set-up time)	250 ns	150 ns
B. Minimum data pulse width	500 ns	300 ns
C. Minimum clock pulse width	1.0 μs	250 ns
D. Minimum time between clock activation and strobe	1.0 μs	400 ns
E. Minimum strobe pulse width	500 ns	300 ns
F. Typical time between strobe activation and output transition	1.0 μs	1.0 μs

(b)

FIG. 12-12 (*a*) Shift register truth table. (*b*) Shift register timing diagram.

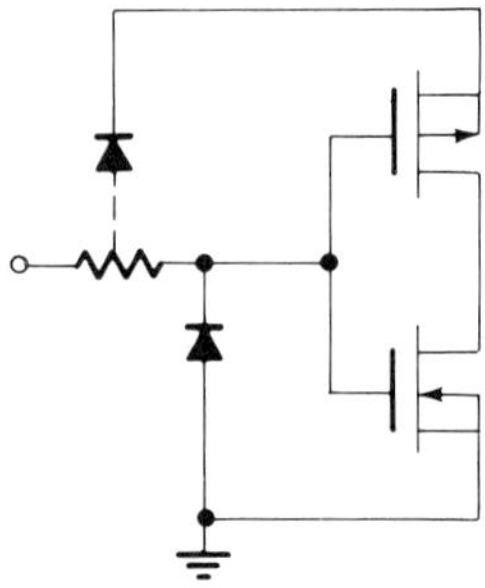

FIG. 12-13 Input protection circuitry.

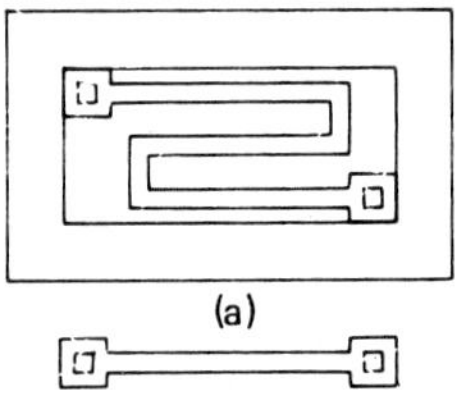

FIG. 12-14 (*a*) Diffused circuit resistor. (*b*) Implanted polysilicon resistor.

BiMOS PIC designs. Figure 12-14*a* and *b* contains two resistors of identical value, but of dramatically different area.

The base and resistor (P^+) diffusion is part of the normal process sequence and is used wherever relatively small values of resistors (less than 10 kΩ, typically) are incorporated into BiMOS circuits. For larger resistor values the ion-implanted polysilicon technique is very useful, although it adds manufacturing (process) steps. Another potential consideration regarding polysilicon resistors is their inherent negative (approximately −2000 ppm/°C) temperature coefficient. For some extended-temperature applications either the resistor-related specification limits must be broadened and/or the process controls tightened when incorporating implanted polysilicon resistors.

12.7 BiMOS POWER INTEGRATED CIRCUITS

A device which represents the composite nature of BiMOS design is shown in Fig. 12-15. The design is the UCN-4820 series IC, and 8-bit serial-in parallel-out (latched) high-current driver. The IC incorporates an 8-bit shift register, an 8-bit D latch, and eight Darlington outputs rated at 350 mA (guaranteed) per channel. To relate this to the earlier discussion on cells and transistors, etc., this design incorporates the Darlington of Fig. 12-6, the latch of Fig. 12-9, the shift register of Fig. 12-11, and additional control circuitry for output enable, clock, strobe, etc.

Another design which uses more random logic is the decoder/driver chip shown in Fig. 12-16. This device (UCN-4805) is a three-line (BCD) to eight-output decoder which includes 60-V output source capability. The various elements of this design include the high-voltage PNP (Fig. 12-3); a 50-mA Darlington (Fig. 12.5); a low-current, high-voltage NPN (Fig. 12-8); and implanted polysilicon resistors (Fig. 12-

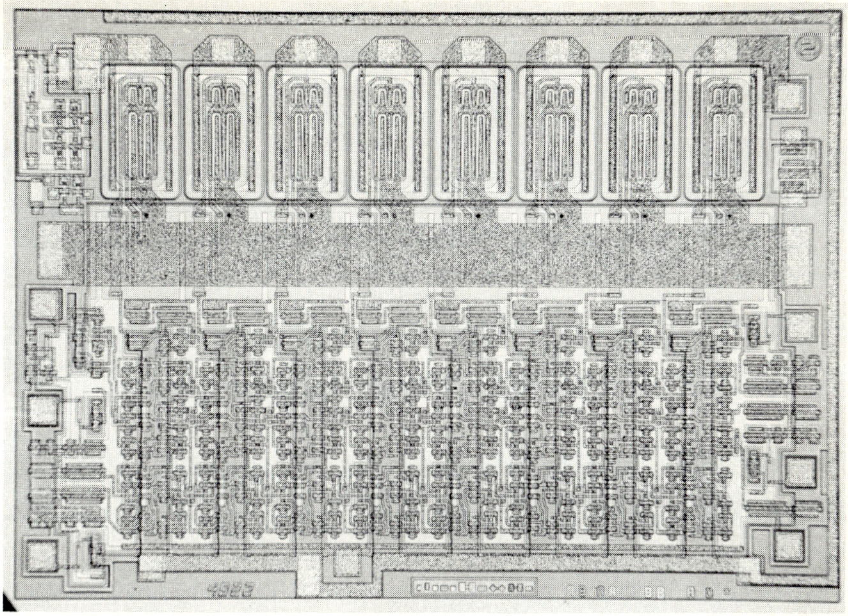

FIG. 12-15 Photomicrograph of the UCN-4822.

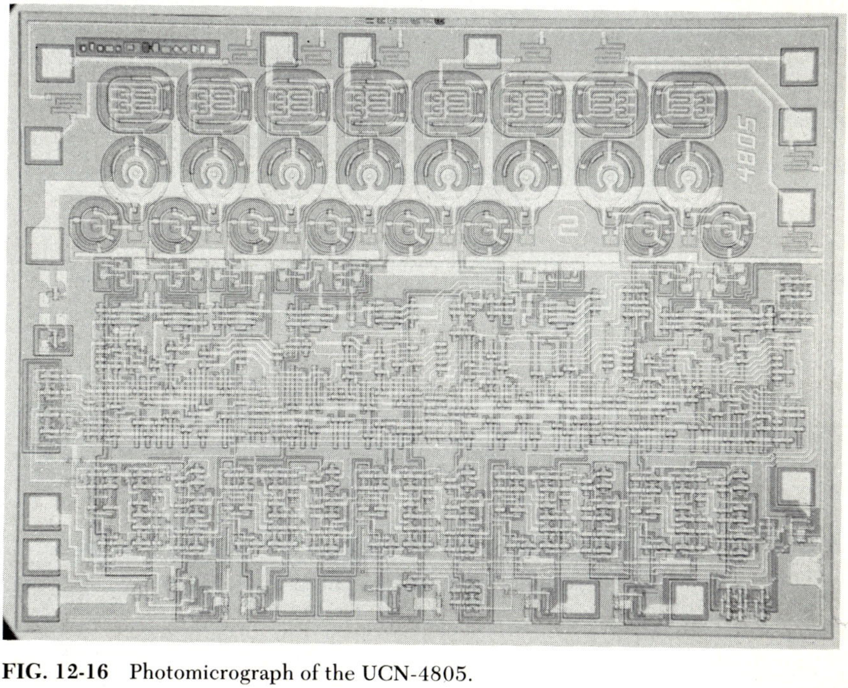

FIG. 12-16 Photomicrograph of the UCN-4805.

14*b*). Obviously, such a random logic decoder is not nearly as easily executed as repetitive designs using shift registers, latches, and output structures only.

Improvements in BiMOS designs are being implemented (as will be shown), particularly in the area of size reductions (which are cost-related) and efforts to enhance dynamic characteristics of the CMOS. Speed increases are not easily obtained, since one of the factors (B&R junction depth) runs counter to improvements in breakdown voltage. Speed requires shallow diffusions with small channel spacings in the MOS, while the high-voltage bipolar section requires deep, contoured diffusions and thus these two concerns are at odds.

12.8 BiMOS APPLICATIONS

The origins of BiMOS interface PICs stem from a need to satisfy customer requirements; an example is the combining of an 8-bit latch and eight NPN Darlingtons (an IC array which became available to satisfy demand). The development of both process and product (the UCN-4401A, a 4-bit latch/driver) began concurrently. Normally, the pioneering of both product and process would not be undertaken concurrently, but the 4-bit latch/driver (which after it was developed was swiftly made into an 8-bit device) went ahead. A vehicle (device) was necessary, and the latch/driver development proved successful.

One of the typical applications of these units is shown in Fig. 12-17;

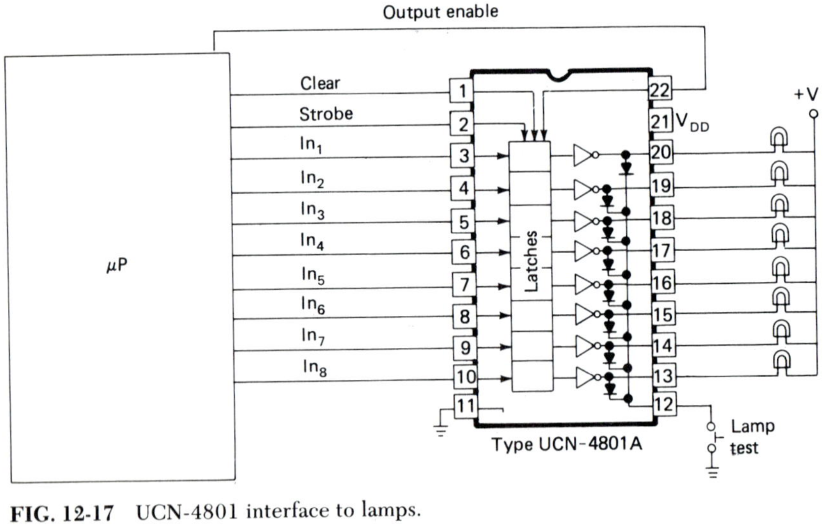

FIG. 12-17 UCN-4801 interface to lamps.

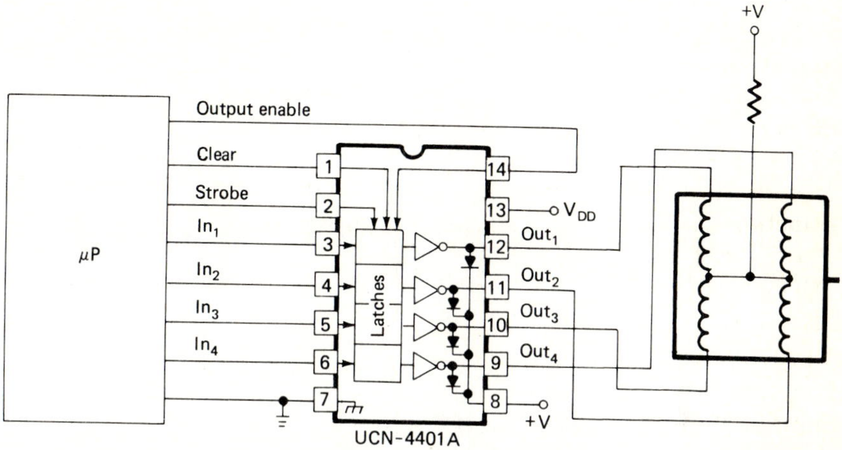

FIG. 12-18 UCN-4401 interface to stepper motor (unipolar).

this represents a microprocessor-to-lamp interface using the latch/driver IC, but could very easily represent a relay, solenoid, LED, thermal printer, or other general interface requiring a device capable of 350 mA at 50 V per output line at reduced duty cycles (350-mA output is not simultaneously allowed as the package power rating prohibits this. The package thermal rating will not allow full current in all outputs at 100% duty cycle.)

The 4-bit device is shown in Fig. 12-18 in a microprocessor-to-stepper-motor interface; a number of variations on this theme may be used. Stepper motors may be driven in a unipolar scheme (Fig. 12-19*a* shows waveforms for single-phase or "wave" drive, and Fig. 12-19*b* shows

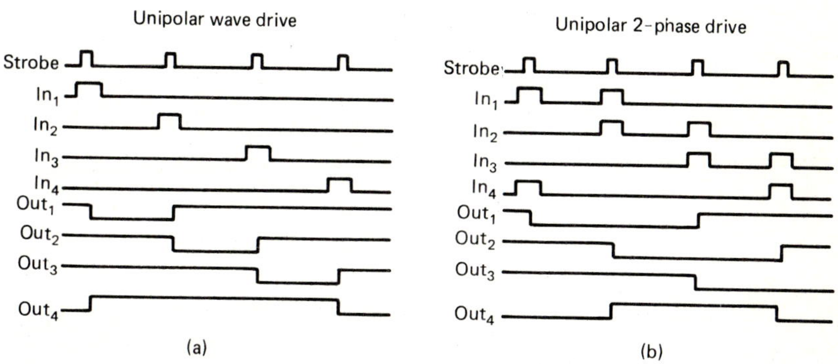

FIG. 12-19 (*a*) Unipolar "wave" (single-phase) drive waveforms. (*b*) Unipolar two-phase drive waveforms.

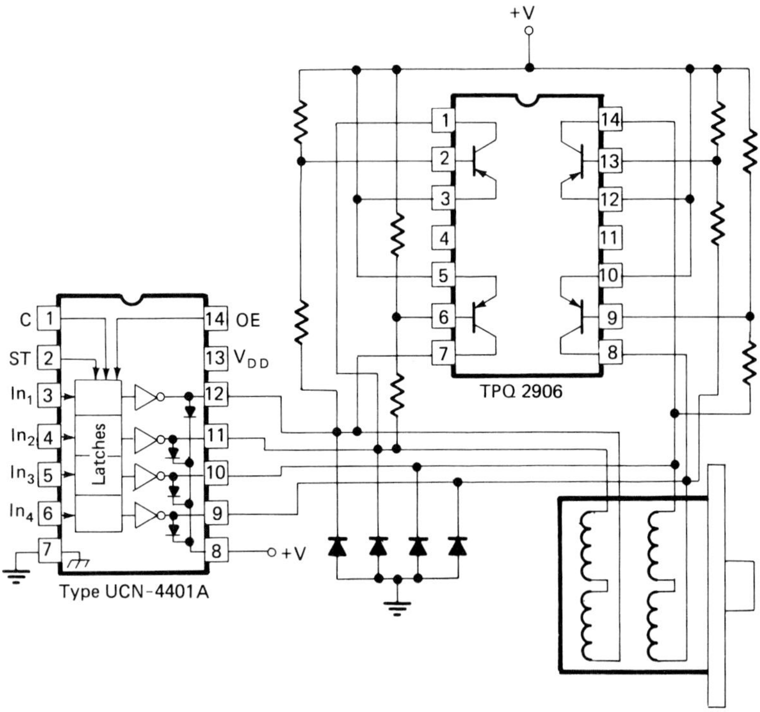

FIG. 12-20 UCN-4401 combined with quad PNP array and diodes to form bipolar stepper motor interface.

waveforms for the higher-torque two-phase drive). Also possible are bipolar schemes such as the combination of the UCN-4401A with a DIP quadruple PNP array and clamp diodes as shown in Fig. 12-20. These diagrams represent the more commonly used techniques, but many others are available, limited mainly by the creative talent of the designer.

Another very common use of BiMOS PICs is in interfacing to flat-panel (planar-style) displays, particularly for high-volume vacuum fluorescent (VF) applications. Both a serial-in/parallel-out (the UCN-4810A, a 10-bit shift register–latch–source driver) and a parallel-in/parallel-out (the UCN-4815, an 8-bit latch–source driver) device are shown in Fig. 12-21. This represents one of the smaller, less complex applications, a 20-character, 16-segment ("British flag" style) alphanumeric system. Other, larger and more complex display systems utilize these devices separately, or in various combinations, and the choice is dependent upon system I/O considerations (serial versus parallel, etc.). Since the serial-in unit is cascadable, the I/O lines are usually

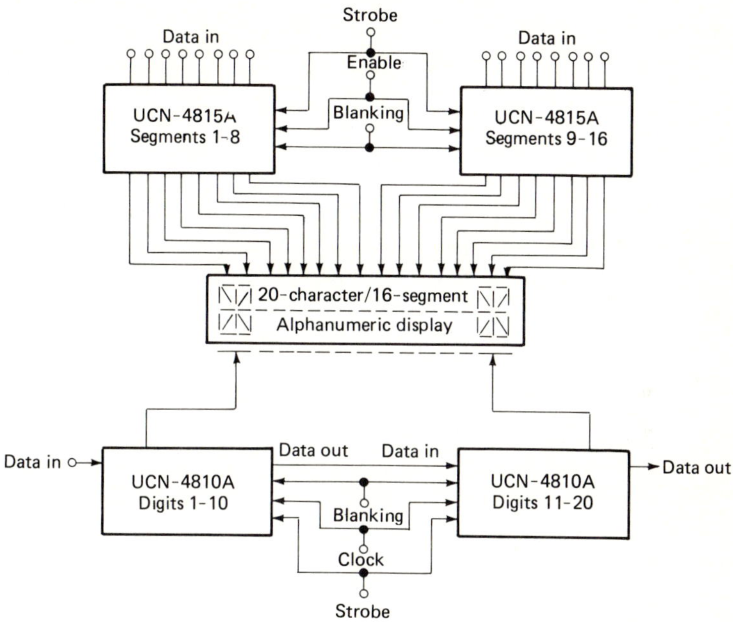

FIG. 12-21 UCN-4810 and 4815 interface to VF display (20 character, 16 segment).

fewer, and wiring (interconnect) is generally simplified. The use of the serial 4810-type device readily allows loading a single bit into the shift register and advancing it (thus activating only a single output in the string) for use as a character or digit driver in multiplexed (MUXed) displays.

Depending upon system requirements, various connection and timing prospects exist. For display applications the dot or segment data may be loaded in a serial 40-bit chain (as shown in Fig. 12-22), with the grid (digit or character) data handled separately. Normal VF display blanking periods are 20 μs, and with the availability of 5-MHz BiMOS the data may be loaded (for many system requirements) during the blanking period. However, since these units contain a latch this is not necessary; the new data may be loaded during the output On period (greater than 100 μs, but dependent upon display size, number of characters, etc.). Another alternative is shown in Fig. 12-23; the grid data are again separate, but the use of four data lines allows much faster input of dot (anode) data, although it uses more I/O lines.

Another early use of these ICs was for interface to the elements of a thermal printer (ceramic hybrid type). Although the UCN-4810 was intended for 60-V, 50-mA applications, it proved to be very successful

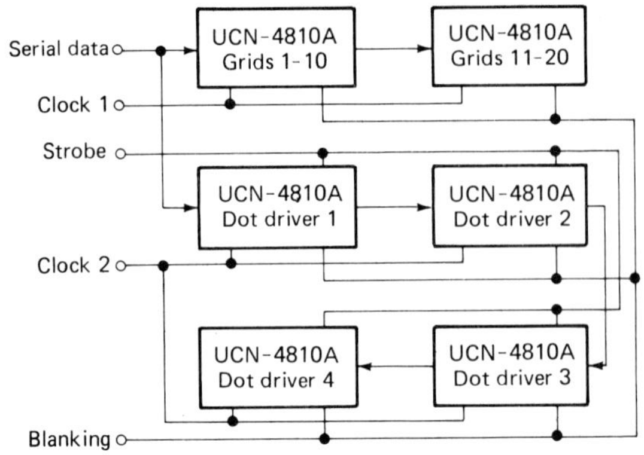

FIG. 12-22 Display block diagram: dot or segment data connected serially.

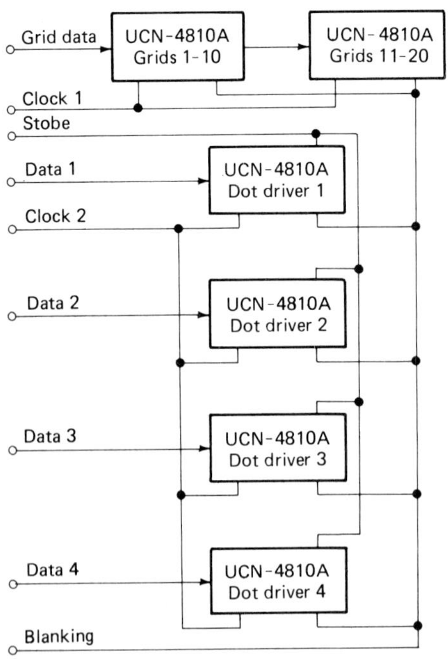

FIG. 12-23 Display block diagram: dot or segment data connected to parallel bus.

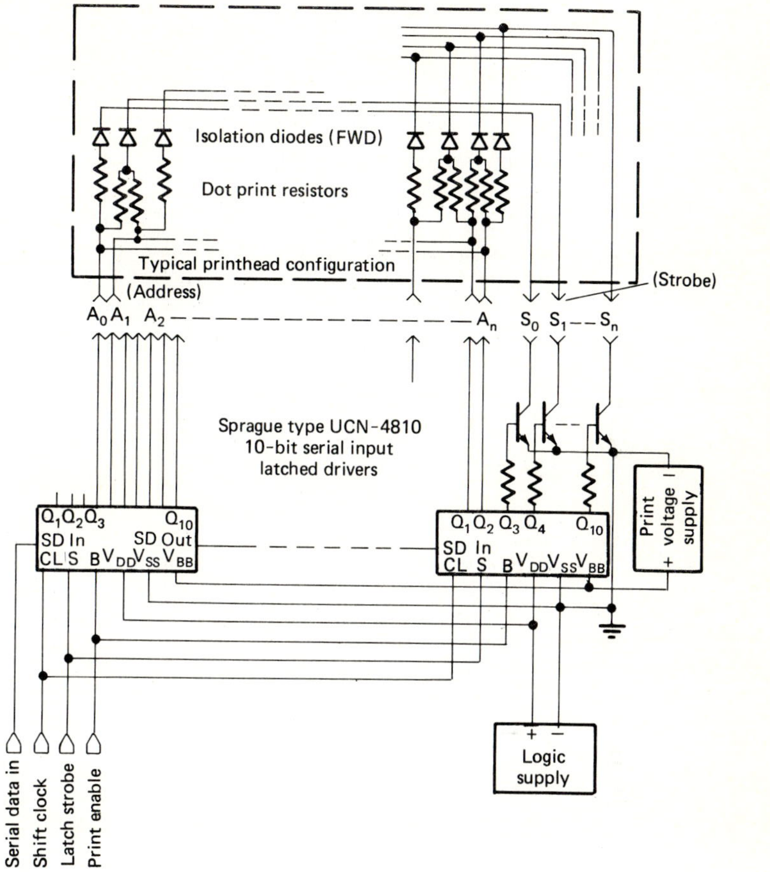

FIG. 12-24 UCN-4810 interface for hybrid thermal printer.

in a lower-voltage, slightly higher-current use (Fig. 12-24). As this is written a 32-bit serial-parallel IC (with 40-V, 100-mA outputs) is complete and will greatly improve the prospects for thermal printer applications.

The serial-to-parallel approach is gaining momentum; an early product entry and type of application are shown in Fig. 12-25. The figure is for an electrosensitive printer, but could easily be for a thermal printer, relays, print hammers, etc. This 8-bit serial-parallel latch/driver has a 500-mA (maximum) output current rating and is available in three voltage ratings (50 V, 80 V, and 100 V), and, in an effort to satisfy additional applications, higher-current products are in the offing, both in parallel and serial form. The printer industry has a considerable need for (and interest in) hammer driver ICs that contain logic

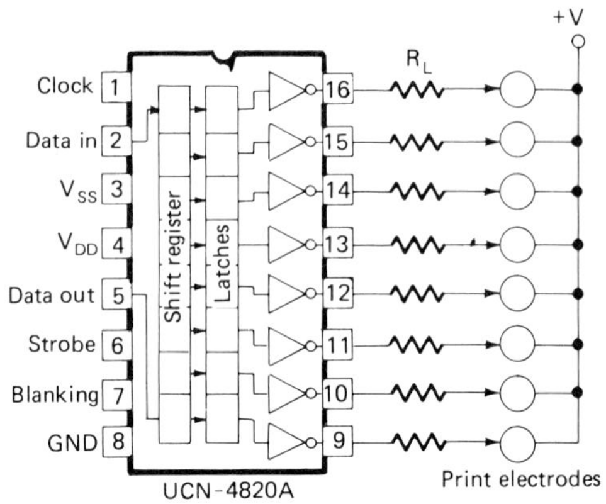

FIG. 12-25 UCN-4820 interface for electrosensitive printer.

plus "muscle." Already under way are a 4-bit serial-to-parallel BiMOS PIC with 60-V, 1.5-A outputs, and a latched, parallel version with the same output ratings. Further evolution will provide increased levels of current, increased breakdown voltage, increased power ($P = V \times I$), and more outputs per chip. This will in turn pose new and challenging problems (particularly in testing, etc.) along with the opportunities that will be created.

12.9 BiMOS TESTING

From its inception BiMOS has faced difficulties associated with testing devices accurately, quickly, and (of course) automatically. Largely because of the rather limited (even as yet) merged-technology product incorporating both low-level circuitry (logic, etc.) and high-current, and/or high-voltage structures within one device, automatic test equipment (ATE) has lagged behind the need to perform both dynamic and power testing upon the same unit. Much in-house test equipment has been necessary, but newer test equipment is becoming available which will more adequately satisfy needs for automatic test equipment capable of determining device timing and such test challenges as measuring the $LV_{CE(sus)}$ of power outputs. The collector-emitter sustaining voltage, $LV_{CE(sus)}$, is a critical concern to any PIC which operates inductive loads. It is also a difficult test to perform and most ATE capable of dynamic device testing has little or no provision for the higher-current and high-voltage combinations of collector-emitter sustaining voltage measurements and/or high-current saturation voltage measurements.

Most early PICs were largely power buffers (linear ICs) and could, in many instances, be tested either with equipment designed for linear use or with modified transistor testers. With the advent of technologies such as BiMOS, most ATE became inadequate since the various input dynamic tests (setup time, hold time, strobe pulse width, etc.) required for shift registers and latches could not easily be satisfied. Hence, in-house implementation (often microprocessor-based and dedicated to a particular product) was necessary until new ATE could satisfy this dynamic and power combination. Obviously, this problem was an immediate concern to merged logic and power IC manufacturers, but with shipment of the product the concern then shifted to the user. A user lacking suitable ATE cannot perform incoming tests in any meaningful, comprehensive manner, nor can many (or most) of the various independent test and reliability operations—an obvious dilemma to many, but one which is (with time) diminishing as new, more powerful ATE is made available.

Another testing concern relates to the use and sale of BiMOS PICs for extended-temperature applications, particularly military and avionics systems. It has always been extremely difficult to guarantee dynamic characteristics at temperature extremes (particularly cold), and much device characterization and added test capability may be required to satisfy such uses.

The low volume and higher unit prices of the military and avionics market should allow much of the needed temperature testing of dc limits. However, for most commercial requirements with extended temperature limits the solution is sound *characterization* of devices and the use (by both supplier and user) of sensible *guard bands*. As noted, testing at low temperature poses significant problems (including icing of sockets, contacts, etc.), although it has become increasingly feasible to test units at elevated temperatures, via automatic "hot rail" techniques.

Guaranteeing device dynamics with requirements for timing conditions of under 100 ns is no easy task; and satisfying timing requirements at extremes of temperature compounds device testing massively. Having sound characterization and a stable, mature process technology and design are preventive solutions to a great many of the problems of dynamic testing (at any temperature).

12.10 BiMOS TEMPERATURE LIMITATIONS

Presently the maximum recommended junction temperature for Sprague BiMOS PICs is +125°C. This limit is predicated on the metal-gate CMOS circuitry and certainly contains a large element of tradition (CMOS ICs with extended temperature range are usually specified to +125°C). The concern for maximum junction temperature is largely related to sodium in the circuit oxides and its potential for drift (accelerated by temperature and bias) and the device threshold shifts which may result.

MOS circuitry requires considerable care relating to surface cleanliness, unlike bipolar circuitry, which requires much less attention pertaining to surface contamination. Any sodium-related contaminants, and their effect on circuit thresholds and drift, are greatly activated by temperatures above +125°C.

Considerable biased life testing has been done (successfully) at +150°C, and more recent activity is oriented toward testing at + 175°C. These higher temperatures are presently being utilized only to accelerate life testing, although it is anticipated that in the future, BiMOS PICs will become available with a recommended maximum junction temperature of +150°C. Extended- and full-temperature-range devices may be purchased which are processed through a product enhancement program which includes burn-in at +150°C for 96 h, temperature cycle, testing at elevated temperature, etc.

Further reductions in sodium content and improved control of thresholds should subsequently allow extending junction temperature (to +150°C) without the concern and need for the added time and cost of a burn-in program. This certainly will further enhance the prospects for BiMOS PICs in both extended- and full-temperature environments such as military and avionics systems.

12.11 MILITARY AND AVIONICS USES

The remarkable acceptability of BiMOS PICs has been very apparent, and is exemplified by the very early interest of designers of military and avionics systems. The interest was so acute that incorporating both extended temperature range and hermetic packaging went ahead, despite some early trepidation (internally). The concerns for the early incorporation of BiMOS PICs into high-reliability, extended-temperature-range applications and hermetic packages proved largely to be "fear and superstition" resulting from the lack of substantiating data (at the time) to aid decision making. The decision to proceed was largely based upon intuition and an assessment of the gain versus risk of aggressively pursuing high-reliability BiMOS ICs.

12.11.1 Circuit Evaluation

Prior to committing BiMOS wafers for assembly in hermetic packages for extended-range use, the wafers are evaluated for electrical characteristics to determine whether they are suitable for high-reliability operation at −55 to +125°C. Bipolar gain at low temperatures, CMOS thresholds at high temperatures, and leakage, output saturation voltages, circuit resistor tolerances, and other measurements over a wide temperature range are correlated with the data for +25°C. This precludes premature assembly of devices into costly, hermetic packages before it has been determined that they are well-suited to the requirements of extended-temperature operation.

12.11.2 Screening

BiMOS PICs for military and avionics applications are screened to MIL STD 883 class B, and are often processed through a 168-h at +125°C burn-in per a standard procedure for class B high-reliability hardware. Devices are usually specified (and tested) at three temperatures, −55°C, +25°C, and +125°C, although some of the display-related BiMOS units are designated for use at −40°C, +25°C, and +85°C (neither the display itself nor the operator is likely to withstand a broader temperature range). The −40 to +85°C is not widely in use and may be diminishing as the BiMOS PICs are more capable and suitable for full (−55 to +125°C) range.

12.12 HERMETIC PACKAGING

The initial development of BiMOS PICs was stimulated by commercial and industrial applications; the first were plastic DIPs. The addition of the military and avionics market required use of hermetic DIPs, and hence the next step was to incorporate side-brazed ceramic DIPs (H-DIPs), which are built domestically. This satisfied the more stringent hermetic requirements, but the expense of the side-brazed package combined with domestic labor and assembly costs required the further addition of a lower-cost "CER-DIP" hermetic package.

The CER-DIP package is coming increasingly into use; it costs considerably less than the side-brazed units, while still affording the hermeticity required by most applications. Thus, BiMOS PICs are available in three package types (plastic, CER-DIP, and side-brazed ceramic DIP), and in three temperature ranges (−20°C to +85°C, −40°C to

+85°C, and −55°C to +125°C, the second and third ranges having guaranteed device limits and conditions). Obviously, the cost of the hermetic, high-reliability BiMOS PICs is a function of temperature, package type, and screening requirements. The commercial hardware, in plastic DIPs, is the lowest-cost; and the size-brazed ceramic package, rated at −55°C to +125°C and combined with the cost of burn-in, is the most expensive (a 10:1 price differential is not unusual).

BiMOS PICs are being incorporated into an increasing number of military and avionics systems, and are also coming into use in telecommunications systems (which require extended-temperature-range performance and/or tightened quality control and screening).

12.13 CHIP SIZE (COST) REDUCTIONS

Design geometries of BiMOS PICs have changed dramatically since the original units appeared, and device designs are being "shrunk" as a result of the improvements which have evolved in the past few years. Size reductions and improved yields constitute important steps in the maturation of BiMOS, and the costs of most of the older product will decline as devices are redesigned.

The dramatic chip reductions afforded by the new design rules and process improvements are shown by comparing the initial UCN-4810A design with its successor. Figure 12-26*a* is the original design, while Fig. 12-26*b* is the UCN-5810A, a new, smaller design photographed here at the same magnification. The newer chip is approximately half the area of the original, providing a significant cost factor relating to the ability of BiMOS to compete with alternative technologies. Device designs will capitalize on the newer, smaller geometries, thus affording improved market penetration against more mature product alternatives (discretes, combinations of ICs, or ICs with discretes) with their (generally) lower unit prices.

In addition to the various competing technologies mentioned earlier, BiMOS PICs must also compete with multipackage alternatives. Thus, early large-volume market penetration may be slowed by the inability to compete effectively with combinations of TTL or CMOS logic with suitable discretes (power transistors or Darlingtons) or power arrays. Two (or more) package design solutions are very much in competition with single-chip BiMOS PICs, and the ability to minimize chip size (and costs) with newer design geometries and yield improvements is of considerable importance. New BiMOS designs must be aggressive, and unit

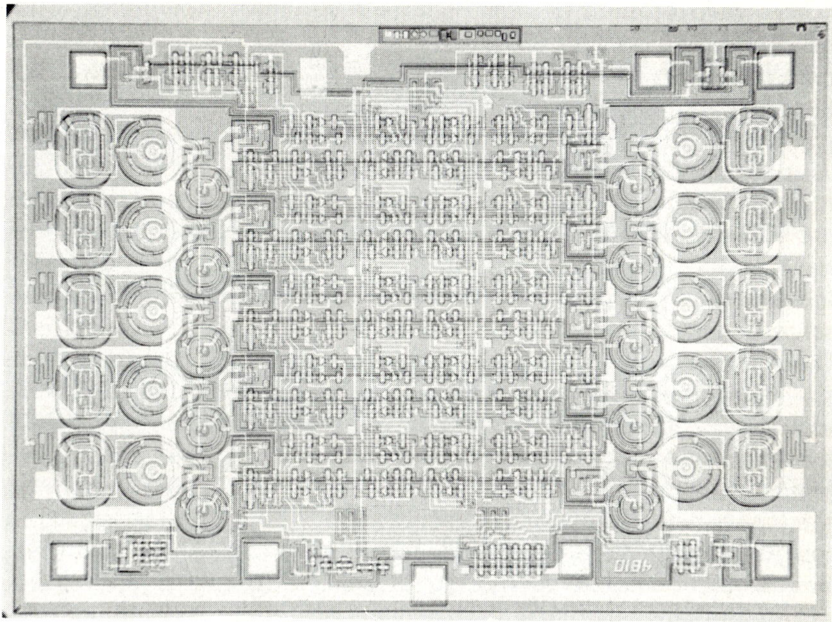

(*a*)

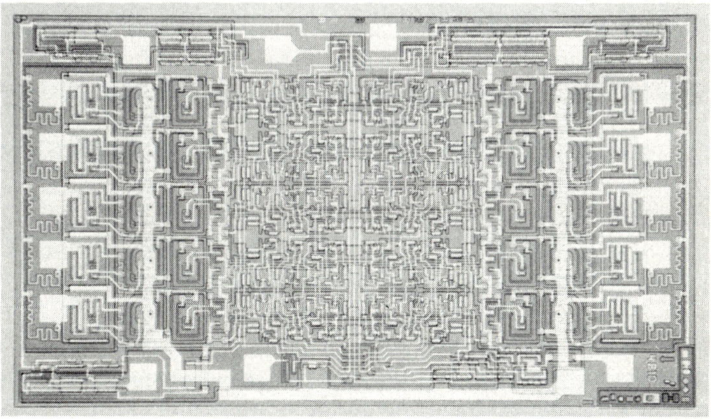

(*b*)

FIG. 12-26 (*a*) Photograph of the original UCN-4810. (*b*) Photograph of the UCN-5810.

pricing must be directly competitive with all other available solutions; otherwise market opportunities will be restricted.

12.14 NEW-PRODUCT EVALUATION

The key ingredient of new-product selection is a continual dialogue with customers and potential customers. Assessing opportunities and providing a circuit solution for customer needs is the secret to success; this, of course, leans very heavily on the applications engineering arm of marketing. "Forcing" new PICs has never worked well, but offering solutions via answers, products, and service to satisfy customer problems seem to be a very successful and accepted solution.

The new-product concepts for the immediate future seem to fall into the "more of the same, but with extension" category. Activity is already under way (or planned) for BiMOS products encompassing (1) higher power (quadruple 1.5 to 2-A latch/driver ICs); (2) higher voltage (improvements in display-related ICs with output voltage increased to 150 V to 200 V); (3) additional outputs per package (a 32-bit display IC with high-voltage outputs); and (4) combinations of increased voltage, current, power, and/or output lines (some specific product plans, and many discussions—including 128 outputs per chip). Most of the forthcoming BiMOS PICs will result from customer needs which extend and/or combine available hardware solutions. An occasional "revolution" should occur, but most of the new-product activity (including BiMOS) is of an evolutionary nature. A recent evolution is the 32-bit serial-to-parallel driver shown in Fig. 12-27 (the UCN-5818); this display IC has 60-V (80-V with test selection) source outputs capable of 40 mA. Also a similar 32-bit thermal printer IC with 40-V, 100-mA outputs is about to be introduced (Fig. 12-28: the UCN-5832).

12.15 ALTERNATE SOURCING

As this is written there are no official or contractual alternate sourcing arrangements for BiMOS PICs; however, some of the ICs presently being manufactured are effectively alternate-sourced via a competitor duplicating function (taking into account parameters, etc.) even though another (similar) technology may be used. Some of the Sprague BiMOS PICs are alternate-sourced in this manner by Texas Instruments and/or Gould AMI; Texas Instruments uses Sprague's BIDFET process, while Gould AMI uses a CMOS process with "stacked" devices to obtain 60-V ratings.

FIG. 12-27 Photograph of the UCN-5818.

Such functional alternates may succeed in many products; however, future, higher-current BiMOS PICs will likely require a much more direct approach. A technology "swap" or licensing may be a necessity (and reality); but broadening the market for BiMOS PICs must occur first. A larger market will attract more suppliers, both as alternate

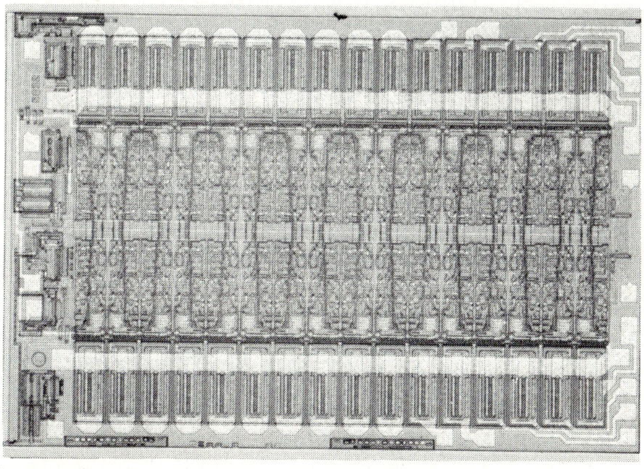

FIG. 12-28 Photograph of the UCN-5832.

sources and as purveyors of new device designs, and will justify resources for new vendor entry into such an activity.

Design resources are (and will be) in heavy demand, and fragmenting a modest market produces little of value to any supplier. Thus, there is a need for a larger market to "grow" the supply and use of BiMOS PICs, but this is somewhat impeded by the lack of alternate suppliers—a classic chicken and egg situation for both supplier and user.

12.16 THE FUTURE OF BiMOS POWER INTEGRATED CIRCUITS

Predicting the future of high-technology products and/or processes is, obviously, tenuous and uncertain. Despite the more complex and costly processing requirements for BiMOS PICs, the technology shows every sign of expansion in the 1980s. As of this time (early 1985) no competing technology appears ready to displace the "appropriate" choice of BiMOS.

However, a constant awareness of competing alternatives is necessary, along with contingent possibilities. BiMOS is a maturing technology and will grow in use, both in systems applications and in breadth of product. Broader use of BiMOS ICs should effect a declining unit price, thus increasing its use still further.

12.17 THE "INFLUENCE"

The microprocessor stands out as "the influence" upon the need for and creation of additional BiMOS PICs. The microprocessor is not capable of driving most peripheral loads, although it is marvelously well-suited for control and calculating functions. Microprocessor I/O lines often dictate use of latches and/or shift registers to allow effective use of the microprocessor logic capabilities, and here increasingly is "the influence" directed toward BiMOS PICs for interface. Logic, control, and power may be included in BiMOS, thus freeing the microprocessor from most "housekeeping" chores. The microprocessor is so pervasive that the growth of interface PICs appears directly tied to the dramatic growth in its use.

The recent past has provided an incredible growth rate for BiMOS PICs (at least those of Sprague Electric), and these increased rates have severely tested the ability to respond to a growing market. The combinations of growth factors at work include (1) broader use of existing products, (2) new BiMOS products, (3) declining BiMOS prices, (4) new customers, (5) increased use in more specialized form (military, telecommunications, etc.), (6) supplanting older hardware combinations with BiMOS single-chip solutions, (7) improved reliability, and (8) space reductions.

With such forces at work it is not difficult to imagine growth rates for BiMOS PICs (at least this interface portion) which could average 50 percent a year through the next few years (at least well into the mid-1980s). The market available for BiMOS PICs is large (although not well-defined), and supplier market data are proprietary; however, a growth rate of 50 percent a year will draw much more attention to this technology within a very short time.

BiMOS represents one of the very exciting and challenging opportunities in the semiconductor world, and most of its success lies ahead. An ending quote from Charles Kettering seems very appropriate: "My interest is in the future because I am going to spend the rest of my life there."

Index

Abrupt junction, **2**.3
AC amplifiers, **6**.9, **6**.11
Adjustable three-terminal regulators, **11**.24, **11**.30
Anisotropic etching, **5**.16
Antimony, **5**.2
Arsenic, **5**.2
Atomic number, **2**.3
Atomic weight, **2**.3
Au-Al intermetals, **8**.27
Auger recombination, **2**.33, **2**.34
Avalanche breakdown, **2**.2, **2**.6, **2**.46, **3**.15
Avalanche effect, **2**.5, **2**.6

Ballast resistors, **1**.6, **2**.45, **4**.5, **6**.25, **11**.8
Band-gap narrowing, **2**.36, **2**.37, **2**.45
Band-gap reference, **11**.3, **11**.12, **11**.26
Base current balance, **2**.52, **2**.54
Base Gummel number, **2**.21
Base transit-time, **4**.6
Base widening, **2**.25
Base width, **5**.4
Beveled junction, **3**.20
Bias chain, **9**.25
Bias network, **9**.24, **9**.26
Bonding:
  die, **7**.2, **7**.7
  wire, **6**.3, **6**.4, **7**.15
Bonding pad, **7**.3
Boron, **5**.2
Breakdown voltages, **2**.6, **5**.17, **6**.28
Bulk parasitics, **4**.19
Buried layer, **2**.7, **2**.8, **2**.47, **4**.2, **4**.8, **4**.20, **5**.2
BV_{CBO} (collector-base breakdown voltage), **4**.13
BV_{CEO} (collector-to-emitter breakdown voltage), **3**.31, **3**.32, **4**.3, **4**.6, **4**.14, **4**.15, **4**.19, **4**.22
BV_{CES} [collector-to-emitter (drain-to-source) breakdown voltage], **3**.31, **3**.32

Case temperature, **7**.8
Centrifuge, **8**.75
Channel stopper, **4**.22–**4**.24
Clamping diode, **4**.11
Class B stages, **10**.14
CMOS, **3**.52
Collector buried layer, **5**.12
Collector series resistance, **2**.24
Collector-to-base avalanche breakdown, **2**.3, **2**.7, **2**.10
Collector width, **2**.7
Common-base current gain, **2**.12
Common-emitter current gain, **2**.14, **5**.7, **5**.14
Commutation speed, **10**.1
Compensation, **11**.20

Complementary transistor, **5.**9
Constant-current regulator, **11.**39
Control circuit, **6.**27
Copper alloy, **1.**15, **7.**7, **7.**12
Copper support, **6.**16
Corrosion:
  electrolytic, **8.**8
  galvanic, **8.**7
Cross talk, **9.**59
Crossover distortion, **10.**25
Crystal structure, **2.**3
Current-boost circuits, **11.**35
Current controllers, **10.**1, **10.**27
Current crowding, **2.**32, **2.**45
Current density, **6.**4, **6.**6
Current gain, **2.**21, **2.**23
Current-induced avalanche breakdown, **2.**46
Current induced base, **2.**25
Current-limit protection, **11.**2, **11.**8, **11.**10, **11.**15, **11.**32
Current limitation circuit, **6.**7
Current overload, **11.**7
Cutoff frequency f_T, **2.**52, **4.**6, **4.**8

Darlington PNP-NPN transistor, **10.**8
DC amplifiers, **6.**9, **6.**11
Dead time, **10.**17
Degree:
  of cure, **8.**14
  of polymerization, **8.**18
Delay time, **2.**48, **2.**49, **2.**52, **4.**19
Density, **2.**3
Dielectric breakdown, **3.**15
Dielectric constant, **2.**3, **2.**4
Dielectric isolation, **5.**15
Differential Scanning Calorimeter (DSC), **8.**14
Diffusion coefficient, **5.**5
Diffusion current, **2.**11, **2.**12, **2.**19
Diffusion length, **2.**18, **2.**36
Diffusion term, **2.**19
Diffusivity, **2.**3, **2.**12, **2.**19, **2.**29, **2.**36
Direct recombination, **2.**33, **2.**34
Dissipated power, **6.**9, **6.**11, **6.**16, **9.**3, **9.**28, **9.**49, **9.**50, **9.**68, **9.**70
Distortion, **9.**12, **9.**22, **9.**35, **9.**42, **9.**50, **9.**58, **9.**59, **9.**66, **9.**67, **9.**71, **9.**72
DMOS (double-diffused), **3.**7, **3.**8, **3.**27, **3.**47, **3.**49–**3.**52
Double-diffusion process, **5.**12
Drift term, **2.**19
Dropout voltage, **11.**8, **11.**10, **11.**30
Dual regulators, **11.**40
Dual tracking, **11.**42

E_{CR}, **2.**6, **2.**46
E_{SB}, **4.**3
Early effect, **2.**23, **2.**37
Efficiency, **6.**13, **6.**14, **9.**2, **9.**5–**9.**7, **9.**12, **9.**17
Einstein's relationship, **2.**20
Electrical switching times, **2.**48
Electromigration, **6.**3, **8.**36
  bulk, **8.**40
  grain boundary, **8.**40
  surface, **8.**40
  vacation motion, **8.**40
Electron-electron scattering, **2.**29
Electronic system, **1.**3
Emitter-base breakdown, **2.**3
Emitter degeneration, **11.**8
Emitter efficiency, **4.**9
Emitter Gummel number, **2.**18, **2.**21, **2.**36
Encapsulant, **7.**2, **7.**7, **7.**14
Energy gap, **2.**3
Epitaxial-base vertical PNP transistor, **5.**2
Epitaxial layer, **4.**3, **5.**2
Equivalent thermal circuit, **2.**39, **2.**40
Error amplifier, **11.**2, **11.**6, **11.**12, **11.**17, **11.**26

Failure rate, **8.**66
Fall time, **2.**49, **2.**50, **2.**55
Fault conditions, **11.**33
Feedback, **11.**7
Field limiting rings, **3.**22, **3.**24, **3.**25
Field plate, **3.**21, **3.**22, **3.**25, **4.**15, **4.**17, **5.**17
Fish-bone emitter, **5.**12
Fixed-negative regulators, **11.**16
Fixed regulators, **11.**11
Fixed three-terminal regulator, **11.**2
Forced current gain, **2.**53
Forward and reverse second breakdown, **4.**24
Forward safe operating area, **2.**43, **2.**45
Forward second breakdown, **2.**43, **6.**8, **6.**25
Full-bridge structures, **10.**20

Gain-bandwidth product, **5.**7, **5.**15, **5.**18
Gate, **4.**20, **4.**21

Gaussian distribution, **5**.5
Generation current, **2**.11
Glass transition temperature, **8**.21
Glue die attach, **8**.58
Graded base, **2**.20, **2**.21
Gradient, **9**.68
Grounded base, **6**.30
Gummel number, **4**.19

h_{fe}, **3**.32, **4**.6
Half-bridge structures, **10**.15
Heat sinking, **8**.56, **11**.34
Heat source, **6**.14, **6**.16
High-current transistor, **5**.7
High injection, **2**.29, **2**.34, **2**.35
High injection in the base, **2**.22, **2**.23
High injection in the collector, **2**.24
High-temperature storage, **8**.67
High-voltage processes, **4**.13, **6**.29
High-voltage transistor, **5**.16
Hole-electron scattering, **2**.29
Homogeneous base, **2**.19
Hysteresis, thermal protection with, **6**.23
Hysteresis voltage, **6**.24

I^2L (integrated injection logic), **4**.18
I^2L NAND gate, **4**.18, **4**.19
I^2L structures, **1**.4
I_{SB}, **4**.3
Impurity profiles, **5**.4, **5**.9
Indirect recombination, **2**.33, **2**.34
Inductive load, **6**.8
Inductive load protection, **1**.7
Industrial applications, **6**.27
Infant mortality, **8**.3
Interfaceless power system (IPS), **1**.10
Intrinsic carrier concentration, **2**.3
Ion implantation, **1**.5
Ionization energy, **2**.6
Isolated-collector vertical PNP, **4**.24
Isotherms, **11**.6, **11**.7

Joule effect, **6**.3
Junction curvature, **2**.3
Junction temperature, **6**.19, **6**.22, **6**.23, **7**.8

Kirkendall vacancies, **8**.31

Latch-up, **4**.20
Latchback, **3**.27, **3**.28
Lateral NPN, **4**.19
Lateral PNP transistor, **2**.8, **2**.15, **4**.5, **4**.8, **4**.18, **4**.24, **5**.2, **8**.47
Lattice constant, **2**.3
LDMOS (lateral DMOS), **3**.8, **3**.13, **3**.27, **3**.28, **3**.33, **3**.41, **3**.46, **3**.47, **3**.54
Lead bending, **8**.74
Leakage current, **2**.6, **2**.11
Life:
  median, **8**.3
  useful, **8**.3, **8**.37
Life test, **8**.66
Lifetime, **2**.18, **2**.33–**2**.36
Limiting circuit, **6**.8
Limiting velocity, **2**.3
Line, **11**.30
Line regulation, **11**.6
Load curve, **9**.19–**9**.21
Load regulation, **11**.6, **11**.10, **11**.30
Logo permanence, **8**.74
Low injection, **2**.19, **2**.20, **2**.23, **2**.37

Maximum junction temperature, **2**.38
Maximum power, **2**.38, **2**.40, **2**.43
Maximum temperature, **2**.38, **6**.9
Mechanical shocks, **8**.75
Mechanical vibration, **8**.75
Merged transistor logic, **4**.18
Moat etch technology, **3**.20
Mobility, **2**.3, **2**.19, **2**.29–**2**.32
Model, **8**.9
Modified epitaxial-base vertical PNP, **5**.5
Moisture, **8**.5
MTBF (mean time between failure), **6**.4
MTL (merged transistor logic), **4**.18
Multiplication coefficient, **2**.6
Multiwatt, **7**.11
Music power, **9**.9, **9**.12

Neutrality condition, **2**.4
Nonsaturation, **2**.25, **2**.55
Nonsaturation region, **2**.24, **2**.26, **2**.51, **2**.53
Nonuniform oxide, **5**.17

Offset comparator, **6**.7
On-chip compensation, **11**.7

On resistance, **3**.21, **3**.27, **3**.34, **3**.37, **3**.38, **3**.42, **3**.44, **3**.50
Operating life, **8**.67
Output impedance, **11**.6, **11**.10, **11**.32
Output music power, **9**.9
Output power, **6**.12, **6**.14, **9**.5, **9**.9, **9**.13, **9**.16, **9**.28, **9**.29, **9**.58
Oversaturation, **2**.28, **2**.54

Package, **6**.16
  hermetic, **7**.16, **8**.12
  plastic, **8**.12
Package construction, **7**.3
Package electrical design, **7**.15
Package mechanical design, **7**.9
Package thermal design, **7**.4
Parasitic MOS, **1**.3, **4**.21–**4**.23, **8**.47
Parasitic NPN, **4**.20
Parasitic PNP, **4**.11
Passivation surface, **7**.3
Peak current gain, **2**.53
Peak value, **2**.17
Permittivity of free space, **2**.2, **2**.4
Phantom gate, **8**.53
Physical switching times, **2**.49, **2**.50
Piezoresistive effects, **1**.6
Pinch-in, **2**.47
Pinch-off, **2**.32
Planar junction, **2**.10, **2**.11
Plane limited junction, **2**.7
Plane unlimited junction, **2**.3
Poisson equation, **2**.3, **2**.10, **2**.39, **2**.45
Polysilicon field plate, **4**.16, **4**.17
Power transistors, **6**.5, **6**.27
Pressure cooker, **8**.7, **8**.73
Pressure pot, **8**.73
Protection, **9**.58, **9**.64, **9**.68, **9**.70, **9**.77
Protection circuit, **6**.24, **6**.27, **11**.2, **11**.8, **11**.32
Pseudosphere, **1**.10
Pulse time, **6**.24
Punch-through, **2**.3, **2**.7, **2**.10, **2**.16, **2**.17, **2**.23, **3**.15, **4**.6
Punch-through breakdown, **3**.15
Purple plague, **8**.31
Push-pull amplifiers, **10**.17

Q_{SS}, **4**.21
Quadruple in-line, **7**.10
Quasi-neutrality condition, **2**.19, **2**.25
Quasi rise time, **2**.49, **2**.53
Quasi saturation, **2**.24, **2**.29, **2**.31, **2**.51, **2**.53, **2**.55
Quasi-saturation current gain, **2**.31
Quasi-saturation region, **2**.24, **2**.26
Quasi storage time, **2**.50, **2**.55
Quiescent current, **11**.26, **11**.39

$r_{bb'}$, **2**.47
Reach-through, **1**.5, **2**.3, **2**.7, **2**.8, **2**.11, **2**.14, **2**.46
Redistribution, **5**.5
Reference voltage, **6**.27
Reich-Hakim model, **8**.9
Reliability, **8**.3
Reliability evaluations, **8**.35
Reliability models, **8**.8
Reliability tests, **8**.27, **8**.64–**8**.74
  HTRB (high-temperature reverse bias), **8**.27, **8**.69
  humidity, **8**.27, **8**.72
  thermal, **8**.27
Resin:
  epoxy, **8**.13
  silicon, **8**.13
Resistance:
  to electrostatic discharge, **8**.75
  soldering heat, **8**.74
  to thermal humidity, **8**.6
Resistivity, **6**.24
Resurf devices, **3**.49
Reverse second breakdown, **2**.43, **2**.45, **10**.4, **10**.5, **10**.25
Ripple, **9**.10, **9**.56, **9**.58, **9**.59
Ripple rejection, **11**.6, **11**.30, **11**.35
Rise time, **2**.48, **2**.49, **2**.52, **2**.53

Safe operating area (SOA), **2**.43, **4**.5, **6**.2, **9**.64, **9**.74, **11**.2, **11**.8, **11**.10, **11**.13, **11**.21, **11**.31
Salt atmosphere, **8**.75
Saturated gain, **2**.27
Saturation, **2**.24, **2**.28, **2**.51, **2**.53, **6**.9
Saturation losses, **10**.10
Saturation region, **2**.24, **2**.25, **2**.27
SCR, **4**.19, **4**.20
Second breakdown, **1**.6, **6**.2, **6**.24
Semi-insulating polycrystalline silicon, **5**.17
Short-circuit protection, **6**.28, **11**.11, **11**.15, **11**.21

Single-point grounding, **11**.35
Sinker, **4**.2, **4**.9, **4**.15, **4**.20
Small-signal transistor, **5**.7
Solder (preform), **8**.57
  hard, **8**.58
  soft, **6**.16, **8**.58
Solderability, **8**.74
Spread of the R_s value, **6**.6
Stability, **11**.12
Start-up circuit, **11**.11, **11**.26
Storage tests, **8**.35
Storage time, **2**.49, **2**.50, **2**.54
Stress, mechanical, **8**.5
Surface current, **2**.11
Surface parasitics, **4**.19
Sustaining, **2**.2
Sustaining field, **2**.13
Sustaining voltage, **2**.12–**2**.14
Switching amplifier, **6**.7, **6**.12, **6**.14, **6**.22, **6**.23
Switching components, **10**.1
Switching frequency, **6**.13
Switching losses, **6**.13
Synthetic PNP, **4**.8

Temperature gradient, **6**.16, **9**.71, **9**.72
Tests, reliability, **8**.27, **8**.64–**8**.74
  HTRB (high-temperature reverse bias), **8**.27, **8**.69
  humidity, **8**.27, **8**.72
  thermal, **8**.27
TH (temperature humidity), **8**.7
THB (temperature humidity bias), **8**.7
Thermal capacitance, **2**.38, **2**.39, **6**.18–**6**.20
Thermal conductivity, **2**.3, **2**.38, **2**.39, **6**.9, **6**.14, **7**.7
Thermal considerations, **11**.34
Thermal crowding, **2**.33, **2**.41
Thermal cycle, **7**.2, **7**.14, **8**.35, **8**.64
Thermal expansion coefficient, **2**.3, **8**.21
Thermal fatigue test, **8**.69
Thermal feedback, **9**.48–**9**.52, **9**.59
Thermal gradients, **9**.25, **9**.26, **9**.50, **9**.70, **9**.72, **11**.8
Thermal hysteresis, **6**.24
Thermal impedance, **2**.40, **6**.19
Thermal limit, **11**.8, **11**.32
Thermal overload protection, **11**.11, **11**.21, **11**.26
Thermal protection, **6**.27, **11**.7, **11**.12
Thermal regulation, **11**.6
Thermal resistance, **2**.38, **2**.39, **2**.42, **6**.9, **6**.20, **7**.4, **7**.15, **8**.70, **8**.75, **10**.14
Thermal resistor, **6**.19
Thermal runaway, **2**.44
Thermal sensors, **6**.22, **6**.23
Thermal shocks, **8**.35, **8**.65
Thermal shutdown, **11**.2, **11**.11
Thermal stability factor, **2**.44
Thermoelastic stress, **7**.13
Thermogravimetrical analysis, **8**.20
Thermomechanical effects, **8**.55
Thickness of the base, **6**.25
Three-terminal voltage regulators, **11**.2, **11**.24, **11**.30
Threshold voltage, **3**.4, **4**.20–**4**.22, **8**.53
Top and bottom isolation, **4**.12, **4**.15, **4**.24
Total base doping, **2**.10
Transfer efficiency, **6**.9
Transient-voltage protection, **6**.26
Transistor geometry, **6**.24
Transit time, **2**.51, **2**.52, **4**.19
Transition capacitance, **2**.5
Transport factor, **2**.18, **2**.21
Transverse base resistance r_{bb}, **2**.32, **2**.47
Triple-diffused PNP transistor, **5**.8
Two-sided PN junction, **2**.5

V_{CER}, **2**.16, **2**.17
V_{CES}, **2**.17
V_{CEV}, **2**.16, **2**.17
Vertical NPN transistor, **2**.7, **2**.10, **2**.14, **2**.15, **2**.18, **2**.22, **2**.36
Vertical PNP transistor, **1**.5, **4**.5, **4**.6, **4**.8, **5**.4
VLAI (very large area integration), **1**.1
VLSI (very large scale integration), **1**.1
VMOS (V-groove DMOS), **3**.9, **3**.11, **3**.12
Voltage and current ratings, **10**.1
Voltage reference, **11**.2, **11**.3, **11**.17

Wear-out, **8**.3

Zener breakdown, **3**.15
Zener breakdown voltages, **2**.6
Zener effect, **2**.5
Zener reference, **11**.3

ABOUT THE EDITOR

Paolo Antognetti is a professor on the faculty of the University of Genoa, Italy. Dr. Antognetti's extensive background in semiconductor modeling and simulation includes teaching and research positions as a post-doctoral fellow at Stanford University and as visiting professor at MIT. He is the author or coauthor of more than thirty papers in the field, and the editor of two books on VLSI aspects of computer-aided design and process modeling. He has also translated several books on electronic engineering into Italian.